W9-CRQ-477

Simulation with Arena

Fourth Edition

McGraw-Hill Series in Industrial Engineering and Management Science

Simulation with Arena

Fourth Edition

W. David Kelton

Professor
Department of Quantitative Analysis and Operations Management
University of Cincinnati

Randall P. Sadowski

Product Manager
Application Programs
Rockwell Automation

David T. Sturrock

Product Manager
Simulation
Rockwell Automation

Boston   Burr Ridge, IL   Dubuque, IA   New York   San Francisco   St. Louis
Bangkok   Bogotá   Caracas   Kuala Lumpur   Lisbon   London   Madrid   Mexico City
Milan   Montreal   New Delhi   Santiago   Seoul   Singapore   Sydney   Taipei   Toronto

Higher Education

SIMULATION WITH ARENA, FOURTH EDITION

Published by McGraw-Hill, a business unit of The McGraw-Hill Companies, Inc., 1221 Avenue of the Americas, New York, NY 10020. Copyright © 2007 by The McGraw-Hill Companies, Inc. All rights reserved. No part of this publication may be reproduced or distributed in any form or by any means, or stored in a database or retrieval system, without the prior written consent of The McGraw-Hill Companies, Inc., including, but not limited to, in any network or other electronic storage or transmission, or broadcast for distance learning.

Some ancillaries, including electronic and print components, may not be available to customers outside the United States.

This book is printed on acid-free paper.

2 3 4 5 6 7 8 9 0 DOC/DOC 0 9 8 7

ISBN-13 978–0–07–352341–5
ISBN-10 0–07–352341–0

Publisher: *Suzanne Jeans*
Senior Sponsoring Editor: *Michael S. Hackett*
Developmental Editor: *Larry Goldberg*
Executive Marketing Manager: *Michael Weitz*
Project Coordinator: *Melissa M. Leick*
Senior Production Supervisor: *Sherry L. Kane*
Media Project Manager: *Laurie Lenstra*
Associate Media Producer: *Christina Nelson*
Senior Designer: *David W. Hash*
Cover Designer: *Peter Alan Kauffman*
Printer: *R. R. Donnelley/Crawfordsville, IN*

Library of Congress Cataloging-in-Publication Data

Kelton, W. David.
  Simulation with Arena / W. David Kelton, Randall P. Sadowski, David T. Sturrock. — 4th ed.
    p. cm. — (McGraw-Hill series in industrial engineering and management science)
  ISBN 978–0–07–352341–5 — ISBN 0–07–352341–0 (hard copy : alk. paper)
  1. Computer simulation. 2. Arena (Computer file). I. Sadowski, Randall P. II. Sturrock, David T. III. Title.
IV. Series.

QA76.9.C65K45   2007
003'.3536—dc22

2006018140
CIP

www.mhhe.com

About the Authors

W. David Kelton is Professor in the Department of Quantitative Analysis and Operations Management at the University of Cincinnati. He received a B.A. in mathematics from the University of Wisconsin-Madison, an M.S. in mathematics from Ohio University, and M.S. and Ph.D. degrees in industrial engineering from Wisconsin.

His research interests and publications are in the probabilistic and statistical aspects of simulation, applications of simulation, and stochastic models. His papers have appeared in *Operations Research, Management Science*, the *INFORMS Journal on Computing, IIE Transactions, Naval Research Logistics*, and the *Journal of the American Statistical Association*, among others.

He is Editor-in-Chief for the *INFORMS Journal on Computing* and has served as Simulation Area Editor for *Operations Research*, the *INFORMS Journal on Computing*, and *IIE Transactions*; Associate Editor of *Operations Research*, the *Journal of Manufacturing Systems*, and *Simulation*; and was Guest Co-Editor for a special simulation issue of *IIE Transactions*. Awards include the TIMS College on Simulation award for best simulation paper in *Management Science*, the IIE Operations Research Division Award, a Meritorious Service Award from *Operations Research*, the INFORMS College on Simulation Distinguished Service Award, and the INFORMS College on Simulation Outstanding Simulation Publication Award. He was President of the TIMS College on Simulation, and was the INFORMS co-representative to the Winter Simulation Conference Board of Directors from 1991 through 1999, serving as Board Chair for 1998. In 1987 he was Program Chair for the WSC, and in 1991 was General Chair. He has worked on grants and consulting contracts from a number of corporations, foundations, and agencies. He recently made it down a black-diamond run, this time on his skis.

Randall P. Sadowski is currently Product Manager for scheduling and data-tracking applications for Rockwell Automation. He was previously director of university relations, chief applications officer, vice president of consulting services and user education at Systems Modeling Corporation.

Before joining Systems Modeling, he was on the faculty at Purdue University in the School of Industrial Engineering and at the University of Massachusetts. He received his bachelor's and master's degrees in industrial engineering from Ohio University and his Ph.D. in industrial engineering from Purdue.

He has authored over 50 technical articles and papers, served as chair of the Third International Conference on Production Research and was the general chair of the 1990 Winter Simulation Conference. He is on the visiting committee for the IE departments at Lehigh University, the University of Pittsburgh, and Ohio University. He is co-author, with C. Dennis Pegden and Robert E. Shannon, of *Introduction to Simulation Using SIMAN*.

He is a Fellow of the Institute of Industrial Engineers and served as editor of a two-year series on Computer Integrated Manufacturing Systems for *IE Magazine* that received the 1987 IIE Outstanding Publication award. He has served in several positions at IIE, including president at the chapter and division levels, and vice president of Systems Integration at the international level. He founded the annual IIE/RA Student Simulation Contest. He collects tools and is the proud owner of a tractor named Dutch, a stocked farm pond, and a recently acquired monster mower.

David T. Sturrock is Simulation Product Manager at Rockwell Automation. He is responsible for the overall success of simulation products in such varied markets as manufacturing, high-speed processing, contact centers, business processes, and real-time testing/control. Dave has applied simulation techniques in the areas of transportation systems, scheduling, plant layout, contact centers, capacity analysis, process design, health care, and real-time control.

Dave received his bachelor's degree in industrial engineering from The Pennsylvania State University with concentrations in manufacturing and automation. During ten years at Inland Steel Company, he worked as a plant industrial engineer and subsequently formed and led a simulation group. He started using SIMAN just after it was first released in 1983. He joined Systems Modeling as a consultant in 1988. His first assignment was to "help out with the next release," and he has been doing that ever since.

He was General Chair for the international 1999 Winter Simulation Conference (WSC) and has participated in several funded research projects, written a handful of papers, and is an active member of the Institute of Industrial Engineers (IIE), AMA, SME, and INFORMS. He does frequent speaking engagements around the world, which, so far, have taken him to about 40 countries.

In his leisure time, Dave enjoys camping, hiking, canoeing, skiing, woodworking, and traveling with his wife and three daughters. Dave also enjoys working with teenagers in many settings, including advising a high-adventure group and conducting "Success Skills" workshops at local public schools.

To those in the truly important arena of our lives:

Albert, Anna, Anne, Christie, and Molly;

Aidan, Charity, Emma, Jenny, Michael, Noah, Sammy, Sean, Shelley, and Tierney;

Diana, Kathy, Melanie, and Victoria.

Preface

This fourth edition of *Simulation with Arena* has the same goal as the first three editions: to provide an introduction to simulation using Arena. It is intended to be used as an entry-level simulation text, most likely in a first course on simulation at the undergraduate or beginning graduate level. However, material from the later chapters could be incorporated into a second, graduate-level course. The book can also be used to learn simulation independent of a formal course (more specifically, by Arena users). The objective is to present the concepts and methods of simulation using Arena as a vehicle to help the reader reach the point of being able to carry out effective simulation modeling, analysis, and projects using the Arena simulation system. While we'll cover most of the capabilities of Arena, the book is not meant to be an exhaustive reference on the software, which is fully documented in its extensive online reference and help system.

Included is a CD with the Arena 10.0 academic software and all the examples in the text. A Web site for the book can be found at http://www.arenasimulation.com/programs/sim_w_Arena_4.asp. We encourage all readers to visit this site to learn of any updates or errata for the book or example files supplied, possible additional exercises, and other items of interest. The site also contains material to support instructors who have adopted the book for use in class, including downloadable lecture slides and solutions to exercises; university instructors who have adopted the book should contact the local McGraw-Hill representative for authorization (see http://www.mhhe.com/catalogs/sem/engineering to locate representatives in the U.S., or call 1-609-426-5793 to locate representatives outside the U.S.).

We've adopted an informal, tutorial writing style centered around carefully crafted examples to aid the beginner in understanding the ideas and topics presented. Ideally, readers would build simulation models as they read through the chapters. We start by having the reader develop simple, well-animated, high-level models, and then progress to advanced modeling and analysis. Statistical analysis is not treated as a separate topic, but is integrated into many of the modeling chapters, reflecting the joint nature of these activities in good simulation studies. We've also devoted more advanced chapters to statistical issues and project planning to cover more advanced issues not treated in our modeling chapters. We believe that this approach greatly enhances the learning process by placing it in a more realistic and (frankly) less boring setting.

We assume neither prior knowledge of simulation nor computer-programming experience. We do assume basic familiarity with computing in general (files, folders, basic editing operations, etc.), but nothing advanced. A fundamental understanding of probability and statistics is needed, though we provide a self-contained refresher of these subjects in Appendices C and D.

Here's a quick overview of the topics and organization. We start in Chapter 1 with a general introduction, a brief history of simulation, and modeling concepts.

Chapter 2 addresses the simulation process using a simple simulation executed by hand and briefly discusses using spreadsheets to simulate very simple models. In Chapter 3, we acquaint readers with Arena by examining a completed simulation model of the problem simulated by hand in Chapter 2, rebuilding it from scratch, going over the Arena user interface, and providing an overview of Arena's capabilities; we also provide a small case study illustrating how knowledge of just these basic building blocks of Arena allows one to address interesting and realistic issues.

Chapters 4 and 5 advance the reader's modeling skills by considering one "core" example per chapter, in increasingly complex versions to illustrate a variety of modeling and animation features; the statistical issue of selecting input probability distributions is also covered in Chapter 4 using the Arena Input Analyzer.

Chapter 6 uses the model in Chapter 5 to illustrate the basic Arena capabilities of statistical analysis of output, including single-system analysis, comparing multiple scenarios (configurations of a model), and searching for an optimal scenario; this material uses the Arena Output and Process Analyzers, as well as OptQuest for Arena.

In Chapter 7, we introduce another "core" model, again in increasingly complex versions, and then use it to illustrate statistical analysis of long-run (steady-state) simulations. Alternate ways in which simulated entities can move around is the subject of Chapter 8, including material-handling capabilities, building on the models in Chapter 7. Chapter 9 digs deeper into Arena's extensive modeling constructs, using a sequence of small, focused models to present a wide variety of special-purpose capabilities; this is for more advanced simulation users and would probably not be covered in a beginning course.

In Chapter 10, we describe a number of topics in the area of customizing Arena and integrating it with other applications like spreadsheets and databases; this includes using VBA (Visual Basic for Applications) with Arena. Chapter 11 shows how Arena can handle continuous and combined discrete/continuous models, such as fluid flow. Chapter 12 covers more advanced statistical concepts underlying and applied to simulation analysis, including random-number generators, variate and process generation, variance-reduction techniques, sequential sampling, and designing simulation experiments. Chapter 13 provides a broad overview of the simulation process and discusses more specifically the issues of managing and disseminating a simulation project.

Appendix A describes a complete modeling specification from a project for *The Washington Post* newspaper. In Appendix B, we give an overview and a link to problem statements for the Arena modeling contest held annually by the Institute of Industrial Engineers (IIE) and Rockwell Automation. Appendix C gives a complete but concise review of the basics of probability and statistics couched in the framework of their role in simulation modeling and analysis. The probability distributions supported by Arena are detailed in Appendix D. Installation instructions for the Arena academic software can be found in Appendix E. All references are collected in a single References section at the end of the book. The index is extensive, to aid readers in locating topics and seeing how they relate to each other; the index includes authors cited.

As mentioned above, the presentation is in "tutorial style," built around a sequence of carefully crafted examples illustrating concepts and applications, rather than in the conventional style of stating concepts first and then citing examples as an afterthought. So it probably makes sense to read (or teach) the material essentially in the order presented. A one-semester or one-quarter first course in simulation could cover all the material in Chapters 1–8, including the statistical material. Time permitting, selected modeling and computing topics from Chapters 9–11 could be included, or some of the more advanced statistical issues from Chapter 12, or the project-management material from Chapter 13, according to the instructor's tastes. A second course in simulation could assume most of the material in Chapters 1–8, then cover the more advanced modeling ideas in Chapters 9–11, followed by topics from Chapters 12 and 13. For self-study, we'd suggest going through Chapters 1–6 to understand the basics, getting at least familiar with Chapters 7 and 8, then regarding the rest of the book as a source for more advanced topics and reference. Regardless of what's covered, and whether the book is used in a course or independently, it will be helpful to follow along in Arena on a computer while reading this book.

The CD included contains the academic version of Arena (see Appendix E for installation instructions), which has all the modeling and analysis capabilities of the complete commercial version but limits model size. All the examples in the book, as well as all the exercises at the ends of the chapters, will run with this educational version of Arena. The CD also contains files for all the example models in the book, as well as other support materials. This software can be installed on any university computer as well as on students' computers. It is intended for use in conjunction with this book for the purpose of learning simulation and Arena. It is not authorized for use in commercial environments.

If you were familiar with the third edition, here are the main changes:

- All the examples have been updated to conform to the current Arena version (10.0). The software is largely consistent with what was discussed in the third edition, but there are several new features and capabilities that we illustrate, including model documentation, enhanced plots, file reading and writing, printing, and animation symbols.

- Section 2.7 in Chapter 2 is new and gives a brief introduction to using spreadsheets to simulate, with an examples of both a static and a dynamic model.

- Section 3.5 in Chapter 3 is new and gives a case study of parallel vs. serial processing, illustrating that just this basic set of Arena tools enables one to address interesting and practical issues.

- In Chapter 5, we have replaced the car-repair model in the third edition with an enhanced version of the call-center model similar to what was in the first two editions. However, we build it up in three stages to enable more manageable teaching chunks.

- Chapter 6 has the same mission as in the third edition (statistical analysis for terminating systems), but uses the enhanced call-center model from Chapter 5.

- The new Chapters 7, 8, 9, 12, and 13 cover the same material as in the third edition, except for updates.

- Chapter 10 has a new Section 10.6 on real-time integration, sharing simulation data with other devices during a run.

- Chapter 11 has a new Section 11.2.4 on the Flow Process panel for continuous-change modeling.

- Appendix B, on the contest problems, has been reduced to a single page, and the (growing) list of problem descriptions has been moved to the public Web site.

- The support materials on the Web site (slides and solutions) have all been updated.

As with any labor like this, there are a lot of people and institutions that supported us in a lot of different ways. First and foremost, Lynn Barrett at Rockwell Automation really made this all happen by reading (and re-reading and re-re-reading, and then fixing) our semi-literate drafts, orchestrating the composing and production, reminding us of what month (and year) it was, and tolerating our tardiness and fussiness and quirky personal-hyphenation habits; her husband Doug also deserves our thanks for putting up with her putting up with us. Rockwell Automation provided resources in the form of time, software, hardware, technical assistance, and moral encouragement; we'd particularly like to thank the Arena development team—Norene Collins, Cory Crooks, Glenn Drake, Tim Haston, Cynthia Kasales, Judy Kirby, Frank Palmieri, Dave Takus, Christine Watson, and Vytas Urbonavicius—as well as Steve Frank, Judy Jordan, Gavan Hood, Scott Miller, Dennis Pegden, Jon Phillips, Darryl Starks, and Nancy Swets. And a special note of thanks goes to Deb Sadowski for her writing and influence as a co-author of the first two editions. The Department of Quantitative Analysis and Operations Management at the University of Cincinnati was also quite supportive.

We are also grateful to Gary Lucke and Olivier Girod of *The Washington Post* for allowing us to include a simulation specification that was developed for them by Rockwell Automation as part of a larger project. Special thanks go to Pete Kauffman for his cover design and production assistance, and to Jim McClure for his cartoon and illustration design. And we appreciate the skillful motivation and gentle nudging by our editor at McGraw-Hill, Michael Hackett. The reviewers, Bill Harper, Mansooreh Mollaghasemi, Barry Nelson, Ed Watson, and King Preston White, Jr., provided extremely valuable input and help, ranging from overall organization and content all the way to the downright subatomic. Thanks are also due to the many individuals who have used part or all of the early material in classes (as well as to their students who were subjected to early drafts), as well as a host of other folks who provided all kinds of input, feedback and help: Christos Alexopoulos, Ken Bauer, Diane Bischak, Sherri Blaszkiewicz, Eberhard Blümel, Mike Branson, Jeff Camm, Colin Campbell, John Charnes, Chun-Hung Chen, Hong Chen, Jack Chen, Russell Cheng, Christopher Chung, Frank Ciarallo, John J. Clifford, Mary Court, Tom Crowe, Halim Damerdji, Pat Delaney, Mike Dellinger, Darrell Donahue, Ken Ebeling, Neil Eisner, Gerald Evans, Steve Fisk, Michael Fu, Shannon Funk, Fred Glover, Dave Goldsman, Byron Gottfried, Frank Grange, Don Gross, John Gum, Tom Gurgiolo, Jorge Haddock, Bill Harper, Joe Heim, Michael Howard, Arthur Hsu, Eric Johnson, Elena Joshi, Keebom Kang, Elena Katok, Jim Kelly, Teri King, Gary Kochenberger, Patrick Koelling, David Kohler, Wendy Krah, Bradley Kramer, Michael Kwinn, Jr., Averill Law, Larry Leemis, Marty Levy, Bob Lipset, Gerald Mackulak, Nancy Markovitch, Deb Mederios, Brian Melloy, Mansooreh Mollaghasemi, Ed Mooney, Jack Morris, Jim Morris, Charles Mosier, Marvin Nakayama, Dick Nance, Barry Nelson, James Patell, Cecil Peterson, Dave Pratt, Mike Proctor, Madhu Rao, James Reeve, Steve Roberts, Paul Rogers, Ralph Rogers, Tom Rohleder, Jerzy Rozenblit, Salim Salloum, G. Sathyanarayanan, Bruce Schmeiser, Carl Schultz, Thomas Schulze, Marv Seppanen, Michael Setzer, David Sieger, Robert Signorile, Julie Ann Stuart, Jim Swain, Mike Taaffe, Laurie Travis, Reha Tutuncu, Wayne Wakeland, Ed Watson, Michael Weng, King Preston White, Jr., Jim Wilson, Irv Winters, Chih-Hang (John) Wu, James Wynne, and Stefanos Zenios.

W. DAVID KELTON
University of Cincinnati
david.kelton@uc.edu

RANDALL P. SADOWSKI
Rockwell Automation
rpsadowski@ra.rockwell.com

DAVID T. STURROCK
Rockwell Automation
dtsturrock@ra.rockwell.com

Contents

CHAPTER 1

What Is Simulation?

Simulation refers to a broad collection of methods and applications to mimic the behavior of real systems, usually on a computer with appropriate software. In fact, "simulation" can be an extremely general term since the idea applies across many fields, industries, and applications. These days, simulation is more popular and powerful than ever since computers and software are better than ever.

This book gives you a comprehensive treatment of simulation in general and the Arena simulation software in particular. We cover the general idea of simulation and its logic in Chapters 1 and 2 (including a bit about using spreadsheets to simulate) and Arena in Chapters 3–9. We don't, however, intend for this book to be a complete reference on everything in Arena (that's what the help systems in the software are for). In Chapter 10, we show you how to integrate Arena with external files and other applications and give an overview of some advanced Arena capabilities. In Chapter 11, we introduce you to continuous and combined discrete/continuous modeling with Arena. Chapters 12-13 cover issues related to planning and interpreting the results of simulation experiments, as well as managing a simulation project. Appendix A is a detailed account of a simulation project carried out for *The Washington Post* newspaper. In Appendix B, we provide a link to statements of fairly complex problems from recent student competitions on Arena modeling held by the Institute of Industrial Engineers and Rockwell Automation (formerly Systems Modeling). Appendix C provides a quick review of probability and statistics necessary for simulation. Appendix D describes Arena's probability distributions, and Appendix E provides software installation instructions. After reading this book, you should be able to model systems with Arena and carry out effective and successful simulation studies.

This chapter touches on the general notion of simulation. In Section 1.1, we describe some general ideas about how you might study models of systems and give some examples of where simulation has been useful. Section 1.2 contains more specific information about simulation and its popularity, mentions some good things (and one bad thing) about simulation, and attempts to classify the many different kinds of simulations that people do. In Section 1.3, we talk a little bit about software options. Finally, Section 1.4 traces changes over time in how and when simulation is used. After reading this chapter, you should have an appreciation for where simulation fits in, the kinds of things it can do, and how Arena might be able to help you do them.

1.1 Modeling

Simulation, like most analysis methods, involves systems and models of them. So in this section, we give you some examples of models and describe options for studying them to learn about the corresponding system.

1.1.1   What's Being Modeled?

Computer simulation deals with models of systems. A *system* is a facility or process, either actual or planned, such as:

- A manufacturing plant with machines, people, transport devices, conveyor belts, and storage space.
- A bank with different kinds of customers, servers, and facilities like teller windows, automated teller machines (ATMs), loan desks, and safety deposit boxes.
- An airport with departing passengers checking in, going through security, going to the departure gate, and boarding; departing flights contending for push-back tugs and runway slots; arriving flights contending for runways, gates, and arrival crew; arriving passengers moving to baggage claim and waiting for their bags; and the baggage-handling system dealing with delays, security issues, and equipment failure.
- A distribution network of plants, warehouses, and transportation links.
- An emergency facility in a hospital, including personnel, rooms, equipment, supplies, and patient transport.
- A field-service operation for appliances or office equipment, with potential customers scattered across a geographic area, service technicians with different qualifications, trucks with different parts and tools, and a central depot and dispatch center.
- A computer network with servers, clients, disk drives, tape drives, printers, networking capabilities, and operators.
- A freeway system of road segments, interchanges, controls, and traffic.
- A central insurance claims office where a lot of paperwork is received, reviewed, copied, filed, and mailed by people and machines.
- A criminal-justice system of courts, judges, support staff, probation officers, parole agents, defendants, plaintiffs, convicted offenders, and schedules.
- A chemical-products plant with storage tanks, pipelines, reactor vessels, and railway tanker cars in which to ship the finished product.
- A fast-food restaurant with different types of staff, customers, and equipment.
- A supermarket with inventory control, checkout, and customer service.
- A theme park with rides, stores, restaurants, workers, guests, and parking lots.
- The response of emergency personnel to the occurrence of a catastrophic event.

People often study a system to measure its performance, improve its operation, or design it if it doesn't exist. Managers or controllers of a system might also like to have a readily available aid for day-to-day operations, like help in deciding what to do in a factory if an important machine goes down.

We're even aware of managers who requested that simulations be constructed but didn't really care about the final results. Their primary goal was to focus attention on understanding how their system worked. Often simulation analysts find that the process of defining how the system works, which must be done before you can start developing the simulation model, provides great insight into what changes need to be made. Part of this is due to the fact that rarely is there one individual responsible for understanding how an entire system works. There are experts in machine design, material handling, processes, and so on, but not in the day-to-day operation of the system. So as you read on, be aware that simulation is much more than just building a model and conducting a statistical experiment. There is much to be learned at each step of a simulation project, and the decisions you make along the way can greatly affect the significance of your findings.

1.1.2   How About Just Playing with the System?

It might be possible to experiment with the actual physical system. For instance:

- Some cities have installed entrance-ramp traffic lights on their freeway systems to experiment with different sequencing to find settings that make rush hour as smooth and safe as possible.
- A supermarket manager might try different policies for inventory control and checkout-personnel assignment to see what combinations seem to be most profitable and provide the best service.
- An airline could test the expanded use of automated check-in kiosks (and employees to urge passengers to use them) to see if this speeds check-in.
- A computer facility can experiment with different network layouts and job priorities to see how they affect machine utilization and turnaround.

This approach certainly has its advantages. If you can experiment directly with the system and know that nothing else about it will change significantly, then you're unquestionably looking at the right thing and needn't worry about whether a model or proxy for the system faithfully mimics it for your purposes.

1.1.3   Sometimes You Can't (or Shouldn't) Play with the System

In many cases, it's just too difficult, costly, or downright impossible to do physical studies on the system itself.

- Obviously, you can't experiment with alternative layouts of a factory if it's not yet built.
- Even in an existing factory, it might be very costly to change to an experimental layout that might not work out anyway.
- It would be hard to run twice as many customers through a bank to see the effect of closing a nearby branch.
- Trying a new check-in procedure at an airport might initially cause a lot of people to miss their flights if there are unforeseen problems with the new procedure.
- Fiddling around with emergency room staffing in a hospital clearly won't do.

In these situations, you might build a *model* to serve as a stand-in for studying the system and ask pertinent questions about what *would* happen in the system *if* you did this or that, or *if* some situation beyond your control were to develop. *Nobody gets hurt, and your freedom to try wide-ranging ideas with the model could uncover attractive alternatives that you might not have been able to try with the real system.*

However, you have to build models carefully and with enough detail so that what you learn about the model will never[1] be different from what you would have learned about

---

[1] Well, hardly ever.

the system by playing with it directly. This is called model *validity*, and we'll have more to say about it later, in Chapter 13.

1.1.4   Physical Models

There are lots of different kinds of models. Maybe the first thing the word evokes is a physical replica or scale model of the system, sometimes called an *iconic* model. For instance:

- People have built *tabletop* models of material handling systems that are miniature versions of the facility, not unlike electric train sets, to consider the effect on performance of alternative layouts, vehicle routes, and transport equipment.
- A full-scale version of a fast-food restaurant placed inside a warehouse to experiment with different service procedures was described by Swart and Donno (1981). In fact, most large fast-food chains now have full-scale restaurants in their corporate office buildings for experimentation with new products and services.
- Simulated control rooms have been developed to train operators for nuclear power plants.
- Physical flight simulators are widely used to train pilots. There are also flight-simulation computer programs, with which you may be familiar in game form, that represent purely logical models executing inside a computer. Further, physical flight simulators might have computer screens to simulate airport approaches, so they have elements of both physical and computer-simulation models.

Although iconic models have proven useful in many areas, we won't consider them.

1.1.5   Logical (or Mathematical) Models

Instead, we'll consider *logical* (or *mathematical*) models of systems. Such a model is just a set of approximations and assumptions, both structural and quantitative, about the way the system does or will work.

A logical model is usually represented in a computer program that's exercised to address questions about the model's behavior; if your model is a valid representation of your system, you hope to learn about the system's behavior too. And since you're dealing with a mere computer program rather than the actual system, it's usually easy, cheap, and fast to get answers to a lot of questions about the model and system by simply manipulating the program's inputs and form. Thus, you can make your mistakes on the computer where they don't count, rather than for real where they do. As in many other fields, recent dramatic increases in computing power (and decreases in computing costs) have impressively advanced your ability to carry out computer analyses of logical models.

1.1.6   What Do You Do with a Logical Model?

After making the approximations and stating the assumptions for a valid logical model of the target system, you need to find a way to deal with the model and analyze its behavior.

If the model is simple enough, you might be able to use traditional mathematical tools like queueing theory, differential-equation methods, or something like linear programming to get the answers you need. This is a nice situation since you might get fairly simple formulas to answer your questions, which can easily be evaluated numerically; working with the formula (for instance, taking partial derivatives of it with respect to controllable input parameters) might provide insight itself. Even if you don't get a simple closed-form formula, but rather an algorithm to generate numerical answers, you'll still have exact answers (up to roundoff, anyway) rather than estimates that are subject to uncertainty.

However, most systems that people model and study are pretty complicated, so that *valid* models[2] of them are pretty complicated too. For such models, there may not be exact mathematical solutions worked out, which is where simulation comes in.

1.2 Computer Simulation

Computer simulation refers to methods for studying a wide variety of models of real-world systems by numerical evaluation using software designed to imitate the system's operations or characteristics, often over time. From a practical viewpoint, simulation is the process of designing and creating a computerized model of a real or proposed system for the purpose of conducting numerical experiments to give us a better understanding of the behavior of that system for a given set of conditions. Although it can be used to study simple systems, the real power of this technique is fully realized when we use it to study complex systems.

While simulation may not be the only tool you could use to study the model, it's frequently the method of choice. The reason for this is that the simulation model can be allowed to become quite complex, if needed to represent the system faithfully, and you can still do a simulation analysis. Other methods may require stronger simplifying assumptions about the system to enable an analysis, which might bring the validity of the model into question.

1.2.1 Popularity and Advantages

Over the last two or three decades, simulation has been consistently reported as the most popular operations research tool:

- Rasmussen and George (1978) asked M.S. graduates from the Operations Research Department at Case Western Reserve University (of which there are many since that department has been around a long time) about the value of methods after graduation. The first four methods were *statistical analysis*, *forecasting*, *systems analysis*, and *information systems*, all of which are very broad and general categories. Simulation was next, and ranked higher than other more traditional operations research tools like linear programming and queueing theory.
- Thomas and DaCosta (1979) gave analysts in 137 large firms a list of tools and asked them to check off which ones they used. Statistical analysis came in first, with 93% of the firms reporting that they use it (it's hard to imagine a large firm that wouldn't), followed by simulation (84%). Again, simulation came in higher

---

[2] You can always build a simple (maybe simplistic) model of a complicated system, but there's a good chance that it won't be valid. If you go ahead and analyze such a model, you may be getting nice, clean, simple answers to the wrong questions. This is sometimes called a Type III Error—working on the wrong problem (statisticians have already claimed Type I and Type II Errors).

than tools like linear programming, PERT/CPM, inventory theory, and nonlinear programming.

- Shannon, Long, and Buckles (1980) surveyed members of the Operations Research Division of the American Institute of Industrial Engineers (now the Institute of Industrial Engineers) and found that among the tools listed, simulation ranked first in utility and interest. Simulation was second in familiarity, behind linear programming, which might suggest that simulation should be given stronger emphasis in academic curricula.

- Forgionne (1983); Harpell, Lane, and Mansour (1989); and Lane, Mansour, and Harpell (1993) all report that, in terms of utilization of methods by practitioners in large corporations, statistical analysis was first and simulation was second. Again, though, academic curricula seem to be behind since linear programming was more frequently *taught*, as opposed to being *used* by practitioners, than was simulation.

- Morgan (1989) reviewed many surveys of the above type and reported that "heavy" use of simulation was consistently found. Even in an industry with the lowest reported use of operations research tools (motor carriers), simulation ranked first in usage.

The main reason for simulation's popularity is its ability to deal with very complicated models of correspondingly complicated systems. This makes it a versatile and powerful tool. Another reason for simulation's increasing popularity is the obvious improvement in performance/price ratios of computer hardware, making it ever more cost effective to do what was prohibitively expensive computing just a few years ago. Finally, advances in simulation software power, flexibility, and ease of use have moved the approach from the realm of tedious and error-prone, low-level programming to the arena of quick and valid decision making.

Our guess is that simulation's popularity and effectiveness are now even greater than reported in the surveys described above, precisely due to these advances in computer hardware and software.

1.2.2   The Bad News

However, simulation isn't *quite* paradise, either.

Because many real systems are affected by uncontrollable and random inputs, many simulation models involve random, or *stochastic*, input components, causing their output to be random too. For example, a model of a distribution center would have arrivals, departures, and lot sizes arising randomly according to particular probability distributions, which will propagate through the model's logic to cause output performance measures like throughput and cycle times to be random as well. So running a stochastic simulation once is like performing a random physical experiment once, or watching the distribution center for one day—you'll probably see something different next time, even if you don't change anything yourself. In many simulations, as the time frame becomes longer (like months instead of a day), most results averaged over the run will tend to settle down and become less variable, but it can be hard to determine how long is "long enough" for this to happen. Moreover, the model or study might dictate that the simulation stop at a particular point (for instance, a bank is open from 9 to 5), so running it longer to calm the output is inappropriate.

Thus, you have to think carefully about designing and analyzing simulation experiments to take account of this uncertainty in the results, especially if the appropriate time frame for your model is relatively short. We'll return to this idea repeatedly in the book and illustrate proper statistical design and analysis tools, some of which are built into Arena, but others you have to worry about yourself.

Even though simulation output may be uncertain, we can deal with, quantify, and reduce this uncertainty. You might be able to get rid of the uncertainty completely by making a lot of over-simplifying assumptions about the system; this would get you a nice, simple model that will produce nice, non-random results. Unfortunately, though, such an over-simplified model will probably not be a *valid* representation of the system, and the error due to such model invalidity is impossible to measure or reliably reduce. For our money, we'd prefer an approximate answer to the right problem rather than an exact answer to the wrong problem (remember the Type III Error?).

1.2.3 Different Kinds of Simulations

There are a lot of ways to classify simulation models, but one useful way is along these three dimensions:

- **Static vs. Dynamic:** Time doesn't play a natural role in static models but does in dynamic models. The Buffon needle problem, described at the beginning of Section 1.3.1, is a static simulation. The small manufacturing model described in Chapters 2 and 3 is a dynamic model. Most operational models are dynamic; Arena was designed with them in mind, so our primary focus will be on such models (except for Section 2.7.1, which develops a static model).

- **Continuous vs. Discrete:** In a continuous model, the state of the system can change continuously over time; an example would be the level of a reservoir as water flows in and is let out, and as precipitation and evaporation occur. In a discrete model, though, change can occur only at separated points in time, such as a manufacturing system with parts arriving and leaving at specific times, machines going down and coming back up at specific times, and breaks for workers. You can have elements of both continuous and discrete change in the same model, which are called *mixed continuous-discrete models*; an example might be a refinery with continuously changing pressure inside vessels and discretely occurring shutdowns. Arena can handle continuous, discrete, and mixed models; our focus will be on discrete models for most of the book, though Chapter 11 discusses continuous and mixed models.

- **Deterministic vs. Stochastic:** Models that have no random input are deterministic; a strict appointment-book operation with fixed service times is an example. Stochastic models, on the other hand, operate with at least some inputs being random—like a bank with randomly arriving customers requiring varying service times. A model can have both deterministic and random inputs in different components; which elements are modeled as deterministic and which as random are issues of modeling realism. Arena easily handles deterministic and stochastic inputs to models and provides many different probability distributions and processes that you can use to represent the random inputs. Since we feel that at least some element of uncertainty is usually present in reality, most of our illustrations involve random inputs somewhere in the model. As noted earlier, though, stochastic models produce uncertain output, which is a fact you must consider carefully in designing and interpreting the runs in your project.

1.3 How Simulations Get Done

If you've determined that a simulation of some sort is appropriate, you next have to decide how to carry it out. In this section, we discuss options for running a simulation, including software.

1.3.1 By Hand

In the beginning, people really *did* do simulations by hand (we'll show you just one, which is painful enough, in Section 2.4).

For instance, around 1733 a fellow by the name of Georges Louis Leclerc (who later was invited into the nobility, due no doubt to his simulation prowess, as Comte de Buffon) described an experiment to estimate the value of π. If you toss a needle of length l onto a table painted with parallel lines spaced d apart (d must be $\geq l$), it turns out that the needle will cross a line with probability $p = 2l/(\pi d)$. So Figure 1-1 shows a simulation experiment to estimate the value of π. (Don't try this at home, or at least not with a big needle.)

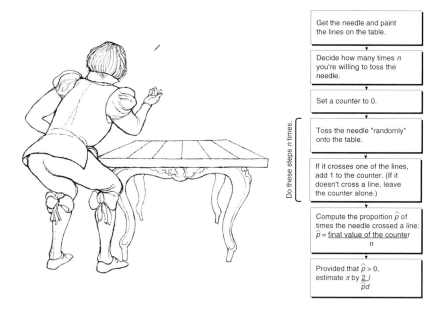

Figure 1-1. The Buffon Needle Problem

Though this experiment may seem pretty simple (probably even silly) to you, there are some aspects of it that are common to most simulations:

- The purpose is to estimate something (in this case, π) whose value would be hard to compute exactly (okay, maybe in 1733 that was true).
- The estimate we get at the end is not going to be exactly right; that is, it has some error associated with it, and it might be nice to get an idea of how large that error is likely to be.
- It seems intuitive that the more tosses we make (that is, the bigger n is), the smaller the error is likely to be and thus the better the estimate is likely to be.
- In fact, you could do a *sequential* experiment and just keep tossing until the probable error is small enough for you to live with, instead of deciding on the number n of tosses beforehand.
- You might be able to reduce the error without increasing the number of tosses if you invest a little up-front work. Weld a second needle to the first so they cross at right angles at their midpoints; such a weapon is called a *Buffon cross*. Leave the lines on the table alone. On each toss, record separately whether each needle crosses a line (it could be that they both cross, neither crosses, or just one but not the other crosses), and get two different estimates of π. It's intuitive (and, happily, true) that whether one needle crosses a line is negatively correlated with whether the other one does, so the two estimates of π will be negatively correlated with each other. The average of the two estimates is also an unbiased estimate of π, but will have less variance than a single-needle estimate from the same number of tosses since it's likely that if one estimate is high, the other one will be low. (This is a physical analog of what's called a *variance-reduction technique*, specifically *antithetic variates*, and is discussed in Section 12.4.2. It seems like some kind of cheat or swindle, but it's really a fair game.)

We'll come back to these kinds of issues as we talk about more interesting and helpful simulations. (For more on the Buffon needle problem, as well as other such interesting historical curiosities, see Morgan, 1984.)

In the 1920s and 1930s, statisticians began using random-number machines and tables in numerical experiments to help them develop and understand statistical theory. For instance, Walter A. Shewhart (the quality control pioneer) did numerical experiments by drawing numbered chips from a bowl to study the first control charts. Guinness Brewery employee W. S. Gossett did similar numerical sampling experiments to help him gain insight into what was going on in mathematical statistics. (To protect his job at Guinness, he published his research under the pseudonym "Student" and also developed the t distribution used widely in statistical inference.) Engineers, physicists, and mathematicians have used various kinds of hand-simulation ideas for many years on a wide variety of problems.

1.3.2 Programming in General-Purpose Languages

As digital computers appeared in the 1950s and 1960s, people began writing computer programs in general-purpose procedural languages like FORTRAN to do simulations of more complicated systems. Support packages were written to help out with routine chores like list processing, keeping track of simulated events, and statistical bookkeeping.

This approach was highly customizable and flexible (in terms of the kinds of models and manipulations possible), but also painfully tedious and error prone since models had to be coded pretty much from scratch every time. (Plus, if you dropped your deck of cards, it could take quite a while to reconstruct your "model.") For a more detailed history of discrete-event simulation languages, see Nance (1996) and Nance and Sargent (2003).

As a kind of descendant of simulating with general-purpose programming languages, people sometimes use spreadsheet software for some kinds of simulation. This has proven popular for static models, perhaps with add-ins to facilitate common operations and to provide higher-quality tools (like random-number generators) than what comes standard with spreadsheets. But for all but the very simplest dynamic models, the inherent limitations of spreadsheets make it at best awkward, and usually practically impossible, to use them for simulations of large, realistic, dynamic models. Section 2.7 briefly illustrates two examples of simulating with spreadsheets.

1.3.3   Simulation Languages

Special-purpose *simulation languages* like GPSS, Simscript, SLAM, and SIMAN appeared on the scene some time later and provided a much better framework for the kinds of simulations many people do. Simulation languages became very popular and are still in use.

Nonetheless, you still have to invest quite a bit of time to learn about their features and how to use them effectively. And, depending on the user interface provided, there can be picky, apparently arbitrary, and certainly frustrating syntactical idiosyncrasies that bedevil even old hands.

1.3.4   High-Level Simulators

Thus, several high-level "simulator" products emerged that are indeed very easy to use. They typically operate by intuitive graphical user interfaces, menus, and dialogs. You select from available simulation-modeling constructs, connect them, and run the model along with a dynamic graphical animation of system components as they move around and change.

However, the domains of many simulators are also rather restricted (like manufacturing or communications) and are generally not as flexible as you might like in order to build valid models of your systems. Some people feel that these packages may have gone too far up the software-hierarchy food chain and have traded away too much flexibility to achieve the ease-of-use goal.

1.3.5   Where Arena Fits In

Arena combines the ease of use found in high-level simulators with the flexibility of simulation languages and even all the way down to general-purpose procedural languages like the Microsoft® Visual Basic® programming system or C if you really want. It does this by providing alternative and interchangeable *templates* of graphical simulation modeling and analysis *modules* that you can combine to build a fairly wide variety of simulation models. For ease of display and organization, modules are typically grouped into *panels* to compose a template. By switching panels, you gain access to a whole different set of simulation modeling constructs and capabilities. In most cases, modules from different panels can be mixed together in the same model.

Arena maintains its modeling flexibility by being fully *hierarchical*, as depicted in Figure 1-2. At any time, you can pull in low-level modules from the Blocks and Elements panel and gain access to simulation-language flexibility if you need to and mix in SIMAN constructs together with the higher-level modules from other templates (Arena is based on, and actually includes, the SIMAN simulation language; see Pedgen, Shannon, and Sadowski 1995 for a complete discussion of SIMAN). For specialized needs, like complex decision algorithms or accessing data from an external application, you can write pieces of your model in a procedural language like Visual Basic or C/C++. All of this, regardless of how high or low you want to go in the hierarchy, takes place in the same consistent graphical user interface.

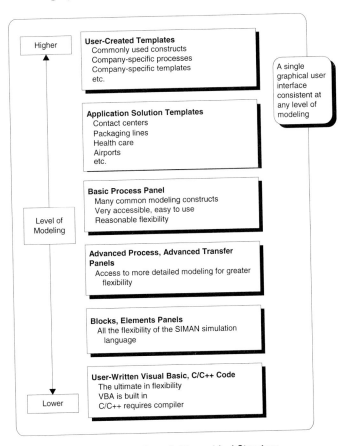

Figure 1-2. Arena's Hierarchical Structure

In fact, the modules in Arena are composed of SIMAN components; you can create your own modules and collect them into your own templates for various classes of systems. For instance, Rockwell Automation (formerly Systems Modeling) has built templates for general modeling (the Arena template, which is the primary focus of this book), high-speed packaging, contact centers, and other industries. Other people have built templates for their company in industries as diverse as mining, auto manufacturing, fast-food, and forest-resource management. In this way, you don't have to compromise between modeling flexibility and ease of use. While this textbook focuses on modeling with the Arena template, you can get a taste of creating your own modules in Chapter 10.

Further, Arena includes dynamic animation in the same work environment. It also provides integrated support, including graphics, for some of the statistical design and analysis issues that are part and parcel of a good simulation study.

1.4 When Simulations Are Used

Just as the capabilities and sophistication of simulation languages and packages have increased dramatically over the last 40 or 50 years, the concept of how and when to use simulation has changed. For a thorough account of the development of computer simulation in general since the earliest days, see Nance and Sargent (2003).

1.4.1 The Early Years

In the late 1950s and 1960s, simulation was a very expensive and specialized tool that was generally used only by large corporations that required substantial capital investments. Typical simulation users were found in steel and aerospace corporations. These organizations would form groups of people, mostly Ph.D.s, who would develop large, complex simulation models using available languages, such as FORTRAN. These models would then be run on large mainframes charging from $600 to $1,000 per hour. Interestingly, the personal computers that reside on most engineers' desks today are much more powerful and faster than the mainframes of the 1960s.

1.4.2 The Formative Years

The use of simulation as we know it today began during the 1970s and early 1980s. Computers were becoming faster and cheaper, and the value of simulation was being discovered by other industries, although most of the companies were still quite large. However, simulation was seldom considered until there was a disaster. It became the tool of choice for many companies, notably in the automotive and heavy industries, for determining why the disaster occurred and, sometimes, where to point the finger of blame.

We recall the startup of an automotive assembly line, an investment of over $100 million, that was not achieving much of its potential. The line was producing a newly released vehicle that was in great demand—far greater than could be satisfied by the existing output of the line. Management appointed a S.W.A.T. team to analyze the problem, and that team quickly estimated the lost potential profit to be in excess of $500,000 per day. The team was told, "Find the problem and fix it." In about three weeks, a simulation was developed and used to identify the problem, which turned out not to have been on the initial suspect list. The line was ultimately modified and did produce according to specifications; unfortunately, by that time the competition was producing similar vehicles, and the additional output was no longer needed. Ironically, a simulation model had been used during the design of the assembly line to determine its feasibility. Unfortunately, many of the processes were new, and engineering had relied on equipment vendors to provide estimates of failures and throughputs. As is often the case, the vendors were extremely optimistic in their estimates. If the original design team had used the simulation to perform a good sensitivity analysis on these questionable data, the problem might have been uncovered and resolved well before implementation.

During this time, simulation also found a home in academia as a standard part of industrial engineering and operations research curricula. Its growing use in industry compelled universities to teach it more widely. At the same time, simulation began to reach into quantitative business programs, broadening the number and type of students and researchers exposed to its potential.

1.4.3 The Recent Past

During the late 1980s, simulation began to establish its real roots in business. A large part of this was due to the introduction of the personal computer and animation. Although simulation was still being used to analyze failed systems, many people were requesting simulations before production was to begin. (However, in most cases, it was really too late to affect the system design, but it did offer the plant manager and system designer the opportunity to spruce up their resumes.) By the end of the 1980s, the value of simulation was being recognized by many larger firms, several of which actually made simulation a requirement before approval of any major capital investment. However, simulation was still not in widespread use and was rarely used by smaller firms.

1.4.4 The Present

Simulation really began to mature during the 1990s. Many smaller firms embraced the tool, and it began to see use at the very early stages of projects—where it could have the greatest impact. Better animation, greater ease of use, faster computers, easy integration with other packages, and the emergence of simulators have all helped simulation become a standard tool in many companies. Although most managers will readily admit that simulation can add value to their enterprise, it has yet to become a standard tool that resides on everyone's computers. The manner in which simulation is used is also changing; it is being employed earlier in the design phase and is often being updated as changes are made to operating systems. This provides a living simulation model that can be used for systems analysis on very short notice. Simulation has also invaded the service industry where it is being applied in many non-traditional areas.

The major impediments preventing simulation from becoming a universally accepted and well-utilized tool are model-development time and the modeling skills required for the development of a successful simulation. Those are probably the reasons why you're reading this book!

1.4.5 The Future

The rate of change in simulation has accelerated in recent years, and there is every reason to believe that it will continue its rapid growth and cross the bridges to mainstream acceptance. Simulation software has taken advantage of new operating systems to provide greater ease of use, particularly for the first-time user. This trend must continue if simulation is to become a state-of-the-art tool resident on every systems-analysis computer. These new operating systems have also allowed for greater integration of simulation with other packages (like spreadsheets, databases, and word processors). It is now becoming possible to foresee the complete integration of simulation with other software packages that collect, store, and analyze system data at the front end along with software that helps control the system at the back end.

The Internet and intranets have had a tremendous impact on the way organizations conduct their business. Information-sharing in real time is not only possible, but is becoming mandatory. Simulation tools are being developed to support distributed model-building, distributed processing, and remote analysis of results. Soon, a set of enterprise-wide tools will be available to provide the ability for employees throughout an organization to obtain answers to critical and even routine questions.

In order to make simulation easier to use by more people, we will see more vertical products aimed at very specific markets. This will allow analysts to construct simulations easily, using modeling constructs designed for their industry or company with terminology that directly relates to their environment. These may be very specialized tools designed for very specific environments, but they should still have the capability to model any system activities that are unique to each simulation project. Some of these types of products are on the market today in application areas such as communications, semiconductors, contact centers, and business-process re-engineering.

Today's simulation projects concentrate on the design or redesign of complex systems. They often must deal with complex system-control issues, which can lead to the development of new system-control logic that is tested using the developed simulation. The next logical step is to use that same simulation to control the real system (Wysk, Smith, Sturrock, Ramaswamy, Smith, and Joshi, 1994). This approach requires that the simulation model be kept current, but it also allows for easy testing of new system controls as the system or products change over time. As we progress to this next logical step, simulations will no longer be disposable or used only once, but will become a critical part of the operation of the ongoing system.

With the rapid advances being made in computers and software, it is very difficult to predict much about the simulations of the distant future, but even now we are seeing the development and implementation of features such as automatic statistical analysis, software that recommends system changes, simulations totally integrated with system operating software, and yes, even virtual reality.

CHAPTER 2

Fundamental Simulation Concepts

In this chapter, we introduce some of the underlying ideas, methods, and issues in simulation before getting into the Arena software itself in Chapter 3 and beyond. These concepts are the same across any kind of simulation software, and some familiarity with them is essential to understanding how Arena simulates a model you've built.

We do this mostly by carrying through a simple example, which is described in Section 2.1. In Section 2.2, we explore some options for dealing with the example model. Section 2.3 describes the various pieces of a simulation model, and Section 2.4 carries out the simulation (by hand), describing the fundamental organization and action. After that, two different simulation-modeling approaches are contrasted in Section 2.5, and the issue of randomness in both input and output is introduced in Section 2.6. Section 2.7 considers using spreadsheets to simulate, with both a static and a (very) simple dynamic model. Finally, Section 2.8 steps back and looks at what's involved in a simulation project, although this is taken up more thoroughly as part of Chapter 13.

By the end of this chapter, you'll understand the fundamental logic, structure, components, and management of a simulation modeling project. All of this underlies Arena and the richer models you'll build with it after reading subsequent chapters.

2.1 An Example

In this section, we describe the example system and decide what we'd like to know about its behavior and performance.

2.1.1 The System

Since a lot of simulation models involve waiting lines or *queues* as building blocks, we'll start with a very simple case of such a model representing a portion of a manufacturing facility. "Blank" parts arrive to a drilling center, are processed by a single drill press, and then leave; see Figure 2-1. If a part arrives and finds the drill press idle, its processing at the drill press starts right away; otherwise, it waits in a First-In First-Out (FIFO) queue. This is the *logical* structure of the model.

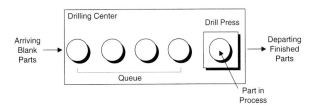

Figure 2-1. A Simple Processing System

You have to specify the *numerical* aspects as well, including how the simulation starts and stops. First, decide on the underlying "base" units with which time will be measured; we'll use minutes for all time measurements here. It doesn't logically matter what the time units are, so pick whatever is most appropriate, familiar, and convenient for your application.[1] You can express input time quantities in different units if it's custom- ary or convenient, like minutes for mean service times but hours for mean machine-up times. However, for arithmetic and calculations, all times must be converted to the base time units if they're not already in it. Arena allows you to express input times in different units, but you must also declare the base time units into which all times are converted during the simulation run, and in which all time-based outputs are reported.

The system starts at time 0 minutes with no parts present and the drill press idle. This *empty-and-idle* assumption would be realistic if the system starts afresh each morning, but might not be so great as a model of the initial situation to simulate an ongoing operation.

The time durations that will make the simulation move are in Table 2-1. The (sequen- tial) part number is in the first column, the second column has the time of arrival of each part, the third column gives the time *between* a part's arrival and that of the next part (called an *interarrival time*), and the service time (required to process on the drill press, not counting any time spent waiting in the queue) is in the last column. All times are in minutes. You're probably wondering where all these numbers came from; don't worry about that right now, and just pretend we observed them in the drilling center or that we brashly made them up.

Table 2-1. Arrival, Interarrival, and Service Times of Parts (in Minutes)

Part Number	Arrival Time	Interarrival Time	Service Time
1	0.00	1.73	2.90
2	1.73	1.35	1.76
3	3.08	0.71	3.39
4	3.79	0.62	4.52
5	4.41	14.28	4.46
6	18.69	0.70	4.36
7	19.39	15.52	2.07
8	34.91	3.15	3.36
9	38.06	1.76	2.37
10	39.82	1.00	5.38
11	40.82	.	.
.	.	.	.
.	.	.	.

[1] Not only should you be sensible about choosing the base time units (e.g., for a simulation of 20 years, don't choose seconds as your base units, and for a simulation of two minutes, don't measure time in days), but you should also choose units that avoid both extremely big and extremely tiny time values in the same model since, even with Arena's double-precision arithmetic, the computer might have trouble with round-off error.

We've decided that the simulation will stop at exactly time 20 minutes. If there are any parts present at that time (in service at the drill press or waiting in the queue), they are never finished.

2.1.2 Goals of the Study

Given a logical/numerical model like this, you next have to decide what output performance measures you want to collect. Here's what we decided to compute:

- The *total production* (number of parts that complete their service at the drill press and leave) during the 20 minutes of operation. Presumably, more is better.
- The *average waiting time in queue* of parts that enter service at the drill press during the simulation. This time in queue records only the time a part is waiting in the queue and not the time it spends being served at the drill press. If WQ_i is the waiting time in queue of the ith part and it turns out that N parts leave the queue during the 20-minute run, this average is

$$\frac{\sum_{i=1}^{N} WQ_i}{N}.$$

(Note that since Part 1 arrives at time 0 to find the drill press idle, $WQ_1 = 0$ and $N \geq 1$ for sure so we don't have to worry about dividing by zero.) This is generally called a *discrete-time* (or *discrete-parameter*) statistic since it refers to data, in this case the waiting times WQ_1, WQ_2, . . ., for which there is a natural first, second, . . . observation; in Arena, these are called *tally* statistics since values of them are "tallied" when they are observed (using a feature of the underlying SIMAN simulation language called *Tally*). From a performance standpoint, small is good.

- The *maximum waiting time in queue* of parts that enter service at the drill press during the simulation. This is a worst-case measure, which might be of interest in giving service-level guarantees to customers. Small is good.
- The *time-average number of parts waiting in the queue* (again, not counting any part in service at the drill press). By "time average," we mean a weighted average of the possible queue lengths (0, 1, 2, . . .) weighted by the proportion of time during the run that the queue was at that length. Letting $Q(t)$ be the number of parts in the queue at any time instant t, this time-average queue length is the total area under the $Q(t)$ curve, divided by the length of the run, 20. In integral-calculus terms, this is

$$\frac{\int_0^{20} Q(t)dt}{20}.$$

Such *time-persistent* statistics are common in simulation. This one indicates how long the queue is (on average), which might be of interest for allocating floor space.

- The *maximum number of parts that were ever waiting in the queue*. Actually, this might be a better indication of how much floor space is needed than is the time average if you want to be reasonably sure to have room at all times. This is another worst-case measure, and smaller is presumably better.
- The *average* and *maximum total time in system* of parts that finish being processed on the drill press and leave. Also called *cycle time*, this is the time that elapses between a part's arrival and its departure, so it's the sum of the part's waiting time in queue and its service time at the drill press. This is a kind of turnaround time, so smaller is better.
- The *utilization* of the drill press, defined as the proportion of time it is busy during the simulation. Think of this as another time-persistent statistic, but of the "busy" function

$$B(t) = \begin{cases} 1 \text{ if the drill press is busy at time } t \\ 0 \text{ if the drill press is idle at time } t \end{cases}.$$

The utilization is the area under $B(t)$, divided by the length of the run:

$$\frac{\int_0^{20} B(t)dt}{20}.$$

Resource utilizations are of obvious interest in many simulations, but it's hard to say whether you "want" them to be high (close to 1) or low (close to 0). High is good since it indicates little excess capacity, but can also be bad since it might mean a lot of congestion in the form of long queues and slow throughput.

There are usually a lot of possible output performance measures, and it's probably a good idea to observe a lot of things in a simulation since you can always ignore things you have but can never look at things you don't have, plus sometimes you might find a surprise. The only downside is that collecting extraneous data can slow down execution of the simulation.

2.2 Analysis Options

With the model, its inputs, and its outputs defined, you next have to figure out how to get the outputs by transforming the inputs according to the model's logic. In this section, we'll briefly explore a few options for doing this.

2.2.1 Educated Guessing

While we're not big fans of guessing, a crude "back-of-the-envelope" calculation can sometimes lend at least qualitative insight (and sometimes not). How this goes, of course, completely depends on the situation (and on how good you are at guessing).

A possible first cut in our example is to look at the average inflow and processing rates. From Table 2-1, it turns out that the average of the ten interarrival times is 4.08 minutes, and the average of the ten service requirements is 3.46 minutes. This looks pretty hopeful, since parts are being served faster than they're arriving, at least on average, meaning that the system has a chance of operating in a stable way over the long term and not "exploding." If these averages were *exactly* what happened for each part— no variation either way—then there would never be a queue and all waiting times in queue would be zero, a happy thought indeed. No matter how happy this thought might be, though, it's probably wrong since the data clearly show that there *is* variation in the interarrival and service times, which could create a queue sometimes; for example, if there happened to be some small interarrival times during a period when there also happened to be some large service times.

Suppose, on the other hand, that the averages in the input data in Table 2-1 had come out the other way, with the average interarrival time being smaller than the average service time. If this situation persisted, parts would, on average, be arriving faster than they could be served, implying heavy congestion (at least after a while, probably longer than the 20-minute run we have planned). Indeed, the system will explode over the long run— *not* a happy thought.

The truth, as usual, will probably be between the extremes. Clearly, guessing has its limits.

2.2.2 Queueing Theory

Since this is a queue, why not use queueing theory? It's been around for almost a century, and a lot of very bright people have worked very hard to develop it. In some situations, it can result in simple formulas from which you can get a lot of insight.

Probably the simplest and most popular object of queueing theory is the *M/M/1 queue.* The first "M" states that the arrival process is *Markovian*; that is, the interarrival times are independent and identically distributed "draws" from an exponential probability distribution (see Appendices C and D for a brief refresher on probability and distributions). The second "M" stands for the service-time distribution, and here it's also exponential. The "1" indicates that there's just a single server. So at least on the surface this looks pretty good for our model.

Better yet, most of our output performance measures can be expressed as simple formulas. For instance, the average waiting time in queue (expected from a long run) is

$$\frac{\mu_S^2}{\mu_A - \mu_S}$$

just where μ_A is the expected value of the interarrival-time distribution and μ_S is the expected value of the service-time distribution (assuming that $\mu_A > \mu_S$ so the queue doesn't explode). So one immediate idea is to use the data to estimate μ_A and μ_S, then plug these estimates into the formula; for our data, we get $3.46^2/(4.08 - 3.46) = 19.31$ minutes.

Such an approach can sometimes give a reasonable order-of-magnitude approximation that might facilitate crude comparisons. But there are problems too (see Exercise 2-6):

- The estimates of μ_A and μ_S aren't exact, so there will be error in the result as well.
- The assumptions of exponential interarrival-time and service-time distributions are essential to deriving the formula above, and we probably don't satisfy these

assumptions. This calls into question the validity of the formula. While there are more sophisticated versions for more general queueing models, there will always be assumptions to worry about.

- The formula is for long-run performance, not the 20-minute period we want. This is typical of most (but not all) queueing theory.
- The formula doesn't provide any information on the natural variability in the system. This is not only a difficulty in analysis but might also be of inherent interest itself, as in the variability of production. (It's sometimes possible, though, to find other formulas that measure variability.)

Many people feel that queueing theory can prove valuable as a first-cut approximation to get an idea of where things stand and to provide guidance about what kinds of simulations might be appropriate at the next step in the project. We agree, but urge you to keep in mind the problems listed above and temper your interpretations accordingly.

2.2.3 Mechanistic Simulation

So all of this brings us back to simulation. By "mechanistic" we mean that the individual operations (arrivals, service by the drill press, etc.) will occur as they would in reality. The movements and changes of things in the simulation model occur at the right "time," in the right order, and have the right effects on each other and the statistical-accumulator variables.

In this way, simulation provides a completely concrete, "brute-force" way of dealing directly with the model. There's nothing mysterious about how it works—just a few basic ideas and then a whole lot of details and bookkeeping that software like Arena handles for you.

2.3 Pieces of a Simulation Model

We'll talk about the various parts of a simulation model in this section, all in reference to our example.

2.3.1 Entities

Most simulations involve "players" called *entities* that move around, change status, affect and are affected by other entities and the state of the system, and affect the output performance measures. Entities are the *dynamic* objects in the simulation—they usually are created, move around for a while, and then are disposed of as they leave. It's possible, though, to have entities that never leave but just keep circulating in the system. However, all entities have to be created, either by you or automatically by the software.

The entities for our example are the parts to be processed. They're created when they arrive, move through the queue (if necessary), are served by the drill press, and are then disposed of as they leave. Even though there's only one kind of entity in our example, there can be many independent "copies," or *realizations* of it in existence at a time, just as there can be many different individual parts of this type in the real system at a time.

Most entities represent "real" things in a simulation. You can have lots of different kinds of entities and many realizations of each kind of entity existing in the model at a time. For instance, you could have several different *kinds* of parts, perhaps requiring

Figure 2-2. A Victorious Breakdown Demon

different processing and routing and having different priority; moreover, there could be several realizations of each kind of part floating around in the model at a time.

There are situations, though, where "fake" (or "logic") entities not corresponding to anything tangible can be conjured up to take care of certain modeling operations. For instance, one way to model machine failures is to create a "breakdown demon" (see Figure 2-2) that lurks in the shadows during the machine's up time, runs out and kicks the machine when it's supposed to break down, stands triumphantly over it until it gets repaired, then scurries back to the shadows and begins another lurking period representing the machine's next up time. A similar example is a "break angel" that arrives periodically and takes a server off duty. (For both of these examples, however, Arena has easier built-in modeling constructs that don't require creation of such fake entities.)

Figuring out what the entities are is probably the first thing you need to do in modeling a system.

2.3.2 Attributes

To individualize entities, you attach *attributes* to them. An attribute is a common characteristic of all entities, but with a specific value that can differ from one entity to another. For instance, our part entities could have attributes called `Due Date`, `Priority`, and `Color` to indicate these characteristics for each individual entity. It's up to you to figure out what attributes your entities need, name them, assign values to them, change them as appropriate, and then use them when it's time (that's all part of modeling).

The most important thing to remember about attributes is that their values are tied to specific entities. The same attribute will generally have different values for different entities, just as different parts have different due dates, priorities, and color codes. Think of an attribute as a tag attached to each entity, but what's written on this tag can differ across entities to characterize them individually. An analogy to traditional computer programming is that attributes are *local* variables—in this case, local to each individual entity.

Arena keeps track of some attributes automatically, but you may need to define, assign values to, change, and use attributes of your own.

2.3.3 (Global) Variables

A *variable* (or a *global* variable) is a piece of information that reflects some characteristic of your system, regardless of how many or what kinds of entities might be around. You can have many different variables in a model, but each one is unique. There are two types of variables: Arena built-in variables (number in queue, number of busy servers, current simulation clock time, and so on) and user-defined variables (mean service time, travel time, current shift, and so on). In contrast to attributes, variables are not tied to any specific entity, but rather pertain to the system at large. They're accessible by all entities, and many can be changed by any entity. If you think of attributes as tags attached to the entities currently floating around in the room, then think of variables as (rewriteable) writing on the wall.

Variables are used for lots of different purposes. For instance, the time to move between any two stations in a model might be the same throughout the model, and a variable called `Transfer Time` could be defined and set to the appropriate value and then used wherever this constant is needed; in a modified model where this time is set to a different constant, you'd only need to change the definition of `Transfer Time` to change its value throughout the model. Variables can also represent something that changes *during* the simulation, like the number of parts in a certain subassembly area of the larger model, which is incremented by a part entity when it enters the area and decremented by a part when it leaves the area. Arena Variables can be vectors or matrices if it is convenient to organize the information as lists or two-dimensional tables of individual values.

Some built-in Arena variables for our model include the status of the drill press (busy or idle), the time (simulation clock), and the current length of the queue.

2.3.4 Resources

Entities often compete with each other for service from *resources* that represent things like personnel, equipment, or space in a storage area of limited size. An entity *seizes* (units of) a resource when available and *releases* it (or them) when finished. It's better to think of the resource as being given to the entity rather than the entity being assigned to the resource since an entity (like a part) could need simultaneous service from multiple resources (such as a machine and a person).

A resource can represent a group of several individual servers, each of which is called a *unit* of that resource. This is useful to model, for instance, several identical "parallel" agents at an airline ticketing counter. The number of available units of a resource can be changed during the simulation run to represent agents going on break or opening up their stations if things get busy. If a resource has multiple units, or a variable number of units, we have to generalize our definition in Section 2.1.2 of resource utilization to be the time-average number of units of the resource that are busy, divided by the time-average number of units of the resource that are available. In our example, there is just a single drill press, so this resource has a single unit available at all times.

2.3.5 Queues

When an entity can't move on, perhaps because it needs to seize a unit of a resource that's tied up by another entity, it needs a place to wait, which is the purpose of a *queue*. In Arena, queues have names and can also have capacities to represent, for instance, limited floor space for a buffer. You'd have to decide as part of your modeling how to handle an entity arriving at a queue that's already full.

2.3.6 Statistical Accumulators

To get your output performance measures, you have to keep track of various intermediate *statistical-accumulator variables* as the simulation progresses. In our example, we'll watch:

- The number of parts produced so far
- The total of the waiting times in queue so far
- The number of parts that have passed through the queue so far (since we'll need this as the denominator in the average waiting-time output measure)
- The longest time spent in queue we've seen so far
- The total of the time spent in the system by all parts that have departed so far
- The longest time in system we've seen so far
- The area so far under the queue-length curve $Q(t)$
- The highest level that $Q(t)$ has so far attained
- The area so far under the server-busy function $B(t)$

All of these accumulators should be initialized to 0. When something happens in the simulation, you have to update the affected accumulators in the appropriate way.

Arena takes care of most (but sometimes not all) of the statistical accumulation you're likely to want, so most of this will be invisible to you except for asking for it in some situations. But in our hand simulation, we'll do it all manually so you can see how it goes and will understand how to do it yourself in Arena if you need to.

2.3.7 Events

Now let's turn to how things work when we run our model. Basically, everything's centered around events. An *event* is something that happens at an instant of (simulated) time that might change attributes, variables, or statistical accumulators. In our example, there are three kinds of events:

- **Arrival:** A new part enters the system.
- **Departure:** A part finishes its service at the drill press and leaves the system.
- **The End:** The simulation is stopped at time 20 minutes. (It might seem rather artificial to anoint this as an event, but it certainly changes things, and this is one way to stop a simulation.)

In addition to the above events, there of course must be an initialization to set things up. We'll explain the logic of each event in more detail later.

Other things happen in our example model, but needn't be separate events. For instance, parts leave the queue and begin service at the drill press, which changes the system, but this only happens because of some other entity's departure, which is already an event.

To execute, a simulation has to keep track of the events that are supposed to happen in the (simulated) future. In Arena, this information is stored in an *event calendar*. We won't get into the details of the event calendar's data structure, but here's the idea: When the logic of the simulation calls for it, a *record* of information for a future event is placed on the event calendar. This event record contains identification of the entity involved, the event time, and the kind of event it will be. Arena places each newly scheduled event on the calendar so that the next (soonest) event is always at the top of the calendar (that is, the new event record is *sorted* onto the calendar in increasing order of event times). When it's time to execute the next event, the top record is removed from the calendar and the information in this record is used to execute the appropriate logic; part of this logic might be to place one or more new event records onto the calendar. It's possible that, at a certain time, it doesn't make sense to have a certain event type scheduled (in our model, if the drill press is idle, you don't want a departure event to be scheduled), in which case there's just no record for that kind of event on the calendar, so it obviously can't happen next. Though this model doesn't require it, it's also possible to have several events of the same kind scheduled on the calendar at once, for different times and for different entities.

In a discrete-event model, the variables that describe the system don't change between successive events. Most of the work in event-driven simulation involves getting the logic right for what happens with each kind of event. As you'll see later, though, modeling with Arena usually gets you out of having to define this detailed event logic explicitly, although you can do so if you want in order to represent something very peculiar to your model that Arena isn't set up to handle directly.

2.3.8   Simulation Clock

The current value of time in the simulation is simply held in a variable called the *simulation clock*. Unlike real time, the simulation clock does not take on all values and flow continuously; rather, it lurches from the time of one event to the time of the next event scheduled to happen. Since nothing changes between events, there is no need to waste (real) time looking at (simulated) times that don't matter.

The simulation clock interacts closely with the event calendar. At initialization of the simulation, and then after executing each event, the event calendar's top record (always the one for the next event) is taken off the calendar. The simulation clock lurches forward to the time of that event (one of the data fields in the event record), and the information in the removed event record (entity identification, event time, and event type) is used to execute the event at that instant of simulated time. How the event is executed clearly depends on what kind of event it is as well as on the model state at that time, but in general could include updating variables and statistical accumulators, altering entity attributes, and placing new event records onto the calendar.

While we'll keep track of the simulation clock and event calendar ourselves in the hand simulation, these are clearly important pieces of any dynamic simulation, so Arena keeps track of them (the clock is a variable called TNOW in Arena).

2.3.9   Starting and Stopping

Important, but sometimes-overlooked, issues in a simulation are how it will start and stop. For our example, we've made specific assumptions about this, so it'll be easy to figure out how to translate them into values for attributes, variables, accumulators, the event calendar, and the clock.

Arena does a lot of things for you automatically, but it can't decide modeling issues like starting and stopping rules. You have to determine the appropriate starting conditions, how long a run should last, and whether it should stop at a particular time (as we'll do at time 20 minutes) or whether it should stop when something specific happens (like as soon as 100 finished parts are produced). It's important to think about this and make assumptions consistent with what you're modeling; these decisions can have just as great an effect on your results as can more obvious things like values of input parameters (such as interarrival-time means, service-time variances, and the number of machines).

You should do *something* specific (and conscious) to stop the simulation with Arena, since it turns out that, in many situations, taking all the defaults will cause your simulation to run forever (or until you get sick of waiting and kill it, whichever comes first).

2.4 Event-Driven Hand Simulation

We'll let you have the gory details of the hand simulation in this section, after outlining the action and defining how to keep track of things.

2.4.1 Outline of the Action

Here's roughly how things go for each event:

- **Arrival:** A new part shows up.
    - Schedule the next new part to arrive later, at the next arrival time, by placing a new event record for it onto the event calendar.
    - Update the time-persistent statistics (between the last event and now).
    - Store the arriving part's time of arrival (the current value of the clock) in an attribute, which will be needed later to compute its total time in system and possibly the time it spends waiting in the queue.
    - If the drill press is idle, the arriving part goes right into service (experiencing a time in queue of zero), so the drill press is made busy and the end of this part's service is scheduled. Tally this part's time in queue (zero).
    - On the other hand, if the drill press is already busy with another part, the arriving part is put at the end of the queue and the queue-length variable is incremented.
- **Departure:** The part being served by the drill press is done and ready to leave.
    - Increment the number-produced statistical accumulator.
    - Compute and tally the total time in system of the departing part by taking the current value of the clock minus the part's arrival time (stored in an attribute during the Arrival event).
    - Update the time-persistent statistics.
    - If there are any parts in queue, take the first one out, compute and tally its time in queue (which is now ending), and begin its service at the drill press by scheduling its departure event (that is, place it on the event calendar).
    - On the other hand, if the queue is empty, then make the drill press idle. Note that in this case, there's no departure event scheduled on the event calendar.
- **The End:** The simulation is over.
    - Update the time-persistent statistics to the end of the simulation.
    - Compute and report the final summary output performance measures.

After each event (except the end event), the event calendar's top record is removed, indicating what event will happen next and at what time. The simulation clock is advanced to that time, and the appropriate logic is carried out.

2.4.2   Keeping Track of Things

All the calculations for the hand simulation are detailed in Table 2-2. Traces of $Q(t)$ and $B(t)$ over the whole simulation are in Figure 2-3. Each row in Table 2-2 represents an event concerning a particular part (in the first column) at time t (second column), and the situation *after* completion of the logic for that event (in the other columns). The other column groups are:

- **Event:** This describes what just happened; Arr and Dep refer respectively to an arrival and a departure.
- **Variables:** These are the values of the number $Q(t)$ of parts in queue and the server-busy function $B(t)$.
- **Attributes:** Each arriving entity's arrival time is assigned when it arrives and is carried along with it throughout. If a part is in service at the drill press, its arrival time is at the right edge of the column and is not enclosed in parentheses. The arrival times of any parts in the queue, in right-to-left order (to agree with Figure 2-1), extend back toward the left and are in parentheses. For instance, at the end of the run, the part in service arrived at time 18.69, and the part that's first (and only) in the queue arrived at time 19.39. We have to keep track of these to compute the time in queue of a part when it enters service at the drill press after having waited in the queue as well as the total time in system of a part when it leaves.
- **Statistical Accumulators:** We have to initialize and then update these as we go along to watch what happens. They are:

  P = the total number of parts produced so far
  N = the number of entities that have passed through the queue so far
  ΣWQ = the sum of the queue times that have been observed so far
  WQ^* = the maximum time in queue observed so far
  ΣTS = the sum of the total times in system that have been observed so far
  TS^* = the maximum total time in system observed so far
  $\int Q$ = the area under the $Q(t)$ curve so far
  Q^* = the maximum value of $Q(t)$ so far
  $\int B$ = the area under the $B(t)$ curve so far

- **Event Calendar:** These are the event records as described earlier. Note that, at each event time, the top record on the event calendar becomes the first three entries under "Just-Finished Event" at the left edge of the entire table in the next row, at the next event time.

Table 2-2. Record of the Hand Simulation

Just-Finished Event: Entity No.	Time t	Event Type	Q(t)	B(t)	Arrival Times: (In Queue)	In Service	P	N	ΣWQ	WQ^*	ΣTS	TS^*	$\int Q$	Q^*	$\int B$	Event Calendar [Entity No., Time, Type]
—	0.00	Init	0	0	( )	—	0	0	0.00	0.00	0.00	0.00	0.00	0	0.00	[1, 0.00, Arr] [−, 20.00, End]
1	0.00	Arr	0	1	( )	0.00	0	1	0.00	0.00	0.00	0.00	0.00	0	0.00	[2, 1.73, Arr] [1, 2.90, Dep] [−, 20.00, End]
2	1.73	Arr	1	1	(1.73)	0.00	0	1	0.00	0.00	0.00	0.00	0.00	1	1.73	[1, 2.90, Dep] [3, 3.08, Arr] [−, 20.00, End]
1	2.90	Dep	0	1	( )	1.73	1	2	1.17	1.17	2.90	2.90	1.17	1	2.90	[3, 3.08, Arr] [2, 4.66, Dep] [−, 20.00, End]
3	3.08	Arr	1	1	(3.08)	1.73	1	2	1.17	1.17	2.90	2.90	1.17	1	3.08	[4, 3.79, Arr] [2, 4.66, Dep] [−, 20.00, End]
4	3.79	Arr	2	1	(3.79, 3.08)	1.73	1	2	1.17	1.17	2.90	2.90	1.88	2	3.79	[5, 4.41, Arr] [2, 4.66, Dep] [−, 20.00, End]
5	4.41	Arr	3	1	(4.41, 3.79, 3.08)	1.73	1	2	1.17	1.17	2.90	2.90	3.12	3	4.41	[2, 4.66, Dep] [6, 18.69, Arr] [−, 20.00, End]
2	4.66	Dep	2	1	(4.41, 3.79)	3.08	2	3	2.75	1.58	5.83	2.93	3.87	3	4.66	[3, 8.05, Dep] [6, 18.69, Arr] [−, 20.00, End]
3	8.05	Dep	1	1	(4.41)	3.79	3	4	7.01	4.26	10.80	4.97	10.65	3	8.05	[4, 12.57, Dep] [6, 18.69, Arr] [−, 20.00, End]
4	12.57	Dep	0	1	( )	4.41	4	5	15.17	8.16	19.58	8.78	15.17	3	12.57	[5, 17.03, Dep] [6, 18.69, Arr] [−, 20.00, End]
5	17.03	Dep	0	0	( )	—	5	5	15.17	8.16	32.20	12.62	15.17	3	17.03	[6, 18.69, Arr] [−, 20.00, End]
6	18.69	Arr	0	1	( )	18.69	5	6	15.17	8.16	32.20	12.62	15.17	3	17.03	[7, 19.39, Arr] [−, 20.00, End] [6, 23.05, Dep]
7	19.39	Arr	1	1	(19.39)	18.69	5	6	15.17	8.16	32.20	12.62	15.17	3	17.73	[−, 20.00, End] [6, 23.05, Dep] [8, 34.91, Arr]
—	20.00	End	1	1	(19.39)	18.69	5	6	15.17	8.16	32.20	12.62	15.78	3	18.34	[6, 23.05, Dep] [8, 34.91, Arr]

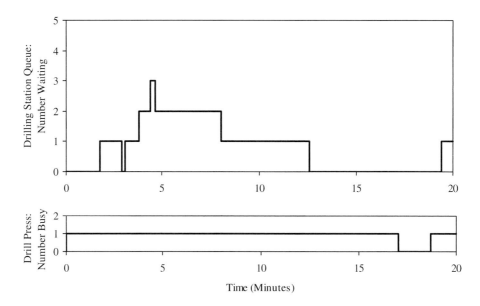

Figure 2-3. Time-Persistent Curves for the Hand Simulation

2.4.3 Carrying It Out

Here's a brief narrative of the action:

- $t = 0.00$, **Init:** The model is initialized, with all variables and accumulators set to 0, the queue empty, the drill press idle, and the event calendar primed with the first arrival happening at time 0.00 and the end of the simulation scheduled for time 20.00. To see what happens next, just take the first event record off the event calendar—the arrival of Entity 1 at time 0.00.

- **Entity 1, $t = 0.00$, Arr:** The next (second) part arrival is scheduled by creating a Part 2 entity, setting its Arrival Time to the current time (0) plus its interarrival time (1.73, from Table 2-1), and placing it on the event calendar. Meanwhile, the part arriving now (Part 1) makes the drill press busy, and the arrival time, 0.00, of this part is stored in its attribute (not in parentheses). The queue is still empty because this part finds the drill press idle and begins service immediately (so the parentheses to contain the times of arrival of parts in queue have nothing between them). Since the entity passed through the queue (with time in queue of 0), N is incremented, ΣWQ is augmented by the waiting time in queue (0), and we check to see if a new maximum waiting time in queue has occurred (no). There has been no production yet, so P stays at 0. No total times in system have yet been observed, so ΣTS and TS^* are unchanged. The time-persistent statistics $\int Q$, Q^*, and $\int B$ remain at 0 since no time has yet passed. Referring to Table 2-1, the service time for Part 1 is set to 2.90 minutes, and the Part 1 entity is returned to the event calendar. Taking the top record off the event calendar, the next event will be the arrival of Entity 2 at time 1.73.

- **Entity 2, $t = 1.73$, Arr:** The next (third) part arrival is scheduled by creating a Part 3 entity, setting its Arrival Time to the current time (1.73) plus its interarrival time (1.35, from Table 2-1), and placing it on the event calendar, which schedules the arrival of this part at time $1.73 + 1.35 = 3.08$. Meanwhile, the part arriving now (Part 2) finds the drill press already busy (with Part 1), so it must queue up. The drill press is still busy, so $B(t)$ stays at 1, but the queue length $Q(t)$ is incremented from 0 to 1. Now there is a queue, and the arrival time (the time now, which is 1.73) of the part in the queue (the currently arriving Part 2) is stored as an attribute of that part in the queue (in parentheses); Part 1, which arrived at time 0.00, is still in service. This event has resulted in no new production and no new time-in-queue observations, so P, N, ΣWQ, WQ^*, ΣTS, and TS^* are unchanged. $\int Q$ is augmented by $0 \times (1.73 - 0.00) = 0$ and $\int B$ is augmented by $1 \times (1.73 - 0.00) = 1.73$. Since the new value of $Q(t)$ is 1, which is greater than the former $Q^* = 0$, we set $Q^* = 1$ as the new maximum queue length observed so far. We don't update the time of the next departure, since it's still (correctly) scheduled as the departure time of Part 1, which arrived earlier and is now partially processed. The next event will be the departure of Part 1 at time 2.90.

- **Entity 1, $t = 2.90$, Dep:** Since this is a departure event, we don't need to schedule the next arrival—it's already on its way, with the (correct) arrival event [3, 3.08, Arr] already on the event calendar. Part 1 is now done being processed by the drill press and ready to depart. Since there is a queue, the drill press will stay busy, so $B(t)$ remains at 1, but $Q(t)$ is decremented (from 1 to 0) and the queue becomes empty (as Part 2, which arrived at time 1.73, leaves it to enter service). Part 2's waiting time in queue of duration $2.90 - 1.73 = 1.17$ has now been completed, which is added into ΣWQ, and N is incremented; this is a new maximum waiting time in queue, so WQ^* is redefined to 1.17. A finished part (Part 1) has been produced so P is incremented and the total time in system for Part 1 is computed as $2.90 - 0.00 = 2.90$, which is added into ΣTS; this is a new maximum total time in system, so TS^* is updated to 2.90. We augment $\int Q$ by $1 \times (2.90 - 1.73) = 1.17$, and $\int B$ is augmented by the same amount. No new maximum for $Q(t)$ has been realized, so Q^* is unchanged at 1. From Table 2-1, the next service time is 1.76, so the next departure (of Part 2, which is now entering service) will be at time $2.90 + 1.76 = 4.66$. The next event is the arrival of Part 3 at time 3.08.

- **Entity 3, $t = 3.08$, Arr:** The next (fourth) part arrival is scheduled by creating a Part 4 entity, setting its Arrival Time to $3.08 + 0.71 = 3.79$ (0.71 is the next interarrival time in Table 2-1), and placing it on the event calendar. Meanwhile, the part arriving now (Part 3) finds the drill press already busy (with Part 2), so Part 3 queues up. The drill press is still busy, so $B(t)$ stays at 1, but the queue length $Q(t)$ is incremented from 0 to 1. The arrival time (3.08) of the part in queue (the currently arriving Part 3) is stored as an attribute of that part; Part 2, which arrived at time 1.73, is still in service. With this event, we're not experiencing any new production or time-in-queue observations, so P, N, ΣWQ, WQ^*, ΣTS, and TS^* are unchanged. $\int Q$ is augmented by $0 \times (3.08 - 2.90) = 0$ and $\int B$ is augmented by

$1\times(3.08 - 2.90) = 0.18$. The new value of $Q(t)$ is 1, which is no more than the former $Q^* = 1$, so we leave $Q^* = 1$ unchanged. We don't update the time of the next departure, since it's still (correctly) scheduled as the departure time of Part 2, which arrived earlier and is now partially processed. The next event will be the arrival of Part 4 at time 3.79.

- **Entity 4, $t = 3.79$, Arr:** The next (fifth) part arrival is scheduled for time $3.79 + 0.62 = 4.41$. The part arriving now (Part 4) queues up since the drill press is now busy; $B(t)$ stays at 1, and the queue length $Q(t)$ is incremented to 2. The arrival time (3.79) of Part 4, now arriving, is stored as an attribute of that part, and placed at the end of the queue (the queue is FIFO); Part 2, which arrived at time 1.73, is still in service. We're not experiencing any new production or time-in-queue observations, so P, N, ΣWQ, WQ^*, ΣTS, and TS^* are unchanged. $\int Q$ is augmented by $1\times(3.79 - 3.08) = 0.71$ and $\int B$ is augmented by $1\times(3.79 - 3.08) = 0.71$. The new value of $Q(t)$ is 2, which exceeds the former $Q^* = 1$, so we now set $Q^* = 2$. We don't update the time of the next departure, since it's still (correctly) scheduled as the departure time of Part 2. The next event will be the arrival of Part 5 at time 4.41.
- **Entity 5, $t = 4.41$, Arr:** The next (sixth) part arrival is scheduled for time $4.41 + 14.28 = 18.69$. The part arriving now (Part 5) joins the end of the queue, $B(t)$ stays at 1, the queue length $Q(t)$ is incremented to 3, and Part 2 continues in service. As in the preceding event, P, N, ΣWQ, WQ^*, ΣTS, and TS^* are unchanged. $\int Q$ is augmented by $2\times(4.41 - 3.79) = 1.24$ and $\int B$ is augmented by $1\times(4.41 - 3.79) = 0.62$. $Q(t)$ has now risen to 3, a new maximum, so we set $Q^* = 3$. The time of the next departure is still (correctly) scheduled as the departure time of Part 2 at time 4.66, which now floats to the top of the event calendar and will be the next event to occur.
- **Entity 2, $t = 4.66$, Dep:** Part 2 now is done and is ready to depart. Since there is a queue, the drill press will stay busy so $B(t)$ remains at 1, but $Q(t)$ is decremented to 2. The queue advances, and the first part in it (Part 3, which arrived at time 3.08) will enter service. A waiting time in queue of duration $4.66 - 3.08 = 1.58$ for Part 3 has now been completed, which is added into ΣWQ, and N is incremented; this is a new maximum waiting time in queue, so WQ^* is changed to 1. A new part has been produced so P is incremented to 2 and the total time in system of Part 2 is computed as $4.66 - 1.73 = 2.93$, which is added into ΣTS; this is a new maximum total time in system, so TS^* is changed to 2.93. $\int Q$ is augmented by $3\times(4.66 - 4.41) = 0.75$ and $\int B$ is augmented by $1\times(4.66 - 4.41) = 0.25$. No new maximum for $Q(t)$ has been realized, so Q^* is unchanged. The next departure (of Part 3, which is now entering service) is scheduled for time $4.66 + 3.39 = 8.05$; no update to the time of the next arrival is needed. The next event is the departure of Entity 3 at time 8.05.
- **Entity 3, $t = 8.05$, Dep:** Part 3 departs. Part 4, first in queue, enters service, adds its waiting time in queue $8.05 - 3.79 = 4.26$ (a new maximum) into ΣWQ, and N is incremented; $B(t)$ stays at 1 and $Q(t)$ is decremented to 1. Part 3's total time in system is $8.05 - 3.08 = 4.97$ (a new maximum), which is added into ΣTS, and P is incremented. $\int Q$ is augmented by $2\times(8.05 - 4.66) = 6.78$ and $\int B$ is augmented by

$1 \times (8.05 - 4.66) = 3.39$; Q^* is unchanged. The next departure (of Part 4, which is now entering service) is scheduled for time $8.05 + 4.52 = 12.57$; no update to the time of the next arrival is needed. The next event is the departure of Entity 4 at time 12.57.

- **Entity 4, $t = 12.57$, Dep:** Part 4 departs. Part 5 enters service, adds its waiting time in queue $12.57 - 4.41 = 8.16$ (a new maximum) into ΣWQ, and N is incremented; $B(t)$ stays at 1 and $Q(t)$ is decremented to 0 (so the queue becomes empty, though a part is in service). Part 4's total time in system is $12.57 - 3.79 = 8.78$ (a new maximum), which is added into ΣTS, and P is incremented. $\int Q$ is augmented by $1 \times (12.57 - 8.05) = 4.52$ and $\int B$ is augmented by the same amount; Q^* is unchanged. The next departure (of Part 5, which is now entering service) is scheduled for time $12.57 + 4.46 = 17.03$; no update to the time of the next arrival is needed. The next event is the departure of Entity 5 at time 17.03.

- **Entity 5, $t = 17.03$, Dep:** Part 5 departs. But since the queue is empty, the drill press becomes idle ($B(t)$ is set to 0) and there is no new waiting time in queue observed, so ΣWQ, WQ^*, and N remain unchanged. Part 5's total time in system is $17.03 - 4.41 = 12.62$ (a new maximum), which is added into ΣTS, and P is incremented. $\int Q$ is augmented by $0 \times (17.03 - 12.57) = 0$ and $\int B$ is augmented by $1 \times (17.03 - 12.57) = 4.46$; Q^* is unchanged. Since there is not now a part in service at the drill press, the event calendar is left without a departure event scheduled; no update to the time of the next arrival is needed. The next event is the arrival of Entity 6 at time 18.69.

- **Entity 6, $t = 18.69$, Arr:** This event processes the arrival of Part 7 to the system, which is now empty of parts and the drill press is idle, so this is really the same scenario as in the arrival of Part 1 at time 0; indeed, the "experience" of Part 7 is (probabilistically) identical to that of Part 1. The next (seventh) part arrival is scheduled for time $18.69 + 0.70 = 19.39$. Meanwhile, the arriving Part 6 makes the drill press busy, but the queue is still empty because this part finds the drill press idle and begins service immediately. The waiting time in queue of Part 6 is 0, so N is incremented to 6, and ΣWQ is "augmented" by 0, which is certainly not a new maximum waiting time in queue. No part is departing, so there is no change in P, ΣTS, or TS^*. Since $Q(t)$ and $B(t)$ have both been at 0 since the previous event, there is no numerical change in $\int Q$, $\int B$, or Q^*. The departure of Part 6 is scheduled for time $18.69 + 4.36 = 23.05$ (so will not occur since this is after the simulation end time, as reflected in the order of the updated event calendar). The next event is the arrival of Part 7 at time 19.39.

- **Entity 7, $t = 19.39$, Arr:** The next (eighth) part arrival is scheduled for time $19.39 + 15.52 = 34.91$, which is beyond the termination time (20), so will not happen. Part 8 queues up, $B(t)$ stays at 1, and $Q(t)$ is incremented to 1 (not a new maximum). Since there is no new waiting time in queue being observed, N, ΣWQ, and WQ^* are unchanged; since no part is departing, P, ΣTS, and TS^* are unchanged. $\int Q$ is augmented by $0 \times (19.39 - 18.69) = 0$ and $\int B$ is augmented by $1 \times (19.39 - 18.69) = 0.70$. The next departure is already properly scheduled. The next event will be the end of the simulation at time 20.

Table 2-3. Final Output Performance Measures from the Hand Simulation

Performance Measure	Value
Total production	5 parts
Average waiting time in queue	2.53 minutes per part (6 parts)
Maximum waiting time in queue	8.16 minutes
Average total time in system	6.44 minutes per part (5 parts)
Maximum total time in system	12.62 minutes
Time-average number of parts in queue	0.79 part
Maximum number of parts in queue	3 parts
Drill-press utilization	0.92 (dimensionless proportion)

- $t = 20.00$, **The End:** The only task here is to update the areas $\int Q$ and $\int B$ to the end of the simulation, which in this state adds $1 \times (20.00 - 19.39) = 0.61$ to both of them.

The bottom row of Table 2-2 shows the ending situation, including the final values of the statistical accumulators.

2.4.4 Finishing Up

The only cleanup required is to compute the final values of the output performance measures:

- The average waiting time in queue is $\Sigma WQ/N = 15.17/6 = 2.53$ minutes per part.
- The average total time in system is $\Sigma TS/P = 32.20/5 = 6.44$ minutes per part.
- The time-average length of the queue is $\int Q/t = 15.78/20 = 0.79$ part (t here is the final value, 20.00, of the simulation clock).
- The utilization of the drill press is $\int B/t = 18.34/20 = 0.92$.

Table 2-3 summarizes all the final output measures together with their units of measurement.

During the 20 minutes, we produced five finished parts; the waiting time in queue, total times in system, and queue length do not seem too bad; and the drill press was busy 92% of the time. These values are considerably different from what we might have guessed or obtained via an oversimplified queueing model (see Exercise 2-6).

2.5 Event- and Process-Oriented Simulation

The hand simulation we struggled through in Section 2.4 uses the *event orientation* since the modeling and computational work is centered around the events, when they occur, and what happens when they do. This allows you to control everything; have complete flexibility with regard to attributes, variables, and logic flow; and to know the state of everything at any time. You easily see how this could be coded up in any programming language or maybe with macros in a spreadsheet (for very simple models only), and people have done this a lot. For one thing, computation is pretty quick with a custom-written, event-oriented code. While the event orientation seems simple enough in principle (although not much fun by hand) and has some advantages, you can imagine that it becomes very complicated for large models with lots of different kinds of events, entities, and resources.

A more natural way to think about many simulations is to take the viewpoint of a "typical" entity as it works its way through the model, rather than the omniscient orientation of the master controller keeping track of all events, entities, attributes, variables, and statistical accumulators as we did in the event-oriented hand simulation. This alternative view centers on the *processes* that entities undergo, so is called the *process orientation*. As you'll see, this is strongly analogous to another common modeling tool—namely, flowcharting. In this view, we might model the hand-simulated example in steps like this (put yourself in the position of a typical part entity):

- Create yourself (a new entity arrives).
- Write down what time it is now on one of your attributes so you'll know your arrival time later for the waiting-time-in-queue and total-time-in-system computations.
- Put yourself at the end of the queue.
- Wait in the queue until the drill press becomes free (this wait could be of 0 duration, if you're lucky enough to find the drill press idle at this time).
- Seize the drill press (and take yourself out of the queue).
- Compute and tally your waiting time in queue.
- Stay put, or *delay* yourself, for an amount of time equal to your service requirement.
- Release the drill press (so other entities can seize it).
- Increment the production-counter accumulator on the wall and tally your total time in system.
- Dispose of yourself and go away.

This is the sort of "program" you write with a process-oriented simulation language like SIMAN, and is also the view of things normally taken by Arena (which actually translates your flowchart-like description into a SIMAN model to be run, though you don't need to get into that unless you want to; see Section 7.1.6). It's a much more natural way to think about modeling, and (importantly) big models can be built without the extreme complexity they'd require in an event-oriented program. It does, though, require more behind-the-scenes support for chores like time advance, keeping track of time-persistent statistics (which didn't show up in the process-oriented logic), and output-report generation. Simulation software like Arena provides this support as well as a rich variety of powerful modeling constructs that enable you to build complicated models relatively quickly and reliably.

Most discrete-event simulations are actually *executed* in the event orientation, even though you may never see it if you do your modeling in the process orientation. Arena's hierarchical nature allows you to get down into the event orientation if you need to in order to regain the control to model something peculiar, and in that case you have to think (and code) with event-oriented logic as we did in the hand simulation.

Because of its ease and power, process-oriented logic has become very popular and is the approach we'll take from now on. However, it's good to have some understanding of what's going on under the hood, so we first made you suffer through the laborious event simulation.

2.6   Randomness in Simulation

In this section, we'll discuss how (and why) you model randomness in a simulation model's input and the effect this can have on your output. We'll need some probability and statistics here, so this might be a good time to take a quick glance at (or a painstaking review of) Appendix C to review some basic ideas, terminology, and notation.

2.6.1   Random Input, Random Output

The simulation in Section 2.4 used the input data in Table 2-1 to drive the simulation recorded in Table 2-2, resulting in the numerical output performance measures reported in Table 2-3. This might be what happened from 8:00 to 8:20 on some particular Monday morning, and if that's all you're interested in, you're done.

But you're probably interested in more, like what you'd expect to see on a "typical" morning and how the results might differ from day to day. And since the arrival and service times of parts on other days would probably differ from those in Table 2-1, the trace of the action in Table 2-2 will likely be changed, so the numerical output performance measures will probably be different from what we got in Table 2-3. Therefore, a single run of the example just won't do since we really have no idea how "typical" our results are or how much variability might be associated with them. In statistical terms, what you get from a single run of a simulation is a *sample of size one*, which just isn't worth much. It would be pretty unwise to put much faith in it, much less make important decisions based on it alone. If you rolled a die once and it came up a four, would you be willing to conclude that all the other faces were four as well?

So random input looks like a curse. But you must often allow for it to make your model a valid representation of reality, where there may also be considerable uncertainty. The way people usually model this, instead of using a table of numerical input values, is to specify *probability distributions* from which observations are *generated* (or *drawn* or *sampled*) and drive the simulation with them. We'll talk in Section 4.4 about how you can determine these input probability distributions using the Arena Input Analyzer. Arena internally handles generation of observations from distributions you specify. Not only does this make your model more realistic, but it also sets you free to do more simulation than you might have observed data for and to explore situations that you didn't actually observe. As for the tedium of generating the input observations and doing the simulation logic, that's exactly what Arena (and computers) like to do for you.

But random input induces randomness in the output too. We'll explore this a little bit in the remainder of this chapter, but will take it up more fully in Chapter 6, Section 7.2, and Chapter 12 and show you how to use the Arena Output Analyzer, Process Analyzer, and OptQuest®, from OptTek Systems, Inc., to help interpret and appropriately cope with randomness in the output.

2.6.2 Replicating the Example

It's time to confess: We generated the input values in Table 2-1 from probability distributions in Arena. The interarrival times came from an exponential distribution with a mean of 5 minutes, and the service times came from a triangular distribution with a minimum of 1 minute, mode of 3 minutes, and maximum of 6 minutes. (You'll see in Chapter 3 how all this works in Arena.) See Appendix D for a description of Arena's probability distributions.

So instead of just the single 20-minute run, we could make several (we'll do five) independent, statistically identical 20-minute runs and investigate how the results change from run to run, indicating how things in reality might change from morning to morning. Each run starts and stops according to the same rules and uses the same input-parameter settings (that's the "statistically identical" part), but uses separate input random numbers (that's the "independent" part) to generate the actual interarrival and service times used to drive the simulation. Another thing we must do is *not* carry over any in-queue or in-process parts from the end of one run to the beginning of the next as they would introduce a link, or correlation, between one run and the next; any such luckless parts just go away. Such independent and statistically identical runs are called *replications* of the simulation, and Arena makes it very easy for you to carry them out—just enter the number of replications you want into a dialog box on your screen. You can think of this as having five replications of Table 2-1 for the input values, each one generating a replication of the simulation record in Table 2-2, resulting in five replications of Table 2-3 for all the results.

We wish you'd admire (or pity) us for slugging all this out by hand, but we really just asked Arena to do it for us; the results are in Table 2-4. The column for Replication 1 is the same as what's in Table 2-3, but you can see that there can be substantial variation across replications, just as things vary across days in the factory.

Table 2-4. Final Output Performance Measures from Five Replications of the Hand Simulation

| Performance Measure | Replication | | | | | Sample | | 95% |
	1	2	3	4	5	Avg.	Std. Dev.	Half Width
Total production	5	3	6	2	3	3.80	1.64	2.04
Average waiting time in queue	2.53	1.19	1.03	1.62	0.00	1.27	0.92	1.14
Maximum waiting time in queue	8.16	3.56	2.97	3.24	0.00	3.59*	2.93*	3.63*
Average total time in system	6.44	5.10	4.16	6.71	4.26	5.33	1.19	1.48
Maximum total time in system	12.62	6.63	6.27	7.71	4.96	7.64*	2.95*	3.67*
Time-average number of parts in queue	0.79	0.18	0.36	0.16	0.05	0.31	0.29	0.36
Maximum number of parts in queue	3	1	2	1	1	1.60*	0.89*	1.11*
Drill-press utilization	0.92	0.59	0.90	0.51	0.70	0.72	0.18	0.23

*Taking means and standard deviations of the "maximum" measures might be of debatable validity—what is the "mean maximum" supposed to, well, mean? It might be better in these cases to take the *maximum* of the individual-replication maxima if you really want to know about the extremes.

The columns in Table 2-4 give the sample mean and sample standard deviation (see Appendix C) across the individual-replication results for the output performance measure in each row. The sample mean provides a more stable indication of what to expect from each performance measure than what happens on an individual replication, and the sample standard deviation measures cross-replication variation.

Since the individual replication results are independent and identically distributed, you could form a confidence interval for the true expected performance measure μ (think of μ as the sample mean across an infinite number of replications) as

$$\overline{X} \pm t_{n-1,1-\alpha/2} \frac{s}{\sqrt{n}}$$

where $\overline{X}$ is the sample mean, s is the sample standard deviation, n is the number of replications ($n = 5$ here), and $t_{n-1,1-\alpha/2}$ is the upper $1 - \alpha/2$ critical point from Student's t distribution with $n - 1$ degrees of freedom. Using the total-production measure, for example, this works out for a 95% confidence interval ($\alpha = 0.05$) to

$$3.80 \pm 2.776 \frac{1.64}{\sqrt{5}}$$

or 3.80 ± 2.04; the half width for 95% confidence intervals on the expectations of all performance measures are in the last column of Table 2-4. The correct interpretation here is that in about 95% of the cases of making five simulation replications as we did, the interval formed like this will contain or "cover" the true (but unknown) expected value of total production. This confidence interval assumes that the total-production results from the replications are normally distributed, which is not really justified here, but we use this form anyway; see Section 6.3 for more on this issue of statistical *robustness*. You might notice that the half width of this interval (2.04) is pretty big compared to the value at its center (3.80); that is, the *precision* is not too good. This could be remedied by simply making more than the five replications we made, which looks like it was not enough to learn anything precise about the expected value of this output performance measure. The great thing about collecting your data by simulation is that you can always[2] go get more by simply calling for more replications.

2.6.3 Comparing Alternatives

Most simulation studies involve more than just a single setup or configuration of the system. People often want to see how changes in design, parameters (controllable in reality or not), or operation might affect performance. To see how randomness in simulation plays a role in these comparisons, we made a simple change to the example model and re-simulated (five replications).

The change we made was just to double the arrival rate—in other words, the mean interarrival time is now 2.5 minutes instead of 5 minutes. The exponential distribution is still used to generate interarrival times, and everything else in the simulation stays the same. This could represent, for instance, acquiring another customer for the facility, whose part-processing demands would be intermingled with the existing customer base.

---

[2] Well, almost always.

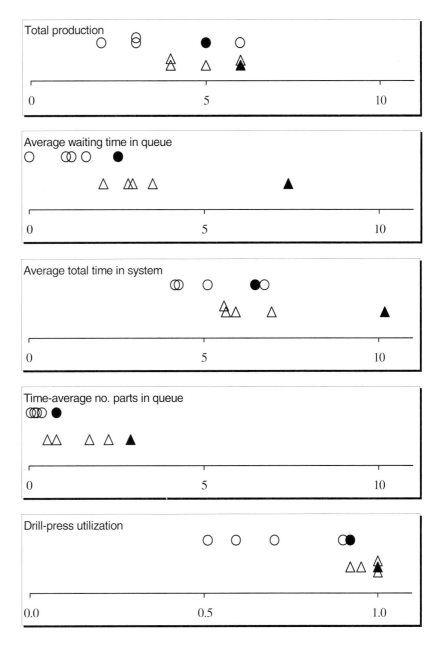

Figure 2-4. *Comparing the Original (Circles) and Double-Time Arrival (Triangles) Systems (Replication 1 Is Filled In, and Replications 2-5 Are Hollow)*

Figure 2-4 indicates what happened to the five "non-extreme" performance measures; the upper row (circles) of each plot indicates the original configuration (and is taken right out of Table 2-4), and the lower row (triangles) is for the new configuration. For each performance measure, the results of the five replications of each model are indicated, and the result from the first replication in each case is filled-in (replications 2-5 are hollow).

It's not clear that total production really increases much with the double-time arrivals, due to limited processing capacity. It does appear, though, that the time in queue and in system, as well as the utilization, tend to increase, although not in every case (due to variability). And despite some overlap, the average queue lengths seem to grow somewhat when the arrival rate is doubled. While formal statistical analyses would be possible here to help sort things out (see Section 6.4), simple plots like these can sometimes be pretty illuminating (though you'd probably want substantially more than just five replications of each scenario).

Moreover, note that relying on just one replication (the first, indicated by the filled-in symbols) could be misleading. For instance, the average waiting time in queue in the first replications of each model variant would have suggested that it is far greater for the double-time arrivals, but looking at the spread across replications indicates that the difference isn't so clear-cut or dramatic. This is exactly the danger in relying on only a single run to make important decisions.

2.7 Simulating with Spreadsheets

You'll agree that simulating by hand doesn't have much of a future. People use spreadsheets for a lot of things, so how about simulation? Well, maybe, depending on the type and complexity of the model, and on what add-in software you might have. In this section, we'll discuss two examples, first a static model and then a dynamic model (recall Section 1.2.3); both, however, are very simple.

2.7.1 A Newsvendor Problem

Rupert sells daily newspapers on a street corner. Each morning he must buy the same, fixed number q of copies from the printer at $c = 55\text{¢}$ each, and sells them for $r = \$1.00$ each through the day. He's noticed that demand D during a day is close to being a random variable X that's normally distributed with mean $\mu = 135.7$ and standard deviation $\sigma = 27.1$, except that D must be a nonnegative integer to make sense, so $D = \max(\lfloor X \rceil, 0)$ where $\lfloor \cdot \rceil$ rounds to the nearest integer (Rupert's not your average newsvendor). Further, demands from day to day are independent of each other. Now if demand D in a day is no more than q, he can satisfy all customers and will have $q - D \geq 0$ papers left over, which he sells as scrap to the recycler on the next corner at the end of the day for $s = 3\text{¢}$ each (after all, it's old news at that point). But if $D > q$, he sells out all of his supply of q and just misses those other $D - q > 0$ sales. Each day starts afresh, independent of any other day, so this is a single-period (single-day) problem, and for a given day is a static model since it doesn't matter when individual customers show up through the day.

How many papers q should Rupert buy each morning to maximize his expected profit? If he buys too few, he'll sell out and miss out on the 45¢ profit on each lost sale, but if he buys too many he'll have some left over to scrap at a loss of 52¢ each. If he had a crystal ball and knew in the morning what demand was going to be that day, he'd of course buy just that many; Rupert's good, but he's not that good, so he needs to decide on a fixed value of q that will apply to every day.

This is a genuine classic in operations research, a single-period, perishable-inventory problem, with more serious applications like food stocks and blood banks. It has many versions, variants, and extensions, and a lot of attention has been paid to getting exact analytical solutions. But it's simple to simulate it; so simple, in fact, that it can be done in a spreadsheet.

First, let's figure out an expression for Rupert's profit on a typical day. Clearly, his daily cost is cq (we won't impose some enigmatic negative-good-will cost for lost sales if $D > q$ and he sells out). If it turns out that $D \leq q$ then he sells D and meets all customer demand, gets revenue rD from customers, and has $q - D$ left over as scrap so gets $s(q - D)$ more revenue, such as it is, from the recycler. On the other hand, if $D > q$ he sells q, gets revenue rq from customers, but gets no scrap revenue. Thus, Rupert's total daily revenue is $r \min (D, q) + s \max (q - D, 0)$; to see this, just play out the two cases $D \leq q$ and $D > q$. So his daily profit, in dollars, is

$$W(q) = r \min (D, q) + s \max (q - D, 0) - cq.$$

We've written this as a function of q since it clearly depends on Rupert's choice of q. In mathematical-programming or optimization parlance, q is Rupert's *decision variable* to try to maximize his *objective function* of expected daily profit, $E(W(q))$. Since D is a random variable, so too is $W(q)$, and we need to take its expectation (see Appendix C for a refresher on probability and statistics) to get a deterministic objective function to try to maximize. Since expected values are long-run averages, in this case, over many days, Rupert is maximizing his long-run average daily profit by finding the best set-it-and-forget-it constant value of q. He's not trying to adjust q each day depending on demand, which is impossible since he has to commit to buying a fixed number q papers in the morning, before that day's demand is realized. Depending on the form of the probability distribution of D, it might be possible in some cases to derive an exact expression for $E(W(q))$ that we could try to maximize with respect to q. But we can always build a simple simulation model and heed the lessons from Section 2.6 about statistical analysis of simulation output to get a valid and precise *estimate* of the optimal q, without making strong and maybe unrealistic assumptions about the distribution of D just to enable an analytical result.

To simulate this, we need to generate numerical values on the demand random variable $D = \max (\lfloor X \rceil, 0)$; such are called *random variates* on D. Thus, we first need to generate random variates on X, which has a normal distribution with mean $\mu = 135.7$ and standard deviation $\sigma = 27.1$. Section 12.2 discusses random-variate generation in general, so we'll trust you to look ahead if you're interested at this point. In this case we can generate a normally distributed random variate as $X = \Phi_{\mu,\sigma}^{-1} (U)$ where U is a *random number* (continuous random variate distributed uniformly between 0 and 1, discussed in Section 12.1), $\Phi_{\mu,\sigma}$ is the cumulative distribution of the normal distribution with mean

μ and standard deviation σ, and $\Phi_{\mu,\sigma}^{-1}$ is the functional inverse of $\Phi_{\mu,\sigma}$, i.e., $\Phi_{\mu,\sigma}^{-1}$ "undoes" whatever $\Phi_{\mu,\sigma}$ does. Another way of looking at the functional inverse is that we need to solve $U = \Phi_{\mu,\sigma}(X)$ for X, where U would be a specific number between 0 and 1 from the random-number generator. As you may know, neither $\Phi_{\mu,\sigma}$ nor $\Phi_{\mu,\sigma}^{-1}$ for the normal distribution can be written as an explicit closed-form formula, so evaluation of $\Phi_{\mu,\sigma}^{-1}(U)$ must be by numerical approximation, but this is built into spreadsheet software, and certainly into simulation software as well.

We used the Microsoft® Excel spreadsheet to create the file `Newsvendor.xls` in Figure 2-5. You'll find this file in the Book Examples folder, which is in turn in the Arena 10.0 folder if you followed the instructions in Appendix E for Arena installation from the CD supplied with this book; a typical complete path to this folder would be `C:\Program Files\Rockwell Software\Arena 10.0\Book Examples`. In cells B4 – B8 (shaded blue in the file), are the input parameters, and in row 2 (pink) are some trial values for the decision variable q (cells H2, K2, N2, Q2, and T2). Column D just gives the day number for reference. We simulated 30 days, though as mentioned earlier, each day is independent so is really an independent, static simulation.

Column E has the demand on each day, via the Excel formula

```
MAX(ROUND(NORMINV(RAND(), $B$7, $B$8), 0), 0)
```

that is the same all the way down the column. Let's dissect this formula from the inside out:

- RAND() is Excel's built-in random-number generator, which returns a number between 0 and 1 that's supposed to be uniformly distributed and independent across draws. See Section 12.1 for a discussion of random-number generators, which are more subtle than you might think, but are clearly essential to good stochastic simulation. Unfortunately, not all random-number generators supplied with various software packages are of high quality (however, the one built into Arena is excellent). Excel's RAND() function is "volatile" by default, meaning that every instance of it re-draws a "fresh" random number upon any recalculation of the spreadsheet, which you can force by hitting the F9 key, by saving the file, or by editing just about anything anywhere in the sheet.
- NORMINV(u, μ, σ) is Excel's built-in numerical approximation to $\Phi_{\mu,\sigma}^{-1}$, so when used with $u = $ RAND() returns a random variate distributed normally with mean μ and standard deviation σ. We have μ in cell B7 and σ in cell B8; the $ in Excel-formula cell references forces what immediately follows it (column letter or row number) not to change when the formula is copied into other cells, and here we want both the column and row to stay the same when this formula is copied into other cells since μ and σ are in these fixed cells.
- ROUND(x, 0) in Excel returns x rounded to the nearest integer. The second argument's being 0 ensures that the rounding is to the nearest integer as opposed to other kinds of rounding (see Excel's Help).
- MAX($a, b, ...$), as you might guess, returns the maximum of its arguments.

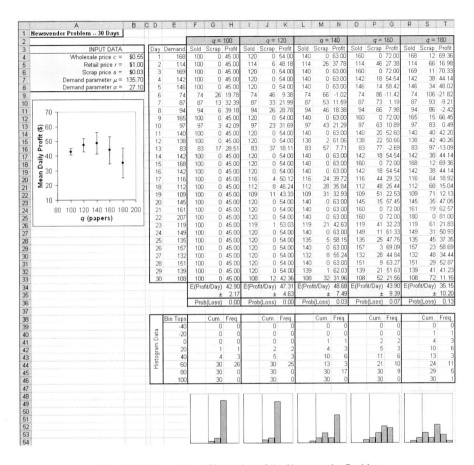

Figure 2-5. Spreadsheet Simulation of the Newsvendor Problem

We entered this formula originally in cell E4, then just copied it to the 29 columns below to get independent demands on 30 days, as RAND() draws a fresh random number each time it appears. Note that if you hit the F9 key or otherwise force the sheet to recalculate, you get a new column of demand numbers, and most other things in the sheet also change accordingly.

For the case of $q = 100$, columns F, G, and H in rows 4 – 33 contain the number of papers sold, scrapped, and the daily profit, for each of days 1 – 30:

- Cell F4 is the number sold on day 1 if $q = 100$. The Excel formula is $=$ MIN($E4, H$2). $E4 is the demand for day 1. We apply the $ to column E since we'll later want to copy this formula for other values of q in columns I, L, O, and R but keep the column reference to E, but no $ for row 4 since we want it to change to rows 5 – 33 to pick up demands for subsequent days. H$2 is $q = 100$. We have no $ for column H since we'll later want it to change for other q values in columns I, L, O, and R, but we apply a $ to row 2 to keep the reference to it for rows below. Thus, this is $\min(D, q)$, as desired.

- Cell G4 is the number of papers scrapped at the end of day 1 if $q = 100$. The Excel formula is $= \mathtt{MAX(H\$2 - \$E4, 0)}$, with $ usage the same as for Cell F4 and for the same reasons. This is $\max(q - D, 0)$.
- Cell H4 is the profit, in dollars, on day 1 if $q = 100$. The Excel formula is $= \mathtt{\$B\$5*F4 + \$B\$6*G4 - \$B\$4*H\$2}$, which is $r \min(D, q) + s \max(q - D, 0) - cq$. We'll let you figure out why the $ signs are present and absent where they are.

We then simply copied cells F4, G4, and H4 into the 29 rows below, allowing the column letters and row numbers to run as appropriate and controlled by the $ placement in the formulas, to get the *simulation table*, as this is sometimes called, for the case $q = 100$.

At the bottom of these columns, shaded green in the file, we added the average profit (cell H34) and the half width of a 95% confidence interval on the expected average profit (cell H35) over the 30 days, with formulae just like those toward the end of Section 2.6.2; the value 2.045 is the upper 0.975 critical point from the t table with 29 degrees of freedom. Note that, as in Section 2.6.2, the form of this confidence interval assumes that the individual profits in a column are normally distributed, which is not quite true, especially for small values of q where there's little variability in profit, but we do so anyway; again, see the end of Section 6.3 for more on statistical robustness.

We also added in row 36 the proportion of days in which a loss was incurred, which is an estimate of the probability of incurring a loss in a day, using the Excel COUNTIF function. With this low value of q a loss is quite unlikely since demand will be at least q on nearly all days, but that will change for higher values of q. And in rows 38 – 46 we collect frequency-distribution data (again using COUNTIF) to construct a histogram of profits directly below; the vertical line in the histograms, which is red in the file, is at zero, so we have an indication to its left of the probability of incurring a loss, as well as the general distribution of the profit data.

To complete the simulation study, we took columns F, G, and H and replicated them into four more blocks of three columns each out to the right, but with q taking the values 120, 140, 160, and 180, and added histograms at the bottom for each value of q. To visualize the mean results, we added the plot on the left edge to show the average profit as a dot, with vertical I-beams showing the confidence intervals, for different values of q on the horizontal axis.

Notice that, for a given day, we chose to use the same realized numerical demand variate (column E) for all trial values of q, rather than generate fresh independent demands for each value of q, which would have been valid as well. We're not lazy (at least about this) and filling more Excel columns with fresh demands wouldn't cost much, but we did it this way deliberately. Our interest is really in *comparing* profit for different values of q, and by using the same realized numerical demands on a given day for all q, the differences we see in profit will be attributable to differences in q (which is the effect we're trying to measure) rather than to chance differences in demand. This is an example of a *variance-reduction technique*, in particular of *common random numbers*, discussed more fully in Section 12.4. Such strategies in stochastic simulation can help you get more precise answers, often with little or no more work (in this case, with *less* work) and are worth thinking about.

If you open the spreadsheet file yourself, you won't see the same numbers as in Figure 2-5 due to the volatility of Excel's RAND ( ). And as you recalculate (tap the F9 key a few times) you'll see all the numbers change, the objects in the mean plot on the left will jump up and down, and the histograms at the bottom will undulate left and right, all reflecting this volatility and the underlying truth that such a stochastic simulation is an experiment with random outcomes; indeed the jumping up and down of the I-beams in the mean plot at the left is a reminder that confidence intervals themselves are random intervals and depend on the particular sample you happened to get. Notice that all five of the dots and I-beams tend to jump up or down together, reflecting the common-random-numbers variance reduction mentioned in the preceding paragraph; see Exercise 2-10.

Despite this randomness, it looks like Rupert should probably set q to be around 140, maybe a bit less, to maximize his expected daily profit. From the confidence-interval I-beams, it also looks like we might want to increase the sample size to considerably more than 30 days to pin this down better (Exercise 2-11). And clearly it would help to consider more values of q in a finer mesh (Exercise 2-12), but this study at least tells us that Rupert should probably buy at least 120 papers each morning, but no more than 160, since otherwise he loses too much profit due to lost sales or scrapping too much at a loss.

But looking beyond averages via the histograms at the bottom, the higher the q, the more variability (risk) there is in the profit, seen by the histograms' broadening out. This is good and bad news. With high q, Rupert has the large inventory to reap a lot of profit on high-demand days, but you also see more negative profits when demand is low, on boring slow-news days when he gets stuck with a lot to scrap at a loss. If Rupert is a risk-taker, he might go for a high value of q (sometimes you eat the bear, sometimes the bear eats you). At the other end, very low q values entail little risk since demand seldom falls short of such small q's, but the downside is that average profit is also low as Rupert cannot cash in on high-demand days since he sells out. Such risk/return tradeoffs in business decisions are commonly quantified via spreadsheet simulations of static models like this.

It's quite easy to simulate the newsvendor problem in Arena; see Exercises 6.22 and 6.23.

2.7.2 A Single-Server Queue

Spreadsheets might also be used to simulate some dynamic models, but only very simple ones. Consider a single-server queue that's logically the same as what we plodded through by hand in Section 2.4, but different on some specifics. As before, customers arrive one at a time (the first customer arrives at time 0 to find the system empty and idle), may have to wait in a FIFO queue, are served one at a time by a single server, and then depart. But we'll make the following changes from the hand-simulated model:

- Interarrival times have an exponential probability distribution with mean $1/\lambda =$ 1.6 minutes, and are independent of each other (see Appendices C and D). The parameter $\lambda = 1/1.6 = 0.625$ is the *arrival rate* (per minute here).

- Service times have a continuous uniform distribution between $a = 0.27$ minute and $b = 2.29$ minutes, are independent of each other, and are independent of the arrival times.

This is called the *M/U/1 queue*, where the U refers to the uniformly distributed service times.

Instead of observing all the outputs in the hand simulation, we'll just watch the waiting times in queue, WQ_1, WQ_2, ..., WQ_{50} of the first 50 customers to complete their queue waits (not including their service times). In the hand simulation, we set up a clock, state variables, an event calendar, and statistical accumulators to simulate this, but *for a single-server FIFO queue* there is a simple recursion for the WQ_i values (if that's all we're interested in by way of output), Lindley's (1952) formula. Let S_i be the time in service for customer i, and let A_i be the interarrival time between customers $i - 1$ and i. Then

$$WQ_i = \max(WQ_{i-1} + S_{i-1} - A_i, 0) \quad \text{for } i = 2, 3, \dots$$

and we initialize by setting $WQ_1 = 0$ since the first customer arrives to an empty and idle system.

To simulate this, we need to generate random variates from the exponential and continuous uniform distributions. Again referring ahead to Section 12.2, there are simple formulas for both (as before, U denotes a random number distributed continuously uniformly between 0 and 1):

- Exponential interarrival times with mean $1/\lambda$: $A_i = -(1/\lambda) \ln(1 - U)$ where ln is the natural (base e) logarithm.
- Continuously uniformly distributed service times between a and b: $S_i = a + (b - a)U$.

Figure 2-6 shows the Excel file `MU1.xls` in the Book Examples folder. Cells B4, B5, and B6 are the parameters, and column D just gives the customer number $i = 1, 2, \dots,$ 50 for a run of the model. Interarrival times are in column E, using the Excel formula `-$B$4*LN(1 - RAND())`, and service times are in column F, generated by the formula `$B$5 + ($B$6 - $B$5)*RAND()`.

Lindley's recursion is in column G, initialized by the 0 in cell G4. A typical entry is in cell G9, `MAX(G8 + F8 - E9, 0)`, which generates WQ_6 from WQ_5 (in cell G8), S_5 (cell F8), and A_6 (cell E9). In this way, each queue wait in column G is linked to the one above it, establishing a correlation between them. Since this is correlation of entries in a sequence with each other, it's called *autocorrelation*. In this case, it's *positive* autocorrelation since a large wait probably means that the next wait will be large too, and vice-versa, which you can understand from looking at Lindley's recursion or just from your own painful experiences standing in line. So, unlike the newsvendor's profit rows in Section 2.7.1, the waiting times in queue here are *not* independent from row to row in column G, which is typical of the output sequence within a run of a dynamic simulation.

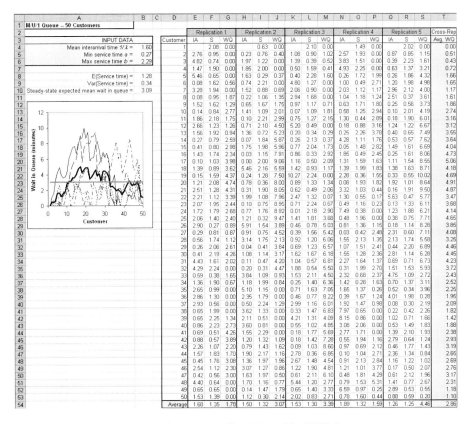

Figure 2-6. Spreadsheet Simulation of the M/U/1 Queue

At the bottom of columns E, F, and G we give the averages. For columns E and F, this is just to confirm that they're close to the expected values of the interarrival times and service times, respectively $1/\lambda = 1.6$ and $(a + b)/2 = 1.28$ (see Appendix D). For column G, this is the average wait in queue for the first 50 customers; i.e., the output result of the simulation.

We could also easily compute the sample standard deviation of the queue waits in column G, and then confidence intervals as in the newsvendor simulation, but we refrain from doing so for two very important reasons:

1. Due to the autocorrelation among the WQ_i values in column G, the sample variance would be *biased* as an estimator of the variance of a "typical" WQ_i, since they are to some extent "pegged" to each other so are not free to vary as much as they would if they were independent. With our positive autocorrelation, the sample variance estimator would be biased low, i.e., would be too small and understate the variance of a "typical" WQ_i. Just because you're able to compute some number, or for that matter plot some graph, you shouldn't if it could be misleading.

2. It's not clear what a "typical" WQ_i even means here, since the initialization $WQ_1 = 0$ makes the early part of the output sequence smaller than later on, so the distribution (and thus variance) of the WQ_i values is changing as i grows. Thus, it's not even clear what it is we'd be estimating by a sample-variance estimate.

We copied columns E, F, and G into five more blocks of three columns each out to the right, creating five independent replications of this run of 50 queue waits. All random numbers, and thus results, are independent across these five sets of three columns each. This is the same model all the way across, just replicated five independent times.

The plot on the left traces the WQ_i sequence in each of the five replications using different dash patterns, and also plots the average of the five values of WQ_i, for a fixed i, across the replications (in column T) as the thicker solid curve. Note the variation within a run, and the average thick curve calms things down a bit. All curves start off at 0 on the left, reflecting the deterministic initialization $WQ_1 = 0$, and rise from there, subject to a lot of wiggling that reflects output variability. The positive autocorrelation is in evidence since the curves, while wiggly, do not break off from themselves abruptly as they progress to the right, and points on them tend to be close to their neighboring points for a little while left and right.

The thick gray horizontal line plots the value in cell B10, which is the steady-state expected wait in queue, $E(WQ_\infty)$. This can be computed exactly, from queueing theory (see, for example Gross and Harris, 1998), for single-server FIFO queues with exponential interarrival times and service times S_i with any distribution, as

$$E(WQ_\infty) = \lambda \frac{\text{Var}(S_i) + (E(S_i))^2}{2(1 - \lambda E(S_i))} .$$

In our case, $\lambda = 0.625$ (the reciprocal of cell B4), $E(S_i) = (a + b)/2 = 1.28$ in cell B8, and $\text{Var}(S_i) = (b - a)^2/12 = 0.34$ in cell B9 (see Appendix D). Though the variation in the plot obscures it, a little imagination might suggest that the thick solid average curve is converging to the gray horizontal line; see Exercise 2-13.

Why no histograms like those at the bottom in newsvendor problem spreadsheet? They could have been constructed in the same way, and could provide useful information about the distribution of the queue waits, beyond just their means. The potential problem with this is the autocorrelation in the queue-wait sequence, so a histogram, like a variance estimate, could be biased if our run were not long enough to ensure that the waiting times in queue eventually visit all possible regions and in the right proportions; the autocorrelation links together nearby observations, which would impede such visitation if the run were too short. However, a very long run could indeed produce a nearly unbiased histogram, which could be useful; in practice, it might be hard to determine how long is long enough. The question is whether to present something that's accurate and useful most of the time, at the risk of its being occasionally biased and misleading.

Simulating a single-server queue in Arena is quite simple, and is basically the central example we use in Chapter 3 to introduce Arena.

2.7.3 Extensions and Limitations

Spreadsheet simulation is popular for static models, many involving financial or risk analysis. Commercial add-in packages to Excel, like @RISK (Palisade Corporation, 2006) and Crystal Ball® (Decisioneering Inc., 2006), facilitate common operations, provide better random-number generators, make it easy to generate variates from many distributions, and include tools for analysis of the results.

However, spreadsheets are not very well suited for simulation of dynamic models. Absent something like Lindley's recursion for very special and simple cases like the single-server FIFO queue in Section 2.7.2, spreadsheets are not very good tools for dynamic models. Our focus in the rest of the book is on dynamic models, for which Arena is specifically designed.

2.8 Overview of a Simulation Study

In deciding how to model a system, you'll find that issues related to design and analysis and representing the model in the software certainly are essential to a successful simulation study, but they're not the only ingredients. We'll take all this up in Chapter 13 in some detail, but we want to mention briefly at this early point what's involved.

No simulation study will follow a cut-and-dried "formula," but there are several aspects that do tend to come up frequently:

- *Understand the system.* Whether the system exists or not, you must have an intuitive, down-to-earth feel for what's going on. This will entail site visits and involvement of people who work in the system on a day-to-day basis.
- *Be clear about your goals.* Realism is the watchword here; don't promise the sun, moon, and stars. Understand what can be learned from the study, and expect no more. Specificity about what is to be observed, manipulated, changed, and delivered is essential. And return to these goals throughout the simulation study to keep your attention focused on what's important, namely, making decisions about how best (or at least better) to operate the system.
- *Formulate the model representation.* What level of detail is appropriate? What needs to be modeled carefully and what can be dealt with in a fairly crude, high-level manner? Get buy-ins to the modeling assumptions from management and those in decision-making positions.
- *Translate into modeling software.* Once the modeling assumptions are agreed upon, represent them faithfully in the simulation software. If there are difficulties, be sure to iron them out in an open and honest way rather than burying them. Involve those who really know what's going on (animation can be a big help here).
- *Verify that your computer representation represents the conceptual model faithfully.* Probe the extreme regions of the input parameters, verify that the right things happen with "obvious" input, and walk through the logic with those familiar with the system.
- *Validate the model.* Do the input distributions match what you've observed in the field? Do the output performance measures from the model match up with those

from reality? While statistical tests can be carried out here, a good dose of common sense is also valuable.

- *Design the experiments.* Plan out what it is you want to know and how your simulation experiments will get you to the answers in a precise and efficient way. Often, principles of classical statistical experimental design can be of great help here.

- *Run the experiments.* This is where you go to lunch while the computer is grinding merrily away, or maybe go home for the night or the weekend, or go on vacation. The need for careful experimental design here is clear. But don't panic—your computer probably spends most of its time doing nothing, so carrying out your erroneous instructions doesn't constitute the end of the world (remember, you're going to make your mistakes on the computer where they don't count rather than for real where they do).

- *Analyze your results.* Carry out the right kinds of statistical analyses to be able to make accurate and precise statements. This is clearly tied up intimately with the design of the simulation experiments.

- *Get insight.* This is far more easily said than done. What do the results mean at the gut level? Does it all make sense? What are the implications? What further questions (and maybe simulations) are suggested by the results? Are you looking at the right set of performance measures?

- *Document what you've done.* You're not going to be around forever, so make it easier on the next person to understand what you've done and to carry things further. Documentation is also critical for getting management buy-in and implementation of the recommendations you've worked so hard to be able to make with precision and confidence.

By paying attention to these and similar issues, your shot at a successful simulation project will be greatly improved.

2.9 Exercises

2-1   For the hand simulation of the simple processing system, define another time-persistent statistic as the total number of parts in the system, including any parts in queue and in service. Augment Table 2-2 to track this as a new global variable, add new statistical accumulators to get its time average and maximum, and compute these values at the end.

2-2   In the preceding exercise, did you really need to add state variables and keep track of new accumulators to get the *time-average* number of parts in the system? If not, why not? How about the *maximum* number of parts in the system?

2-3   In the hand simulation of the simple processing system, suppose that the *queue discipline* were changed so that when the drill press becomes idle and finds parts waiting in queue, instead of taking the first one, it instead takes the one that will require the *shortest processing time* (this is sometimes called an *SPT* queue discipline). To make this work, you'll need to assign a second attribute to parts in the system when they arrive, representing what their service time at the drill press will be. Re-do the hand simulation. Is this a better rule? From what perspective?

2-4   Suppose that, in the hand simulation of the simple processing system, a constant setup time of 2 minutes was required once a part entered the drill press but before its service could actually begin. When a setup is going on, regard the drill press as being busy. Re-do the hand simulation and discuss the results.

2-5   Suppose the drill press can work on two parts simultaneously (and they enter, are processed, and leave the drill press independently). There's no difference in processing speed if there are two parts in the drill press instead of one. Redefine $B(t)$ to be the number of parts in service at the drill press at time (so $0 \leq B(t) \leq 2$), and the drill press utilization is redefined as

$$\frac{\int_0^T B(t)\,dt}{2T}$$

where $T = 20$. Re-run the original simulation to measure the effect of this change.

2-6   In Section 2.2.2, we used the M/M/1 queueing formula with the mean interarrival time μ_A and the mean service time μ_S estimated from the data in Table 2-1 as 4.08 and 3.46, respectively. This produced a "predicted" value of 19.31 minutes for the average waiting time in queue. However, the hand-simulation results in Section 2.4.4 produced a value of 2.53 minutes for this measure.

  (a)   If these two (very different) numbers are supposed to estimate or approximate the same thing (average waiting time in queue), why are they so apparently different? Give at least three reasons.

  (b)   We carried out the simulation for a million minutes (rather than 20 minutes), using the same sources of interarrival and service times as in Table 2-1, and got a value of 3.60 for the average waiting time in queue (as you can imagine, it really took us a long time to do this by hand, but we enjoyed it). Why is this yet different from the values 19.31 and 2.53 being compared in part (a)?

  (c)   We consulted an oracle, who sold us the information that the interarrival times are actually "draws" from an exponential distribution with mean 5 minutes, and the service times are actually draws from a triangular distribution with minimum 1 minute, mode 3 minutes, and maximum 6 minutes (see Appendix D for more information on these distributions, and Appendix C to brush up on your probability). With this (exact) information, the mean interarrival time is actually $\mu_A = 5$ minutes, and the mean service time is actually $\mu_S = (1 + 3 + 6)/3 = 3.33$ minutes. Using these values, use the M/M/1 formula in Section 2.2.2 to predict a value for the average waiting time in queue. Why is it yet different from the previous three values (19.31, 2.53, and 3.60) for this same quantity discussed previously in this exercise?

(**d**) Using the information from the oracle in part (c), we treated ourselves to another million-minute hand simulation (just as enjoyable), but this time drew the service times from an exponential (not triangular) distribution with mean 3.33 minutes, and got a result of 6.61 for the mean waiting time in queue. Compare and contrast this with the previous four values for this quantity discussed previously in this exercise.

2-7 Here are the actual numbers used to plot the triangles for the double-time-arrival model in Figure 2-4:

| | Replication | | | | |
Performance Measure	1	2	3	4	5
Total production	6	4	6	4	5
Average waiting time in queue	7.38	2.10	3.52	2.81	2.93
Average total time in system	10.19	5.61	5.90	6.93	5.57
Time-average no. parts in queue	2.88	0.52	1.71	0.77	2.25
Drill-press utilization	1.00	0.92	1.00	0.95	1.00

For each of these five performance measures, compute the sample mean, sample standard deviation, and half widths of 95% confidence intervals on the expected performance measures, as done in Table 2-4. Comparing these confidence intervals to those in Table 2-4, can you clarify any differences beyond the discussion in Section 2.6.3 (which was based on looking at Figure 2-4)?

2-8 In the hand simulation of the simple processing system, suppose the drill press is taken down for burn-in maintenance at time 3 minutes, and it takes 3 minutes to do this work and bring it back up; it then stays up for the rest of the 20-minute simulation length. If there's a part in service when the drill press goes down, it just sits there, its processing suspended until the drill press comes back up, when it's resumed and needs just its remaining processing time (its processing does not have to start over from scratch). During the down time, the arrival process of new parts continues without regard to down time. Regard the drill press as being busy when it's down, and define the time in system of a part to include the time it sits in the drill press but not being worked on if it's unlucky enough to be in service when the drill press is taken down. Using the same input data as in Table 2-1, carry out this simulation by hand and note any differences in the output performance measures from the original model. (See Exercise 6-18 for the question of whether this model change induces statistically significant changes in the output performance measures.)

2-9 In Exercise 2-8, refine the output performance measure on the status of the drill press from just the two states (busy vs. idle, where in Exercise 2-8 the drill press was to be regarded as "busy" during its down time) to three states:

1. busy working on a part and not down,
2. idle and not down, and
3. down, with or without a part being stuck in the machine for the down time.

Give the output performance measures from this simulation under this refined definition of the drill-press state.

2-10 Modify the newsvendor problem in Section 2.7.1 to use independent demands for each trial value of q, and note the difference in the behavior of the output, particularly in the mean plot on the left.

2-11 Increase the sample size on the newsvendor problem in Section 2.7.1 from 30 days to 120. As in the original problem, use the same realized demands in column E for all values of q. Don't forget to change the statistical-summary and histogram cells appropriately at the bottom of the columns, as well as the cells referred to in the various graphics. What is the effect on the results? Based on your results, about what sample size (in days) would be needed to bring the maximum half-width of all five confidence intervals down to under $\pm$ \$2.00? Don't actually carry this out, but just get an estimate; see Section 6.3.

2-12 Based on your results from Exercise 2-11, try a different set of trial values of q to try to home in more precisely on the optimal q. Keep the sample size at 120 days. Depending on what you choose to try, you may need to rescale the axes on the mean plot on the left.

2-13 Modify the M/U/1 queue spreadsheet simulation from Section 2.7.2 to have exponential service times with mean 1.28 minutes, rather than the original uniform service times, resulting in the M/M/1 queue. Change the steady-state expected wait in queue appropriately; see the exponential-distribution entry in Appendix D.

2-14 Modify the M/U/1 queue spreadsheet simulation from Section 2.7.2 to make ten replications rather than 5, and to run for 200 customers rather than 50. To avoid clutter, plot only the cross-replication average queue-wait curve. Compare with the original in Section 2.7.2.

CHAPTER 3

A Guided Tour Through Arena

As we were honest enough to admit in Chapter 2, we really used Arena to carry out the "hand" simulation in Section 2.4, as well as the multiple replications and the modified model with the double-time arrivals in Section 2.6. In this chapter, we'll lead you on a tour through Arena by having you start up Arena, browse through the existing model we built for the hand simulation, run it, and then build it from scratch. Then we'll use just these basic building blocks in a case study exploring a question of real interest, whether it's better to have specialized serial processing or generalized parallel processing, and the effect of variability on such a decision. We'll also explore the Arena user interface, get you into the help and documentation systems, discuss different ways to run your simulation, and illustrate some of the drawing and graphics tools.

Section 3.1 gets you to start Arena on your computer, and in Section 3.2, you'll open an existing model and look around. In Section 3.3, you'll go through the model in some detail, browsing the dialog boxes and animation, running the model, and taking a look at the results; in Section 3.4, you'll construct this model from scratch. Section 3.5 contains the case study mentioned above. In Section 3.6, we'll briefly go over many of Arena's pieces and capabilities, including what's available in the menus and toolbars and the drawing and printing capabilities. There's a broad and deep help system in Arena, with all of the detailed technical documentation, which is the subject of Section 3.7. There are a lot of options for running and controlling simulations, which are discussed in Section 3.8.

By the end of this chapter, you'll have a good feel for how Arena works and have an idea of the things you can do with it. You'll be able to work effectively with Arena to build simple models and maybe take a stab at doing some not-so-simple things as well by cruising the menus and dialogs on your own, with the aid of the help and documentation systems. While you can probably make some sense out of things by just reading this chapter, you'll be a lot better off if you follow along in Arena on your computer. More information on building your own models with Arena is discussed in Chapter 4 and beyond.

3.1 Starting Up

Arena is a true Microsoft® Windows® operating system application, so its look and feel will already be familiar to you, and all the usual features and operations are there. In addition, Arena is fully compatible with other Windows software, like word processors, spreadsheets, and CAD packages, so you can easily move things back and forth (Chapter 10 goes into detail about Arena's interaction and communication with other software).

By the way, we're assuming you're already comfortable with the basics of working with Windows, such as:

- Disks, files, folders, and paths.
- Using the mouse and keyboard, including clicking, double-clicking, and right-clicking.
- Operating on windows, like moving, resizing, maximizing, minimizing, and closing.
- Accessing things from menus. We'll use notation like "*M > C > S > T*" to mean open the *M* menu, choose *C* from it, then choose *S* from the submenu menu (if any), then choose the tabbed page labeled *T* (if any), etc.
- Using the *Control*, *Alt*, and *Shift* keys. By "*Ctrl+whatever*," we'll mean to hold down the *Ctrl* key and press "*whatever*" (this will also apply for *Alt+whatever* and *Shift+whatever*). If "whatever" is a keyboard key, it's not case-sensitive. "Whatever" could also be a mouse click, like *Ctrl+Click* to extend a selection to include additional items.
- *Cut* (or the menu command *Edit > Cut* or the shortcut key combination *Ctrl+X*), *Copy* (or *Edit > Copy* or *Ctrl+C*), and *Paste* (or *Edit > Paste* or *Ctrl+V*) of text and other items.
- Filling out dialog boxes by entering and editing text entries, pressing buttons, selecting and clearing (that is, unchecking) check boxes, clicking exactly one from a list of option buttons (also called radio buttons), and selecting items from drop-down list boxes.

If any of these things are unfamiliar to you, it would probably be a good idea for you to go through a tutorial on Windows before moving on.

Go to your computer, on which Arena is already installed per the instructions that came with it (see Appendix E for instructions on installing the academic version of Arena, which is what's on the CD packaged with this book). Approach the computer cautiously but with confidence—if it senses you're afraid, it could attack. Locate the Arena icon, or a shortcut to it, and double-click on it (or launch Arena by starting Windows and clicking the *Start* button, then *Programs > Rockwell Software > Arena 10.0*, and finally the *Arena 10.0* icon). In a moment, the Arena copyright window will come up; if you're running an academic version (which is what's on the CD with this book) or an evaluation version, you'll get a message box to this effect, which you should read and then click *OK* (or just click the *Enter* key on your keyboard since the *OK* button is already selected by default).

At the top left of the Arena window are the *File*, *View*, *Tools*, and *Help* menus (in addition to several other menus if a blank model file was automatically opened when Arena started up). You'll also see toolbars with various buttons, only a few of which are available unless you have a model file open:

Create a *New* blank model file. This is equivalent to the menu command *File > New* and to the keyboard operation *Ctrl+N*.

Display a dialog box to open a previously saved model; equivalently *File > Open* or *Ctrl+O*. You may need to navigate around to other folders or disks to find what you want.

   ☞   *Template Attach* (*Templates*, of which there are several, contain the modeling elements); equivalently *File > Template Panel > Attach*. The template files (with file name extension.*tpo*) are in the Template folder, which in turn is in the Arena 10.0 folder. You can also right-click in the Project Bar on the left (see Figure 3-1), then *Template Panel > Attach* from the pop-up that appears.

   ☞   *Template Detach* (when you don't need the modeling elements in the active panel any more); equivalently *File > Template Panel > Detach* or right-click in the active panel (which you want to detach) the Project Bar on the left, then *Template Panel > Detach* in the pop-up.

   ▶?   *Context Help* to provide help on a menu or toolbar command. Click on it to add the question mark to your mouse arrow and then click on a toolbar button or menu command to get help on it; closing that help window returns the mouse pointer to arrow-only.

   Tooltips provide another source of quick and brief help (even quicker and briefer) on toolbar buttons. If your mouse remains motionless over a button for a second or two, a little box shows up with the name of the button. If you want to know more about that button, you could use ▶? as just described, or maybe look it up (now that you at least know its name) in Arena's help system; more on that in Section 3.7. If you get tired of being pestered by tooltips at every turn, you can turn them off via *View > Toolbars > Toolbars* and clear (uncheck) the *Show Tooltips* option.

   When you're done with your Arena session and want to get out, click ✕ at the upper-right corner of the Arena window, or *File > Exit*, or *Alt+F4,* or right-click in the Arena window bar at the very top and select Close from the pop-up.

3.2   Exploring the Arena Window

In this section, we'll open an existing model, use it to look around the Arena window so you can get familiar with where things are, and introduce some basic Arena terminology.

3.2.1   *Opening a Model*

The ready-made model for the hand simulation can be found via *File > Open* (or just click 📂 to bring up the Open dialog box). File names appear in a scrolling box, and you can also navigate to other folders or drives. Find the file named `Model 03-01.doe`; the file name extension .*doe*[1] is the default for Arena files. In a typical installation using the CD that came with this book, it will be in the Book Examples folder, which is in turn in the Arena 10.0 folder. Click on this file name (highlighting it), and then click the *Open* button (or just double-click the file name).

   You should get an Arena window that looks something like Figure 3-1 (you might see different toolbars and buttons on your computer or see some things in different places). We'll call this Model 3-1.

---

[1] In its early development, Arena was code-named "Bambi." We're not making this up.

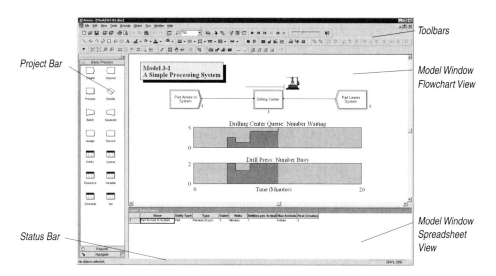

Project Bar

Status Bar

Toolbars

Model Window
Flowchart View

Model Window
Spreadsheet
View

Figure 3-1. Arena Window for the Simple Processing System, Model 3-1

3.2.2 Basic Interaction and Pieces of the Arena Window

As shown in Figure 3-1, the Arena window with this model open is divided into several pieces.

On the right, taking up most of the screen, is the *model window*, which is actually inside the Arena window. If you had several Arena models open at once you'd have a separate model window for each of them, all inside the Arena window, just as in word-processing or spreadsheet software. Switch between model windows by just clicking in them (if the one you want is visible), or use the Arena Window menu to select from the entire list. If you have a lot of models open, you can cycle among them via *Ctrl+Tab*, or you might want to minimize some of them to icons with the ▬ button in each one. The Window menu also has commands (*Cascade*, *Tile*, etc.) for how you'd like to arrange the open models or their minimized icons. Create a new (blank) model window via ▯ (or *File > New* or *Ctrl+N*), save the active model window via 🖫 (or *File > Save* or *Ctrl+S*) or *File > Save As,* and open a previously saved model window via 📂 (or *File > Open* or *Ctrl+O*). Resizing and repositioning a model window works just like any Microsoft® Windows® operating system application.

The familiar cut, copy, and paste operations work within Arena as well as between Arena and other applications. For instance, you might have several Arena model windows open, and you might want to copy some objects from one to another. Just select the objects with the mouse (*Ctrl+Click* to extend the selection, or drag a box across them if they're positioned that way), copy them to the Clipboard (*Ctrl+C* or 🗎 or *Edit > Copy*), switch to the other window, and paste them in (*Ctrl+V* or 🗎 or *Edit > Paste*). After choosing the paste operation, the mouse pointer changes to cross hairs that you click where you want the northwest corner of the selection to land. Or, you might have Arena open simultaneously with a spreadsheet in which there's a long number you want to put into an Arena dialog text box. Copy the number from the spreadsheet cell, switch to Arena (either via the Windows® Taskbar or by using *Alt+Tab* to cycle through the open applications), position the insertion pointer in the Arena dialog box where you want the number, and paste it in. If you're writing a report in a word processor and want to paste in a "snapshot" of an Arena screen, go to Arena and get it into the state you want to photograph, press the *Prnt Scrn* (Print Screen) key, switch over to your word-processing document, and paste the shot where you want it; if you want just the active window (like a dialog box you want to document), press *Alt+Prnt Scrn* instead, then paste it into the word-processing document.

The model window can be split into two regions, or *views*: the *flowchart view* and the *spreadsheet view*. Often it's helpful to see both the flowchart and spreadsheet views of the model window at the same time. But you can choose to see only one of the views and thus devote all of the real estate in the model window to it by clearing the menu command *View > Split Screen* or clicking ▣ so that it does not appear to be pushed in; in this case, to see one or the other view, just single-click on either a flowchart (▭) or spreadsheet (▦) module in the Project Bar on the left of your screen. The flowchart view contains the model's graphics, including the process flowchart, animation, and other drawing elements. The spreadsheet view can display model data such as times and other parameters, and allows you to enter or edit them (right now it happens to be showing details about something called "Create – Basic Process"). Many model parameters can be viewed and edited in either the flowchart view or the spreadsheet view, but the spreadsheet view gives you access to lots of parameters at once, arranged in compact groups of similar parameters convenient for editing, especially in large models. The horizontal line splitting the flowchart and spreadsheet views (if both views are visible) can be dragged up or down to change the proportion of the model window allocated to the two views.

Down the left edge of the Arena window in Figure 3-1 is the *Project Bar*, which hosts *panels* containing the objects with which you'll be working, displaying one panel at a time. Right now the Project Bar is displaying the Basic Process panel, which contains fundamental building blocks, called *modules*, that are useful in a wide variety of simulation models.

Below the Basic Process panel on the Project Bar is a horizontal button labeled "Reports," which will display another panel containing a road map to the results of a simulation after it's run; click on this button to make this panel visible, and then click on the Basic Process button to make that panel visible again.

The Navigate panel allows you to display different views of a model, including different submodels in a hierarchical model (Model 3-1 doesn't have submodels so the only view in the Navigate panel is Top-Level, though if you click on the + to its left, you open a tree that has three entries for our model, which we'll discuss in Section 3.2.3 below). If the small button (▨) on the right of the horizontal Navigate button is pressed, you'll also get a "thumbnail" of the model window in the top of the Navigate panel, with a translucent blue box showing the location and zoom size of the active window's current view. Clicking anywhere in the thumbnail changes the current view to that spot. Drag the blue box around to pan to other regions of the window, or resize the blue box (hover the pointer over an edge and then click-drag) to change the "altitude" of the zoom. The +/– toggle in the circle in the thumbnail's upper right determines whether the blue box is shown relative to the entire window's capacity (the "+" choice), or takes into account what region in the window actually has content (the "–" choice).

The Project Bar is usually docked to the left edge of the Arena window, but it can be torn off and "floated" anywhere on your screen, or it can be docked to the right edge of the model window if you prefer. You'll usually need the Project Bar to be visible while working on a model, but if you'd like more room just to look through things, you can push the small ✗ button at the upper right of the Project Bar, or clear *View > Project Bar* to hide it (re-check *View > Project Bar* to display it again).

There are several other panels that come with Arena, perhaps depending on what you licensed. These include Advanced Process (with different and "smaller" building blocks for more detailed modeling), Advanced Transfer (containing many options for moving entities around), and Blocks and Elements (which together give you full access to the SIMAN simulation language that underlies Arena; see Pegden, Shannon, and Sadowski, 1995). Yet more panels contain constructs for specialized applications, like modeling contact centers and high-speed packaging lines. As mentioned earlier, to make the elements in a panel available for use in your model, you need to *Attach* the panel to your model via *File > Template Panel > Attach* or the *Template Attach* button ( ), or by right-clicking in a panel and selecting *Template Panel > Attach* in the pop-up. Panel files have the file name extension *.tpo* and are typically in the Template folder inside the Arena 10.0 folder. If you want Arena to attach certain panels to each new model you start, do *Tools > Options > Settings* and type the file names of those *.tpo* panel files into the *Auto Attach Panels* box there.

At the very bottom of the Arena window is the *Status Bar*, which displays various kinds of information on the status of the simulation, depending on what's going on at the moment. Right now the only thing it shows are the (x, y) coordinates in the world space (see Section 3.2.3) of the location of the mouse pointer. While the simulation runs, the Status Bar will display, for instance, the simulation clock value, the replication number being executed, and the number of replications to be run. You can hide the Status Bar by clearing (unchecking) *View > Status Bar*.

3.2.3 Panning, Zooming, Viewing, and Aligning in the Flowchart View

The particular flowchart view of the model window you see in Figure 3-1 is just one of many possible *views* of the model and the big *world space* in which the flowchart depiction of a model lives. The world space's center has (x, y) coordinates $(0, 0)$, and it extends for thousands of units in all four directions from there; these units are just positional and don't have any particular physical meaning (call them furlongs or youdels[2] if you like). To maximize the size of the model window within the Arena window, click ▢ if it's visible in the upper right corner of the model window. Likewise, to maximize the Arena window itself to consume your entire screen, click its ▢ button.

---

[2] Apologies to Gene Woolsey.

To see different parts of the flowchart view, you can pan around using the scroll bars on the lower and right edges, or the arrow keys (try it; to navigate via the keyboard, you must first make the model window active by clicking in it). You can also zoom in (with the ⌗ button or the + key or *View* > *Zoom In*), or zoom out (with the ⌗ button or the – key or *View* > *Zoom Out*) to see parts of the model from different "altitudes." To pan/ zoom automatically to see all the model at the closest possible zoom, click ⌗ (or *View* > *Views* > *All*, or the * key). If you want to go back to the preceding view (maybe you messed up), click ⌗ (or *View* > *Previous*). If you're at a relatively high altitude but spy a region that you'd like to view up close, select *View* > *Views* > *Region* (or hit the "[" key) to change the mouse pointer to cross hairs, click on one corner of the rectangular region you want to see, then again on the opposite corner—Arena will pan and zoom to see all of that region at the closest possible zoom (that is, lowest possible altitude). Another way to pan and zoom is via the thumbnail option in the Navigate toolbar, described in Section 3.2.2.

If you get to a view you like (and to which you'd like to be able to return instantly), you can save it as a *Named View* and assign a hot key to it. Pan and zoom to the view you want to save, then select *View* > *Named Views* (or hit the the **?** key or the ▼ button), and then click *Add*. You must give the view a descriptive Name, and you can optionally assign a hot key to it as well. To jump back to this view at any time, select *View* > *Named Views* (or hit the **?** key or the ▼ button), click on the view you want, and press the *Show* button. You can also access your Named Views in the Navigate panel of the Project Bar by click- ing the + to the left (in this case of Top-Level) to open up a tree of the Named Views; just click on an entry to go to that view. Yet another way (probably the fastest way) to get to a Named View is to hit the hot key assigned to it; you'll have to remember what the hot keys are, or maybe document them in the model with some text, as described in Section 3.6.3. Hot keys for Named Views are one of the few places in Arena where characters are case-sensitive (for example, "a" and "A" are different). Named Views can be accessed at any time, even while the simulation is running. We've set up three Named Views for Model 3-1: all (hot key **a**), logic (hot key **l**), and plots (hot key **p**). Try them out.

New Arena models start out in a specific "Home" pan/zoom configuration, just to the southeast of the (0, 0) position in the world space, to which you can return by pressing the Home key on your keyboard (or *View* > *Views* > *Home*). To see the largest possible area of the world space (from the maximum altitude), select *View* > *Views* > *Max*.

To get your visual bearings, you can display a background grid of little dots by checking *View* > *Grid* (or by clicking ⌗). If you further want to cause newly placed items to snap to this grid, check *View* > *Snap* (or click ⌗). Both of these actions are toggle keys; that is, you just repeat the action to undo it. To snap existing items to the grid, first select them (maybe using *Ctrl+Click* to keep extending your selection, or dragging a rectangle across them if they're arranged that way) and then *Arrange* > *Snap Object to Grid* to adjust their positions to align with the grid points. To customize the spacing of the grid points, select *View* > *Grid & Snap Settings*; the units are (x, y) values in the measurement units of the world space. You can also display Rulers on the top and left edges, with units' being Arena world units, by pushing the ⌗ button or checking *View* > *Rulers*.

You might want to align objects in the flowchart view precisely, either horizontally or vertically, with respect to their edges or centers, and there are several options for doing this. One option is to establish horizontal and vertical *Guides* in the flowchart view; make sure Rulers are displayed, then click the *Guides* button (┼) or select *View > Guides*, and then drag down from the horizontal ruler on the top or drag to the right from the vertical ruler on the left to place a dashed blue Guide line (you can have several of each). Then, click the *Glue to Guides* button (⌐) or select *View > Glue to Guides*, and then drag objects toward the guides until a red positional square lights up for edge or center positioning. Once objects are glued to a guide, you can drag the guide, and all objects glued to it will move together and stick with their alignment to the guide. Another option is to select the objects you want to align and then use the Arrange menu; select Align to line the selected objects up on their Top, Bottom, Left, or Right edges; select Distribute to space them evenly horizontally or vertically with spacing specified in Settings. *Arrange > Flowchart Alignment* lines up flowchart modules (see Section 3.2.4 next) in the current set of selected objects.

3.2.4 Modules

The basic building blocks for Arena models are called *modules*. These are the flowchart and data objects that define the process to be simulated and are chosen from panels in the Project Bar. Modules come in two basic flavors: *flowchart* and *data*.

Flowchart modules describe the dynamic processes in the model. You can think of flowchart modules as being nodes or places through which entities flow, or where entities originate or leave the model. To put an instance of a flowchart module of a particular type into your model, drag it from the Project Bar into the flowchart view of the model window (you can drag it around later to reposition it). Flowchart modules are typically connected to each other in some way. In the Basic Process panel, the kinds of flowchart modules available are Create, Dispose, Process, Decide, Batch, Separate, Assign, and Record; other panels have many additional kinds of flowchart modules. Each type of flowchart module in the Basic Process panel has a distinctive shape, similar to classical flowcharting (see Schriber, 1969) and suggestive of what it does. But in other panels (such as the Advanced Process panel), there are many more flowchart-module types than there are reasonable shapes, so they're all represented by simple rectangles. Some panels (like Advanced Transfer) use colors in the rectangles to distinguish different types of flowchart modules, and some panels (like the specialized ones for contact centers and packaging) use more elaborate graphics for them. One way to edit a flowchart module is to double-click on it once it's been placed in the flowchart view of the model window, to bring up a dialog box pertaining to it. Another way to edit flowchart modules is to select a module type (for example, click on a Create or a Process module), either in the Project Bar or in the flowchart view of the model window, and a line for each flowchart module of that type in the model shows up in the spreadsheet view of the model window (if it's visible), where you can edit the entries. This gives you a compact view of all the instances of flowchart modules of that type in your model, which is useful in large models where you might have many such instances.

Data modules define the characteristics of various process elements, like entities, resources, and queues. They can also set up variables and other types of numerical values and expressions that pertain to the whole model. Icons for data modules in the Project Bar look like little spreadsheets. The Basic Process panel's data modules are Entity, Queue, Resource, Variable, Schedule, and Set (other panels contain additional kinds of data modules). Entities don't flow through data modules, and data modules aren't dragged into the model window; rather, data modules exist "behind the scenes" in a model to define different kinds of values, expressions, and conditions. You don't double-click on a data module to edit it, but just single-click it in the Project Bar and a spreadsheet for that type of module will appear in the spreadsheet view of the model window (which must be visible), which you can then edit or extend by double-clicking where indicated to add additional rows. While the default is to edit modules from the spreadsheet view, if you double-click on the number in the left-hand column (or right-click and select Edit via Dialog), you can also edit in the dialog mode. Unlike flowchart modules, you don't have more than one instance of a data module in a model; however, there could be many rows in the spreadsheet for a data module, each typically representing a separate object of that type (for instance, if your model has three different queues, the Queue data module will display three rows, one for each queue, in its spreadsheet).

Flowchart and data modules in a model are related to each other by the names for objects (like queues, resources, entity types, and variables) that they have in common. Arena keeps internal lists of the names you give to these kinds of objects as you define them, and then presents these names to you in drop-down lists in the appropriate places in both flowchart and data modules, which helps you remember what you've named things (and protects you from your own inevitable typos...or, at least, keeps you consistent in your lousy typing so your model will still run).

3.2.5 *Internal Model Documentation*

If you rest your mouse pointer on a module symbol or other object, you'll see a Data Tip. Data Tips have two parts, a *default description* and a *user-defined description*. The default tip will describe some generic information about the object, such as its name and type. The user-defined part will display exactly what is entered in the Object Properties Description field. You can enter text in this field by right-clicking on an object and selecting Properties. Display of these Data Tips can be toggled via *View > Data Tips*. By default, both are enabled, but either the default or the user-defined Data Tips may be disabled.

In addition to the *module* descriptions, you can also enter a *Project Description* that provides some context to the entire model. This is a good place to document what the model does, why it was created, assumptions you have made, and similar information. This is entered in *Run > Setup > Project Parameters* (see Figure 3-15 later in this chapter).

While Data Tips are a useful feature, particularly when your models get large, you might wonder if there is another way to make use of this information. There is. *Tools > Model Documentation Report* creates a custom report summarizing all your model data. It provides some options about what information to include, then generates an HTML report.

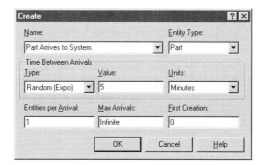

Figure 3-2. The Create Property Dialog Box for Model 3-1

3.3   Browsing Through an Existing Model: Model 3-1

To see how Model 3-1 is set up, we'll now walk you through the flowchart and data modules in the model window and indicate how they're related. Then we'll run this model and look at the results. After that, in Section 3.4, we'll show you how to build this model from scratch.

3.3.1   The Create Flowchart Module

We'll start with the Create module, which we named `Part Arrives to System`, at the left of the flowchart view of the model window. Note that this module is an instance of the general Create module, which we've specialized for our needs in this particular model.

The Create module is the "birth" node for arrival of entities to our model's boundary, representing parts in this case, into the model from outside. Double-click it to open a dialog box like the one in Figure 3-2.

In the Name box, we've typed `Part Arrives to System` as the name of this particular Create module (rather than accepting the bland default Name), which is what appears inside its shape in the flowchart view and in its data tip. We entered `Part` as the Entity Type; there's only one entity type in this model but in general there could be many, and naming them separately keeps them straight and allows you to customize them in useful ways (like separating times or numbers in system by entity type).

Across the center of the dialog box is a bordered area called Time Between Arrivals, where we specify the nature of the time separating consecutive arrivals of `Part` entities originating in this module. In the Type box, we selected `Random (Expo)` (using the list box arrow ▾) so that the interarrival times will be generated as draws on a random variable; in particular, from the exponential distribution (see Appendix C if you need to brush up on your probability and Appendix D for definition of the exponential and other probability distributions). In the Value box, we typed 5, and in the Units box, selected `Minutes` to tell Arena that we mean 5 minutes rather than 5 seconds or 5 hours or 5 days. While the number we typed in was "5," we could have typed "5." or "5.0" since Arena is generally quite robust and forgiving about mixing up integers and real numbers.

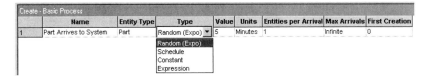

Figure 3-3. The Create Spreadsheet for Model 3-1

In the bottom row of boxes, we said that the number of Entities per Arrival is 1 (the default, so that Parts arrive one at a time rather than in a batch of several), that we don't want to put a cap on the maximum number of arrivals (if we did, this Create module would be "turned off" after that), and that the first Part should arrive right away, at time 0 (rather than after an initial time period that might or might not have the same definition as times between successive arrivals).

To close this Create dialog box, click *Cancel* or ✗ at the upper right; if you'd made any changes that you wanted to retain, you'd click *OK* instead.

An alternative way to edit the Create flowchart module is via the spreadsheet view in the model window. If you click the Create module in the flowchart view of the model window (or on any instance of the Create module there if there were several of them in your model), or on the general Create module shape in the Project Bar, a spreadsheet for your Create module(s) shows up in the spreadsheet view of the model window, as in Figure 3-3. By clicking or double-clicking in each of the fields in this spreadsheet, you can edit the entry or select from options; Figure 3-3 shows the list of options for Type accessed via the drop-down list there (drop-down lists for fields will be offered wherever they make sense). If you had multiple Create modules in your model, each representing a separate source of incoming entities, there would be a separate row in the Create spreadsheet for each. This is convenient in large models to edit many things quickly, or just to get an overview of all the Create modules at once. Selecting a particular Create module in either the flowchart or spreadsheet view selects that module in the other view. By right-clicking in a row of the spreadsheet, you're given the option of editing that module via the dialog box as in Figure 3-2. If you right-click in a numerical field (here or pretty much anywhere in Arena), you can also choose Build Expression to get very useful assistance in putting together a perhaps-complicated algebraic expression that could use many different Arena variables, probability distributions, mathematical functions, and arithmetic operations (we'll discuss the Expression Builder in Sections 3.4.10 and 4.2.4). You can change the field widths in a spreadsheet by dragging left and right the solid vertical bars separating them.

3.3.2   The Entity Data Module

One of the things we did in our Create module was to define an Entity Type that we called `Part`. By selecting the Entity data module in the Project Bar, the Entity spreadsheet for your model shows up in the spreadsheet view of the model window, as in Figure 3-4. Here you can see and edit aspects of the types of entities in your model. In Figure 3-4, the drop-down list for the Initial Picture field is shown, indicating that we decided that our `Part` entities will be animated as blue balls when the simulation runs. There are several fields

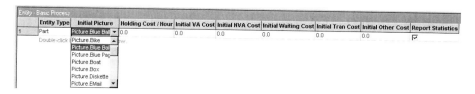

Figure 3-4. The Entity Spreadsheet for Model 3-1

for defining the costing data for entity types. A check box at the end lets you ask for Report Statistics on this entity type, including the average and maximum time in system observed for these types of entities during the run. We only have one entity type in our model, but if you had several, each would have its own row in the Entity spreadsheet.

3.3.3   The Process Flowchart Module

Our Process module, which we named `Drilling Center`, represents the machine, including the resource, its queue, and the entity delay time there (part processing, in our case). Open it by double-clicking on its name, and you should see the dialog box in Figure 3-5.

After entering the Name `Drilling Center`, we selected `Standard` as the type, meaning that the logic for this operation will be defined here in this Process module rather than in a hierarchical submodel. Skipping to the bottom of the dialog box, the Report Statistics check box allows you a choice of whether you want output statistics like utilizations, queue lengths, and waiting times in queue.

The Logic area boxes take up most of the dialog and determine what happens to entities in this module.

The Action we chose, `Seize Delay Release`, indicates that we want this module to take care of the entity's seizing some number of units of a Resource (after a possible wait in queue), then Delay for a time representing the service time, and then Release unit(s) of the Resource so that other entities can seize it. Other possible Actions are simply to Delay the entity here for some time (think of it like a red traffic light, after which the entity proceeds), Seize the Resource and then Delay (but not Release the Resource), or Delay and then Release the Resource that previously had been Seized; several Process modules could be strung together to represent a wide range of processing activities.

You can specify different Priorities for entities to Seize the Resource. Here and elsewhere in Arena, lower numbers mean higher priority.

Define the Resource(s) to be Seized or Released in the Resources box; click *Add* to add a Resource to this list. You can define or edit a particular Resource line by double-clicking its line, or selecting it and then clicking *Edit*, to bring up the Resources dialog box, as in Figure 3-6. Here you define the Resource Name and the Quantity of units (e.g., individual servers) that Seizing entities will Seize and that Releasing entities will Release (this is *not* where you specify the number of units of the Resource that exist—that's done in the Resource data module's Capacity field and will be discussed later). Listing more than one Resource means that Seizing entities must Seize the specified Quantity of each Resource before starting to be processed, like a machine and two operators, and Releasing entities will Release the specified Quantity of the corresponding Resources.

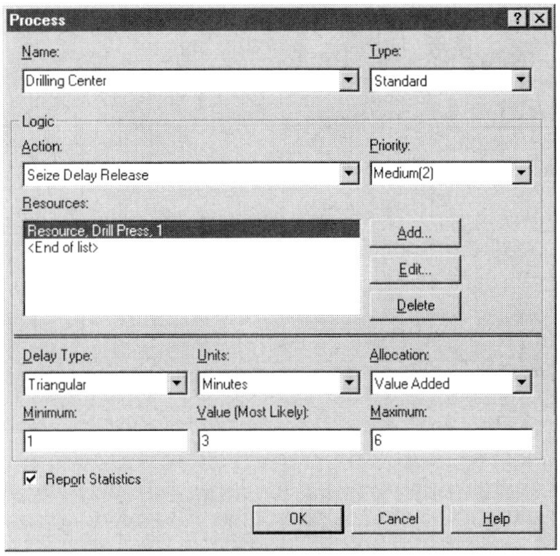

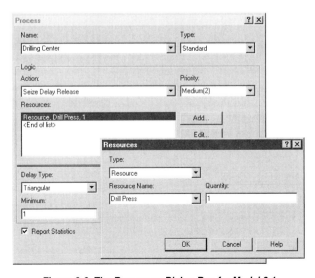

Figure 3-5. The Process Property Dialog Box for Model 3-1

Returning to the Process dialog box in Figure 3-5, the Delay Type drop-down list box offers three probability distributions (Normal, Triangular, and Uniform), a Constant, or a general Expression. The Units field determines the time units for the numerical Delay duration, and the Allocation field relates to how this delay is to be charged. The prompts on the next line change to match your choice of Delay Type. Note that the Expression option for the Delay Type allows you great flexibility in defining the Delay duration, including any other Arena probability distribution; right-clicking in the Expression field lets you bring up the Expression Builder (see Sections 3.4.10 and 4.2.4) to help you.

Figure 3-6. The Resources Dialog Box for Model 3-1

Figure 3-7. The Process Spreadsheet for Model 3-1

Close the Process dialog box with the *Cancel* button; again, if you had made changes that you wanted to retain, you'd click *OK*.

Figure 3-7 illustrates the Process spreadsheet, seen if you select any Process module instance in the flowchart view of the model window, or the general Process module in the Project Bar, with the drop-down box for Delay Type shown (where we've selected `Triangular`). If you had multiple Process modules in your model, there would be a row for each one in the Process spreadsheet. As with Create modules, this provides an alternative way to view simultaneously and edit the fields for your Process module(s). If you click on the "1 Rows" button in the Resources field, a secondary spreadsheet appears (see Figure 3-8) that allows you to edit, add, and delete resources equivalent to the Resources dialog box of Figure 3-6 (you need to click the ✗ button at the upper right of the Resources secondary spreadsheet to close it before you can go on).

3.3.4   The Resource Data Module

Once you've defined a Resource as we've done in this Process module (in our case, we named the resource `Drill Press`), an entry for it is automatically made in the Resource data module; click on it in the Project Bar to view the Resource spreadsheet in Figure 3-9. This spreadsheet allows you to determine characteristics of each Resource in your model, such as whether its Capacity is fixed or varies according to a Schedule (that drop-down list is shown in Figure 3-9, where `Fixed Capacity` has been selected). You can also cause the Resource to fail according to some pattern; try clicking on the "0 Rows" button under the Failures column heading to bring up a secondary spreadsheet for

Figure 3-8. The Resources Secondary Spreadsheet in the Process Spreadsheet for Model 3-1

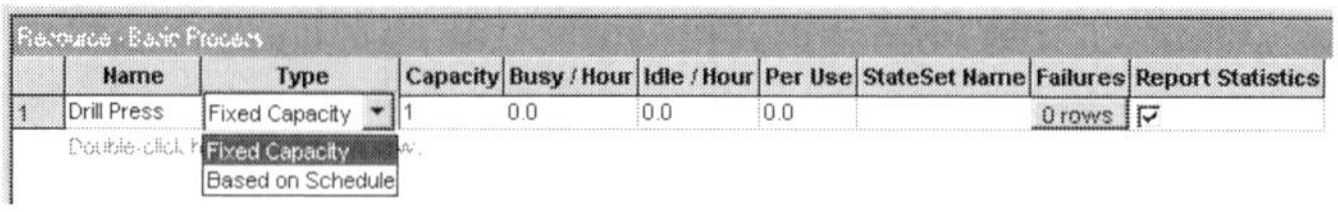

Figure 3-9. The Resource Data Module Spreadsheet for Model 3-1

this (the failure pattern is defined in the Failure data module in the Advanced Process panel, which you might have to attach to the Project Bar for your model).

3.3.5 The Queue Data Module

If the `Drill Press` resource is busy when an entity gets to the Process module, the entity will have to queue up. The Queue spreadsheet, seen in Figure 3-10, appears in the spreadsheet view if you select the Queue data module in the Project Bar. Here you can control aspects of the queues in your model (we only have one, named `Drilling Center.Queue`), such as the discipline used to operate it, as shown in the Type list in Figure 3-10 (`First In First Out` is the default and is selected). You could, for instance, rank the queue according to some attribute of entities that reside in it; if you chose Lowest Attribute Value, the queue would be ranked in increasing order of some attribute, and an additional field would show up in the line for this Queue in which you would have to specify the Attribute to be used for ranking.

3.3.6 Animating Resources and Queues

Speaking of queues, you might have noticed the ——————⊣ just above the Process module in the flowchart view. This is where the queue will be animated, and the Process module acquired this graphic when we specified that we wanted entities to Seize a Resource there.

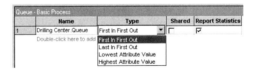

Figure 3-10. The Queue Data Module Spreadsheet for Model 3-1

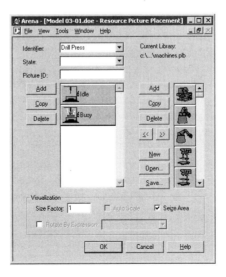

Figure 3-11. The Resource Picture Placement Dialog Box for Model 3-1

And while we're on the subject of animation, you've no doubt noticed the ⬛ above and to the right of the Process module and positioned at what will be the head of the queue animation. This is a Resource animation and will change appearance during the simulation depending on whether the `Drill Press` Resource is Idle or Busy. This did not come "free" with the Resource specified in the Process module; rather, we added it to our model via the *Resource* button (⬛) in the *Animate* toolbar. Double-click on the ⬛ icon to get the Resource Picture Placement dialog, as in Figure 3-11. This allows us to pick pictures from libraries (files with extension *.plb* to their name, usually found in the Arena 10.0 folder) to cause the Resource to be animated differently depending on the state it's in. While you can get an idea of how this works at this point, we'll discuss Resource animation in Sections 3.4.8 and 4.3.3.

3.3.7   The Dispose Flowchart Module

The Dispose module represents entities leaving the model boundaries; double-click its name to bring up the dialog box in Figure 3-12; the Dispose spreadsheet is in Figure 3-13. There's not much to do here—just give the module a descriptive Name and decide if you want output on the Entity Statistics, which include things like average and maximum time in system of entities that go out through this module and costing information on these entities.

3.3.8   Connecting Flowchart Modules

The Create, Process, and Dispose modules are connected (in that order, going left to right) by lines called *Connections*. These establish the sequence that all parts will follow as they progress from one flowchart module to another. To make the Connections, click *Connect* (⬛) or equivalently select *Object > Connect*, which changes the mouse pointer to cross hairs. Click on the *exit point* (▶) from the source module and finally on the *entry point* (■) on the destination module (you can make intermediate clicks if you want this connection to be a series of line segments). To help you hit these exit and entry points, which might appear quite small if you're zoomed out to a high altitude, Arena lights up a

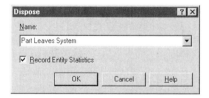

Figure 3-12. The Dispose Property Dialog Box for Model 3-1

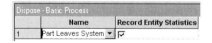

Figure 3-13. The Dispose Spreadsheet for Model 3-1

green box to indicate that you can click now and hit an exit point, and a red box for an entry point.

If you have many connections to make (maybe you placed a lot of flowchart modules to rough out your model), after each one you can right-click in a blank spot of the flowchart view and select Repeat Last Action from the pop-up menu to keep connecting. And if you have *really* a lot of connections to make, you can double-click on the Connect button (or do *Object > Connect* twice in a row) and not even have to bother with the right-click pop-up; when you're done and you want to get out of this, right-click or hit the Escape key (*Esc*).

If *Object > Auto-Connect* is checked, Arena will automatically connect the entry point on a newly placed module to whichever other connect-out module is selected when you place the new module.

If *Object > Smart Connect* is checked, then new Connections are automatically laid out to follow horizontal and vertical directions only, rather than following free-form diagonal directions according to where the connected modules are (unless you make intermediate clicks while drawing a connection, in which case you get the intermediate points and diagonals). This is pretty much a matter of taste and has no bearing on the model's operation or results.

If *Object > Animate Connectors* is checked (or, equivalently, *Animate Connectors* (⊡) is pushed in), then Arena will show entity icons (in our case, the blue balls) running down the connections as the transfers happen when the simulation runs. This is just to let you know during the animation that these transfers are occurring—as far as the simulation and statistics collection are concerned, they are happening in zero simulated time (or, equivalently, at infinite speed). We'll show you how to model non-zero travel times between model locations in Section 4.4, including how they're animated.

If for some reason you want to move a connection point (entry or exit) to somewhere else relative to its module, you can do so but you must first right-click on it and select Allow Move from the pop-up.

3.3.9 *Dynamic Plots*

The two plots were created via the *Plot* button (⊞) from the *Animate* toolbar. They'll dynamically draw themselves as the simulation runs, but then disappear when it's over (we'll show you how to make more detailed plots, which also stick around after the run ends, in Section 7.2.1).

Double-click on the top plot (the one for the queue length) to get the Plot dialog box on the left side of Figure 3-14. In the Expressions window, we have just one entry, so we'll get just one curve on this plot. This entry got there by clicking *Add* on the Plot dialog box to bring up a Plot Expression dialog box (which looks like the right side of Figure 3-14 after it's filled out), where we entered in the Expression box there `NQ(Drilling Center.Queue)`, the number of entities in this queue, which Arena will automatically update as the simulation proceeds. Right-clicking in this Expression box allowed us to use the Arena Expression Builder to help us enter the correct text here (more on the Expression Builder in Sections 3.4.10 and 4.2.4).

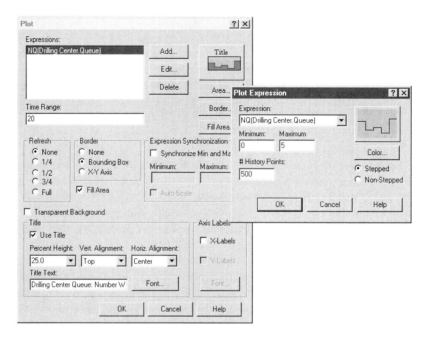

Figure 3-14. The Plot and Plot Expression Dialog Boxes for the Queue-Length Plot for Model 3-1

In the Plot dialog box, go ahead and double-click this Expression (or select it and then click *Edit*) to get the filled-out Plot Expression dialog box shown on the right side of Figure 3-14. The Minimum is the smallest *y*-axis value for this curve, and the Maximum is (you guessed it) the maximum *y*-axis value we want to allow for this curve. Here, we know that the model will start off empty and idle, so it's clear that the Minimum should be 0. However, the Maximum is *a priori* pretty much a guess, which might have to be adjusted after you make your run and see what the actual maximum queue length turns out to be; if your guess on the Maximum is too big, you'll squish your plot toward the bottom, and if you underguess the Maximum, you'll decapitate it (Arena can scale automatically, however, as discussed below). The # History Points is the maximum number of corners on the plot you want to allow for at any given time; if you see your plot dissolving from the left as the simulation runs, you should increase this value. Since this is a queue-length plot, it will be piecewise-constant, so the Stepped appearance is appropriate. The *Color* button in the Plot Expression dialog box allows you to change the color of the curve (we chose black), which might be useful if you're plotting several curves for several different Expressions on the same set of axes. Close the Plot Expression dialog with its *Cancel* button to get back to the Plot dialog box.

Back in the Plot dialog box, we entered 20 for the Time Range to allow room on the *x*-axis for a plot over the whole 20-minute simulation run (the units, which we want to be minutes, are in the Base Time Units for the model, as discussed in Section 3.3.11). Since this is wide enough for the whole run, we selected None for Refresh (the fractions under

Refresh are the portion of the plot that shifts off the left edge as needed to make room for that much in the near future on the right edge). We feel safer with a Bounding Box around the whole Border, and we checked Fill Area to flood the area under the curve with a color. If we were plotting several curves in the same graph, we could scale them on a common *y*-axis by checking Synchronize Min and Max and enter our guesses for the Minimum and Maximum here for all curves; in this case, if we next checked Y-Labels, we could then check Auto Scale to get Arena to change the *y*-axis scale as needed for all curves, and then our guesses on the extremes would just be the initial values (this works even if you're plotting just one curve). You can add a Title to your plot; the fields in that area should be self-explanatory (except maybe Percent Height, which is the percent of the total plot height consumed by the Title). The X-Labels option would label the extreme values of the *x*-axis, but we'll do our own custom labeling in Section 3.3.10; the Y-Labels option would display on the vertical axis the lowest and highest values reached by the curve(s) in the plot. The *Area*, *Border*, and *Fill Area* buttons under the plot thumbnail on the right allow you to select colors for those elements (we chose light gray for the background, dark gray for the fill area flooded under the curve, and black for the border—okay, call us boring). Click *Cancel* to close this Plot dialog box.

The size of the plot is determined by dragging the handles on its borders. Click (once) on the plot and try this (don't worry, there's an Undo). Actually, you have to specify an initial size of the plot after you fill out the dialogs, but you can change this later, as well as drag it around to relocate it.

The Plot and Plot Expression dialog boxes for the Drill Press: Number Busy plot are similar, so we won't go through them in detail (but go ahead and open them to look at them). The only really different thing is that the Expression whose value we want to plot on the *y*-axis is NR(Drill Press),which we know will always be either 0 or 1, so we specified the Maximum in the Plot Expression dialog box to be 2 to make for an attractive and tasteful graph. As before, we used the Expression Builder in the Expression box of the Plot Expression dialog box to figure out that this is the right name and syntax.

3.3.10 *Dressing Things Up*
The various labels in the model window, like the title at the upper left and axis labels for the plot, were done via the *Text* button (**A**) on the *Draw* toolbar. You can control the usual things like font, size, and style from there. To go to a new line in the text use *Ctrl+Enter*. To change the text color, select the text (single-click on it), and use the *Text Color* button (**A** ▾) to select either the color on the underline there (click on the **A** in this case) or to choose a different color (click on the ▾ in this case) that will become the new underline color for future text. You can also resize or rotate text by selecting it and dragging the underline bar.

The *Draw* toolbar also has things like boxes, ellipses, polygons, and lines, as well as the means to control their colors and styles, which you can use to decorate your model window, depending on your artistic creativity and talent (as you can see, ours is severely limited). This is how we made the simple shadow box behind the model title in the upper left of the model window. The *Arrange* toolbar and menu have buttons and commands that allow you to manipulate objects, such as grouping, flipping, sending a draw object to

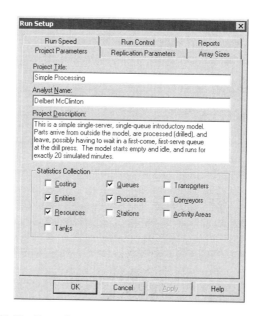

Figure 3-15. The Run > Setup > Project Parameters Dialog Box for Model 3-1

the back or front of a stack of objects, and so on. We'll talk more about artwork in Sections 3.6.2 and 3.6.3.

3.3.11 Setting the Run Conditions

Things like run length and number of replications are set via *Run > Setup*, which brings up a dialog box with five tabbed pages. Figure 3-15 shows the tab for Project Parameters, where we specify a Project Title, Analyst Name, and Project Description, as well as select what kind of output performance measures we want to be told about afterwards. We also chose to document our model internally via entering a brief Project Description.

Figure 3-16 shows the *Replication Parameters* tab of Run Setup, which controls a number of aspects about the run(s). We default the Number of Replications field to 1 (which we'll accept for now since we're only concerned with modeling at the moment, although you know better, from Section 2.6.2). We'll default (that is, not use) the Start Date and Time field, which is for associating a specific calendar date and time with a simulation time of zero. You can also specify a Warm-up Period at the beginning of each replication, after which the statistical accumulators are all cleared to allow the effect of possibly atypical initial conditions to wear off. We specify the Length of Replication to be 20 and select the time unit for that number to be Minutes. The Hours Per Day box defaults to 24 (to answer the question you obviously have about this, it could be convenient to define a day to have, say, 16 hours in the case of a two-shift manufacturing operation, if it's customary to think of time in days). The Base Time Units box specifies the "default" time units in which time-based outputs will be reported, as well as how Arena will interpret some time-based numerical inputs that don't have an accompanying Time Units box (such as the Time Range in the Plot dialog box in Figure 3-14). The

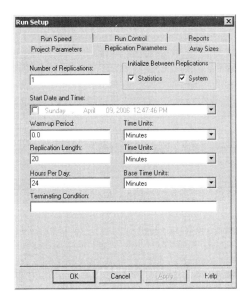

Figure 3-16. The Run > Setup > Replication Parameters Dialog Box for Model 3-1

Terminating Condition box allows you to establish complex or state-dependent termination rules; see Section 12.5.2 for an example where we want the simulation to keep running until the results achieve the statistical precision we'd like. Model 3-1, however, will simply terminate at time 20 minutes. Close the Run Setup dialog box by pressing *Cancel*.

Speaking of termination, you must specify in every Arena model how you want it to terminate. This is really part of modeling. Arena can't know what you want, so does not include any kind of "default" termination. In fact, in most cases, your simulation will just continue running forever or until you intervene to stop it, whichever comes first. In Section 3.8, we'll show you how to pause and then kill your run if you need to.

3.3.12 Running It

To run the model, click the *Go* button (▶) in the *Standard* toolbar (or *Run > Go* or press the *F5* key); note that the buttons in this group are similar to those on a video player. The first time you run a model (and after you make changes to it) Arena checks your model for errors (you can do this step by itself with the ✓ button on the *Run Interaction* toolbar, or *Run > Check Model* or the *F4* key); if you have errors, you'll be gently scolded about them now, together with receiving some help on finding and correcting them. Then you can watch the model animation run, but you'll have to look fast for a run this short unless your computer is pretty laid back. During the animated run, you see the Part entities (the blue balls) arriving and departing, the Resource Picture changing its appearance as the Resource state changes between Idle and Busy, the Queue changing as Part entities enter and leave it, the digital simulation clock in the Status Bar advancing, and the plots being drawn. The counters next to the flowchart modules display different quantities depending

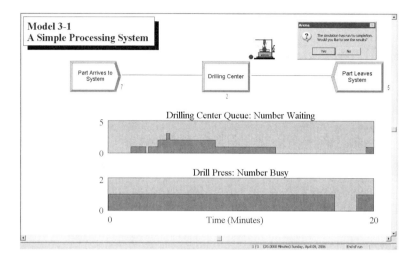

Figure 3-17. Ending Animation State of Model 3-1

on the module type. For the Create module, it's the number of entities that have been created. For the Process module, it's the number of entities that are currently in process there (in service plus in queue), and for the Dispose module, it's the number of entities that have left the system. There are other ways to run your model, and we'll discuss some of them in Section 3.8.

The final state of things should look something like Figure 3-17, except that we've moved the Arena dialog box (asking about seeing the results) out of the way. The plots display the same information as in Figure 2-3 from the hand simulation. The clock in the Status Bar is frozen at its final value, and at that point we see that the Resource is Busy operating on one part, with one part waiting in queue (in agreement with the final state in the hand simulation in Section 2.4.3 and the bottom row of Table 2-2). The final values of the counters next to the flowchart modules are also as they were at the end of the hand simulation in Section 2.4.3.

The Arena box that appears at the end of the run asks if you'd like to see the summary results, which we'll do next in Section 3.3.13. After you look at those reports (or if you choose not to), your model window will appear to be "hung" and you can't edit anything. That's because you're still in *run mode* for the model, which gives you a chance to look at the plots and the final status of the animation. To get out of run mode and back to being able to edit, you have to click *End* (■), just like on a video player.

3.3.13 Viewing the Reports

If you'd like to see the numerical results now, click *Yes* in the Arena box that appears at the end of your simulation, as shown near the top right of Figure 3-17. This opens a new reports window in the Arena window (separate from your model window). The Project Bar now displays the Reports panel, which lists several different Reports you can view, such as Category Overview, Category by Replications, and Resources. Clicking on each of these reports in the Project Bar opens a separate report window in the Arena window (use the Arena Windows menu to see what you have open). Don't forget to close these report windows when you're done viewing them since they don't go away on their own if you simply go back to your model window; if you change your model and then re-run it, you might wind up with several different report windows open and it could get confusing to figure out which one goes with which variant of your model. Actually, when making changes in your model to investigate the effects of different parameter settings or assumptions, you probably should change the name of your *.doe* file slightly, since Arena will simply overwrite previous results to the same report file name if you don't, and you'll lose your previous results. (The Arena Process Analyzer, discussed in Sections 3.6.1 and 6.5, provides a far better way to manage the activity of running multiple variants or scenarios of your model and keeping track of the results for you.)

The default Arena installation automatically brings up the Category Overview Report, which gives you access to most of the results; the other reports listed in the Project Bar repeat a lot of this, but provide more detail. Down the left edge of the report window itself is a tree, which you can expand by clicking the + signs in it (and re-contract with the − signs), giving you a kind of hyperlinked outline to this entire report. The report itself is organized into pages, through which you can browse using the ▶, ▶|, |◀, and ◀ buttons at the top left of the report window. If you want to print some or all of the pages in the report being displayed, click the 🖨 button in the report window (not the similar-looking button above it in the Arena window, which will be dimmed and thus inactive anyway if the report window is active). If you'd like to export the report to a different file, including several common spreadsheet and word-processor formats, click 📖 in the report window and follow the directions there.

But if you're looking for just a few specific results, it's better to click around on the +'s and −'s in the tree outline in the report window. For instance, to see what happened with the queue during our simulation, we clicked down a sequence of + signs into the Queue section of the report (specifically, Simple Processing → Queue → Time → Waiting Time → Drilling Center.Queue), eventually getting to the Waiting Time information, as shown in Figure 3-18. What's selected in the tree is displayed and outlined in the report to the right, and we see from that line that the average waiting time in queue was 2.5283 (the report reminds us that the Base Time Units are minutes), and the maximum waiting time was 8.1598 minutes (both of which agree with the hand-simulation results in Section 2.4.4). A little further down in this view of the report window, under "Other" (to which we could jump directly by clicking on its entry in the tree), we see that the average number waiting (i.e., length) of the queue was 0.7889 part, and the maximum was 3, both of which agree with our hand simulation in Section 2.4.4.

Browse through this report and note that the output performance measures in Table 2-3 are all here, as well as a lot of other stuff that Arena collected automatically (we'll talk more about these things later). By following the branches of the tree as indicated below, you'll find, for example:

- Simple Processing → Entity → Time → Total Time → Part: The average total time in system was 6.4397 minutes, and the maximum was 12.6185 minutes.

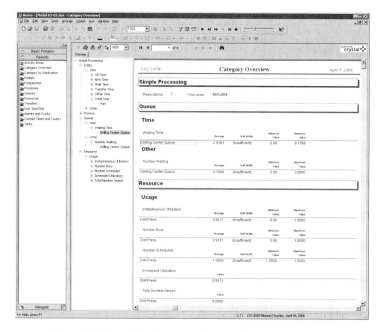

Figure 3-18. Part of the Category Overview Report for Model 3-1

- Simple Processing → Resource → Usage → Instantaneous Utilization → Drill Press: The utilization of the drill press was 0.9171 (i.e., it was busy 91.71% of the time during the simulation). The different measures of utilization are discussed in Section 4.2.5.

- Simple Processing → Process → Other → Number In → Drilling Center: During the simulation, seven entities entered the drilling center Process module.

- Simple Processing → Process → Other → Number Out → Drilling Center: During the simulation, five entities left the drilling center Process module (two fewer than entered, which is how many parts were in the drilling center at termination). This value of 5 also represents the total production in this model, since parts exit the system immediately after leaving the drilling center.

- Simple Processing → Entity → Time → Wait Time → Part: Of the five parts that exited the system, their average wait time in all queues (of which there's only one in this model) was 3.0340 minutes, and the maximum was 8.1598 minutes. The reason the average here differs from the average waiting time in queue = 2.5283 is that the 3.0340 here counts the waiting times of only those five parts that exited the system, while the 2.5283 counted the waiting times of all 6 parts that left the queue. The two maxima, however, are the same in this run since that maximum was achieved earlier in the simulation (the maxima would not necessarily always be equal).

- Simple Processing → Entity → Other → WIP → Part: The work in process (WIP) averaged 1.7060 parts and hit a maximum of four parts at some point(s) in time.

Many of the numbers in the reports (as in our hand simulation in Chapter 2) can be classified as *tally, time-persistent,* or *counter* statistics:

- *Tally statistics* are those that result from taking the average, minimum, or maximum of a list of numbers. For example, the average and maximum total time in system (6.4397 and 12.6185 minutes, respectively) are tally statistics since they're respectively the average and maximum of the total times in system of the five parts that left the system during the simulation. Tally statistics are sometimes called *discrete-time statistics* since their "time" index (1, 2, 3, ...) is a discrete indexing of the time order in which the observations were made.

- *Time-persistent statistics* are those that result from taking the (time) average, minimum, or maximum of a plot of something during the simulation, where the *x*-axis is *continuous* time. Time-persistent averages involve the accumulated area under the plotted curve (i.e., an integral). The average and maximum of the number of parts in queue (0.7889 and 3 parts, respectively) are time-persistent statistics, as is the instantaneous utilization of the drill press (0.9171). Time-persistent statistics are also known as *continuous-time statistics.*

- *Counter statistics*, as the name suggests, are accumulated sums of something. Often, they are simply nose counts of how many times something happened, like the number of parts that left the drilling center (five in our simulation) or that entered the drilling center (seven). But they could be accumulations of numbers that are not all equal to 1; for instance, the accumulated waiting time at the drilling center was 15.1700 minutes, representing the sum (not average) of the waiting times observed in queue there. In the Category Overview report, this number can be found via Simple Processing → Process → Accumulated Time → Accum Wait Time → Drilling Center. Another counter statistic is at Simple Processing → Resource → Usage → Total Number Seized → Drill Press, where we see that the drill press resource was used (either to completion for the part or just started) six times.

If you close a report window, you can view it later as long as you don't delete (or overwrite) the Microsoft® Access database file that Arena creates as it runs. This Access file is named `model_filename.mdb`, where `model_filename` is what you named your *.doe* model file (so in our case, the Access file is named `Model 03-01.mdb`). On the Project Bar, select Reports and then click on the report you want (such as Category Overview) to view it again. The way this works is that Arena uses third-party software called Crystal Reports® from Business Objects (http://www.businessobjects.com) to read the Access database file, extract the useful stuff from it, and then display it to you in the report-window format we've described above.

Right now you might be thinking that this report structure is pretty serious overkill just to get a handful of numbers out of this small model, and it probably is. However, in large, complicated models it's quite helpful to have this structure to organize the myriads of different output numbers and to help you find things and make some quick comparisons and conclusions.

In addition to the reports described above, Arena produces a very compact (to the point of being cryptic) report of many of the simulation results, as a plain ASCII text file named `model_filename.out` (so `Model 03-01.out` for us), as shown in Figure 3-19. Some of the labels are a little different; e.g., "DISCRETE-CHANGE VARIABLES" are the same as time-persistent statistics. You'll find in this file many of the numbers in the reports we talked about above (minor roundoff-error discrepancies are possible), and even a few that are not in the reports above (like the number of observations used for tally statistics). For some purposes, it might be easier and faster to take a quick look at this rather than the report structure we described above. However, the order and arrangement and labeling of things here is decidedly user-hostile; this format

```
Project: Simple Processing
Analyst: Delbert McClinton

Replication ended at time     : 20.0 Minutes
Base Time Units: Minutes


                        TALLY VARIABLES

Identifier              Average   Half Width  Minimum    Maximum    Observations

Drilling Center.WaitTi  3.0340    (Insuf)     .00000     8.1598     5
Drilling Center.TotalT  6.4396    (Insuf)     2.8955     12.618     5
Drilling Center.VATime  3.4056    (Insuf)     1.7641     4.5167     5
Part.VATime             3.4056    (Insuf)     1.7641     4.5167     5
Part.NVATime            .00000    (Insuf)     .00000     .00000     5
Part.WaitTime           3.0340    (Insuf)     .00000     8.1598     5
Part.TranTime           .00000    (Insuf)     .00000     .00000     5
Part.OtherTime          .00000    (Insuf)     .00000     .00000     5
Part.TotalTime          6.4396    (Insuf)     2.8955     12.618     5
Drilling Center.Queue.Wa 2.5283   (Insuf)     .00000     8.1598     6


                     DISCRETE-CHANGE VARIABLES

 Identifier              Average   Half Width  Minimum    Maximum    Final Value

Part.WIP                 1.7059    (Insuf)     .00000     4.0000     2.0000
Drill Press.NumberBusy   .91709    (Insuf)     .00000     1.0000     1.0000
Drill Press.NumberSchedu 1.0000    (Insuf)     1.0000     1.0000     1.0000
Drill Press.Utilization  .91709    (Insuf)     .00000     1.0000     1.0000
Drilling Center.Queue.Num .78890   (Insuf)     .00000     3.0000     1.0000


                        OUTPUTS
  Identifier                          Value

Drilling Center Number Out            5.0000
Drilling Center Accum VA Time         17.028
Drilling Center Number In             7.0000
Drilling Center Accum Wait Time       15.170
Part.NumberIn                         7.0000
Part.NumberOut                        5.0000
Drill Press.NumberSeized              6.0000
Drill Press.ScheduledUtilization      .91709
System.NumberOut                      5.0000


Simulation run time: 0.02 minutes.
Simulation run complete.
```

Figure 3-19. SIMAN Summary Report File (Model 03-01.out) *for Model 3-1*

is actually a leftover from earlier versions of Arena and in fact goes back to the early 1980s and the underlying SIMAN simulation language. However, if, out of nostalgia or a mania for compactness, you want this to be the default report that Arena gives you if you ask for one after the run ends, select *Run > Setup > Reports* and in the Default Report list scroll down and pick SIMAN Summary Report (*.out* file).

The exact meaning of report labels like Average, Minimum, Maximum, and Time-Average should be clear to you by now. But the reports also refer here and there to Half Widths (though we never get numbers for them in this model and are just scolded that we're somehow Insufficient). As you might guess, these will be half widths of confidence intervals (they'll be at level 95%) on the expected value of the corresponding performance measure, provided that our simulation produces adequate data to form them.

If we were doing more than one replication (which we're not in this case), Arena would take the summary results for an output performance measure from each replication, average them over the replications, compute the sample standard deviation from them, and finally compute the half width of a 95% confidence interval on the expected value of this performance measure. This is exactly what we did by hand in Section 2.6.2 for several of the output performance measures there, as given in Table 2-4. Exercise 3-1 asks you to do this with Arena to reproduce (except maybe for round-off) the results in Table 2-4.

If we're interested in the *long-run* (or *steady-state*) behavior of the system after any initialization effects have worn off, we might choose to make just one (really) long replication or run; this issue is taken up in Section 7.2.3. If we do so, we might see half-width numbers in the reports if our run is long enough, even though we're doing only one replication. Arena tries to compute these half widths by breaking the single long run into batches whose means serve as stand-ins for truly independent summary results over replications for making confidence intervals. We'll talk about this more in Section 7.2.3, including a discussion of exactly how these half widths are computed from the simulation output data. The reason you might get "Insufficient" (or, sometimes, "Correlated") instead of a numerical value for the half width is that the run must be long enough to provide data in sufficient quantity and of adequate quality to justify the validity of the half widths; if your run is not long enough for this, Arena simply declines to deliver a value on the theory that a wrong answer is worse than no answer at all.

3.4 Building Model 3-1 Yourself

In this section, we'll lead you through the construction of Model 3-1 from scratch. What you end up with might not look exactly like our Model 3-1 cosmetically, but it should be functionally equivalent and give you the same results.

Before embarking on this, we might mention a couple of little user-interface functions that often come in handy:

- Right-clicking in an empty spot in the flowchart view of the model window brings up a small pop-up box of options, one of which is to repeat the last action (like placing a module in your model, of which you may need multiple instances). This can obviously save you time when you have repetitive actions (though it won't make them any more interesting). Other options here include some views, as well as running or checking the model.

- *Ctrl+D* or pressing the *Ins* key duplicates whatever is selected in the flowchart view of a model window, offsetting the copy a little bit. You'll then probably want to drag it somewhere else and do something to it.

3.4.1   New Model Window and Basic Process Panel

Open a new model window with ▯ (or *File > New* or *Ctrl+N),* which will automatically be given the default name Model1, with the default extension *.doe* when you save it. You can change this name when you decide to save the contents of the model window. Subsequent new model windows during this Arena session will get the default names Model2, Model3, and so on. You might want to maximize your new model window within the Arena window by clicking ▢ near the model window's upper-right corner (if it's not already maximized).

Next, attach the panels you'll need if they're not already there in the Project Bar. For this model, we need only the Basic Process panel, which is a file called BasicProcess.tpo, typically in the Template folder under the Arena 10.0 folder. Click 🗐 (or *File > Template Panel > Attach* or right-click on the Project Bar and select *Attach)* to open the Attach Template Panel dialog in Figure 3-20, where you open the Basic Process panel BasicProcess.tpo (click on it, then click the *Open* button, or just double-click it). You can tell Arena to attach certain panels automatically to the Project Bar of new model windows via *Tools > Options > Settings* by typing the panels' file names (including the *.tpo* extension) into the Auto Attach Panels box.

The attached panel will appear in the Project Bar, with icons representing each of the modules in this panel. Right-click in the panel to change the icon size or to display text only for the modules. Right-clicking in a panel also allows you to detach it if you accidentally attached the wrong one (you can also detach the visible panel via 🗐 or *File > Template Panel > Detach).* You can detach a panel even if you've placed modules from

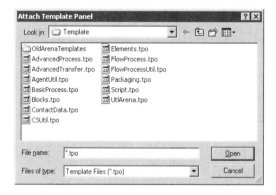

Figure 3-20. The Attach Template Panel Dialog

that panel in your model. If your display isn't tall enough to show the panel's full height, use the scroll bar at its right to get to it all.

3.4.2 Place and Connect the Flowchart Modules

This model requires one instance of each of three flowchart modules: Create, Process, and Dispose. To add an instance of a flowchart module to your model, drag its icon from the Project Bar into the flowchart view of the model window and drop it about where you want it (you can always drag things around later). To help you line things up, remember Grid (⬚), Snap (⬚), Snap to Grid (⬚), Rulers (⬚), Guides (╋), Glue (⬚), *Arrange >* *Align*, and *Arrange > Distribute* from Section 3.2.3.

If you have *Object > Auto-Connect* checked and you dragged the modules into your model in the order mentioned above (without de-selecting a module after dropping it in), Arena will connect your modules in the correct order; if you have *Object > Smart Connect* checked, those connections will be oriented horizontally and vertically.

Figure 3-21 shows how these modules look in the flowchart view of the model window just after we placed them (both Connect toggles were checked), with the Dispose model selected since it was the last one we dragged in. If you did not have *Object > Auto-Connect* checked, you'll need to connect the modules yourself; to do so, use *Connect* (⬚) on the modules' exit (►) and entry (■) points, as described in Section 3.3.8. Recall as well from Section 3.3.8 that, during the animation, you'll see entities running along the connections if the *Animate Connectors* button (⬚) is pressed (or equivalently, *Object > Animate Connectors* is checked) just to let you know that they're moving from one flowchart module to the next. However, this movement is happening in zero simulated time; Section 4.4 describes how to represent positive travel times in your model.

3.4.3 The Create Flowchart Module

Open the "raw" Create module by double-clicking it to get the dialog box in Figure 3-22, where we need to edit several things. We first changed the Name of this instance of the Create module from the default, Create 1, to `Part Arrives to System`, and we changed the Entity Type from the default to `Part`. In the Time Between Arrivals area of the dialog box, we accepted the default Random (Expo) for the Type, changed the default

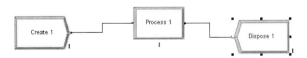

Figure 3-21. Initial Placement of the Flowchart Modules

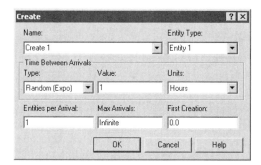

Figure 3-22. The Create Dialog Box

Value of 1 to 5, and selected `Minutes` as the Units from the drop-down list there (note that the default is Hours). We accepted the defaults for the three boxes in the bottom row of the dialog box and clicked *OK* to save our changes; at this point, the Create dialog box should look like Figure 3-2. Recall that we can also view and edit flowchart modules via their spreadsheet view, as detailed in Section 3.3; the completed spreadsheet for this Create module was shown earlier in Figure 3-3. Note that the new name of this Create module now appears in its shape in the flowchart view of the model window.

3.4.4   Displays

As we introduce new modules and new concepts, we'll try to lead you through each dialog box (or, equivalently, spreadsheet). Even though the Create module is fairly simple, the above description was fairly lengthy. To convey this more compactly, we'll use visuals, called *Displays*, as shown in Display 3-1. Note that there are three parts to this display. The upper-right portion has the filled-in dialog box (which in Display 3-1 is the same as Figure 3-2). In some cases, it may show several related dialogs. The upper left shows the module with which the dialog box is associated. Later, this portion may also show buttons we clicked to get the dialog box(es) shown on the upper right. The bottom portion of the display is a table showing the actions required to complete the dialog box(es). The left column of the table defines the dialog boxes or prompts, and the right column contains the entered data or action (italics) like checking boxes or pressing command buttons. In general, we'll try to provide the complete display when we introduce a new module or a new secondary dialog box of a module we've already covered. For modules that aren't new, we'll normally give you only the table at the bottom of the display, which should allow you to recreate all the models we develop easily. There may be more items or prompts in a dialog box than we show in a display; for those we accept the defaults.

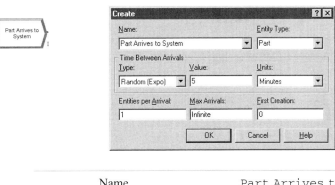

| Name | Part Arrives to System |
| Entity Type | Part |

Time Between Arrivals area	
Type	Random (Expo)
Value	5
Units	Minutes

Display 3-1. The Completed Create Dialog Box

This might be a good time to save your model; choose a name different from ours (which was `Model 03-01.doe`), or put it in a different folder.

3.4.5   The Entity Data Module

Now that you've defined the `Part` Entity Type in your Create module, you might want to say some things about it. The only thing we did was change its initial animation picture from the default "Report" (⊞) to "Blue Ball" (●). To do this, click the Entity data module in the Project Bar to display it in the spreadsheet view of the model window (see Figure 3-4 in Section 3.3.2). In the Initial Picture list, click `Picture.Blue Ball` (clicking the arrow opens the list for scrolling).

3.4.6   The Process Flowchart Module

Display 3-2 indicates what's needed to edit the Process flowchart module. Since the Action you specify includes Seizing a Resource, you *must* hit the *Add* button to define which resource is to be Seized; this brings up the Resources secondary dialog box, which is shown in the display as well. You also might want to make the area for the queue animation longer; click the queue animation (the ⊢————⊣), then drag its left end back to the left (hold down the *Shift* key to constrain its angle to horizontal, vertical, or 45° diagonal).

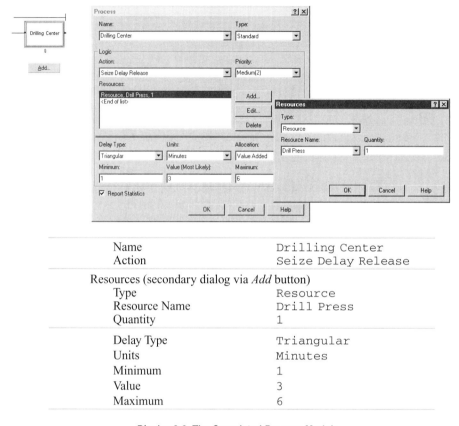

Name	Drilling Center
Action	Seize Delay Release

Resources (secondary dialog via *Add* button)	
Type	Resource
Resource Name	Drill Press
Quantity	1

Delay Type	Triangular
Units	Minutes
Minimum	1
Value	3
Maximum	6

Display 3-2. The Completed Process Module

3.4.7   The Resource and Queue Data Modules

Once you've defined this Process module, your model has both a Resource and a Queue, with names you specified (the name for a Queue is `whatever.Queue`, where `whatever` is the Name you gave to the Process module for this Queue). If you need to specify non-default items for a Resource or a Queue (which we don't in this model), you'd use the Resource and Queue data modules, as described in Sections 3.3.4 and 3.3.5.

3.4.8   Resource Animation

While not necessary for the simulation to work, it's usually nice to animate your Resources. This lets you show their state (just Idle vs. Busy for this model), as well as show entities "residing" in them during processing.

Click on the *Resource* button (⌐) in the *Animate* toolbar to bring up the Resource Picture Placement dialog box. To associate this picture to the Resource, click the Identifier arrow to choose the Resource Name, `Drill Press`. In the list of pictures on the left side of the dialog box, select `Inactive` and then press the *Delete* button to its left; do the same for `Failed`.

Now, if your Drill Press really looks like a white square when it's Idle and a green square when it's Busy, you're all set...but probably it doesn't. If you double-click on one square you get into the Picture Editor where you can try your hand at drawing your drill press in both its corresponding states. Or, if you have a graphics file somewhere depicting your drill press (maybe from a digital camera), you could copy and paste it into the Picture Editor.

Instead, let's pick some attractive artwork out of one of Arena's picture libraries, on the theory that if you were any good at art or photography you probably wouldn't be reading *this* book. So go ahead and close the Picture Editor to get back to the Resource Picture Placement dialog box. To open an Arena picture library, click *Open* along the right column and navigate to the Arena 10.0 folder where you'll see a list of files with *.plb* file name extensions. Open `Machines.plb` to view a gallery on the right of stunning creations from the rustbelt collection. Scroll down a bit and click ▦, then click the white-square Idle button on the left (depressing it), and then click ≤< to copy this picture to the left where it becomes the Idle picture instead of the white square. Similarly, copy ▦ on the right to become the Busy picture on the left. Finally, check the Seize Area box at the bottom so that the Part being processed will show up in the animation. Your Resource Picture Placement dialog box should look like Figure 3-11 in Section 3.3.6 at this point. We'll have more to say about Resource pictures in Section 4.3.3.

3.4.9 The Dispose Flowchart Module

The final flowchart module is the Dispose; Display 3-3 shows how to edit it for this model (the only thing to do is improve on the default Name).

3.4.10 Dynamic Plots

We described most of the entries and properties of the two animated plots earlier in Section 3.3.9. To make such a plot from scratch, press the *Plot* button (▱) on the *Animate* toolbar to get a blank Plot dialog box, and then proceed as indicated in Display 3-4. Remember that initially you might have to guess at the Maximum y-axis value in the Plot Expression dialog box, and perhaps adjust it after you have a feel for the results. Also, when you're done filling in the dialog and click *OK*, your mouse pointer becomes cross hairs; click to determine the location of one corner of the plot, then again to determine the opposite corner (of course, you can resize and reposition the plot later).

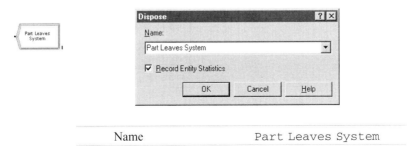

Name	Part Leaves System

Display 3-3. The Completed Dispose Module

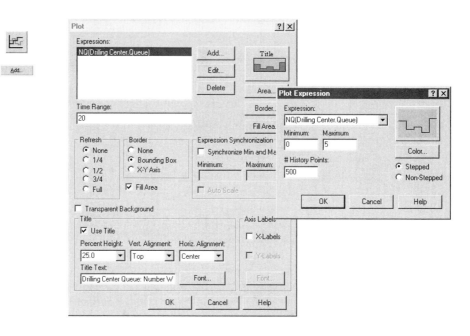

Plot Expressions (secondary dialog via *Add* button)

Expression	`NQ(Drilling Center.Queue)`
Maximum	5
Color	*black*

Plot

Time Range	`20`
X-Labels	*clear* (i.e., *uncheck*)
Title - Use Title	*select*
Horiz. Alignment	*Center*
Title Text	`Drilling Center Queue:`
	`Number Waiting`

Display 3-4. The Completed Plot Dialog Box for the Queue-Length Plot

While it's perfectly legal just to type in NQ(Drilling Center.Queue) manually for the Expression in the Plot Expressions secondary dialog box, you'd first have to know that this is the right thing to type in, which you very well might *not* know. This is one of many places in Arena where you can enter a general algebraic expression, and to do so correctly, you often need to know the exactly correct names of various objects in your model (like Drilling Center.Queue) and built-in Arena functions (like NQ, which returns the current number of entities in the queue named in its argument). To help you out with this memory-intensive task (that's your memory, not your computer's), Arena provides something called the Expression Builder, which you can access by right-clicking in any box that calls for some kind of Expression, and then

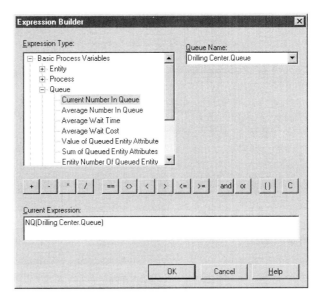

Figure 3-23. The Expression Builder for a Queue-Length Expression

selecting Build Expression. Figure 3-23 shows the Expression Builder window after we expanded the Expression Type tree on the left to get at what we want for the queue-length plot. The label on the box in the upper right (now Queue Name) will change depending on what we select in the Expression Type tree; here it provides a drop-down list where we can specify the queue for which we want to know the Current Number in Queue (we only have one Queue in this model so it's a short list). The Current Expression box at the bottom is the Expression Builder's answer, and clicking *OK* at this point pastes this text back into the Expression box from where we started with our right-click. You can still edit and modify the expression there, as you can in the Current Expression box of the Expression Builder, perhaps using its calculator-type buttons for arithmetic operations, and using the tree at the left to look up the names of other quantities.

The plot for the number busy at the drill press is quite similar, with only two differences from Display 3-4, both of which are in the Plot Expressions secondary dialog box. First, make the Expression NR(Drill Press), an expression you can discover with the Expression Builder via the Expression Type path Basic Process Variables → Resource → Usage → Current Number Busy, and select Drill Press (the only entry) under Resource Name. Finally, make the Maximum 2 since we know this curve will always be at height zero or one.

To make the two plots visually harmonious, we made them the same size and aligned them vertically.

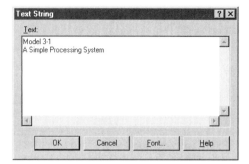

Figure 3-24. The Text String Dialog Box

3.4.11 Window Dressing

Model 3-1 has several text labels in the flowchart view of the model window to help document things as well as indicate what's what during the animation. These were produced via the *Text* button (**A**) on the *Draw* toolbar, which opens a Text String dialog box like the one in Figure 3-24. Type in your text (use *Ctrl+Enter* to go to a new line), perhaps change its font (Times Roman, Arial, etc.), font style (Italics, Bold, etc.), and size via the *Font* button, and then click *OK*. Your mouse pointer becomes cross hairs, which you click to place the northwest corner of the text entry in the flowchart view of your model window. You can drag it around later, as well as resize and reorient it using the underline below it that appears when you select it (hold down the shift key while reorienting it to constrain it to be horizontal or vertical or on a 45° line). To change the text color, select the text and use the *Text Color* button (**A** ▾) to select either the color of the underline there (click on the **A** in this case) or to choose a different color (click on the ▾ in this case and select a color from the palette), which will also become the new underline color.

The yellow backdrop box behind the model label and its shadow were made with the *Box* button (□) on the *Draw* toolbar; clicking this button turns your mouse pointer to cross hairs, after which you click once to determine one corner of the box and again to determine the opposite corner. You can change the fill color by first selecting the box and then clicking the Fill Color button (▨ ▾); change the border color with the Line Color button (▨ ▾). To create faux three-dimensional effects like shadows, you can cleverly "stack" and offset objects (like a yellow box on top of a slightly shifted black box) using options like Send to Back from the *Arrange* toolbar and menu.

If you'd like to drop a graphics file (*.gif, .jpg,* etc.) into the flowchart view of your model window, use *Edit > Insert New Object*, select *Create from File*, next *Browse* to choose the file you want, press *Open*, then *OK*, and finally drop it in with the cross hairs.

Clearly, you could spend a ruinous amount of time on this kind of stuff. Without implying anything about anybody, we offer it as a simple empirical observation that the higher up in an organization you go to present your simulation, the more effort is probably justified on graphics.

3.4.12 The Run > Setup Dialog Boxes

To establish the run conditions, use the *Run > Setup* menu command, where you'll find tabbed pages that control various aspects of how your simulation will execute. You'll need to edit just two of these pages.

The first one is the Project Parameters tab, where you should enter a Project Title, Analyst Name (that's you), and Project Description. You might also need to modify the selection of which statistics will be collected and reported, depending on how you have your defaults set in *Tools > Options > Project Parameters* (we want Entities, Resources, Queues, and Processes). The completed dialog was shown in Figure 3-15 in Section 3.3.11.

The other tab that you need to edit is Replication Parameters. These edits were also discussed in Section 3.3.11, and were shown in Figure 3-16 there.

3.4.13 Establishing Named Views

To set up a Named View for your model, first pan and zoom to the scene you want to remember, then *View > Named Views* (the ❧ button or or type ?), press *Add*, then pick a name and maybe a hot key (case-sensitive). If you want to change this view's definition later, press *Edit* instead once you've panned and zoomed to the new view you want to associate with this name and possible hot key; to delete it from the list, select it there and press *Delete*. We discussed using Named Views in Section 3.2.3.

At first blush, setting up Named Views might seem like a frill, but trust us—you'll want some of these when your models grow.

3.5   Case Study: Specialized Serial Processing vs. Generalized Parallel Processing

A classic operational question is whether to have specialized or generalized workers when processing involves multiple different tasks; a related question is how processing-time variability affects the decision. Based on a paper by Harrison and Loch (1995)[3], we can investigate this using only the Arena modules introduced in Model 3-1. Arguments can be made for both specialized and generalized work, depending on the setting, and experiments have been conducted on real systems, sometimes with disappointing results.

A careful simulation study might help avoid such disappointments. Consider a loan-application office (as did Harrison and Loch), where applications arrive with exponentially distributed interarrival times with mean 1.25 hours; the first application arrives at time zero. Processing each application requires four steps: first a credit check (this takes time but everyone passes), then preparing the loan covenant, then pricing the loan, and finally disbursement of funds. For each application, the steps have to be done in that order. The time for each step is exponentially distributed with mean 1 hour, independent of the other steps and of the arrival process. Initially, the system is empty and idle, and we'll run it for 160 hours (about a work month); see Chapter 5 for other kinds of stopping rules based on entity counts and other conditions. Output performance

---

[3] Thanks to Professor Jim Morris of the University of Wisconsin-Madison for recommending this paper to us.

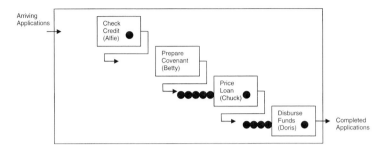

Figure 3-25. Specialized Serial Processing

measures include the average and maximum total number of applications in process, and the average and maximum total time, from entry to exit, that applications spend in the system, as well as their time waiting for the next processing step to begin. There are four employees available (Alfie, Betty, Chuck, and Doris), all equally qualified for any of the four steps, and the question is how best to deploy them.

3.5.1 *Model 3-2: Serial Processing – Specialized Separated Work*
A first thought might be to specialize the employees and assign, say, Alfie to check credit for all applications, Betty to prepare all covenants, Chuck to price all loans, and Doris to disburse all funds. So each application would first have to go through Alfie, then Betty, then Chuck, and finally Doris. A layout might look like Figure 3-25, where there are currently 12 applications (the filled circles) in process, Alfie is busy but with no more applications in queue, Betty is idle (so of course with no queue), Chuck is busy with five more in queue, and Doris is busy with four more in queue. All queues are FIFO. Though Betty might disagree, it's too bad that, under these operational rules, she can't lend a hand to Chuck or Doris right now.

An Arena simulation of this, Model 3-2, is fairly straightforward, and essentially just involves adding three Process modules to Model 3-1, along with three more Resources. The completed model is in Figure 3-26, with the first Process module and its Resources dialog open.

Since the modules are so similar to those in Model 3-1, we'll just highlight the differences, and will trust you to browse through the model:

- The Create module is the same as in Model 3-1, except for the module Name, the name for the Entity Type, the Value of the mean of the exponential interarrival times, and the time Units (now Hours).
- The four Process modules have, as before, the Action of `Seize Delay Release`, but of course have different names for the modules and Resources. Since exponential process times are not among the Delay Type choices, instead choose `Expression` there, and then use the Expression Builder (go to "`Random Distributions`") to fill in the Expression field, as seen in Figure 3-26. The time Units here are also Hours.

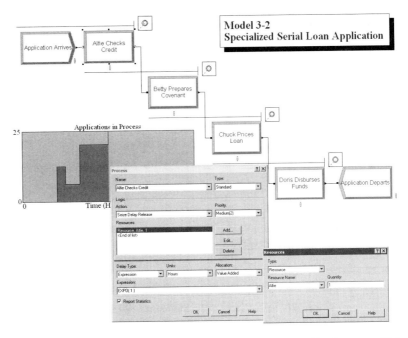

Figure 3-26. Completed Model 3-2 with the First Process Module and its Resources Dialog

- Other than its Name, the Dispose module is the same as in Model 3-1.
- The default Entity Picture (Entity data module, `Picture.Report`) is okay here since the entities are, well, reports of the loan applications.
- The default Resource animations are almost okay, but we made the Idle picture also have a white fill rather than the default green (double-click on the Resource animation icon to get to the Resource Picture Placement window, then double-click on the Idle and Busy icons in turn and use the Fill Color tool). Remember to select in the Identifier field the correct Resource name for each of the four cases. Also, check the Seize Area box.
- The Queue and Resource data modules are automatically filled in correctly if the Process modules were set up first. The defaults for everything (e.g., the Type and Capacity of each of the four Resources) work in this model.
- The dynamic Plot traces the total number of applications in process, which is the Arena Expression `EntitiesWIP(Application)`, and can be found in the Expression Builder under `Basic Process Variables`, then `Entity`, and finally `Number in Process`. "WIP" is an acronym for work in process. The other entries in the Plot dialog are similar those in Model 3-1, adjusted of course for the different time frame and maximum y-axis value.
- In *Run > Setup*, in the Project Parameters tab check the Processes box in Statistics Collection, and enter some documentation elsewhere; in the Replication

Parameters tab enter 160 for the Replication Length and make sure the Time Units and Base Time Units are both Hours.

■ After some initial runs and noting the maximum queue lengths, we lengthened their animations a bit out to the left to accommodate.

Run this model and look at the Category Overview report to find, among other things:

■ The average and maximum total number of applications in process were, respectively, 12.3931 and 21 (via Loan Application → Entity → Other → WIP).

■ The average and maximum total time, from entry to exit, that applications spent in the system were, respectively, 16.0831 hours and 27.2089 hours (Loan Application → Entity → Time → Total Time). Note that this counts only those applications that had left the system when the simulation stopped, not those that were still in process at that time (since, of course, their total time in system had not yet been completed).

■ The average and maximum total time that applications spend waiting for the next processing step to begin were, respectively, 11.9841 hours and 22.2732 hours (Loan Application → Entity → Time → Wait Time). This includes only time "wasted" waiting in queue, and not "value-added" time spent undergoing processing at the four steps, so is a good measure of system inefficiency. Arena counts in this statistic the total waiting time in the four queues only for those applications that completed all four steps and exited the system; waiting times in the individual queues are under Loan Application → Queue → Time and, as noted for Model 3-1, may include more entities than those included for the overall system-wide average and maximum.

■ During the 160 hours, 117 applications were completed (Loan Application → Entity → Other → Number Out), a measure of productivity.

■ Alfie, Betty, Chuck, and Doris were busy, respectively, 82.33%, 70.34%, 80.44%, and 80.80% of the time.

■ Alfie, Betty, Chuck, and Doris processed, respectively, 128, 128, 122, and 117 applications (Loan Application → Process → Other → Number Out). In this model, all applications visit these people in this order, so these numbers in this order must decrease or stay the same. (Under what conditions would they all be the same? Is that possible in this model?)

The main inefficiency in this system is the waiting in queue that the applications must endure, and there are four different places where that could happen. Also, there is the possibility, as in Figure 3-25, that applications could be queued at some stations while other employees are idle, which seems like a waste of resources.

3.5.2 Model 3-3: Parallel Processing – Generalized Integrated Work

Would it be better to "generalize" or "integrate" the work so that each employee completely processes all four steps of one application at a time, and with a single queue of applications "feeding" the group? From the applications' viewpoint, this would present just one opportunity for wasting time waiting in queue, but then again the

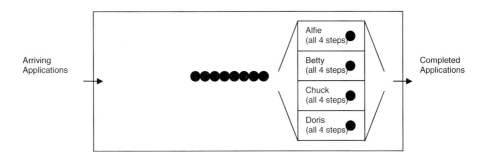

Figure 3-27. Generalized Parallel Processing

processing would be longer. Figure 3-27 shows how this would work, where each employee processes all four steps before moving on to the next application in queue. Like the specialized serial processing snapshot in Figure 3-25, there are currently 12 applications in process, but note that the total number in queue is now eight rather than nine, since Betty is now busy (we don't know how she feels about that, but this looks like better service).

The Arena model for this, Model 3-3, is actually simpler than Model 3-2, and is shown in its completed form in Figure 3-28 with the (sole) Process module and its Resources dialog showing. The Create and Dispose modules are identical to Model 3-2, as is the dynamic Plot and the *Run > Setup* dialog (except for labeling).

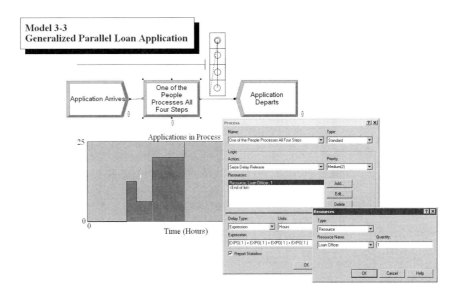

Figure 3-28. Completed Model 3-3 with the Process Module and its Resources Dialog

The main change is that the four Process modules in Model 3-2, each of which represented a single employee, are replaced by a single Process module representing all four employees. We replaced the four single-unit resources in Model 3-2 with a single four-unit resource here, which we named `Loan Officer` (look at the Resource data module in Model 3-3). Note that, in the Resources dialog of the Process module shown in Figure 3-28, the Quantity field is still 1, representing the number of units of the `Loan Officer` resource that are seized and released by each `Application` entity; this is *not* where we specify that there are four units of this resource in existence, but rather in the Capacity column of the Resource data module.

And since each employee must now complete all four tasks in tandem before taking the next application out of the queue, we must make the Delay time longer in the Process module. Each of the four steps requires `EXPO(1)` hours to complete, so we simply add four of these up in the Expression field in the Delay Type area of the Process module, as seen in Figure 3-28, each representing one of the four processing steps (it's okay to do arithmetic inside an Expression field anywhere in Arena). Now we're guessing that, if you're reading a book like this, you probably did well in seventh-grade algebra, so you know that adding up four of the same thing is the same as multiplying that thing by 4, so you may be wondering why we didn't save ourselves some typing and enter something like `4*EXPO(1)` instead. Well, the problem with that is that it would make just a single "draw" from the exponential distribution and multiply that single resulting number by 4, meaning in our case that each of the four steps took exactly the same amount of time, not what we want. So, ugly as it is, `EXPO(1) + EXPO(1) + EXPO(1) + EXPO(1)` is what we need since this makes four separate, independent draws from the exponential distribution to represent four separate, independent times for the four steps.

Finally, a change in the Resource animation would make things a little more realistic (but still not completely right), though would not be necessary for the model or its results to be correct. First, change the icon by double-clicking on the Resource animation and then the Idle icon to get to the Resource Picture Placement window, and duplicate the white square three times and line them all up vertically. Then select and copy this whole thing, go back (just close the window), open the Busy icon, and paste it in to replace what's already there. Back out in the model window, double-click on the two concentric circles representing the Seize Area to get to the Seize Area dialog, select Points, and then Add three times to get three more related seize areas for this multi-unit resource; close the Seize Area dialog and drag these around to where they belong inside the boxes (you may need to turn off Snap to Grid to get them just right). In the animation, then, you'll see as many entity icons in the Resource animation as there are entities in service at any time. What makes even this not completely right is that the icons for the entities in service will always shove themselves toward the original Seize Area (the double concentric circles) rather than staying where they "belong," if you view the boxes from top to bottom as being for Alfie, Betty, Chuck, and Doris (you can really get this right by using Resource Sets on single-unit resources, rather than a single multi-unit resource, as discussed in Model 5-2 in Chapter 5).

This parallel, integrated-work configuration appears to provide better service than the serial, specialized-work setup. The average and maximum total number of applications in process here are, respectively, 4.6118 and 10, compared to 12.3931 and 21 for the serial configuration. The average and maximum total time in system of applications here are, respectively, 5.3842 and 13.7262 hours, down from 16.0831 and 27.2089 hours for the serial model. The average and maximum time wasted waiting in queue here are, respectively, 1.3282 and 6.8231 hours, down markedly from 11.9841 and 22.2732 hours for the serial setup. Productivity went up, from 117 to 135 applications completed, though such improvement is of course limited by the arrival rate. All these improvements are due to less idle time among the employees; the utilization of the `Loan Officer` resource, defined as the time-average number of units busy (which came out to be 3.48) divided by 4, was 87%, compared to the average of 78.48% over the four individual employees in the serial configuration. We just never have a situation like that of Figure 3-25 where jobs are queued yet resources are idle. Of course, these comparisons are from just one replication of each of Models 3-2 and 3-3, so we're on thin statistical ice, as discussed in Section 2.6 (see Exercise 6-19).

3.5.3   Models 3-4 and 3-5: The Effect of Task-Time Variability

Harrison and Loch (1995) pose a further question about the comparison between the above two layouts—will the second one always be a lot better than the first one, as seemed to happen above? While many other aspects of the situation could speak to this question, they looked at variability of the task times.

Each of the four tasks for an application has an average duration of one hour, but these durations have an exponential probability distribution around that mean, so there is variability in the task time; let's get an idea of how much variability. For example, the probability that a task can be done in under ten minutes (1/6 hour) is $F(1/6) = 1 - e^{-1/6} \cong$ 0.15, where F denotes the cumulative distribution function of an exponential random variable with mean one hour (see Appendix C for a refresher on probability and statistics, and Appendix D for definition of the exponential and other probability distributions). But some tasks will take a long time; the probability that a task will take more than two hours is $1 - F(2) = e^{-2} \cong 0.14$, about the same as the 0.15 probability of doing a task in less than ten minutes. Put another way, about 15% of the tasks are very quick (under ten minutes), but about the same number (14%) take a very long time (over two hours). So there's quite a lot of variation in task times with their exponential distribution, and in particular the long task times contribute significantly to congestion and queueing delays, and thus to inefficiency. In the serial configuration of Model 3-2, when just one of these long task times pops up, it creates a big backup at that station (maybe that's what's happening to Chuck in Figure 3-25). In the parallel configuration of Model 3-2, a long task time will also increase congestion, but probably not by as much as in the serial configuration since it's unlikely to be accompanied by similarly long times for the other three tasks for that application, which somewhat dampens its effect.

But what if there were less variability among task times, even while holding the mean task time to the same one hour? Let's take that to the extreme and look at what happens if there's *no* variability, i.e., each task time takes *exactly* one hour (and thus each application will require *exactly* four hours of processing time in total). Maybe we could move toward this with better support for the tasks, either computer-based, clerical, or via better pre-screening. Since it's not under our control in reality, we'll keep the arrival

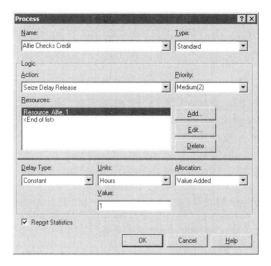

Figure 3-29. Completed Process Module for Alfie in Model 3-4 with Constant Task Times

process the same; i.e., with interarrival times that have an exponential distribution with mean 1.25 hours, so this model still has some stochastic (random) input.

Now how do the serial and parallel configurations compare? We modified Model 3-2 to get Model 3-4, where each of the four Process modules was modified to make the corresponding task time exactly one hour. This was very easy to do; in each of the Process module dialogs just change Delay Type to Constant in the pull-down menu, set Value to 1, and make sure the Units are still Hours. Figure 3-29 shows this for Alfie, and the other three work the same way.

Figure 3-30. Completed Process Module for Model 3-5 with Constant Task Times

Similarly, we created Model 3-5 from Model 3-3 by just getting rid of the ugly `EXPO(1) + EXPO(1) + EXPO(1) + EXPO(1)` in the Process module for the Expression Delay type, changing the Delay Type to Constant, and entering a Value of 4 Hours; Figure 3-30 shows the completed Process module dialogue.

Table 3-1 summarizes the results from the constant-service-time Models 3-4 and 3-5 (bottom half) and repeats those from the exponential-service-times Models 3-1 and 3-2 (top half), rounding to two decimals. For the constant-service models (3-4 and 3-5) there might still be some improvement in going from the serial to parallel operation, but only a very slight one (and might not even be statistically significant, per Exercise 6-19). Evidently, when service times have little or no variability there's not much difference between the specialized serial and generalized parallel work designs.

Of course, you could raise some questions, like:

- In the parallel integrated-work model, wouldn't you expect these generalized people to be at least a little less efficient than if they specialized in doing just one thing? Our Model 3-3 doesn't allow for this possibility, but Exercises 3-13 and 6-20 do.
- When dealing with human resources, don't they need a little idle time, maybe via scheduled breaks, to catch their breath and stay fresh? Resource Schedules, which can be used to model such breaks, will be discussed in Chapters 4 and 5.
- Are the improvements from going from the serial to parallel designs statistically significant, or could they be just spurious, the result of mere sampling fluctuation (after all, we made just a single replication of each configuration)? This is taken up in Chapter 6, and in Exercises 6-19 and 6-20.

3.6   More on Menus, Toolbars, Drawing, and Printing

In this section, we'll briefly mention some miscellaneous information that we haven't covered on the Arena menus and toolbars, and mention a few more things about its drawing and printing capabilities. As we go through the examples in later chapters, we'll give more detail on these things as needed.

3.6.1 Menus

Here we give a quick overview of what's in the Arena menus. Some options in some menus will be dimmed (meaning that you can't select them) if they don't apply in your particular situation or status at the moment. For more information about menu entries,

Table 3-1. Summary Results from All Four Scenarios of the Loan-Processing Model

	Model	Total WIP		Total Time in System		Total Waiting Time		Number Processed	Avg. Utilization
		Avg.	Max.	Avg.	Max.	Avg.	Max.		
Expo service	3-2 (serial)	12.39	21	16.08	27.21	11.98	22.27	117	0.78
	3-3 (parallel)	4.61	10	5.38	13.73	1.33	6.82	135	0.87
Constant service	3-4 (serial)	3.49	12	5.32	11.38	1.32	7.38	102	0.65
	3-5 (parallel)	3.17	11	4.81	10.05	0.81	6.05	102	0.66

remember that you can click ▸? and then use that to click on any menu item (dimmed or not) to get complete documentation on it, including hyperlinks to related topics.

File menu. This is where you create new Arena model files, open existing ones, close windows, and save your models. This is also where you attach and detach Project Bar panels. You can also import CAD drawings from AutoCAD®

(and from other CAD programs in standard DXF format) for use as Arena "backdrops" and, in some cases, active elements (like paths for wire-guided vehicles) to allow you to use existing detailed drawings of facilities. Another type of graphics file you can import is a Microsoft® Visio® drawing file in *.vsd* format. If you change the colors Arena uses, you can save them as a color palette (you can do some of this with Windows® as well); you can also open previously saved color palettes. The Arena printing functions are accessible from this menu. The Send command allows you to send mail from within Arena and attaches any active model to your message. Arena remembers the most recent documents, and you can open them quickly. The Exit command is one of the ways to quit Arena.

Edit menu. Here you'll find the usual options as applied to objects in Arena models. You can Undo previous actions or Redo your Undos. You can Cut or Copy a selected object (or group of objects) to the Clipboard for placement

elsewhere in the current model, to other models, or in some cases, to other applications. Paste allows you to insert the Clipboard contents into a model, and Paste Link creates an OLE link to the source document that's currently in the Clipboard. Duplicate makes a copy of what's selected and places it nearby in the current model, and Delete permanently removes whatever you have selected. You can Select All objects in a model as well as Deselect All. With Entity Pictures you can change what's in the list presented in the Entity data module, as well as the appearance of those pictures; you can copy pictures into this list from Arena's picture libraries (*.plb* files). Calendar Schedules allows you to describe complex time patterns with hierarchies (weeks are made of days, which are made of shifts, and so on), define exceptions like holidays and vacations, and show the net effect of all this in a composite view. Arena's Find function searches all modules and animation objects in the active model for a text string with the usual control over whole-word searches and case sensitivity (this can come in handy for finding entries that you didn't think you had but about which you're getting some error message). You can display additional object Properties, such as its unique object tag. If you have links in your model to other files, such as a spreadsheet or sound file, Links tells you about them and allows you to modify them. Insert New Object lets you make placements from other applications, like graphics and multimedia, and Insert New Control allows insertion of a VBA®/ActiveX® control. Object lets you edit something you've brought into the model from another application.

View menu. From this menu, you can control how your model appears on the screen, as well as which toolbars you want to display. Zooming lets you view the model from different "altitudes" so you can see the big picture or smaller sections in more detail. The Zoom Factor allows you to set how much you zoom in or out each time. Views (whose submenu is shown) offers certain "canned" views of your model; and Named Views lets you define, change, and use your own views. Rulers, Grid, Guides, Glue to Guides, and Snap to Grid are useful if you want to line things up geographically; Grid & Snap Settings gives you control over the spacing for Grid and Snap. Page Breaks displays your model indicating how it will be broken into pages for printing. Data Tips allows you options for displaying object properties in tool tips if you let the mouse hover over that object. Connector Arrows lets you put arrowheads on connectors to visualize the direction of entity flow. Layers lets you control what kinds of objects show up during the edit or run mode. If Split Screen is checked (toggled on), your model window will show both the flowchart and spreadsheet views simultaneously. Runtime Elements Bar, if checked, displays a window to allow you to choose what is displayed during execution. Toolbars is one way you can designate which sets of buttons are displayed on your screen (see Section 3.6.2), Project Bar toggles for whether the Project Bar (which hosts the panel) is visible, and the Status Bar entry lets you decide whether you want to see the horizontal bar at the very bottom of the screen, which tells you what's going on and indicates the world coordinates of the mouse pointer in the Arena workspace. Debug Bar, if checked, shows a window of debugging tools during a run.

Tools menu. Arena News Flash is an Internet-based news feed for updates, etc. Arena comes not only with the modeling capability with which we've spent our time so far, but also contains a suite of related tools, possibly depending on what you've licensed. The Arena Symbol Factory provides a large collection of

graphics objects in many categories from which you can create graphical symbols for animation of things like entities and resources. The Input Analyzer fits probability distributions to your observed real-world data for specifying model inputs (see Section 4.6.4). The Process Analyzer organizes efficient ways for you to make multiple simulation runs, which might represent different model configurations, and keep track of the results; it also helps you carry out proper statistical analyses of your simulation's results, such as a reliable way to select the best from among several different model configurations. Another application with additional statistical capabilities, called the Output Analyzer, comes with Arena but must be launched separately (it's in the Arena 10.0 folder). We'll show you how to use the Process Analyzer and the Output Analyzer in Sections 6.5 and 6.4, respectively. Report Database exports summary statistics from a run to a CSV (Comma-Separated Values) file to be read into a spreadsheet for post-processing analysis. ContactCenter (shown with its submenu open) provides special functions to model contact/call centers. Model Documentation Report produces a compact but complete set of documentation, including run conditions, modules used, and any submodels. Export Model to Database allows you to save the details of your model to an Access or Excel database; Import Model from Database allows you to bring in those details from such a database to construct or update a model quickly. OptQuest for Arena is an application that decides how to change model inputs that you select and then runs a sequence of simulations to search for a combination of these inputs that optimizes (maximizes or minimizes) an output performance measure that you designate; we'll give an example of its use in Section 6.6. AVI Capture enables recording many actions, including editing and animation, into a video (*.avi*) file for playback. The Macro item provides the tools to record and run Visual Basic (VB) macros and to open the VB Editor for writing custom logic. These VB topics are discussed in Section 10.2. Finally, the Options item lets you change and customize a lot of how Arena works and looks to suit your needs (or tastes).

Arrange menu. The items here pertain to the position of modeling modules and graphics objects; some apply only to graphics objects. Bring to Front and Send to Back position the selected object(s) on the top and bottom, respectively, of a "stack" of objects that may overlap. Group and Ungroup, respectively, put together and subsequently take apart objects logically, without affecting their physical appearance; Grouping is useful if you

want to move or copy a complex picture built from many individual objects. The Flip entries invert the selected object(s) around a line in the indicated direction, and Rotate spins the selection clockwise 90°. Align lines up the selected objects along their top, bottom, left, or right edges. Distribute arranges the selected objects evenly in either the horizontal or vertical direction, and Flowchart Alignment arranges the selected flowchart modules evenly both vertically and horizontally. Snap Object to Grid forces the selection to align to the underlying grid of points, and Change Object Snap Point lets you alter the exact point on the selected object that gets snapped.

Object menu. These items relate to the model's logical structure, its flowchart modules, and the connections between logical pieces of the model. Connect changes the pointer to cross hairs and lets you establish graphically a connection between modules for entities to follow; selecting it twice sets up a Connect "session" for repeated use (right-click or hit *Esc* to end the session). Auto-Connect is a toggle that allows you automatically to connect a newly placed module to one that's already selected. Smart Connect causes newly added connections to be drawn in horizontal/vertical segments instead of possibly diagonal lines, unless you make intermediate clicks if you're placing the Connection by hand. Animate Connectors causes entities to show up as they move along Connections between flowchart modules, even if this movement is occurring instantly in the simulation. Animate At Desktop Color Depth, if checked, runs the animation with all the colors on your desktop, which could slow down the simulation; if not checked, the animation runs at 8-bit color depth with no slowdown. Submodel, whose submenu is shown, lets you define and manage hierarchical submodels.

Run menu. This menu contains the *Run > Setup* dialog boxes that we discussed in Sections 3.3.11 and 3.4.12, which control the manner in which the current model will be run (including possibly its run length). It also contains entries for running the simulation in different ways, as well as several options to watch the execution, check it (and view any errors), and to set up and control how the run goes and is displayed on your screen. We'll describe these capabilities further in Section 3.8. You can also access the code that Arena actually generates in the underlying SIMAN simulation language.

Window menu. If you have several models open at once, you can arrange them physically in an overlapping Cascade, or in a non-overlapping Tile arrangement. If you have several models minimized (via the – button in each window), select Arrange Icons to organize them. The Use System Background Color entry causes this model to use whatever background color is selected at the Windows® operating-system level rather than what is set internal to Arena; to return your model to Arena's internal color, select this item again. (This menu item changes/toggles between "System" and "Custom" each time you select it.) Finally, you can activate a model that's already open by selecting it at the bottom of the menu.

Help menu. This is one of several ways to access Arena's online Help system (see Section 3.7 for more help on Help). If you select Arena Help, you'll get to the table of

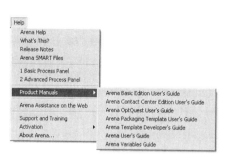

Contents, an Index, and a Search utility for getting to the topic you want. What's This changes the pointer to ▶? that you can then use to click on a menu entry or toolbar button to find out a little bit about it. Release Notes has information on recent changes and system requirements. Arena SMART Files produces a subject-based index to a couple of hundred little Arena models that illustrate a wide variety of specific modeling techniques. Next on this menu, you'll find general help topics on whatever modeling panels you have attached at the moment. Product Manuals takes you to detailed documents on the various components of the Arena software and related products. Arena Assistance on the Web takes you directly to a Web site with the latest help and software information (of course, you have to be online for this to work). The last three entries on this menu provide information on support and training; the copy protection and activation procedures used for commercial and other versions of Arena; and detailed version information.

3.6.2   Toolbars

Arena has several *toolbars* with groups of buttons and drop-down list boxes to facilitate quick access to common activities. Some of these buttons are just faster ways to get at menu items (discussed above), and some represent the only way to do something.

Select the menu option *View > Toolbars* (or right-click in a toolbar) to choose which toolbars will be displayed. As in many applications, you can tear off toolbars and float them in the interior of the Arena window as palettes, or dock them to an edge (if you want it near the edge but not docked, hold down the *Ctrl* key while approaching the shoreline). You won't have to set your toolbar configuration every time you use Arena as it will remember your last configuration. You can also have different configurations for when you're editing your model, when the simulation is running, and when various other Arena window types are active (such as the Picture Editor), and again, Arena will remember what each was.

Customize how toolbars are displayed via *View > Toolbars > Customize* or right-clicking in a toolbar, then selecting Customize, and then Customize again. We'll mention each toolbar in turn below, but you have the option to rearrange what buttons are on which toolbars by this customization capability.

- While you could choose to hide the *Standard* toolbar, it's a little hard to see how you could do much without it:

It starts with buttons to create a New model, Open an existing one, and Save the active model, as on the File menu; also from that menu are buttons to Attach a panel or Detach the visible one, and to Print and do a Print Preview. From the Edit menu are Cut, Copy, and Paste, as well as Undo and Redo. Next is the Toggle Split Screen button for a split model window, and then the magnifying glass to View a Region that you select with it in the flowchart view of the model window at the closest possible zoom. You can choose a Zoom percent from a drop-down list. The Layers button lets you control which types of objects show up in the flowchart view of the model window in both edit and run mode. You can add a Submodel and Connect flowchart modules. You can create and manage calendar time patterns and their exceptions with Edit Time Patterns and Edit Exceptions. Display Composite View lets you manage the capacity and efficiency data associated with specific system elements (like scheduling a resource to be available for two shifts instead of one). The next six buttons are run controls, and will be discussed in Section 3.8. The slider bar allows you to speed up or slow down the animation during a run. The Standard toolbar ends with the context-sensitive Help button; click on it (note the **?** that gets added to the mouse pointer), then click on a toolbar button or a menu command or a module in the Project Bar to learn about it.

- Buttons on the *Draw* toolbar have no corresponding menu options, so drawing can be done only by toolbar access:

This is how you can draw static Lines, Polylines, Arcs of ellipse boundaries, Bézier Curves, Boxes, Polygons, and Ellipses to dress up your model, as well as add Text to annotate it. There are also controls for changing the color of Lines (including borders of shapes), Fill of objects, Text, and the Background of the flowchart view of the model window. You can alter the Line Width and Style (thickness as well as whether it's there or not, and dash patterns), and put arrowheads on lines. Line Patterns provides different pre-drawn line appearances, like roads, tracks, and belts. Fill Pattern provides hatch patterns for shapes. Show Dimensions lets you display the sizes of shapes and lines for precise drawing. You've probably used draw features in other applications, so Arena's capabilities will be familiar. By far, the best way to familiarize yourself with these things is to open up a "test" model window and just try them out; see Section 3.6.3 for more on drawing.

- The *Animate* toolbar contains capabilities to allow you to animate your model or enhance the animation that is inherent in some Arena modules:

Typically, click one of these buttons, enter a dialog box to describe exactly what you want, then place the animation in your model. There are a lot of different capabilities here, and we'll illustrate most of them as we progress through building models in later chapters (we've already used Plot and Resource). For now, hover your mouse above each button to show the Tooltip with its name.

- The *Integration* toolbar contains buttons related to Arena's Module Data Transfer wizard and VBA (the Visual Basic Editor and VBA Design Mode button):

Chapter 10 discusses the use of VBA, which augments Arena's standard modeling features with a complete Visual Basic programming interface.

- The *View* toolbar has buttons to control how you view the flowchart view of the model window:

From here you can Manage Named Views, Zoom In and Out, or choose to View All the model or View Previous. You can also reveal the Rulers and Grid, create Guides, show page breaks for printing, and Snap new objects to the Grid, as well as Glue objects to Guides.

- The *Arrange* toolbar corresponds closely to the Arrange menu:

You can Bring a selected object to the Front or Send it to the Back. A selection of multiple drawing objects can be made into a logical Group and Ungrouped later. Drawing objects can also be Flipped around a Vertical or Horizontal line on their midpoint or Rotated clockwise 90°. You can also Align the selected objects on their Top, Bottom, Left, or Right edges, as well as Space them evenly either horizontally or vertically. Finally, you can Snap selected objects to the Grid.

- The *Run Interaction* toolbar has buttons corresponding to the Check Model, Command, Break, Watch, and Break on Module entries from the Run menu (see Section 3.8 for more). The last button corresponds to Animate Connectors from the Object menu:

- The *Record Macro* toolbar has buttons to Start/Pause/Resume as well as Stop recording a VB macro. A macro is a series of Visual Basic statements stored in a subroutine in a Visual Basic module (macros are discussed in Chapter 10):

■ The *Animate Transfer* toolbar gives you tools to add animation objects to your model:

These include Storage, Seize, Parking, Transporter, Station, Intersection, Route, Segment, Distance, Network, and Promote Path. We'll discuss these capabilities in subsequent chapters as we develop models whose animations can benefit from them.

3.6.3 Drawing

The *Draw* toolbar, mentioned in Section 3.6.2, has a variety of shapes, text tools, and control features to allow you to enhance the model by placing static (no participation in the simulation or animation) objects in the model window to help document things or to make the animation seem more "real" by adding items like walls, aisles, and potted plants. This isn't intended to be a complete, full-featured CAD or artwork capability, but it usually proves adequate; of course, you can always import graphics from other packages, as mentioned in Section 3.4.11. Arena's drawing tools work a lot like other drawing packages, so we'll just point out what's there and let you play with things to get used to them:

■ *Line,* ⟍: Click once on this button, changing the mouse pointer to cross hairs, then click where you want the line to start and again where you want it to end. To constrain the line to be vertical or horizontal or on a 45° angle, hold down the *Shift* key while moving to the end of the line.

■ *Polyline,* ⭧: This lets you draw a jagged line with an unlimited number of points. After selecting this button, click where you want to start and then again for each new point; double-click for the endpoint. Hold down the *Shift* key during a segment to constrain it to vertical, horizontal, or 45°.

■ *Arc,* ⌒: Draw part of the border of an ellipse. Click first for the center of the ellipse, then move the mouse and follow the bounding outline, clicking again when it's the size and shape you want (hold down the *Shift* key to constrain the ellipse to be a circle). At this point, the mouse pointer becomes the end of a line emanating from the ellipse's center; click to define one end of the arc, then again for the other end. To edit the arc later, select it and use the lines to change what part of the arc is shown and use the disconnected handle to change the ellipse size or shape.

■ *Bézier Curve,* ⤳: These have become popular due to their ability to assume a lot of different shapes yet maintain their smoothness and inherent beauty. Click for one endpoint, then make intermediate clicks (up to 30) for the interior "attractor" points; double-click to place the other endpoint. Holding down the *Shift* key while moving to the next point causes the (invisible) lines connecting them to be horizontal, vertical, or at 45°. To change the curvature, select the curve and drag the interior attractor points around; dragging the endpoints anchors the curve to different places. Move the curve by dragging it directly.

- *Box,* □: Click first for one corner, then again for the opposite corner. Hold down the *Shift* key to constrain it to a square. This object, like the next two, has a border regarded as a "line*"* for color and style, as well as a "fill" for color or pattern choices.
- *Polygon,* ⌂: Click for the first point, then move to new locations to click the others; double-click for the final point, which you want to be connected back to the first one. Hold down the *Shift* key to force line segments to be horizontal, vertical, or at 45°. This object has a line border and a fill like a box.
- *Ellipse,* ◯: First click for the center, move the mouse and follow the bounding outline to the size and shape you want, and finally click again. Hold the *Shift* key to force it to a circle. This object has a line border and a fill like a box.
- *Text,* **A**: This is how you add annotation to your model to label things or provide documentation. Clicking the button brings up a dialog box where you type in your text; use *Ctrl+Enter* to go to a new line and *Ctrl+Tab* for a tab. The Font button lets you change the font, style, and size. Closing this dialog box changes the mouse pointer to cross hairs, which you click where you want to position the northwest corner of your text. Use the underline to move, resize, or reorient the text to a different angle (hold the *Shift* key to constrain the angle to horizontal, vertical, or 45°).
- *Line Color,* ✒ ▾: If a line object (a line, polyline, arc, Bézier curve, or the border of a shape) is selected, clicking the paintbrush part of the button changes that object to the color underlining the paintbrush. Clicking the drop-down arrow changes the selected line object to the color you then select from the palette, and changes the underline color to this as well. New line objects will be in the underline color. Arena will remember this line color not only for future line objects in this window, but also for new windows and future Arena sessions, until you change it again.
- *Fill Color,* ▲ ▾: This operates on the interior of a shape (box, polygon, or ellipse) just as Line Color operates on line objects.
- *Text Color,* **A** ▾: This operates on Text drawing objects just as Line Color operates on line objects.
- *Window Background Color,* ▦ ▾: This sets the background color of the flowchart view of the model window to the color you select from the palette.
- *Line Width,* ≡ ▾: Operates on the width of line objects just as Line Color operates on their color.
- *Line Style,* ▤ ▾: Provides dash patterns for lines. The No Line option makes the line invisible but it is still logically there (this might make sense for a border of a shape).
- *Arrow Style,* ⇄ ▾: Provides various arrowheads for lines.
- *Line Pattern,* ▱ ▾: Provides different pre-drawn line appearances, like roads, tracks, and belts.
- *Fill Pattern,* ▨ ▾: Operates on the pattern for the interior of a shape just as Fill Color operates on its color.

- *Show Dimensions,* ⟨x⟩ ▾: Shows sizes of shapes and lengths of lines and line segments, for precise drawing.

3.6.4 Printing

All or parts of the flowchart view of the active model window can be printed directly from Arena (in color, if you have it). *File > Print Preview* (or ⌕) lets you preview what's coming, *File > Print* (or 🖨 or *Ctrl+P*) lets you print it, and *File > Print Setup* lets you select a print driver or do a setup of your printer.

If your model is big, the print will extend across several pages. And if you have Named Views, you'll get a print of the current view, followed by a separate print of each named view. If you don't want all this, use Print Preview to see what's on which page, then selectively print only the pages you want. The menu option *View > Page Breaks* will show in the flowchart view where the pages will end if the model is printed.

As an alternative to printing directly from Arena, remember that you can get to a view you like (maybe even during a pause in the animation), press the *Prnt Scrn* (Print Screen) key, switch over to your word-processing document, and paste the shot in where you want it. Then you can print this document when you'd like. You could also paste the Prnt Scrn image into a paint program and perhaps crop it or otherwise clean it up (in fact, that's exactly how we got all the pieces of the Arena window into this book).

3.7 Help!

Arena has an extensive and comprehensive Help system to serve as a reference, guide you through various operations, and supply examples of modeling facets as well as complete projects. The help system is carefully integrated and provides extensive hyperlinks to other areas to aid in getting the information you need quickly and easily. There are several different ways to access the help system, which we'll describe briefly in this section. However, you may find that the best way to learn about help is just to get in and start exploring.

At any time, you can access the full help system from the *Help* menu. The contents of this menu were described above, at the end of Section 3.6.1.

The ⟨?⟩ button invokes *context-sensitive help*. To use it, just click the button, then click on whatever you're curious about—a toolbar button, a command from a menu, a module in the Project Bar—and you can get to the information you need via a visual path.

Most Arena dialog boxes have a *Help* button that you can press. This is a good way to get direct information on what that part of the software is about, what your options are, how relevant things are defined, related concepts (via hyperlinks to other parts of the Help system), and examples. You can also use the **?** button at the top of the dialog box to access What's This? help information on individual items in a dialog box. Simply click the **?** and then click on the selected item.

In case you forget what a particular button does, you can let your pointer stay motionless on it for a second or two; a little boxed *Tooltip* will appear to remind you what it is.

SMART files, via the Help menu, are subject-categorized Arena models (a couple of hundred) that illustrate, via small working models, a wide variety of Arena modeling capabilities.

Online help is available in the form of technical notes and updates, at http://support.rockwellautomation.com. You can set up your own account there (it's free) to download the latest in such items.

In the Examples folder inside the Arena 10.0 folder are several detailed Example models that you can open, browse, copy, edit (to be safe, edit a copy), and run. If you have an academic or other limited version of Arena, you may not be able to run your models to completion due to the limit on the number of concurrent entities. If you open a large model (one that exceeds the academic version limits), Arena will enter a runtime mode. Runtime mode allows you to make minor changes and run the model, but does not allow significant changes like adding and deleting modules. The Example models illustrate different aspects of building and studying models. The Help topic[4] "Example Models" has a description of what's in each one.

3.8   More on Running Models

Usually you'll just want to run your model to completion as you have it set up, but there are times when you might like to control how the run is done. Entries from the *Run* menu, as well as corresponding buttons from the *Standard* and *Run Interaction* toolbars, let you do this. (See Section 4.5 for details and examples of using these capabilities.)

- *Run > Setup* gives you access to many options for how runs are made for the current model, like deciding whether to see the animation and perhaps running in full-screen mode. These selections and specifications are saved with the current model rather than going into global effect. Click the *Help* button inside any *Run > Setup* dialog box and browse through the hyperlinked topics to get familiar with what the (numerous) possibilities are.

- *Run > Go* (or ▶ from the *Standard* toolbar or the F5 function key) just does it (or resumes it after a pause). If you've made changes to the model since the last check (see below), it gets checked before it's run.

- *Run > Step* (or ▶∣ or F10) executes the model one action at a time so you can see in detail what's going on. This gets really boring so is useful primarily as a debugging or demonstration tool. As with the *Go* button, use of *Step* causes the model to be checked if Arena detects changes since the last check was performed.

- *Run > Fast-Forward* (or ▶▶) disables the animation and executes the run at a much faster rate. You can pause at any time during the run to view the animation. *Fast-Forward* causes the model to be checked if Arena detects changes since the last check was performed.

- *Run > Pause* (or ∣∣ or *Esc*) interrupts the run so you can look at something. Press ▶, ▶∣, or ▶▶ to resume it.

---

[4] To go to a specific Help topic, select *Help > Arena Help > Index*, type the topic name into the first region, and then open (double-click, or single-click and then click Display) the index topic in the second region below.

- *Run > Start Over* (or ⏮ or *Shift+F5*) goes back to the beginning of the simulation and reruns the model. *Start Over* causes the model to be checked if Arena detects changes since the last check was performed.
- While Arena is running your model, it's in what's called run mode, and most of the model-building tools are disabled. So when the run is over, you need to select *Run > End* (or ■ or *Alt+F5*) to get out of run mode and enable the modeling tools again. If you've paused your run before it terminated, this will kill it at that point.
- Use *Run > Check Model* (or ✓ or *F4*) to "compile" your model without running it. If Arena detects errors at this stage, you're told about them (gently, of course) in an Errors/Warnings window; the buttons at the bottom of this window help you find the problem (for example, by selecting the offending module in the flowchart view of the model window).
- *Run > Review Errors* recalls the most recent Errors/Warnings window containing whatever Arena found wrong during that check.
- *Run > Run Control > Command* (or 🔳) gets you to an interactive command-line window that allows control over a lot of how the run is done—like interrupts and altering values. This also checks the model, if required, and starts the run if it's not yet started.
- *Run > Run Control > Breakpoints* (or ✋) lets you set times or conditions to interrupt the model in order to check on or illustrate something.
- *Run > Run Control > Watch* (or 👓) establishes a window in which you can observe the value of a variable or expression as the run progresses. *Run > Setup > Run Control* lets you determine whether this is concurrent with the run or only when the Watch window is the active window (the latter will speed things up).
- *Run > Run Control > Break on Module* (or 🖐) either sets or clears a break put on the selected module. A break on a module halts execution when an entity starts or resumes execution of the logic for the module.
- *Run > Run Control > Highlight Active Module* causes a flowchart module to be highlighted when it is being executed, which provides a visual indication of the action during the animation.
- If *Run > Run Control > Batch Run (No Animation)* is checked, the model will be run without any animation. This is even faster than Fast-Forward, and is usually used when the project is in production mode to produce adequate statistics for precise analysis.
- *Run > SIMAN* allows you to view or write to a file the model file (*.mod*) and experiment file (*.exp*) for the code in the underlying SIMAN simulation language that your Arena model generates. To understand this, you'll of course need to know something about SIMAN (see Pegden, Shannon, and Sadowski, 1995).

3.9 Summary and Forecast

After this guided tour through Arena, you should be a pretty knowledgeable (and tired) tourist. In Chapters 4 through 9, we'll set you out to explore on your own, but with a pretty detailed road map from us. In those chapters, you'll build a sequence of progressively more complex models that illustrate many of Arena's modeling capabilities and sometimes require you to perform a few creative modeling stunts. We'll also integrate at the end of Chapter 4, throughout Chapter 6, at the end of Chapter 7, and in Chapter 12 some information on the statistical aspects of doing simulation studies with Arena, on both the input and output sides.

3.10 Exercises

3-1   Make five replications of Model 3-1 by just asking for them in *Run > Setup > Replication Parameters*. Look at the output and note how the performance measures vary across replications, confirming Table 2-4. To see the results for each of the five replications individually, you'll need to open the Category by Replication report; the confidence-interval half-widths can be seen in the Category Overview report, however.

3-2   Implement the double-time arrival modification to Model 3-1 discussed in Section 2.6.3 by opening the Create module and changing the Value 5 to 2.5 for the mean of the exponential distribution for Time Between Arrivals (don't forget to click *OK*, rather than *Cancel*, if you want this change to happen). Make five replications and compare the results to what we got in the hand simulation (see Figure 2-4). To see the results for each of the five replications individually, you'll need to open the Category by Replication report.

3-3   Lengthen the run in Model 3-1 to 12 hours for a more interesting show. If you want the plots to be complete, you'll have to open them and extend the Time Range (mind the units!), as well as possibly the Maximum value for the *y* axis in the Number in Queue plot. You might also need to increase # History Points in the Number in Queue Plot. Make just one replication.

3-4   Implement the change to Model 3-1 described in Exercise 2-4 from Chapter 2. Open the Process module, change the Delay Type to Expression, and enter the appropriate Expression (use the Expression Builder if you like). Run the model for 20 minutes and check against your hand-simulation results. If they're different, what might be the explanation (assuming you did everything right)?

Next, try running this for 24 hours and watch the queue-length plot (for both plots, change the Time Range to 1440 and the # History Points to 1000, and in the queue-length plot, change the Maximum for the *y*-axis to 50). To allow more room in the queue animation, click on the line for the queue and drag its left end to the left. What's happening? Why?

3-5   Implement the additional statistic-collection function described in Exercise 2-1, and add a plot that tracks the total number of parts in the system (also called *work in process*, abbreviated as WIP) over time. Note that at any given point in simulated time, WIP is the number in queue plus the number in service; you might also use the Expression Builder and check the Help topic EntitiesWIP Variable.

3-6   Modify Model 3-1 with all of the following changes:

- Add a second machine to which all parts go immediately after exiting the first machine for a separate kind of processing (for example, the first machine is drilling and the second machine is washing). Processing times at the second machine are the same as for the first machine. Gather all the statistics as before, plus the time in queue, queue length, and utilization at the second machine.

■ Immediately after the second machine, there's a pass/fail inspection that takes a constant 5 minutes to carry out and has an 80% chance of a passing result; queueing is possible at inspection, and the queue is first-in, first-out. All parts exit the system regardless of whether they pass the test. Count the number that fail and the number that pass, and gather statistics on the time in queue, queue length, and utilization at the inspection center. (HINT: Try the Decide flowchart module.)

■ Include plots to track the queue length and number busy at all three stations. Configure them as needed.

■ Run the simulation for 480 minutes instead of 20 minutes.

3-7 In Exercise 3-6, suppose that parts that fail inspection after being washed are sent back and re-washed, instead of leaving; such re-washed parts must then undergo the same inspection, and have the same probability of failing (as improbable as that might seem). There's no limit on how many times a given part might have to loop back through the washer. Run this model under the same conditions as Exercise 3-6, and compare the results for the time in queue, queue length, and utilization at the inspection center. Of course, this time there's no need to count the number of parts that fail and pass, since they all eventually pass (or do they?). You may have to allow for more room in some of the queue animations and plots' *y*-axes.

3-8 In Exercise 3-7, suppose the inspection can result in one of three outcomes: pass (probability 0.80, as before), fail (probability 0.09), and re-wash (probability 0.11). Failures leave immediately, and re-washes loop back to the washer. The above probabilities hold for each part undergoing inspection, regardless of its past history. Count the number that eventually fail and the number that eventually pass, and gather statistics on the time in queue, queue length, and utilization at the inspection center. (HINT: Explore the Decide flowchart module and contemplate a higher-dimensional coin to flip.)

3-9 In Model 3-1, suppose that instead of having a single source of parts, there are three sources of arrival, one for each of three different kinds of parts that arrive: Blue (as before), Green, and Red. For each color of arriving part, interarrival times are exponentially distributed with a mean of 15 minutes. Run the simulation for 480 minutes, and compute the same performance measures as for Model 3-1. Once the parts are in the system, they retain their correct color (for the animation) but are not differentiated for collection of statistics on time in queue, queue length, or utilization (that is, they're lumped together for purposes of processing and statistics collection on these output performance measures); however, collect statistics separately by part color for total time in system. Processing times at the drilling center are the same as in Model 3-1 and are the same regardless of the color of the part.

3-10 In science museums, you'll often find what's called a *probability board* (also known as a *quincunx*):

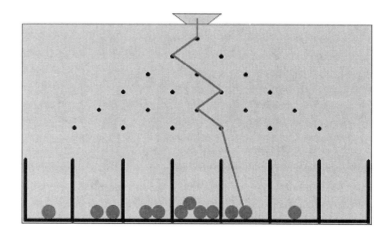

This is like a big, shallow, tilted baking pan with a slot at the midpoint of the top edge through which marbles roll, one at a time, from a reservoir outside the board; let's say the reservoir has k marbles in it. Just below the slot is a fixed peg, which each incoming marble hits and causes the marble to roll left or right off of; assume that you've tilted the board so that there's an equal chance that the marble will roll left vs. right (interpret "left" and "right" from your viewpoint as you look at the board from in front of it, which is the opposite from the marbles' viewpoint as they dive down the board nose first on their stomachs). Below this peg is a row of two pegs, parallel to the top edge of the board but offset horizontally from the first peg so that the two pegs in this second row are diagonally arranged below the first peg, as in the picture. Assume that the board's tilt angle, the peg spacing, the marbles' mass, and the gravitational field of the host planet are just right so that each marble will next hit exactly one of the two pegs in the second row (which peg it hits is determined by whether it rolled left or right off of the first peg). The marble will next roll left or right off of whichever peg it hits in the second row (again, assume a 50-50 chance of rolling left vs. right). The next parallel row of pegs has three pegs in it, again offset so that each marble will hit exactly one of them and roll left or right, again with equal probabilities. This continues through the last row; let's say that the number of rows is n so that the last row has n pegs in it ($n = 6$ in the picture, counting the first peg at the top as its own row). After rolling off of a peg in the last row, the marble will land in exactly one of $n + 1$ bins arranged diagonally under the last row of pegs. Create an Arena simulation model to simulate a probability board with $n = 6$ rows of pegs and $k = 1000$ marbles in the reservoir. Animate the marbles bouncing down the board, and also animate the number of marbles accumulating in the bins at the bottom with Level animation objects from the *Animate* toolbar. In addition, count the number of marbles that land in each of the 7 bins. The proportion of marbles landing in each bin estimates the probabilities of what distribution? What if somebody opens a window to the left of the board and a wind comes in to blow the marbles toward the right as they roll, so that there's a 75% (rather than 50%) chance that they'll roll to the right off of each peg?

3-11 In Exercise 3-9, as part of the processing of parts, an inspection is included (there's no extra processing time required for the inspection, and during the inspection, the part entity continues to occupy the resource). Each part passes this inspection with probability 0.93, regardless of its color. If it passes, it just leaves the system, as before. If it fails, it must loop back and be re-processed, going to the end of the queue if there is one and taking an amount of time for re-processing that is an independent draw from the same processing-time distribution; parts undergoing re-processing must still be inspected (with the same pass probability) and can fail multiple times (there's no limit on the number of failures for a given part) before finally passing and leaving. Adjust the plots if needed so they show the entire curve. Compare the average time of parts in the processing queue, the average length of the queue, the resource utilization, and the average time in system by part color with what you got from Exercise 3-9 ... does it make sense? (HINT: Explore the Decide module, with "Type" chosen to be "2-way by Chance.")

3-12 In Exercise 3-11, after (finally) passing inspection, parts' paint need to be touched up, so they are sent to one of three separate touch-up-paint booths, one for each color, with each part being directed to the booth for its color (of course); each touch-up booth has its own FIFO queue and its own (single) server. Touch-up times are TRIA(3, 9, 18) minutes, regardless of color. Add a plot that tracks, on a single set of axes, the queue lengths of each of the touch-up booths (with appropriate colors for each of the three curves). Collect the same output statistics as for Exercise 3-9, in addition to the time in queue, number in queue, and utilization for each touch-up booth. (HINT: Explore the Decide module some more, this time with "Type" chosen to be "N-way by Condition" and then "Add" conditions based on Entity Type.)

3-13 In Model 3-3, time studies showed that moving to this integrated work entailed an average increase of 18% in the time it takes to complete each of the four tasks to process an application since employees are no longer specialized in just one task, as they were in Model 3-2. Model this as increasing each service time by 18% from what it was in Model 3-3. If this happens, is the generalized parallel integrated-work scheme still advisable, in comparison with the specialized serial organization of Model 3-2?

3-14 Five identical machines operate independently in a small shop. Each machine is up (i.e., works) for between six and ten hours (uniformly distributed) and then breaks down. There are two repair technicians available, and it takes one technician between one and three hours (uniformly distributed) to fix a machine; only one technician can be assigned to work on a broken machine even if the other technician is idle. If more than two machines are broken down at a given time, they form a (virtual) FIFO "repair" queue and wait for the first available technician. A technician works on a broken machine until it is fixed, regardless of what else is happening in the system. All uptimes and downtimes are independent of each other. Starting with all machines at the beginning of an "up" time, simulate this for 160 hours and observe the time-average number of machines that are down (in repair or in queue for repair), as well as the utilization of the repair technicians as a group. Animate the machines when they're either undergoing repair or in queue for a repair technician, and plot the total number of machines down (in repair plus in queue) over time. (HINT: Think of the machines as "customers" and the repair technicians as "servers" and note that there are always five machines floating around in the model and they never leave.)

━━━━━━━━
CHAPTER 4

Modeling Basic Operations and Inputs

In Chapters 2 and 3, we introduced you to a simple processing system (Model 3-1), conducted a hand simulation (Chapter 2), and examined an Arena model (Chapter 3). We also considered a couple of other models, in Sections 2.7 and 3.5, but all were quite simple and stylized. In this chapter, we'll work with a more realistic system and develop complete models of several versions of that system, with each version adding complexity and new modeling concepts. We'll also talk about how you can realistically specify input probability distributions that represent the actual system under study.

Section 4.1 describes this more complicated system—a sealed electronic assembly and test system. We then discuss how to develop a modeling approach, introduce several new Arena concepts, build the model, and show you how to run it and view the results. By this time, you should start to become dangerous in your modeling skills. In Section 4.2, we'll enhance the model by enriching the scheduling, failure, and states of resources, and give you alternate methods for studying the results. Section 4.3 shows you how to dress up the animation a little bit. In Section 4.4, we'll generalize how entities move around, introducing the notions of Stations, Routes for non-zero travel times, and animation of transfers. Section 4.5 addresses how to find and correct errors in your models. Finally, in Section 4.6, we take up the issue of how you specify quantitative inputs, including probability distributions from which observations on random variables are "generated" to drive your simulation. When you finish this chapter, you should be able to build some reasonably elaborate models of your own, as well as specify appropriate and realistic distributions as input to your models.

4.1 Model 4-1: An Electronic Assembly and Test System

This system represents the final operations of the production of two different sealed electronic units, shown in Figure 4-1. The arriving parts are cast metal cases that have already been machined to accept the electronic parts.

The first units, called Part A, are produced in an adjacent department, outside the bounds of this model, with interarrival times to our model being exponentially distributed with a mean of 5 (all times are in minutes). Upon arrival, they're transferred (instantly) to the Part A Prep area, where the mating faces of the cases are machined to assure a good seal, and the part is then deburred and cleaned; the process time for the combined operation at the Part A Prep area follows a TRIA(1, 4, 8) distribution. The part is then transferred (instantly, again) to the sealer.

The second units, called Part B, are produced in a different building, also outside this model's bounds, where they are held until a batch of four units is available; the batch is

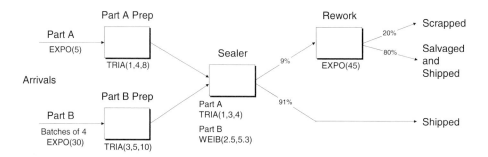

Figure 4-1. Electronic Assembly and Test System

then sent to the final production area we are modeling. The time between the arrivals of successive batches of Part B to our model is exponential with a mean of 30 minutes. Upon arrival at the Part B Prep area, the batch is separated into the four individual units, which are processed individually from here on, and the individual parts proceed (instantly) to the Part B Prep area. The processing at the Part B Prep area has the same three steps as at the Part A Prep area, except that the process time for the combined operation follows a TRIA(3, 5, 10) distribution. The part is then sent (instantly) to the sealer.

At the sealer operation, the electronic components are inserted, the case is assembled and sealed, and the sealed unit is tested. The total process time for these operations depends on the part type: TRIA(1, 3, 4) for Part A and WEIB(2.5, 5.3) for Part B (2.5 is the scale parameter β and 5.3 is the shape parameter α; see Appendix D). Ninety-one percent of the parts pass the inspection and are transferred immediately to the shipping department; whether a part passes is independent of whether any other parts pass. The remaining parts are transferred instantly to the rework area where they are disassembled, repaired, cleaned, assembled, and re-tested. Eighty percent of the parts processed at the rework area are salvaged and transferred instantly to the shipping department as reworked parts, and the rest are transferred instantly to the scrap area. The time to rework a part follows an exponential distribution with mean of 45 minutes and is independent of part type and the ultimate disposition (salvaged or scrapped).

We want to collect statistics in each area on resource utilization, number in queue, time in queue, and the cycle time (or total time in system) separated out by shipped parts, salvaged parts, or scrapped parts. We will initially run the simulation for four consecutive 8-hour shifts, or 1,920 minutes.

4.1.1   Developing a Modeling Approach

Building a simulation model is only one component of a complete simulation project. We will discuss the entire simulation project in Chapter 13. Presume for now that the first two activities are to state the study objective and define the system to be studied. In this case, our objective is to teach you how to develop a simulation model using Arena. The system definition was given above. In the real world, you would have to develop that definition, and you may also have to collect and analyze the data to be used to specify the input parameters and distributions (see Section 4.5). We recommend that the next activity be the development of a modeling approach. For a real problem, this may require the definition of a data structure, the segmentation of the system into submodels, or the development of control logic. For this problem, it requires only that we decide which Arena modules will provide the capabilities we need in order to capture the operation of the system at an appropriate level of detail. In addition, we must decide how we're going to model the different processing times at the sealer operation. To simplify this task, let's separate the model into the following components: part arrival, prep areas, sealer operation, rework, part departure, and part animation. Also, we'll assume that all entities in the system represent individual parts that are being processed.

Because we have two distinct streams of arriving entities to our model, each with its own timing pattern, we will use two separate Create modules (one for each part type) to generate the arriving parts.

We also have different processing times by part type at the sealer operation, so we'll use two Assign modules to define an attribute called `Sealer Time` that will be assigned the appropriate sealer processing time after the parts are generated by the Create modules. When the parts are processed at the sealer operation, we'll use the time contained in the `Sealer Time` attribute for the processing time there, rather than generating it on the spot as we did in the models in Chapter 3.

Each of the two prep areas and the sealer operation will be modeled with its own Process module, very much like the Process module used in the models in Chapter 3. An inspection is performed after the sealer operation has been completed, which results in parts going to different places based on a "coin flip" (with just the right bias in the coin). We'll use a Decide module with the pass or fail result being based on the coin flip. The rework area will be modeled with Process and Decide modules, as it also has a pass or fail option. The part departures will be modeled with three separate Record and Dispose modules (shipped, salvaged, and scrapped) so we can keep corresponding cycle-time statistics sorted out by shipped vs. salvaged vs. scrapped. All of these modules can be found on the Basic Process panel.

4.1.2 Building the Model

To build the model, you need to open a new model window and place the required modules on the screen: two Create, two Assign, four Process, two Decide, three Record, and three Dispose modules.

Your model window should now look something like Figure 4-2, assuming you've made the Connections or used the Auto-Connect feature (Object menu) while placing the modules in the appropriate sequence (the numbers inside your module shapes might be different if you placed your modules in a different order, but that doesn't matter since they're all "blanks" at this point). You might want to use the *File > Save* function now to save your model under a name of your choosing.

Now let's open each module and enter the information required to complete the model. Start with the Create 1 module that will create the arriving Part A entities. Display 4-1 (the "Display" device was described in Section 3.4.4) provides the information required to complete this module. Note that this is very similar to the Create

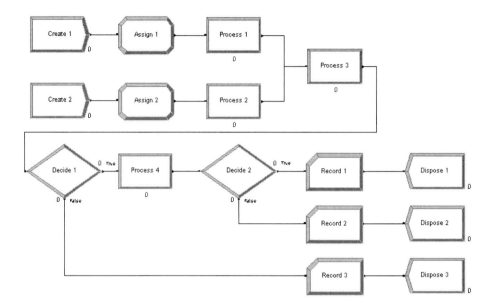

Figure 4-2. Model Window of Placed Modules

module used in Model 3-1. We've given the module a different Name and specified the Entity Type as Part A. The Time Between Arrivals is Random (i.e., an exponential distribution) with a Value (i.e., mean) of 5, and the units are set to Minutes. The remaining entries are the default options. We can now accept the module by clicking *OK*.

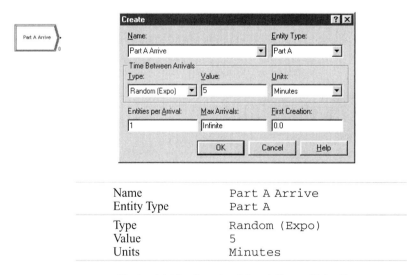

Name	Part A Arrive
Entity Type	Part A
Type	Random (Expo)
Value	5
Units	Minutes

Display 4-1. The Completed Part A Create Dialog Box

Name	Part B Arrive
Entity Type	Part B
Type	Random (Expo)
Value	30
Units	Minutes
Entities per Arrival	4

Display 4-2. The Completed Part B Create Dialog Box Entries

The Create module for the Part B arrivals is very similar to that for Part A, as shown in Display 4-2 (we'll skip the graphics since they're almost the same as what you just saw), except we have filled in one additional field (Entities per Arrival) to reflect the batch size of 4. Recall that the Part B entities arrive in batches of four. Thus, this entry will cause each arrival to consist of four separate entities rather than one.

Having created the arriving parts, we must next define an attribute Sealer Time and assign it the sealer processing time, which is different for each part type. We'll assign these values in the Assign 1 and Assign 2 modules that we previously placed. The Part A assignment is shown in Display 4-3. We've defined the new attribute and assigned it a value from a TRIA(1, 3, 4) distribution. We've also defined an attribute, Arrive Time, which is used to record the arrival time of the entity. The Arena variable TNOW

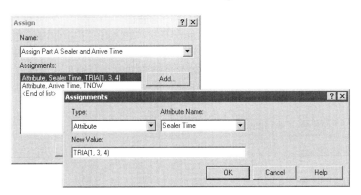

Name	Assign Part A Sealer and Arrive Time
Type	Attribute
Attribute Name	Sealer Time
New Value	TRIA(1, 3, 4)
Type	Attribute
Attribute Name	Arrive Time
New Value	TNOW

Display 4-3. Assigning the Part A Sealer Time and Arrival Time

Name	Assign Part B Sealer and Arrive Time
Type	Attribute
Attribute Name	Sealer Time
New Value	WEIB(2.5, 5.3)
Type	Attribute
Attribute Name	Arrive Time
New Value	TNOW

Display 4-4. Assigning the Part B Sealer Time and Arrival Time

provides the current simulation time, which in this case is the time the part arrived or was created (a good way to discover TNOW is to right-click in the New Value field of the Assign module's Assignments dialog box, select Build Expression, guess Date and Time Functions, and it's first in the list, described as Current Simulation Time).

The assignment to the Sealer Time and Arrive Time attributes for Part B is shown in Display 4-4. Although four entities are created in the previous module for each arrival, they'll each be assigned a different (independent) value from the sealer-time distribution in the following Assign module.

Having completed the two part-arrival modules and the assignment of the sealer times, we can now move to the two prep areas that are to be modeled using the two Process modules previously placed. The completed dialog box for the Prep A Process area is given in Display 4-5.

The Process module has four possible Actions: Delay, Seize Delay, Seize Delay Release, and Delay Release. The Delay action will cause an entity to undergo a specified time Delay. This Action does not require a Resource. This implies that waiting will occur and that multiple entities could undergo the Delay simultaneously. Since our prep area requires the use of a machine or Resource, we need an Action that will allow for waiting, queueing until the prep resource is available, and delaying for the processing time. The Seize Delay Action provides the waiting and delaying, but it does not release the Resource at the end of processing to be available for the next entity. If you use this Action, it is assumed that the Resource would be Released downstream in another module. The Seize Delay Release option provides the set of Actions required to model our prep area accurately. The last Action, Delay Release, assumes that the entity previously Seized a Resource and will undergo a Delay here, followed by the Release of the Resource.

You might notice that when you select one of the last three options, a list box appears in the empty space below the Action selection box. Click *Add* to enter the Resource information.

In entering data, we strongly urge you to make use of drop-down lists whenever possible. The reason for this caution is that once you type a name, you must always match what you typed the first time. Arena names are not case-sensitive, but the spelling and any embedded blanks must be identical. Picking the name from the list assures that it is the same. If you type in a slightly different name, Arena will give you an error message

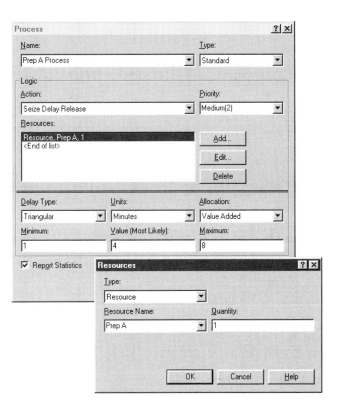

Name	Prep A Process
Action	Seize Delay Release
Resources	
Type	Resource
Resource Name	Prep A
Quantity	1
Delay Type	Triangular
Units	Minutes
Minimum	1
Value (Most Likely)	4
Maximum	8

Display 4-5. Prep A Process Dialog Box

the first time you check or attempt to run the model (or, worse yet, your model might run but it will be wrong).

Also note that when you place a module, Arena automatically provides default names and values. These default names are the object name (module, resource, etc.) with an appended number. The appended number is incremented for each additional name, if a unique name is required; for example, Process 1, Process 2, and so on. There are two reasons for this. The first is a matter of convenience—you can accept the default resource name, or you can change it. The second reason is that all names for any objects in Arena must be unique, even if the object type is different. Otherwise, Arena could not determine which object to associate with a name that had been used more than once.

To help you, Arena does a lot of automatic naming, most of which you won't even notice. For example, if you click on the Queue data module, you'll see that Arena also assigned the name `Prep A Process.Queue` to the queue at this prep area. In most cases, you can assign your own names rather than accepting the default names.

You might also notice that when you select either of the two actions that include a Seize and then accept the module, Arena will automatically place an animated queue (a horizontal line with a small vertical line at the right) near the associated Process module. This will allow you to visualize entities waiting in the queue during the simulation run. If you click on this queue, the queue name will be displayed.

The second Process module is filled out in an almost-identical fashion, with the exception of the name (`Prep B Process`), the resource name (`Prep B`), and the parameters for the process time (3, 5, 10). We have not included a display for this module.

The next step is to enter data for the sealer operation, which is the third Process module we placed. The entries for the dialog box are shown in Display 4-6. Note that in the upstream Assign modules we defined the attribute `Sealer Time` when the arriving parts were created. When an entity gains control of, or *seizes*, the resource, it will undergo a process delay equal to the value contained in its `Sealer Time` attribute.

The inspection following the sealer operation is modeled using the first Decide module. We'll accept the default Type, `2-way by Chance`, as we have only a pass or fail option, and it will be decided by chance. The dialog box requires that we enter a Percent True, and it provides two ways for entities to leave the module—True or False. In this case, we will enter the Percent True as 9%. This will result in 9% of the entities (which we'll treat as the failed items) following the True branch, and 91% (the passed items) following the False branch.[1] Parts that pass are sent to Shipping, and parts that fail are sent to Rework. The data for this Decide module are shown in Display 4-7. By the

Name	Sealer Process
Action	Seize Delay Release
Resources	
Resource Name	Sealer
Quantity	1
Delay Type	Expression
Units	Minutes
Expression	Sealer Time

Display 4-6. The Sealer Dialog Box

[1] We could have just as well reversed the interpretation of True as fail and False as pass, in which case the Percent True would be 91 (and we'd probably change the module Name to `Passed Sealer Inspection`).

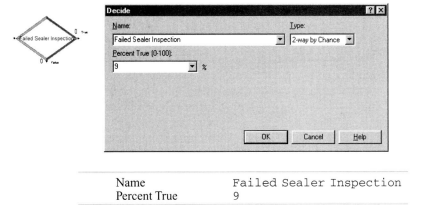

Name	Failed Sealer Inspection
Percent True	9

Display 4-7. The Sealer Inspection Dialog Box

way, if you've been building this model as we've moved through the material, now would be a good time to click the *Save* button—you never know when somebody might bump the power switch!

The remaining Process module will be used to model the rework activity. The data for this Process module are in Display 4-8. This module is very similar to the Prep A and B Process modules with the exceptions of the Name, Resource Name, and Expression.

The final Decide module is used to separate the salvaged and scrapped parts following rework. The data for this Decide module are in Display 4-9. We've chosen the True branch to represent the scrapped parts (20%) and the False branch to represent the salvaged parts.

Name	Rework Process
Action	Seize Delay Release
Resources	
Resource Name	Rework
Quantity	1
Delay Type	Expression
Units	Minutes
Expression	EXPO(45)

Display 4-8. The Rework Process Dialog Box

Name	Failed Rework Inspection
Percent True	20

Display 4-9. The Rework Inspection Dialog Box

Having defined all of the operations, we now need to fill in the Record and Dispose modules. Remember that as part of the simulation output, we wanted to collect statistics on resource utilization, number in queue, and time in queue at each of the operations. These three statistics are automatically collected whenever you use a Process module with an Action option that requires a Resource (assuming that the Report Statistics box for the module is checked and that the Processes box is checked in *Run > Setup > Project Parameters*). We also wanted statistics on the cycle time separated by shipped parts, salvaged parts, and scrapped parts. The Record module provides the ability to collect these cycle times in the form of Tallies. The completed dialog box for the scrapped parts tally is shown in Display 4-10. We picked the Type `Time Interval` from the drop-down list. The Tally Name defaults to the module name. This will cause Arena to record as a Tally statistic the time between the attribute carrying the arrival time (`Arrive Time`) of the part to the system and the time that it arrived at this Record module, which will be the entity's time in system.

The remaining two Record modules are named `Record Salvaged Parts` and `Record Shipped Parts`. We have not bothered to include displays on these modules as they are completely analogous to the `Record Scrapped Parts` module in Display 4-10.

The final three modules Dispose of the entities as they leave the system. For this model, we could have directed all entities to a single Dispose module. However, one of the features of a Dispose module is the inclusion of an animation variable, which appears near the lower right-hand corner of the module. This animation variable will display the current count for the total number of entities that have passed through this module during the run and allow the viewer to discern the proportion of entities that have taken each of the three possible paths through the system.

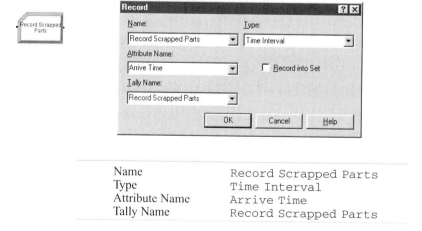

Name	Record Scrapped Parts
Type	Time Interval
Attribute Name	Arrive Time
Tally Name	Record Scrapped Parts

Display 4-10. The Record Scrapped Parts Dialog Box

Name	Scrapped

Display 4-11. The Scrapped Dispose Dialog Box

The data for the `Scrapped` Dispose module are shown in Display 4-11. We have defaulted on the check in the box entitled Record Entity Statistics. However, if we had wanted to keep entity flow statistics only on the parts that were shipped, including the salvaged parts, then we could have cleared this box (leaving checks in the remaining two Dispose modules). Doing so would have caused only the selected parts to be included in the automatic entity-flow statistics. Of course, you need to make sure the Entities box in *Run > Setup > Project Parameters* is checked in order to get any of these results.

The two other Dispose modules, `Salvaged` and `Shipped`, are filled out in a similar way.

You're nearly ready to run the model. Although it has taken you some time to get this far, once you get accustomed to working with Arena, you'll find that you could have accomplished these steps in only a few minutes.

The model could actually run at this point, but once started, it would continue to run forever because Arena doesn't know when to stop the simulation. You establish the run parameters for the model by selecting *Run > Setup*. The Run Setup dialog box has six tabbed pages that can be used to control the simulation run. The data for the first tab, Project Parameters, are shown in Display 4-12. We've entered the project title and analyst name so they will appear on the output reports, and also a brief synopsis under Project Description for documentation purposes. In the statistics collection area, we have *cleared* the Entities selection as we don't need those data for our analysis (so we won't get statistics on time in system sorted out by entity type). You might try running the simulation with this box checked in order to see the difference in the output reports.

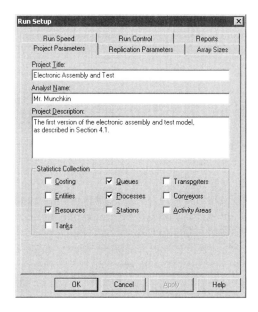

Project Title	Electronic Assembly and Test
Analyst Name	Mr. Munchkin
Project Description	The first version of the elect-ronic assembly and test model, as described in Section 4.1.
Statistics Collection Entities	*clear*

Display 4-12. The Run Setup Project Parameters

You also need to specify the run length, which is done under the Replication Parameters tab. We've set the Replication Length to `32` hours (four consecutive 8-hour shifts), the Base Time Units to `Minutes`, and defaulted the remaining fields. The completed dialog box is shown in Display 4-13. We've also accepted the defaults for the remaining four tabs in *Run > Setup*: Array Sizes, Run Speed, Run Control, and Reports. You might want to look at these tabs to get an idea of the options available.

Before we run our newly created model, let's give it one final tweak. Since we have two different part types, it might be nice if we could distinguish between them in the animation. Click on the Entity data module found in the Basic Process panel and note that the initial picture for both parts is `Picture.Report`. When we run our model, all of our parts will use this same icon for displaying entities on the animation.

Now click on the Initial Picture cell for Part A, and use the list to select a different picture. We've chosen the blue ball for Part A and the red ball for Part B, as shown in Display 4-14. This will allow us to distinguish easily between the two parts in the

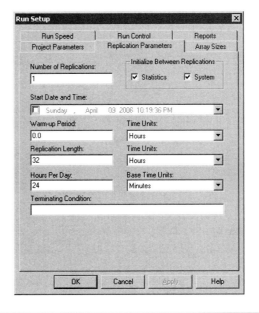

| Replication Length | 32 |
| Base Time Units | Minutes |

Display 4-13. The Run Setup Replication Parameters

animation. If you're interested in seeing what these icons look like, you can select *Edit > Entity Pictures* from the menu bar at the top of your screen to open the Entity Picture Placement window, which will allow you to see the icons currently available down the left column. We'll show you later how to use this feature in more detail.

Your final model should look something like Figure 4-3.

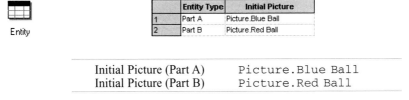

| Initial Picture (Part A) | Picture.Blue Ball |
| Initial Picture (Part B) | Picture.Red Ball |

Display 4-14. The Entity Data Module

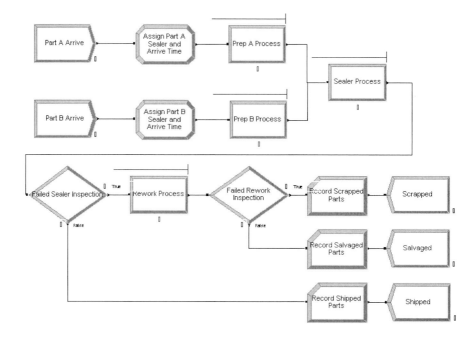

Figure 4-3. The Final Model 4-1

4.1.3 Running the Model

Before running your model, you might want to check it for errors. You can do this by clicking the *Check* button (✔) on the Run Interaction toolbar, the *Run > Check Model* command, or the *F4* key on the keyboard. With a little luck, the response will be a small window with the message "No errors or warnings in model." If you have no luck at all, an error window will open with a message describing the error. If this occurs, you might want to select the *Find* option, if the button is enabled. This feature attempts to point you to where Arena thinks the error might be. We suggest that you intentionally insert an error into your model and try these features. As you build more complex models, you might just find yourself using these features quite often.

If your model check results in no errors, you're now ready to run the simulation. There are four ways to run a simulation, but we'll only talk about three of them here. The first way is to run the simulation with the animation. Use the *Go* button (►) on the Standard toolbar, the *Run > Go* command, or the *F5* key. If you've not yet checked your model or if you've made a change since the last check, Arena will first check the model, then initialize the model with your data, and finally run it. You'll notice during the run that Arena hides some of the graphics so that your focus is on the animation. Don't worry, though. They'll return when you end the run (or you can check to see that they're still there by using the *View > Layers* command).

If you leave the status bar active (at the bottom of the screen), you can tell what Arena is doing. Toward the right of this bar are three pieces of information: the replication number, the current simulation time, and the simulation status.

After the simulation starts to run, you may want to speed up or slow down the animation. You can do this while the model is running with the slider bar on the right end of the Standard toolbar, or by pressing the "<" key to slow it down or the ">" key to speed it up. If you do either, the current Animation Speed Factor is displayed at the far left of the status bar. You can also increase or decrease the Animation Speed Factor from the Run Speed tab of the Run Setup dialog box. This option can also be used to enter an exact speed factor.

During the simulation run, you can also pause the simulation using the *Pause* button (❚❚) on the Standard toolbar, *Run > Pause*, or the *Esc* key. This temporarily suspends the simulation, and the message "User interrupted" will appear on the status bar.

While you're in Pause mode, you might want to double-click on one of the entities that is visible in the animation. An Entity Summary dialog box lists the values of each of the entity's attributes. This can be a very useful feature when trying to debug a model. You can also use the *Step* button (▶❙) on the Run toolbar to move entities through the system one step at a time. You can continue the simulation run "normally" at any time with the *Go* button.

This method of running a simulation provides the greatest amount of information, but it can take a long time to finish. In this case, the time required to complete the run depends on the Animation Speed Factor. You can skip ahead in time by selecting the *Fast-Forward* button (▶▶) on the Run toolbar, or *Run > Fast-Forward*. This will cause the simulation to run at a much faster speed by not updating the animation graphics. At any time during the Fast-Forwarding run, you can select the *Go* button again and return to the animation mode. During an animation-mode run, you can Zoom In (+), Zoom Out (−), or move about in the simulation window (arrow keys or scroll bars).

Using Fast-Forward will run the simulation in much less time, but if you're only interested in the numerical simulation results, you might want to disable the animation (computation and graphics update) altogether. You do this with the *Run > Run Control > Batch Run (No Animation)* option.

The *Run > Run Control* option also allows you to configure a number of other runtime options. For now, select the *Batch Run (No Animation)* option. Note that if you revisit this option, there is a check to the left. Accept this option and click the *Run* button. Note how much faster the simulation runs. The only disadvantage is that you must terminate the run and reset the animation settings in order to get the animation back. If you have large models or long runs and you're interested only in the numerical results, this is the option to choose since it is even faster than Fast-Forward.

While you're building a model, you should probably have most of the toolbars visible and accessible. However, when you're running a model, many of the toolbars simply consume space because they are not active during runtime. Arena recognizes this and will save your toolbar settings for each mode. To take advantage of this, pause during the run and remove the toolbars that you want to inactivate while the simulation is running. When you end the run, these toolbars will reappear. You could also go to *Run > Setup > Run*

Control before the run and check the box for Run in Full-Screen Mode (since the *End* button will be gone at the end, you need to use the menu option *Run > End* instead).

4.1.4   Viewing the Results

If you selected the *Run > Go* menu option (or the *Go* button, ►), you might have noticed that, in addition to the blue and red balls (our Part A and B entities) moving through the system, there are several animated counters being incremented during the simulation run. There is a single counter for each Create, Process, and Dispose module and two counters for each Decide module. The counters for the Create, Dispose, and Decide modules are incremented each time an entity exits the module. In the case of the Process modules, the counter is the number of entities that are currently at that module, including any entities in the queue waiting for the resource plus any entities currently in process. If you selected the *Run > Fast-Forward* menu option (or the *Fast-Forward* button, ▶▶), these counters (and the entities in the queues) will be updated at the end of the run and whenever you pause or change views of the model. The final numbers resulting from our simulation are shown in Figure 4-4.

If you run the model to completion, Arena will ask if you want to see the results. If you select *Yes*, you should get a window showing the Category Overview Report (the default report). When the first page of the report appears, you might find it strange that the message "No Summary Statistics Are Available" is displayed. The system summary statistics are entity and costing statistics, which we elected not to collect when we *cleared* the Entities selection on the Project Parameters tab of the Run Setup dialog box (see Display 4-12). At some point, you might want to change these selections and view the difference in the reports.

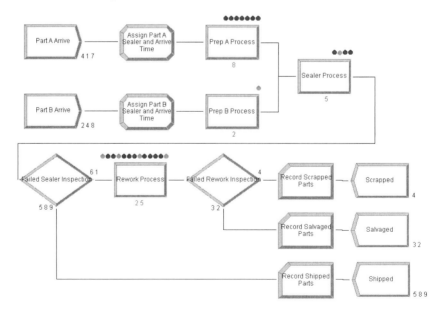

Figure 4-4. The Animated Results for Model 4-1

Recall that you can navigate through the report using the tree listing under the Preview tab at the left side of the report window or by using the arrow buttons in the upper-left corner of the report window. This report will provide statistics by the categories selected in the Run Setup dialog box (Project Parameters tab, Statistics Collection area). For our model, you will find sections on Process, Queue, Resource, and User Specified. The User Specified section is there because we included Record modules in our model to collect statistics on the cycle times sorted by departure type.

As in Chapter 3, you will find three types of statistics in our report: *tally*, *time-persistent*, and *counter*. A fourth statistic (*outputs*) is available when multiple replications are made (we'll talk about them in Chapter 6, when we do multiple replications). The tally statistics here include several process times, queue times, and the interval times collected by our Record modules. The time-persistent statistics include number waiting in queue, resource usage, and resource utilization. Counters include accumulated time, number in, number out, and total number seized.

The tally and time-persistent statistics provide the average, 95% confidence-interval half width, and the minimum and maximum observed values. With the exception of the half-width column, these entries are as described in Chapters 2 and 3, and should be self explanatory.

At the end of a single-replication simulation run, Arena attempts to calculate a 95% confidence-interval half width for the steady-state (long-run) expected value of each observed statistic, using a method called *batch means* (see Section 7.2.3). Arena first checks to see if sufficient data have been collected to test the critical statistical assumption (uncorrelated batches) required for the batch-means method. If not, the annotation "Insufficient" appears in the report and no confidence-interval half width is produced, as can be seen for several of the results. If there are enough data to test for uncorrelated batches, but the test is failed, the annotation "Correlated" appears and once again there's no half width, which is the case for several of the results. If there was enough data to perform the test for uncorrelated batches *and* the test were passed, the half width (the "plus-or-minus" amount) of a 95% confidence interval on the long-run (steady-state) expected value of the statistic would be given (this happens for a few of the results in this run). In this way, Arena refuses to report unreliable half-width values even though it could do so from a purely computational viewpoint. The details and importance of these tests are further discussed in Section 7.2.3.

Trying to draw conclusions from this single short run could be misleading because we haven't yet addressed issues like run length, number of replications, or even whether long-run steady-state results are appropriate (but we will, in Section 7.2). However, if you look at the results for the waiting time in queue and number waiting in queue (Figure 4-5), you see that the rework station is plagued by waiting times and queue lengths that are much longer than for the other stations (you can corroborate this imbalance in congestion by looking at the Process and Resources sections of the Category Overview report as well). This implies either that the rework area doesn't have enough capacity to handle its work, or there is a great deal of variability at this station. We'll address this issue in a minute in Section 4.2. (By the way, note that in some places in the Category Overview Report, you get a color-coded graphic for some of the results.)

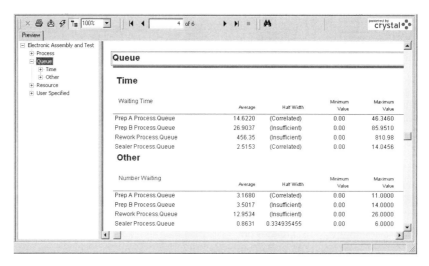

Figure 4-5. The Queue Section of the Category Overview Report: Model 4-1

4.2   Model 4-2: The Enhanced Electronic Assembly and Test System

Having constructed and run our model, the next activity would be to verify that the "code" (Arena file) is free of bugs[2] and also to validate that the conceptual model really represents the system being studied (see Sections 2.8 and 13.6). For this example, that's fairly easy. We can examine the logic constructs we selected from the modules we used and compare them to the problem definition. With much larger and more complex systems, this can become a challenging task. An animation is often useful during the verification and validation phases because it allows you to view the entire system being modeled as it operates. If you ran the model we developed and viewed its animation, you should have noted that it appeared to operate quite similarly to the way we described the system. If verification can be very difficult, complete validation (the next activity) can sometimes be almost impossible. That's because validation implies that the simulation is behaving just like the real-world system, which may not even exist, so you can't tell. And even if the system does exist, you must have output performance data from it, as well as convince yourself and other nonbelievers that your model can really capture and predict the events of the real system. We'll discuss both of these activities in much more detail in Chapter 13.

For now, let's assume that as part of this effort you showed the model and its accompanying results to the production manager. Her first observation was that you didn't have a complete definition of how the system works. Whoever developed the problem

---

[2] Of course, you're already familiar with this term "bug" in reference to an error in a computer program, but do you know the etymology of this entomological neology? We do. The original computer "bug" was a luckless moth electrocuted in a relay panel of an early computer used by the venerable Grace Murray Hopper (who coined the term) at Vassar College, thus shorting circuits and causing errors; you can see its sad postmortem photo at http://www.cs.vassar.edu/images/Bug.GIF.

definition looked only at the operation of the first shift. But this system actually operates two shifts a day, and on the second shift, there are two operators assigned to the rework operation. This would explain our earlier observation when we thought the rework operation might not have enough capacity. The production manager also noted that she has a failure problem at the sealer operation. Periodically, the sealer machine breaks down. Engineering looked at the problem some time ago and collected data to determine the effect on the sealer operation. They felt that these failures did not merit any significant effort to correct the problem because they didn't feel that the sealer operation was a bottleneck. However, they did log their observations, which are still available. Let's assume that the mean uptime (from the end of one failure to the onset of the next failure) was found to be 120 minutes and that the distribution of the uptime is exponential (which, by the way, is often used as a realistic model for uptimes if failures occur randomly at a uniform rate over time). The time to repair also follows an exponential distribution with a mean of 4 minutes.

In addition, the production manager indicated that she was considering purchasing special racks to store the waiting parts in the rework area. These racks can hold ten assemblies each, and she would like to know how many racks to buy. Our next step is to modify the model to include these three new aspects, which will allow us to use some additional Arena features.

In order to incorporate these changes into our model, we'll need to introduce several new concepts. Changing from a one- to a two-shift operation is fairly easy. In Model 4-1, we set our run length to four 8-hour shifts and made no attempt to keep track of the day/ shift during the run. We just assumed that the system conditions at the end of a shift were the same at the start of the next shift and ignored the intervening time. Now we need to model explicitly the change in shifts, because we have only one operator in the first shift and two in the second shift for the rework process.

We'll add this to our model by introducing a Resource Schedule for the rework resource, which will automatically change the number of rework resources throughout the run by adjusting the resource capacity. While we're making this change, we'll also increase the run length so that we simulate more than just two days of a two-shift operation. We'll model the sealer failures using a Resource Failure, which allows us to change the available capacity of the resource (much like the Resource Schedule), but has additional features specifically designed for representing equipment failures. Finally, we'll use the Frequencies statistic to obtain the type of information we need to determine the number of racks that should be purchased.

4.2.1 Expanding Resource Representation: Schedules and States

So far we've modeled each of our resources (prep area, sealer, and rework) as a single resource with a fixed capacity of 1. You might recall that we defaulted all of this information in the Resource data module. To model the additional rework operator, we could simply change the capacity of the rework resource to 2, but this would mean that we would *always* have two operators available. What we need to do is to schedule one rework operator for the first shift (assume each shift is 8 hours) and two rework operators for the second shift. Arena has a built-in construct to model this, called a *Schedule*, that allows you to vary the capacity of a resource over time according to a fixed pattern. A resource Schedule is defined by a sequence of time-dependent resource capacity changes.

We also need to capture in our model the periodic random breakdowns (or failures) of the sealer machine. This could be modeled using a Schedule, which would define an available resource capacity of 1 for the uptimes and a capacity of 0 for the time to repair. However, there is a built-in construct designed specifically to model failures. First, let's introduce the concept of *Resource States*.

Arena automatically has four Resource States: *Idle*, *Busy*, *Inactive*, and *Failed*. For statistical reporting, Arena keeps track of the time the resource was in each of the four states. The resource is said to be Idle if no entity has seized it. As soon as an entity seizes the resource, the state is changed to Busy. The state will be changed to Inactive if Arena has made the resource unavailable for allocation; this could be accomplished with a Schedule's changing the capacity to 0. The state will be changed to Failed if Arena has placed the resource in the Failed state, which also implies that it's unavailable for allocation.

When a failure occurs, Arena causes the entire resource to become unavailable. If the capacity is 2, for example, both units of the resource will be placed in the Failed state during the repair time.

4.2.2   Resource Schedules

Before we add our resource schedule for the rework operation, let's first define our new 16-hour day. We do this in the Replication Parameters tab of the *Run > Setup* option, by changing the Hours Per Day from 24 to `16` (ignore the warning about Calendars that you might get here). While we're in this dialog box, let's also change our Time Units for the Replication Length to `Days` and the Replication Length itself to `10` days.

You can start the definition of a resource schedule in either the Resource or Schedule data module. We'll start with the Resource data module. When you click on this module from the Basic Process panel, the information on the current resources in the model will be displayed in the spreadsheet view of the model window along the bottom of your screen. Now click in the Type column for the Rework resource row and select `Based on Schedule` from the list. When you select this option, Arena will add two new columns to the spreadsheet view—Schedule Name and Schedule Rule. Note that the Capacity cell for the Rework resource has been dimmed because the capacity will instead be based on a schedule. In addition, the cells for two new columns are also dimmed for the other three resources because they still have a fixed capacity. Next you should enter the schedule name (e.g., `Rework Schedule`) in the Schedule Name cell for the Rework resource.

Finally, you need to select the Schedule Rule, which can affect the specific timing of when the capacity defined by the schedule will actually change. There are three options for the Schedule Rule: Wait (the default), Ignore, and Preempt. If a capacity decrease of x units is scheduled to occur and at least x units of the resource are idle, all three options immediately cause x units of the resource unit(s) to become Inactive. But if fewer than x units of the resource are idle, each Schedule Rule responds differently: (The three options are illustrated in Figure 4-6.)

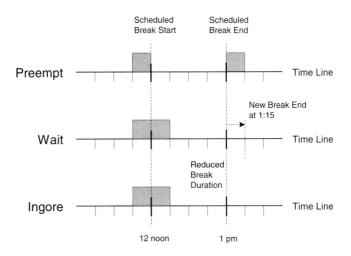

Figure 4-6. The Preempt, Wait, and Ignore Schedule Rules

- The *Ignore* option immediately decreases the resource capacity, ignoring the fact that the resource is currently allocated to an entity, but "work" on the in-service entities continues unabated. When units of the resource are released by the entity, these units are placed in the Inactive state. However, if the resource capacity is increased again (that is, the scheduled time at the lower capacity expires) before the entities release the units of the resource, it's as if the schedule change never occurred. The net effect is that the time for which the resource capacity is scheduled to be reduced may be shortened with this option.
- The *Wait* option, as its name implies, will wait until the in-process entities release their units of the resource before starting the actual capacity decrease. Thus the reduced capacity time will always be of the specified duration, but the time between these reductions may increase.
- The *Preempt* option attempts to preempt the last unit of the resource seized by taking it away from the controlling entity. If the preempt is successful and a single unit of capacity is enough, then the capacity reduction starts immediately. The preempted entity is held internally by Arena until the resource becomes available, at which time the entity will be reallocated the resource and continue with its remaining processing time. This provides an accurate way to model schedules and failures because, in many cases, the processing of a part is suspended at the end of a shift or when the resource fails. If the preempt is unsuccessful or if more than one unit is needed, then the Ignore rule will be used for any remaining capacity.

So when should you use each of the rules? Generally, we recommend that you examine closely the actual process and select the option that best describes what actually occurs at the time of a downward schedule change or resource failure. If the resource under consideration is the bottleneck for the system, your choice could significantly

Resource

	Name	Type	Capacity	Schedule Name	Schedule Rule
1	Prep A	Fixed Capacity	1	1	Wait
2	Prep B	Fixed Capacity	1	1	Wait
3	Sealer	Fixed Capacity	1	1	Wait
4	Rework	Based on Schedule	Rework Schedul	Rework Schedule	Ignore

Rework Resource
        Type   Based on Schedule
        Schedule Name   Rework Schedule
        Schedule Rule   Ignore

Display 4-15. The Resource Data Module: Selecting a Resource Schedule

affect the results obtained. But sometimes it isn't clear what to do, and, while there are no strict guidelines, a few rules of thumb may be of help. First, if the duration of the scheduled decrease in capacity is very large compared to the processing time, the Ignore option may be an adequate representation. If the time between capacity decreases is large compared to the duration of the decrease, the Wait option could be considered. For this model, we've selected the Ignore option because, in most cases, an operator will finish his task before leaving and that additional work time is seldom considered.

The final spreadsheet view of the first four columns is shown in Display 4-15 (there are actually additional columns to the right, which we don't show here). There is also a dialog form for entering these data, which can be opened by right-clicking on the rework cell in the Name column and selecting *Edit via Dialog*.

Now that you've named the schedule and indicated the schedule rule, you must define the actual schedule the resource should follow. One way to do this is by clicking on the Schedule data module and entering the schedule information in the spreadsheet view. A row has already been added that contains our newly defined Rework Schedule. Clicking in the Durations column will open the Graphical Schedule Editor, a graphical interface for entering the schedule data. The horizontal axis is the calendar or simulation time. (Note that a day is defined as 16 hours based on our day definition in the Run Setup dialog box.) The vertical axis is the capacity of the resource. You enter data by clicking on the *x-y* location that represents day one, hour one, and a capacity value of one. This will cause a solid blue bar to appear that represents the desired capacity during this hour; in this case, one. You can complete the data entry by repeatedly clicking or by clicking, holding, and then dragging the bar over the first eight hours. Complete the data schedule by entering a capacity of 2 for hours nine through 16. It's not necessary to add the data for Day 2, as the data entered for Day 1 will be repeated automatically for the rest of the simulation run. The complete schedule is shown in Figure 4-7. You might note that we used the *Options* button to reduce our maximum vertical axis (capacity) value from ten to four; other things can be changed in the Options dialog box, like how long a time slot lasts, the number of time slots, and whether the schedule repeats from the beginning or remains forevermore at a fixed-capacity level.

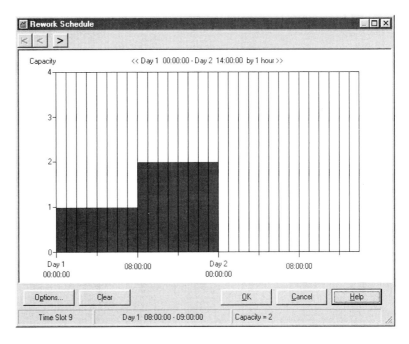

Figure 4-7. The Graphical Schedule Editor: The Rework Schedule

You can also enter these data manually by right-clicking in the Durations column in the Schedule module spreadsheet view and selecting *Edit via Dialog*. If you select this option, first enter the schedule name, then click the *Add* button to open the Durations window. Here you define the (Capacity, Duration) pairs that will make up the schedule. In this case, our two pairs are 1, 8 and 2, 8 (Display 4-16). This implies that the capacity will be set to 1 for the first 480 minutes, then 2 for the next 480 minutes. This schedule will then repeat for the duration of the simulation run. You may have as many (Capacity, Duration) pairs as are required to model your system accurately. For example, you might want to include operator breaks and the lunch period in your schedule. There is one caution, or feature,[3] of which you should be aware. If, for any pair, no entry is made for the Duration, it will default to infinity. This will cause the resource to have that capacity for the entire remaining duration of the simulation run. As long as there are positive entries for all durations, the schedule will repeat for the entire simulation run.

If you use the Graphical Schedule Editor to create the schedule and then open the dialog box, you will find that the data have been entered automatically. Note that you cannot use the Graphical Schedule Editor if you have any time durations that are not integer, or if any entries require an Expression (for example, a time duration that's a draw from a random variable).

---

[3] This is called a feature if you do it intentionally and an error if you do it accidentally.

Schedule

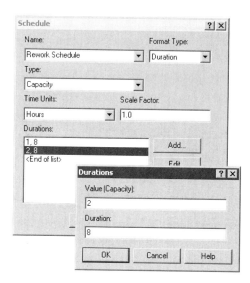

Name	Rework Schedule
Value	1
Duration	8
Value	2
Duration	8

Display 4-16. The Schedule Data Module Dialog Box

4.2.3 Resource Failures

Schedules are intended to model the planned variation in the availability of resources due to shift changes, breaks, vacations, meetings, etc. Failures are primarily intended to model random events that cause the resource to become unavailable. You can start your failure definition in either the Resource or the Failure data module. The Failure data module can be found in the Advanced Process panel (which you might need to attach at this point via the 📂 button or *File > Template Panel > Attach* if it's not already accessible in the Project Bar). Since we started with the Resource module in developing our schedule, let's start with the Failure data module for our Sealer failure.

If you need to make the Advanced Process panel visible in the Project Bar, click on its name, and then click on the Failure data module. The spreadsheet view for this module will show no current entries. Double-click where indicated to add a new row. Next select the default name in the Name column and replace it with a meaningful failure name, like Sealer Failure. Then select whether the failure is Count-based or Time-based using the list in the Type cell. A Count-based failure causes the resource to fail after the specified number of entities have used the resource. This count may be a fixed number or may be generated from any expression. Count-based activities are fairly common in industrial models. For example, tool replacement, cleaning, and machine adjustment are

	Name	Type	Up Time	Up Time Units	Down Time	Down Time Units	Uptime in this State only
1	Sealer Failure	Time	EXPO(120)	Minutes	EXPO(4)	Minutes	

Double-click here to add a new row.

Failure

Name	Sealer Failure
Type	Time
Up Time	EXPO(120)
Up Time Units	Minutes
Down Time	EXPO(4)
Down Time Units	Minutes

Display 4-17. The Sealer Failure Spreadsheet View

typically based on the number of parts that have been processed rather than on elapsed time. Although these may not normally be viewed as "failures," they do occur on a periodic basis and prevent the resource from producing parts. On the other hand, we frequently model failures as Time-based because that's the way we've collected the failure data. In our model, the problem calls for a Time-based failure. So click on the Type cell and select the Time option. When you do this, the column further to the right in the spreadsheet will change to reflect the different data requirements between the two options. Our Up Time and Down Time entries are exponential distributions with means of 120 minutes and 4 minutes, respectively. We also need to change the Up Time Units and Down Time Units from Hours to Minutes.

The last field, Uptime in this State Only, allows us to define the state of the resource that should be considered as "counting" for the uptimes. If this field is defaulted, then all states are considered. Use of this feature is very dependent on how your data were collected and the calendar timing of your model. Most failure data are simply logged data; for example, only the time of the failure is logged. If this is the case, then holidays, lunch breaks, and idle time are included in the time between failures, and you should default this field. Only if your time between failures can be linked directly to a specific state should this option be chosen. Many times equipment vendors will supply failure data based on actual operating hours; in this case, you would want to select this option and specify the Busy state. Note that if you select this option you must also define the Busy state using the StateSet data module found in the Advanced Process panel.

The final spreadsheet view for our Sealer Failure is shown in Display 4-17.

Having completed the definition of our Sealer Failure, we now need to attach it to the Sealer resource. Open the Resource data module (back in the Basic Process panel) and click in the Failures column for the Sealer resource row. This will open another window with the Failures spreadsheet view. Double-click to add a new row and, in the Failure Name cell, select Sealer Failure from the list. We must also select the Failure Rule—Ignore, Wait, or Preempt. These options are the same as for schedules, and you respond in an identical manner. Returning to our rules of thumb for choosing the Failure Rule option, because our expected uptime (120 minutes) is large compared to our failure duration (4 minutes), we'll use the Wait option. The final spreadsheet view is shown in

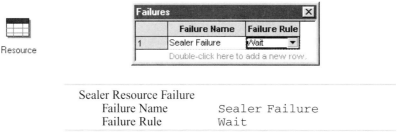

Sealer Resource Failure
 Failure Name   Sealer Failure
 Failure Rule   Wait

Display 4-18. The Sealer Resource Data Module: Failures Spreadsheet View

Display 4-18. If you have multiple resources with the same failure profile, they can all reference the same Failure Name. Although they will all use the same failure profile, they will each get their own independent random samples during the simulation run.

4.2.4 Frequencies

Frequencies are used to record the time-persistent occurrence frequency of an Arena variable, expression, or resource state. We can use the *Frequencies* statistic type to obtain the information we need to determine the number of racks required at the Rework area. We're interested in the status of the rework queue—specifically, how many racks of 10 should we buy to ensure that we have sufficient storage almost all of the time. In this case, we're interested in the amount of time the number in queue was 0 (no racks needed), greater than 0 but no more than 10 (one rack needed), greater than 10 but no more than 20 (two racks needed), etc.

Frequency statistics are entered using the Statistic data module, which can be found in the Advanced Process panel. Clicking on this data module will open the spreadsheet view, which is initially empty; double-click to add a new row. We'll first enter the name as Rework Queue Stats. Next, select Frequency in the Type list and default on the Value entry for the Frequency Type. You might note that when we selected the Type, the Report Label cell was automatically given the same name as the Name cell, Rework Queue Stats, which we'll accept to label this output in the reports.

We now need to develop an Expression that represents the number in the rework queue. To request this information, we need to know the Arena variable name for the number in queue, NQ. We can get the name of the queue, Rework Process.Queue, from the Queue data module found in the Basic Process panel. Thus, the Expression we want to enter is NQ(Rework Process.Queue). At this point, you should be asking, "Did you expect me to know all that?" To an experienced (and thus old) SIMAN user, this is obvious. However, it is clearly *not* obvious to the new user. Fortunately, Arena provides an easy way to develop these types of expressions without the need to know all the secret words (e.g., NQ). Place your pointer in the blank Expression cell, right-click, and select Build Expression. This will open the Arena Expression Builder window shown in Display 4-19. Under the Expression Type category Basic Process Variables, you'll find the sub-category Queue. Click on the + sign to expand the options and then select Current Number In Queue. When you do this, two things will happen: the Queue

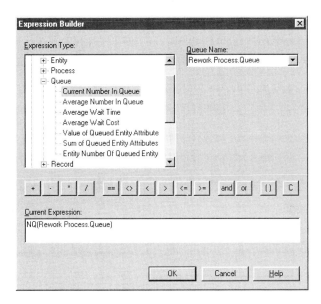

Display 4-19. The Expression Builder Dialog Box

Name field will appear at the right and the Current Expression at the bottom will be filled in using the queue name shown. In our case, the Current Expression was `NQ(Prep A Process.Queue)`, which is not yet what we want (it's the wrong queue). Now use the drop-down list arrow for Queue Name to view and select the `Rework Process.Queue`. Now when you click *OK*, that expression will automatically be entered into the field from which the Expression Builder was opened.

You can right-click on just about any field in which an expression can be entered to open the Expression Builder. For example, we could have used the Expression Builder to find the expression for the current simulation time (`TNOW`). You can also build complex expressions by using the function buttons in the Expression Builder, as well as typing things (the old-fashioned way, assuming that you know what to type) into the Current Expression field at the bottom.

The spreadsheet view to this point (we're not done yet) is shown in Display 4-20.

	Name	Type	Frequency Type	Expression	Report Label	Output File	Categories
1	Rework Queue Stats	Frequency	Value	NQ(Rework Process.Queue)	Rework Queue Stats		0 rows

Double-click here to add a new row.

Statistic

Name	`Rework Queue Stats`
Type	`Frequency`
Frequency Type	`Value`
Expression	`NQ(Rework Process.Queue)`

Display 4-20. Partial Frequency Entry in the Statistic Data Module

Categories					☒
	Constant or Range	Value	High Value	Category Name	Category Option
1	Constant	0		0 Racks	Include
2	Range	0	10	1 Rack	Include
3	Range	10	20	2 Racks	Include
4	Range	20	30	3 Racks	Include
5	Range	30	40	4 Racks	Include
	Double-click here to add a new row.				

Constant or Range	Constant
Value	0
Category Name	0 Racks
Constant or Range	Range
Value	0
High Value	10
Category Name	1 Rack
Constant or Range	Range
Value	10
High Value	20
Category Name	2 Racks

Display 4-21. Categories for Rework Queue Stats Frequency Statistic

The last step in setting up the rework queue statistics is to build the categories that define how we want the values displayed, done in the Categories column at the far right (click the "0 Rows" button to open a spreadsheet to which you add a row for each category). Display 4-21 shows the entries for the first three categories. The first entry is for a queue size of a Constant 0 (in which case, we'd need 0 racks); the following entries are for one rack, two racks, etc. For now, we'll only request this information for up to four racks. If the queue ever exceeds 40 parts, Arena will create an out-of-range category on the output report. In the case of a Range, note that Value is not included in the range, but High Value is. Thus, for instance, Value = 10 and High Value = 20 defines a range of numbers (10, 20]; that is, (strictly) greater than 10 and less than or equal to 20.

Before leaving the Statistic data module, we might also want to request additional information on the Sealer resource. If we run our current model, information on the utilization of the Sealer resource will be included in our reports as before. However, it will not specifically report the amount of time the resource is in a failed state. We can request this statistic by adding a new row in our Statistic data module as shown in Display 4-22. For this statistic, we enter the Name and Type, `Sealer States` and `Frequency`, and select `State` for the Frequency Type. Finally, we select the `Sealer` resource from the list in the Resource Name cell. This will give us statistics based on all the states of the Sealer resource—Busy, Idle, and Failed.

Before you run this model, we recommend that you check the *Run > Run Control > Batch Run (No Animation)* option, which will greatly reduce the amount of time required to run the model. Although slower, an alternative is to select *Run > Fast-Forward* (▸▸).

Statistic - Advanced Process								
Name	Type	Frequency Type	Expression	Resource Name	Report Label	Output File	Categories
1 Rework Queue Stats	Frequency	Value	NQ(Rework Process.Queue)		Rework Queue Stats		5 rows
2 Sealer States	Frequency	State	Expression 1	Sealer	Sealer States		0 rows
Double-click here to add a new row.							

Statistic

Name	Sealer States
Type	Frequency
Frequency Type	State
Resource Name	Sealer

Display 4-22. The Statistic Data Module for the Sealer States

This would also be a good time to save your work. Note that you can still pause the run at any time to determine how far you've progressed.

4.2.5 Results of Model 4-2

Table 4-1 gives some selected results from the Reports for this model (rightmost column), as well as for Model 4-1 for comparison. We rounded everything to two decimals except for the two kinds of utilization results, which we give to four decimals (in order to make a particular point about them).

Table 4-1. Selected Results from Model 4-1 and Model 4-2

Result	Model 4-1	Model 4-2
Average Waiting Time in Queue		
Prep A	14.62	19.20
Prep B	26.90	51.42
Sealer	2.52	7.83
Rework	456.35	116.25
Average Number Waiting in Queue		
Prep A	3.17	3.89
Prep B	3.50	6.89
Sealer	0.86	2.63
Rework	12.95	3.63
Average Time in System		
Shipped Parts	28.76	47.36
Salvaged Parts	503.85	203.83
Scrapped Parts	737.19	211.96
Instantaneous Utilization of Resource		
Prep A	0.9038	0.8869
Prep B	0.7575	0.8011
Sealer	0.8595	0.8425
Rework	0.9495	0.8641
Scheduled Utilization of Resource		
Prep A	0.9038	0.8869
Prep B	0.7575	0.8011
Sealer	0.8595	0.8425
Rework	0.9495	0.8567

The results from this model differ from those produced by Model 4-1 for several reasons. We're now running the simulation for ten 16-hour days (so 160 hours) rather than the 32 hours we ran Model 4-1. And, of course, we have different modeling assumptions about the Sealer and Rework Resources. Finally, all of these facts combine to cause the underlying random-number stream to be used differently (more about this in Chapter 12).

Going from Model 4-1 to Model 4-2 didn't involve any changes in the Prep A or Prep B parts of the model, so the differences we see there are due just to the differences in run length or random bounce. The difference for the Prep B queue results are pretty noticeable, so either this area becomes more congested as time goes by, or else the results are subject to a lot of uncertainty—we don't know which (all the more reason to do statistical analysis of the output, which we're not doing here).

For the Sealer, the queue statistics (both average waiting time and average length) display considerably more congestion for Model 4-2. This makes sense, since we added the Failures to the Sealer in this model, taking it out of action now and then, during which time the queue builds up. The utilization statistics for the Sealer are not much different across the two models, though, since when it's in the Failed state the Sealer is not available, so these periods don't count "against" the Sealer's utilizations.

Unlike the Sealer, the Rework operation seems to be going much more smoothly in Model 4-2. Of course, the reason for this is that we added a second unit of the Rework resource during the second eight-hour shift of each 16-hour day. This increases the capacity of the Rework operation by 50% over time, so that it now has a time-average capacity of 1.5 rather than 1. And accordingly, the utilization statistics of the Rework operation seem to have decreased substantially (more on these different kinds of utilizations below).

Looking at the average time in system of the three kinds of exiting parts, it seems clear that the changes at the Sealer and Rework operations are having their effect. All parts have to endure the now-slower Sealer operation, accounting for the increase in the average time in system of shipped parts. Salvaged and scrapped parts, however, enjoy a much faster trip through the Rework operation (maybe making them feel better after failing inspection), with the net effect being that their average time in system seems to decrease quite a lot.

Now we need to discuss a rather fine point about utilizations. For each Resource, Arena reports two utilization statistics, called *Instantaneous Utilization* and *Scheduled Utilization*.

- *Instantaneous Utilization* is calculated by computing the utilization at a particular instant in time (that is, [number of Resource units busy] / [number of Resource units scheduled] at that point in time), and then calculating a time-weighted average of this over the whole run to produce the value shown in the reports. If there are no units of the Resource scheduled at a particular instant in time, the above ratio is simply defined to be zero, and it is counted as a zero in the time-weighted average reported. This can be a useful statistic for tracking utilization over time (for example, a utilization plot). However, it should be used with caution as it can provide confusing results under certain conditions (more on this shortly). If you like math (and even if you don't), we can express all this as follows. Let $B(t)$ be the number of units of the Resource that are busy at time t, and let $M(t)$ be the number of units of that Resource that are scheduled (busy or not) at time t. Let $U(t) = B(t)/M(t)$ whenever $M(t) > 0$, and simply define $U(t)$ to be 0 if $M(t) = 0$. If the run is from time 0 to time T, then the reported Instantaneous Utilization is

$$\int_0^T U(t)\,dt/T,$$

which is just the time average of the $U(t)$ function as defined above. It's important to note that in reporting this, Arena accounts for time periods when there's no capacity scheduled as time periods with a utilization of zero.

- *Scheduled Utilization* is the time-average number of units of the Resource that are busy (taken over the whole run), divided by the time-average number of units of the Resource that are scheduled (over the whole run). In the above notation, the reported Scheduled Utilization is

$$\frac{\int_0^T B(t)dt \,/\, T}{\int_0^T M(t)dt \,/\, T} = \frac{\int_0^T B(t)dt}{\int_0^T M(t)dt}.$$

    There's no problem with dividing by zero here as long as the Resource was ever scheduled at all to have units available (the only situation that makes sense if the Resource is even present in the model).

Under different conditions, each of these reported utilizations can provide useful information. If the capacity of a Resource stays constant over the entire simulation run (the Resource is never in an Inactive or Failed state), the two reported utilizations will be the same (you're asked to prove this in Exercise 4-19, and you can verify from Table 4-1 that this is so in Model 4-2 for the Resources that have fixed capacity). If a Schedule is attached to the Resource and the scheduled capacities are either zero or a single positive constant (for example, 1), the reported Instantaneous Utilization tells you how busy the Resource was over the entire run (counting zero-capacity periods as zero-utilization periods); a plot of $U(t)$ as defined above would tell you how closely your Schedule tracks what's being asked of the Resource, so it could be useful to investigate staffing plans. The reported Scheduled Utilization tells you how busy the Resource was over the time that it was available at all (in other words, capacity was non-zero).

For example, consider the case where a Resource is scheduled to be available 2/3 of the time with a capacity of 1, and unavailable the other 1/3 of the time (capacity of 0). Suppose further that during the time the Resource was available, it was busy half the time. For this example, the reported Instantaneous Utilization would be 0.3333 and the reported Scheduled Utilization would be 0.5000. This tells you that over the entire run the Resource was utilized 1/3 of the time, but during the time that it was scheduled to be available, it was utilized 1/2 the time. In this case, the reported Instantaneous Utilization is less than or equal to the reported Scheduled Utilization.

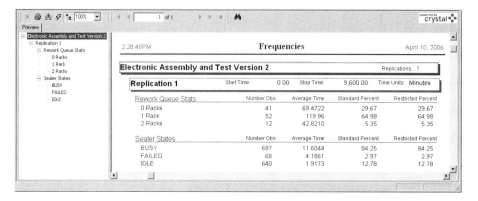

Figure 4-8. The Arena Frequencies Report: Model 4-2

We need to modify the above with an even finer point. If you select a Schedule Rule other than Wait, it is possible to get a reported Scheduled Utilization greater than 1. For example, if you select the Ignore option and the Resource is allocated when it is scheduled to become unavailable (capacity of 0), the Resource capacity is decreased to zero, but the Resource stays allocated until it is released. Say you have a ten-hour run time and the Resource is scheduled to be available for only the first five hours. Let's assume that the Resource was allocated at time 0 to an activity with duration of 6 hours. At time 5, the capacity of the Resource would be reduced to 0, but the Resource would remain allocated until time 6. The reported Instantaneous Utilization would be 0.6 (it was utilized 6 of the 10 hours), while the reported Scheduled Utilization would be 1.2 (it was utilized 6 hours, but only scheduled to be available 5 hours).

A further (and, we promise, final) complication to the above arises when a Schedule is attached to the Resource and the scheduled capacities vary among different positive values over time (e.g., 1, 7, 4, etc.) rather than varying only between 0 and a single positive constant. If you have this situation, we strongly recommend that you not use Instantaneous Utilization even though it will still be reported. Depending on the capacities, the time durations, and the usage, this utilization can be less than, equal to, or greater than the Scheduled Utilization (you're asked to demonstrate this in Exercise 4-20). In this case, we suggest that you instead take advantage of the Frequencies statistic, which will give you detailed and precise information on how the Resource was utilized; see the Help topic "Resource Statistics: Instantaneous Utilization vs. Scheduled Utilization" for more on this. So, for example, in Model 4-2 for the Rework Resource, we'd suggest that you use the Scheduled Utilization of 0.8567 in Table 4-1 rather than the Instantaneous Utilization of 0.8641.

The new frequency statistics are not part of the normal Category Overview report. You must click on the Frequencies report in the Reports panel of the Project Bar. The results are given in Figure 4-8.

The first section shows the statistics we requested to determine the number of racks required at the rework process. In this particular run, there were never more than 20 in the rework queue (that is, there are no data listed for three or four racks), and there were more than ten only 5.35% of the time. This might imply that you could get by with only one rack, or at most, two. The Sealer States statistics give the percentage of times the Sealer spent in the Busy, Failed, and Idle states.

One last note is worth mentioning about frequency statistics. For our results, the last two columns, Standard and Restricted Percent, have the same values. It is possible to exclude selective data resulting in differences between these columns. For example, if you exclude the Failed state for the sealer Frequency, the Standard Percent would remain the same, but the Restricted Percent column would have values only for Busy and Idle, which would sum to 100%.

4.3 Model 4-3: Enhancing the Animation

So far in this chapter we've simply accepted the default animation provided with the modules we used. Although this base animation is often sufficient for determining whether your model is working correctly, you might want the animation to look more like the real system before allowing decision makers to view the model. Making the animation more realistic is really very easy and seldom requires a lot of time. To a large extent, the amount of time you decide to invest depends on the level of detail you decide to add and the nature of the audience for your efforts. A general observation is that, for presentation purposes, the higher you go in an organization, the more time you should probably spend on the animation. You will also find that making the animation beautiful can be a lot of fun and can even become an obsession. So with that thought in mind, let's explore some of what can be done.

We'll modify Model 4-2 into what we'll call Model 4-3, and we'll start by looking at the existing animation. The current animation has three components: entities, queues, and variables. The entities we selected using the Entity data module can be seen when they travel from one module to another or when they're waiting in a queue. For each Process module we placed, Arena automatically added an animation queue, which displays waiting entities during the run. Variables were also placed automatically by Arena to represent the number of entities resident in or that have exited a module.

As suggested by how they came to exist in the model—added when you placed the module—the animation constructs are "attached" to the module in two respects. First, their names, or *Identifiers*, come from values in the module dialog box; you can't change them directly in the animation construct's dialog box. Second, if you move the module, the animation objects that were automatically supplied move with it; however, if you want the animation to stay where it is, just hold the *Shift* key when you move the module. On the second point, if for some reason you replace the animation objects (e.g., queues, counters) that automatically came with the module, they do not move with the module.

Sometimes it's helpful to "pull apart" the animation to a completely different area in the model window, away from the logic. If you do so, you might consider setting up some Named Views (see Section 3.4.13) to facilitate going back and forth. If you want to disconnect an animation construct completely from the module it originally accompanied, *Cut* it to the *Clipboard* and *Paste* it back into the model. It will retain all of its characteristics, but no longer will have any association with the module. An alternative method is to delete the animation construct that came with the module and add it back from scratch using the constructs from the Animate toolbar.

You might even want to leave some of the automatically placed animation constructs with the modules and just make copies of them for the separate animation. If you decide to do this, there are some basic rules to follow. Any animation construct that provides information (e.g., variables and plots) can be duplicated. Animation constructs that show an activity of an entity (e.g., queues and resources) should not be duplicated in an animation. The reason is quite simple. If you have two animated queues with the same name, Arena would not know which animated queue should show a waiting entity. Although Arena will allow you to duplicate an animated queue, it will generally show all waiting entities in the last animated queue that you placed.

We'll use the "pull apart" method to create our enhanced animation. Let's start by using the zoom-out feature, *View > Zoom Out* (![] or the – key), to reduce the size of our model. Now click on a queue and use *Edit > Cut* (✂ or *Ctrl+X*) to cut or remove it from the current model and then *Edit > Paste* (📋 or *Ctrl+V*) to place the queue in the general area where you want to build your animation (click to place the floating rectangle). Repeat this action for the remaining queues, placing them in the same general pattern as in the original model. You can now zoom in on the new area and start building the enhanced animation. We'll start by changing our queues. Then we'll create new pictures for our entities and add resource pictures. Finally, we'll add some plots and variables.

4.3.1 Changing Animation Queues

If you watched the animation closely, you might have noticed that there were never more than about 14 entities visible in any of the queues, even though our variables indicated otherwise. This is because Arena restricts the number of animated entities displayed in any queue to the number that will fit in the drawn space for the animation queue. The simulation may have 30 entities in the queue, but if only 14 will fit on the animation, only the first 14 will show. Then, as an entity is removed from the queue, the next undisplayed entity in the queue will be shown. While the output statistics reported at the end will be correct, this can be rather deceptive to the novice and may lead you to assume that the system is working fine when, in fact, the queues are quite large. There are three obvious ways (at least to us) to avoid this problem: one is to watch the animation variable for the number in queue, the second is to increase the size of the animation queue, and the third is to decrease the size of the entity picture (if it remains visually accurate).

Let's first increase the size of the queue. Figure 4-9 shows the steps we'll go through as we modify our queue. We first select the queue (View 1) by single-clicking on the Rework queue, `Rework Process.Queue`. Notice that two handles appear, one at each end. You can now place your pointer over the handle at the left end, and it will change to cross hairs. Drag the handle to stretch the queue to any length and direction you want (View 2). If you now run the simulation, you should occasionally see a lot more parts waiting at Rework.

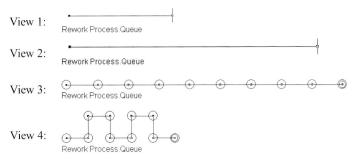

Figure 4-9. Alternate Ways to Display a Queue

We can also change the form of the queue to represent the physical location *point* of each entity in it. Double-click on the selected queue and the Queue dialog box will appear as in Display 4-23.

Select the Point Type of the queue and click the *Points* button. Then add points by successively clicking *Add*. We could change the rotation of the entity at each point, but for now we'll accept the default of these values. When you accept these changes, the resulting queue should look something like the one shown in View 3 of Figure 4-9. Note that the front of the queue is denoted by a point surrounded by two circles. You can then drag any of these points into any formation you like (View 4). If you want all these points to line up neatly, you may want to use the Snap option discussed in Chapter 3. Arena will now place entities on the points during an animation run and move them forward, much like a real-life waiting line operates.

For our animation, we've simply stretched the queues (as shown in View 2 of Figure 4-9) for the Prep A, Prep B, and Sealer areas. We did get a little fancy with the Rework queue. We changed the form of the queue to points and added 38 points. This allowed us to align them in four rows of ten to represent four available racks as shown in Figure 4-10 (this was much easier to do with the Snap option on).

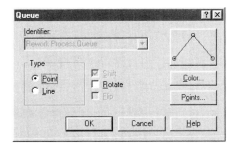

Type		
Point		*select*

Display 4-23. The Queue Dialog Box

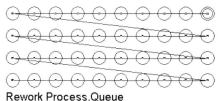

Rework Process.Queue

Figure 4-10. The Rework Queue with 40 Points

4.3.2 Changing Entity Pictures

Now let's focus our attention on our animation entities. In our current animation, we arbitrarily selected blue and red balls for the two kinds of entities. Let's say we want our entities to be similar to the balls but have the letter "A" or "B" displayed inside the ball. You create new pictures in the Entity Picture Placement window, as shown in Figure 4-11, which is opened using *Edit > Entity Pictures*. The left side of this window contains entity pictures currently available in your model, displayed as a list of buttons with pictures and associated names. The right side of this window is used for accessing *picture libraries*, which are simply collections of pictures stored in a file. Arena provides several of these libraries with a starting selection of icons; you might want to open and examine them before you animate your next model (their file names end with *.plb*).

There are several ways to add a new picture to your animation. You can use the *Add* button (on the left) to draw a new picture for the current list, or you can use the *Copy* button (on the left) to copy a picture already on the current list. If you use the Add function, your new entry will not have a picture or a name associated with it until you draw it and give it a name in the Value field above the list. If you use the Copy function, the new picture and name will be the same as the picture selected when you copied.

To add a picture from a library to your current entity picture list, highlight the picture you want to replace on the left, highlight the new selection from a library on the right, and click on the left arrow button ($\leq<$) to copy the picture to your picture list. You can also build and maintain your own picture libraries by choosing the *New* button, creating your own pictures, and saving the file for future use. Or you can use clip art by using the standard copy and paste commands.

For this example, as for most of our examples, we'll keep our pictures fairly simple, but you can make your entity and resource pictures as fancy as you want. Since the blue and red balls were about the right size, let's use them as our starting point. Click on the `Picture.Blue Ball` icon in the current list on the left and then click *Copy*. Now select one of these two identical pictures and change the name (in the Value dialog box) to `Picture.Part A` (now wasn't that obvious?). Note that as you type in the new name it will also change on the selected icon. To change the picture, double-click on the picture icon. This opens the Picture Editor window that will allow you to modify the picture drawing. Before you change this picture, notice the small gray circle in the center of the square; this is the *entity reference point*, which determines the entity's relation with the other animation objects. Basically, this point will follow the paths when the entity is moving, will reside on the seize point when the entity has control of a resource, and so on.

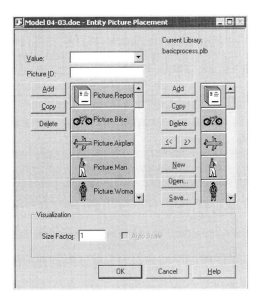

Figure 4-11. The Entity Picture Placement Window

We'll change this picture by inserting the letter "A" in the center of the ball and changing to a lighter fill color so the letter will be visible. When you close this window, the new drawing will be displayed beside the Picture.Part A name. Now repeat the same procedure to make a new picture for Part B. Your final pictures should look something like Figure 4-12.

If you click on one of the pictures, the full name will be displayed in the Value field at the top. Also note that there is a Size Factor field in the lower left-hand portion of the window (see Figure 4-11). You can increase or decrease your entity picture size by changing this value. For our animation, we increased the Size Factor from 1 to 1.3.

The final step is to assign these new pictures to our parts so they will show up in the animation. You do this by clicking on the Entity data module and entering the new names in the Initial Picture cell for our two parts. You might note that your new names will not appear on the drop-down list, so you will need to type them in. However, once you have entered the new names and accepted the data, they will be reflected on the list.

Figure 4-12. The Final Entity Pictures

4.3.3   Adding Resource Pictures

Now that we've completed our animated queues and entities, let's add resource pictures to our animation. You add a resource picture by clicking the *Resource* button (⬛) found in the Animate toolbar. This will open the Resource Picture Placement window, which looks very similar to the Entity Picture Placement window. There's very little difference between an entity picture and a resource picture other than the way we refer to them. Entities acquire pictures by assigning a *picture name* somewhere in the model. Resources acquire pictures depending on their state. In Section 4.2.1, we discussed the four automatic resource states (Idle, Busy, Failed, and Inactive). When you open a Resource Picture Placement window, you might notice that there is a default picture for each of the four default states. You can, however, change the drawings used to depict the resource in its various states, just like we changed our entity pictures.

First we need to identify which resource picture we are creating. You do this by using the drop-down list in the Identifier box to select one of our resources (e.g., Prep A). Now let's replace these pictures as we did for the entity pictures. Double-click on the Idle picture to open the Picture Editor window. Use the background color for the fill, make the line thickness 3 points (from the Draw toolbar), and change the line color. Note that the box must be highlighted in order to make these changes. The small circle with the cross is the *reference point* for the resource, indicating how other objects (like entity pictures) align to its picture; drag this into the middle of your box. Accept this icon (by closing the Picture Editor window) and return to the Resource Picture Placement window. Now let's develop our own picture library. Choosing the *New* button from the Resource Picture Placement window opens a new, empty library file. Now select your newly created icon, click *Add* under the current library area, and then click the right arrow button. Click the *Save* button to name and save your new library file (e.g., Book.plb).

We'll now use this picture to create the rest of our resource pictures. Highlight the Busy picture on the left and the new library picture on the right and use the left arrow button to make your busy picture look just like your idle picture. When the animation is running, the entity picture will sit in the center of this box, so we'll know it's busy. However, you do need to check the Seize Area toggle at the bottom of the window for this to happen. Now copy the same library picture to the inactive and failed states. Open each of these pictures and fill the box with a color that will denote whether the resource is in the failed or inactive state (e.g., red for failed and gray for inactive). Now copy these two new pictures to your library and save it. Your final resource pictures should look something like those shown in Figure 4-13. We also increased our Size Factor to 1.3 so the resource size will be consistent with our entity size.

When you accept the resource pictures and return to the main model window, your pointer will be a cross hair. Position this pointer in the approximate area where you want to place the resource and click. This places the new resource on your animation. (By the way, have you remembered to save your model recently?) Your new resource icon may be larger than you want, so adjust it appropriately by dragging one of its corner handles. The resource picture also contains an object that appears as a double circle with a dashed line

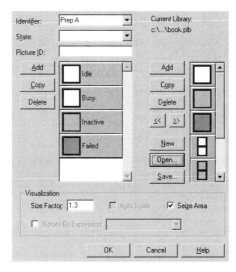

Figure 4-13. The Resource Pictures

connected to the lower left portion of the resource picture—the *Seize Area*. This Seize Area is where your entity will sit when it has control of the resource; drag it to the center of your resource picture if necessary. Now run your animation to see if your entity and resource pictures are the sizes you want. If not, you can adjust the position of the seize area by pausing the run, displaying seize areas (using the *View > Layers* menu option), and dragging it to the desired location. After you've fine-tuned its position, you'll probably want to turn off the display of the seize area layer before ending the run.

Once you're satisfied with your animation of the Prep A resource, you'll want to add animations for the rest of the resources. You could repeat the above process for each resource, or you can copy and paste the Prep A resource for the Prep B and Sealer pictures. Once you've done this, you'll need to double-click on the newly pasted resource to open the Resource Picture Placement window and select the proper name from the drop-down list in the Identifier field.

The picture of the rework resource will have to be modified because it has a capacity of 2 during the second shift. Let's start by doing a copy/paste as before, but when we open the Resource Picture Placement window, edit the Idle picture and add another square (*Edit > Duplicate* or *Edit > Copy* followed by *Edit > Paste*) beside or under the first picture. This will allow room for two entities to reside during the second shift. Copy this new picture to your library and use it to create the remaining rework pictures. Rename the resource (Rework) and close the window.

The original resource animation had only one seize area, so double-click on the seize area, click the *Points* button, and add a second seize area. Seize areas are much like point queues and can have any number of points, although the number of points used depends on the resource capacity. Like a queue, seize areas can be shown as a line. Close this window and position the two seize-area points inside the two boxes representing the resource.

You should now have an animation that's starting to look more like your perceived system. You may want to reposition the resources, queues, stations, and so forth, until you're happy with the way the animation looks. If you've been building this model yourself and based it on Model 4-2, you'll need to clear (uncheck) *Run > Run Control > Batch Run (No Animation)* in order to enjoy the fruits of your artistic labors to create Model 4-3. You also could add text to label things, place lines or boxes to indicate queues or walls, or even add a few potted plants.

4.3.4 Adding Variables and Plots

The last thing we'll do is add some additional variables and a plot to our animation. Variables of interest are the number of parts at each process (in service plus in queue) and the number of parts completed (via each of the three exit possibilities). We could just copy and paste the variables that came with the flowchart modules. First copy and paste the four variables that came with our process modules; put them right under the resource picture that we just created. Then resize them by highlighting the variable and dragging one of the handles to make the variable image bigger or smaller. You can also reformat, change font, change color, etc., by double-clicking on the variable to open the Variable window shown in Figure 4-14. We then repeated this process for the three variables that came with our Dispose modules. Finally, we used the Text tool from the Animate toolbar to label these variables.

Now let's add a plot for the number in the rework queue. Clicking the *Plot* button ( ) from the Animate toolbar opens the Plot window. Use the *Add* button to enter the expression `NQ(Rework Process.Queue)`. Recall that we used this same expression when we created our Frequencies data. We also made a number of other entries as shown in Display 4-24.

Figure 4-14. The Variable Window

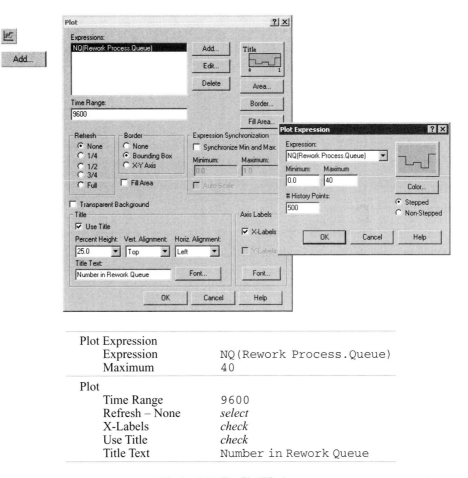

Plot Expression		
	Expression	NQ(Rework Process.Queue)
	Maximum	40
Plot		
	Time Range	9600
	Refresh – None	*select*
	X-Labels	*check*
	Use Title	*check*
	Title Text	Number in Rework Queue

Display 4-24. The Plot Window

After you've accepted these data, you may want to increase the size of the plot in the model window and move it somewhere else on the animation. Finally, we used the Text tool from the Animate toolbar to add the *y*-axis limits for our plot.

Your animation should now be complete and look something like the snapshot shown in Figure 4-15, which was taken at simulation time 5453.0640 minutes. The final numerical results are, of course, the same as those from Model 4-2 since we changed only the animation aspects to create the present Model 4-3.

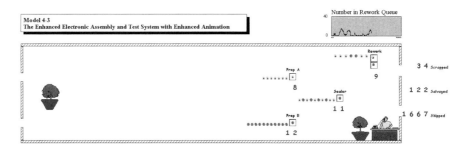

Figure 4-15. The Animation at Time 5469.0640 Minutes: Model 4-3

4.4   Model 4-4: The Electronic Assembly and Test System with Part Transfers

So far in this chapter, we've developed successive models of the electronic assembly and test system with the assumption that all part transfers between operations occurred instantaneously. Let's generalize that assumption and now model the system with all part transfers taking two minutes, regardless of where they're coming from or going to. This includes the transfer of arriving parts to the prep areas and transfer of the departing parts from either the Sealer or Rework station to be scrapped, salvaged, or shipped. We'll modify Model 4-3 to create Model 4-4.

4.4.1   Some New Arena Concepts: Stations and Transfers

In order to model the two-minute transfer times and to show the part movement, we need to understand two new Arena concepts: *Stations* and *Station Transfers*. Arena approaches the modeling of physical systems by identifying locations called *Stations*. Stations may be thought of as a place at which some process occurs. In our example, stations will represent the locations for the part arrivals, the four manufacturing cells, and part departures. Each station is assigned a unique name. In the model, stations also appear as entry points to sections of model logic, working in conjunction with our other new topic, *Station Transfers*.

Station Transfers allow us to send an entity from one station to another without a direct connection. Arena provides several different types of station transfers that allow for positive transfer times, constrained movement using material-handling devices, and flexible routings that depend on the entity type. The station transfer we'll use here is called a *Route*, which allows the movement of entities from one station to another. Routes assume that time may be required for the movement between stations, but that no additional delay is incurred because of other constraints, such as blocked passageways or unavailable material-handling equipment. The route time can be expressed as a constant, a sample from a distribution, or, for that matter, any valid expression.

We often think of stations as representing a physical location in a system; however, there's no strict requirement that this be so, and in fact, stations can be used effectively to serve many other modeling objectives. Stepping back for a moment from their intended use in representing a system, let's examine what happens in Arena when an entity is transferred (e.g., routed) to a station. First, we'll look at the model logic—moving entities from module to module during the run. Underneath the hood, as we discovered (in painstaking detail) in Chapter 2, a simulation run is driven by the entities—creating them, moving them through logic, placing them on the event calendar when a time delay is to be incurred, and eventually destroying them. From this perspective, a station transfer (route) is simply another means of incurring a time delay.

When an entity leaves a module that specifies a Route as the transfer mechanism, Arena places the entity on the event calendar with an event time dictated by the route duration (analogous to a Delay module). Later, when it's the entity's turn to be removed from the event calendar, Arena returns the entity to the flow of model logic by finding the module that defines its destination station, typically a Station module. For example, let's assume a Route module is used to send an entity to a station named `Sealer`. When the entity comes off the event calendar, Arena finds the module that defines the `Sealer` station, and the entity is directed or sent to that module. This is in slight contrast to the direct module connections we've seen so far, where the transfer of an entity from module to module occurred without placing the entity on the event calendar and was represented graphically in the model by a Connection line between two modules. While the direct connections provide a flowchart-like look to a model, making it obvious how entities will move between modules, station transfers provide a great deal of power and flexibility in dispatching entities through a model, as we'll see when we cover Sequences in Section 7.1.

Stations and station transfers also provide the driving force behind an important part of the model's animation—displaying the movement of entities among stations as the model run progresses. The stations themselves are represented in the flowchart view of the model using station marker symbols. These stations establish locations on the model's drawing where station transfers can be initiated or terminated. The movement of entities between the stations is defined by route path objects, which visually connect the stations to each other for the animation and establish the path of movement for entities that are routed between the stations.

You'll soon see that the station markers define either a destination station (for the ending station of a route path) or an origin station that allows transfer out of the module via a station transfer (for the beginning of a route path). You add animation stations via the Station object from the Animate Transfer toolbar. You add route paths by using the Route object from the Animate Transfer toolbar and drawing a polyline that establishes the graphical path that entities follow during their routes. When the simulation is running, you'll see entity pictures moving smoothly along these route paths. This begs the question: How does this relate to the underlying logic where we just learned that an entity resides on the event calendar during its route time delay? The answer is that Arena's animation "engine" coordinates with the underlying logic "engine," in this case, the event calendar. When an entity encounters a route in the model logic and is placed on the event calendar, the animation shows the entity's picture moving between the stations on the route path. The two engines coordinate so that the timing of the entity finishing its animation movement on the route and being removed from the event calendar to continue through model logic coincide, resulting in an animation display that is representative of the model logic at any point in time.

Figure 4-16. The Part Arrival Logic Modules

4.4.2   Adding the Route Logic

Since the addition of stations and transfers affects not only the model, but also the animation, let's start with our last model (Model 4-3). Open this model and use *File > Save As* to save it as Model 04-04.doe. Let's start with the part arrivals. Delete the connections between the two Assign modules and the Part A Prep and Part B Prep Process modules. Now move the two Create/Assign module pairs to the left to allow room for the additional modules that we'll add for our route logic. To give you a preview of where we're headed here, our final modification to this section of our model, with the new modules, is shown in Figure 4-16.

The existing Create and Assign modules for the Part A and Part B arrivals remain the same as in the original model. In order to add our stations and transfers, we first need to define the station where the entity currently resides and then route the entity to its destination station. Let's start with Part A. We place a Station module (from the Advanced Transfer panel, which you may need to attach to your model) to define the location of Part A, as seen in Display 4-25. We've entered a Name (Part A Arrival Station), defaulted the Station Type to Station, and defined our first station (Part A Station). This module defines the new station (or location) and assigns that location to the arriving part. Note that the Name box in the Station module simply provides the text to appear in the model window and does not define the name of the station itself. The Station Name box defines the actual station name to be used in the model logic—Part A Station.

We'll next add the Route module (from the Advanced Transfer panel), which will send the arriving part to the Prep A area with a transfer time of two minutes. We'll provide a module name, enter a Route Time of 2, in units of Minutes, default the Destination Type (Station), and enter the destination Station Name as Prep A Station, as shown in Display 4-26. This will result in any Part A part arrival being sent to the yet-to-be-established Prep A Station (though we just defined its name here).

We've added the same two modules to the Part B arrival stream. The data entries are essentially the same, except that we use Part B in place of all Part A occurrences (modelers who are both lazy and smart would duplicate these two modules, make the few required edits, and connect as needed). You might note that there are no direct connects exiting from the two Route modules. As discussed earlier, Arena will take care of getting the entities to their correct stations.

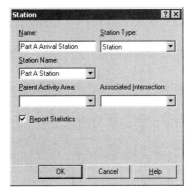

Name	Part A Arrival Station
Station Name	Part A Station

Display 4-25. The Station Module

Now let's move on to the prep areas and make the necessary modifications, the result of which is depicted in Figure 4-17 (almost...we'll make one slight modification here before we're done). The two modules preceding the prep areas' Process modules are Station modules that define the new station locations, `Prep A Station` and `Prep B Station`. The data entries are basically the same as for our previous Station modules (Display 4-25), except for the module Name and the Station Name. The parts that are transferred from the part arrival section using the Route modules will arrive at one of these stations. The two Process modules remain the same. We then added a single Route module, with incoming connectors from both Prep areas. Although the two prep areas are at different locations, the parts are being transferred to the same station, `Sealer`

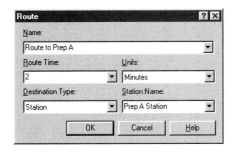

Name	Route to Prep A
Route Time	2
Units	Minutes
Station Name	Prep A Station

Display 4-26. The Route Module

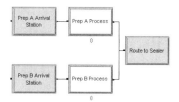

Figure 4-17. The Prep Area Logic Modules (Almost...)

`Station`. Arena will keep track of where the parts originated, the Prep A or Prep B station. Again, the only change from the previously placed Route modules are the module Name, `Route to Sealer`, and the destination Station Name, `Sealer Station`.

By now you can see what is left to complete our model changes. We simply need to break out our different locations and add Station and Route modules where required. The model logic for the Sealer and Rework areas is shown in Figure 4-18. We've used the same notation as before with our new stations named `Sealer Station`, `Rework Station`, `Scrapped Station`, `Salvaged Station`, and `Shipped Station`.

The model logic for our Scrapped, Salvaged, and Shipped areas is shown in Figure 4-19.

At this point, your model should be ready to run (correctly), although you probably will not notice much difference in the animation, except maybe that the queues for Part B Prep and Rework seem to get a lot longer (so we made more room in these queue animations and rescaled the *y*-axis on the Rework Queue plot). If you look at your results, you might observe that the part cycle times are longer than in Models 4-3 and 4-2, which is caused by the added transfer times; however, if you were to make multiple replications of each of these models (see Section 2.6.2), you'd notice quite a lot of variability in these output performance measures across replications, so concluding much from single replications of this model vs. Model 4-3 (or, equivalently, Model 4-2) is risky business.

Maybe you're wondering about a simpler approach to modeling non-zero transfer times—instead of all these Station and Route modules all over the place, why not just

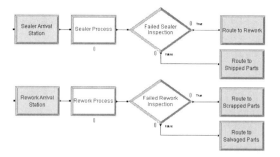

Figure 4-18. The Sealer and Rework Area Logic Modules

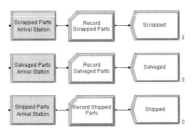

Figure 4-19. The Scrapped, Salvaged, and Shipped Logic Modules

stick in a Process module interrupting each arc of the flowchart where the transfer time occurs with Action = Delay and a Delay Type that's Constant at 2 Minutes? Actually, this *would* work and would produce valid numerical output results. However, it would not allow us to animate the entities moving around, which we'd like to do (for one thing, it's fun to watch...briefly). So we'll now add the route transfers to our animation.

4.4.3   Altering the Animation

In the process of modifying the animation, we moved a lot of the existing objects. If you want your animation to look like ours, you might want to first take a look ahead at Figure 4-20. (Of course, you could always just open `Model 04-04.doe`.)

We'll now show you how to add stations and routes to your animation. Let's start by adding the first few animation station markers. Click the *Station* button (▭) found in the Animate Transfer toolbar to open the Station dialog box shown in Display 4-27 (note the little station-marker symbol out to its left). If you don't have the Animate Transfer toolbar open, you can use *View > Toolbars* (or right-click in any toolbar area) to choose to display it. Now use the drop-down list to select from the already-defined stations our first station, `Part A Station`. When you close the dialog box, your pointer will have changed to cross hairs with a station-marker symbol attached. Position this new pointer to the left of the Prep A queue and click to add the station marker. Alternatively, you don't need to close the dialog box, but just move the pointer (which will be cross hairs with a station-marker symbol attached) to where you want the station marker to be and click (the dialog box will close itself).

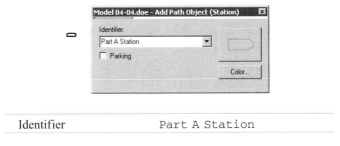

Identifier	Part A Station

Display 4-27. The Station Dialog Box

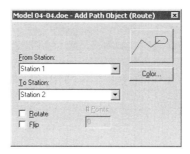

Display 4-28. The Route Dialog Box

Repeat the same set of steps to place the `Prep A Station` marker just above the Prep A queue. Now that we have our first two stations placed, let's add the route path. Click the *Route* button (🖼) in the Animate Transfer toolbar. This will open the Route dialog box shown in Display 4-28. For our animation, we simply accepted the default values, even ignoring the default (and wrong) "From" and "To" station names. After you close the Route dialog box (or even if you don't close it), the pointer changes to cross hairs. Place the cross hairs inside a station marker (`Part A Station`) at the start of a path and click; this will start the route path. Move the pointer and build the remainder of the path by clicking where you want "corners," much like drawing a polyline. The route path will automatically end when you click inside the end station marker (`Prep A Station`).

The above two-step process (first place the Station markers, and then place the Route animations) could be replaced by a one-step process involving only the Route animation tool. Try clicking it on the toolbar, then click the crosshairs in the flowchart view of your model, make a few more clicks for corners, and finally double-click. You'll notice that you got a station marker "for free" at each end of the Route animation. Then double-click on the Station marker at the beginning, use the pull-down list to associate it to the correct logical Station (assumed to be already in your model...if not you could name it here and forget the pull-down list); do the same for the ending Station marker.

If this is your first animated route, you might want to go ahead and run your model so you can see the arriving parts move from the `Part A Station` to the `Prep A Station`. If you're not happy with the placement of the routes, they can be easily changed. For example, if you click on the station marker above the Prep A.Process queue (or, if you have the options checked under *View > Data Tips*, merely hover your pointer over the station marker), the name of the station (`Prep A Station`) will appear below the marker. You can drag the station marker anywhere you want (well, almost anywhere, as long as you stay in the model window). Note that if you move the station marker, the routes stay attached. Once you've satisfied your curiosity, place the stations and route path for the Part B arrival and transfer to the Prep B station.

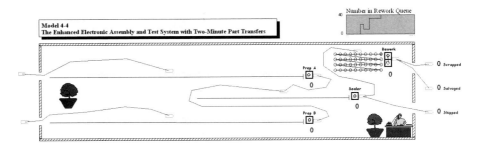

Figure 4-20. The Completed Animation for Model 4-4

If you add the remaining stations and routes, your animation will be complete. However, from a visual point of view you may not have an animation that accurately captures the flow of parts. Recall that we placed our prep stations just above the prep queues, which are to the left of the prep areas. Let's assume that in the physical facility the parts exit from the right side of the area. If you had used the existing prep stations, the parts would have started their routes from the left side of the areas. The solution to this problem is simple—just add a second station marker at the prep areas (placed on the right side) with the same Identifier for the logical station. When Arena needs to route an entity, it will look for the route path, not just the station. So as you build the remaining routes, you'll need to have two station markers (with the same name) at each of the prep areas, the sealer area, and the rework area. Your final animation should look something like Figure 4-20.

Finally, there's one subtle point that you may not have noticed. Run your animation and watch the Part B entities as they travel from arrival to the Part B Prep area. Remember that we're creating four entities at a time, but we see only one moving. Now note what happens as they arrive at the queue—suddenly there are four entities. As it turns out, there are really four entities being transferred, it's just that they're all on top of one another during the transfer—they're all moving at precisely the same speed because the transfer times are a constant for all entities. You can check this out by opening the Route to Prep B Route module and changing Route Time from a constant 2 to EXPO(2), causing the entities' individual travel times to differ. Now when you run the animation, you should see all four entities start out on top of one another and then separate as they move toward the queue. You can exaggerate this by clicking on the route path and making it much longer. Although this will show four entities moving, it may not represent reality since, if you believe the modeling assumptions, this travel time was supposed to be a constant 2 minutes (but your boss will probably never know).

Here's a blatant sleight of hand to keep the model assumptions yet make the animation look right. If we show the B parts arriving, with each part animated as a "big batch" picture, we'll get the desired visual results. There will still be four moving across to the Prep B Station, but only the top one will show. We need only convert the picture back to a single entity when they enter the queue at the prep area. In order to implement this concept, we'll need to create a new "big batch" picture and make two changes to the model.

Figure 4-21. The Batch B Entity Picture

First use *Edit > Entity Pictures* to open the Entity Picture Placement window. Next make a copy of the `Picture.Part B` picture and rename it `Picture.Batch B` in the Value field above it. Edit this new picture's drawing by double-clicking on its button in the list. In the Picture Editor, use Copy and Paste to get four Part B circles as shown in Figure 4-21; move the entity reference point to be near the center of your four-pack. Then close the Picture Edit window and click *OK* to return to the model window.

Now that we have a batch picture, we need to tell Arena when it should be used. Open the `Assign Part B Sealer and Arrive Time` Assign module and Add another assignment (in this case, it doesn't matter where in the list of assignments it's added since there's no dependency in these assignments of one on another). This new assignment will be of Type `Entity Picture`, with the Entity Picture being `Picture.Batch B` (which you'll need to type in here for the first time). This will cause Arena to show the batch picture when it routes the Part B arrivals to the Prep B station. Now we need to convert the picture back to a single entity when it arrives at the Prep B area. You'll need to insert a new Assign module (`Assign Picture`) between the Prep B station, `Prep B Arrival Station`, and the `Prep B Process` Process module. This assignment will have `Entity Picture` Type, with the Entity Picture being `Picture.Part B` (which will be on the drop-down list for Entity Picture). Thus, when the entities enter the queue or service at Prep B Process, they will be displayed as individual parts.

Now if you run the simulation (with the animation turned on), you'll see an arriving batch being routed to the Prep B area and individual parts when the batch enters the queue. Defining different entity pictures allows you to see the different types of parts move through the system. Note the batch of four B parts when they enter the system. Although there are still four of these pictures, each on top of the other, the new Batch B picture gives the illusion of a single batch of four parts entering.

Now that you understand the concept of routes, we'll use them in Chapter 7 in a more complex system.

4.5   Finding and Fixing Errors

If you develop enough models, particularly large ones, sooner or later you'll find that your model either will not run or will run incorrectly. It happens to all of us. If you're building a model and you've committed an error that Arena can detect and that prevents the model from running, Arena will attempt to help you find that error quickly and easily. We suspect that this may already have happened to you, so consider yourself lucky if that's not the case.

Typical errors that prevent a model from running may include: undefined variables, attributes, resources, unconnected modules, duplicate use of module names, misspelling of names (or, more to the point, inconsistent spelling whether it's correct or not), and so on. Although Arena tries to prevent you from committing these types of errors, it can't give you maximum modeling flexibility without the possibility of an occasional error occurring. Arena will find most of these errors when you attempt to check or run your newly created model. To illustrate this capability, let's intentionally insert some errors into the model we just developed, Model 4-4. Unless you're really good, we suggest that you save your model under a different name before you start making these changes.

We'll introduce two errors. First, delete the connector between the first Create module, `Part A Arrive`, and the following Assign module. Next, open the dialog box for this Assign module, `Assign Part A Sealer and Arrive Time`, and change the New Value of the second assignment from `TNOW` to `TNO`. These two errors are typical of the types of errors that new (and sometimes experienced) modelers make. After making these two changes, use the *Run > Check* option or the *Check* button (✔) on the Run Interaction toolbar to check your model. An Arena Errors/Warnings window should open with the message

```
ERROR:
Unconnected exit point
```

This message is telling you that there is a missing connector (the one we just deleted). Now look for the *Find* and *Edit* buttons at the bottom of the window. If you click the *Find* button, Arena will take you to and highlight the offending module. If you click the *Edit* button, Arena will take you directly to the dialog box of that offending module. Thus, Arena attempts to help you find and fix the discovered error. So, click *Find* and add the connector that we deleted.

Having fixed this problem, a new check of the model will expose the second error:

```
ERROR:

A linker error was detected at the following block:

*    3 0$              ASSIGN:
                             Sealer Time=TRIA(1,3,4):
                             Arrive Time=TNO:
                             NEXT(14$);
Undefined symbol : TNO
Possible cause: A variable or other symbol was used
               without first being defined.
Possible cause: A statistic was animated, but statistics collection
               was turned off.
To find the problem, search for the above symbol name using
Edit-Find from the menu.
```

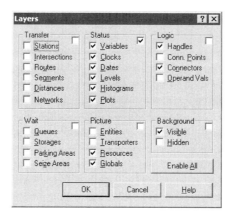

Figure 4-22. The Layers Dialog Box

This message tells you that the symbol TNO (the misspelled Arena variable TNOW) is undefined. Click *Edit* and Arena will take you directly to the dialog box of the Assign module where we planted the error. Go ahead and fix the error. During checking, if you find that you've misused a name or just want to know where a variable is used, use the *Edit > Find* option to locate all occurrences of a string of characters. You might try this option using the string TNOW. If for some reason you forgot what the error was, use *Run > Review Errors* to reopen the error window with the last error message.

There are a number of ways to check the model accuracy or find logic errors. The most obvious is through the use of an animation that shows what the model logic is doing. However, there are times when you must dig a little deeper in order to resolve a problem. Let's first look at the Highlight Active Module option by selecting *Run > Run Control > Highlight Active Module*.

Now run your model. If your modules were not visible at the start of the run, Arena will change the view so you can see modules highlighted as entities pass through. However, on some computers, this happens so fast that you'll typically see only the modules highlighted where the entity stops; e.g., Delay and Dispose modules. If you don't believe us, use the zoom option to zoom in so that only one or two modules fill the entire screen. Now click either the *Go* or *Step* button and you'll see that Arena jumps around and tries to show you only the active modules. Again, it's generally working so fast that not all the modules are highlighted, but it does at least try to show them. When you finally get bored, stop your model and clear the Highlight Active Module option.

You also have some options as to what you see when the model is running. Select *View > Layers* to bring up the dialog box shown in Figure 4-22. You can do this while you're still in Run mode. This dialog box will allow you to turn on (check) or turn off (clear) different items from displaying during the model run.

Now let's assume that you have a problem with your model and you're suspicious of the Process module in the Sealer logic, Sealer Process. It would be nice if the simulation would stop when an entity reaches this module, and there are ways to cause this to happen. You can use the *Run > Run Control > Break on Module* option or the *Module Break* button in the Run Interaction toolbar. You first need to highlight the module or modules on which you want to break. Then select *Run > Run Control > Break on Module* or press the *Module Break* button. This will cause the selected options to be outlined in a red box. Now click the *Go* button. Your model will run until an entity arrives at the selected module—it may take a while! The module will pause when the next entity arrives at the selected module. You can now attempt to find your error (for instance, by repeatedly clicking the *Step* button until you notice something odd) or click *Go*, and the run will continue until the next entity arrives at this module. When you no longer want this break, highlight the module and use *Run > Run Control > Break on Module* to turn off the break.

Now we'll introduce you to the Debug Bar. You can activate this window by selecting the *View > Debug Bar* or the *Run > Run Control > Breakpoints* option. This window has several tabs at the bottom that alter the window contents and allow you to perform different functions.

- The Breakpoints window provides the ability to manipulate the execution to examine operations and logic, evaluate changes in system status, or make presentations to others. Breaks in execution can be established at specific times, when a specific entity number becomes active, on a module, or on a conditional statement.
- The Calendar window allows you to view all future events scheduled on Arena's event calendar for the running simulation. Event times, entity numbers, and event descriptions are displayed in a table format.
- The Active Entity window displays the attribute values of the active entity in a tree-view organization. The value of user-defined and other assignable attributes can be changed via that tree view.
- The Watch windows provide three identical Watch windows that allow you to monitor the values of any expression in the simulation.

Now let's assume that you think you have a problem that occurs at about time 29 and you would like to stop the simulation at that time to see if you can locate the perceived problem. You could slow the model way down and try to hit the *Pause* button at time 29, but chances are fairly good that you will find this rather tedious and hard to do. The easy way is to select the Breakpoints tab in the now-opened Debug window and then press the *New Breakpoint* button in the upper left corner of that window. The Breakpoint Properties dialog will appear, which allows you to set a new breakpoint. We'll accept the defaults except for the Time field where we'll enter a value of 29, as shown in Display 4-29.

Break On	Time
Time	29
Units	Base Time Units

Display 4-29. The Breakpoints Option

When you click *OK*, the window should resemble Figure 4-23.

Now click *Go* and you'll see your simulation run and the stop at time 29. Before we proceed, let's take a peek at the other windows. Select the Calendar tab and you should see the data as displayed in Figure 4-24. This shows that the next event (shown in brackets) is scheduled to occur at time 29.2390. This is the simulation time in minutes. The information to the right shows the time based on how you set your calendar. We didn't use this feature, so it shows the current day and date. You might note that this time is listed as 00:29:14, which appears to be a slightly different time. They are the same because the first time is expressed as fractional units of the base time units (minutes), and the second is expressed as hours:minutes:seconds. It also tells you that it's entity 17 and it will enter an Assign block that is part of the Process module named `Prep A Process` (we'll explain later where this magical Assign block came from). It also shows you the remaining events scheduled to occur, including the last entry, which is the end of the replication at time 9600 minutes.

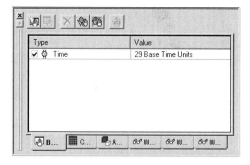

Figure 4-23. The Breakpoints Window

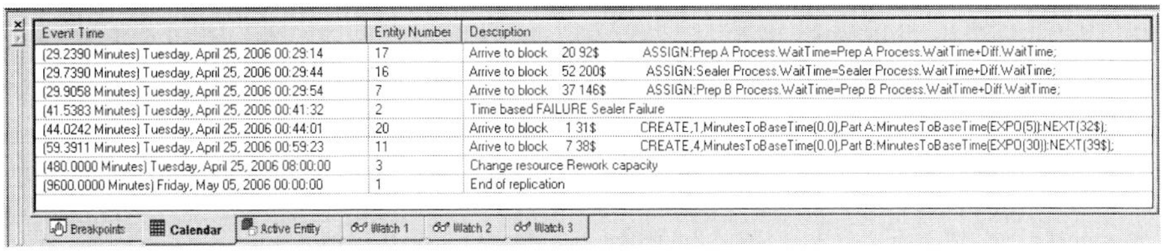

Event Time	Entity Number	Description
(29.2390 Minutes) Tuesday, April 25, 2006 00:29:14	17	Arrive to block 20.92$ ASSIGN:Prep A Process.WaitTime=Prep A Process.WaitTime+Diff.WaitTime;
(29.7390 Minutes) Tuesday, April 25, 2006 00:29:44	16	Arrive to block 52.200$ ASSIGN:Sealer Process.WaitTime=Sealer Process.WaitTime+Diff.WaitTime;
(29.9058 Minutes) Tuesday, April 25, 2006 00:29:54	7	Arrive to block 37.146$ ASSIGN:Prep B Process.WaitTime=Prep B Process.WaitTime+Diff.WaitTime;
(41.5383 Minutes) Tuesday, April 25, 2006 00:41:32	2	Time based FAILURE Sealer Failure
(44.0242 Minutes) Tuesday, April 25, 2006 00:44:01	20	Arrive to block 1.31$ CREATE,1.MinutesToBaseTime(0.0),Part A:MinutesToBaseTime(EXPO(5)):NEXT(32$);
(59.3911 Minutes) Tuesday, April 25, 2006 00:59:23	11	Arrive to block 7.38$ CREATE,4.MinutesToBaseTime(0.0),Part B:MinutesToBaseTime(EXPO(30)):NEXT(39$);
(480.0000 Minutes) Tuesday, April 25, 2006 08:00:00	3	Change resource Rework capacity
(9600.0000 Minutes) Friday, May 05, 2006 00:00:00	1	End of replication

Breakpoints   Calendar   Active Entity   Watch 1   Watch 2   Watch 3

Figure 4-24. The Calendar Window

Just out of curiosity, you might want to press the *Step* button and you'll see the simulation take the next step and stop at time 29.2390, just as predicted. You might also want to check the contents of the other windows, but you should find that they currently do not contain any entries.

Now let's open the Runtime Elements window by selecting the *View > Runtime Elements Bar* option. This window allows you to browse the current property values of simulation elements easily (e.g., resources, queues, transporters, etc.) during a running simulation. As additional elements are defined, they will automatically appear as a new tab under this bar. Our window contains six tabs; Variables, Queues, Resources, Statistics, Active Areas, and Processes. The Variables tab shows a window with all of the variables and their values that were used in this model. In this case, it's not very interesting since we didn't define any user variables and it only contains those defined by Arena.

Now let's examine the contents of the Queues window. After you open this window, make sure that the animation shows in the upper portion of your screen. If it's not there, reposition the display. You should see four queues, one for each of the four Process modules in our model. You might also note that the current number in each queue matches the number shown in the animation. Now press the "+" sign at the left of Prep B Process.Queue. Next expand the Entities in Queue section, then the second entity (Entity 13), and finally the Attributes and User-Defined entries for that entity. The resulting window is shown in Figure 4-25.

This window now shows the details for the second entity in the Prep B queue. You could also get the same information by going to the animation and double-clicking on the second entity in that queue. Most of this information should be self explanatory. Having said that, let's explain the difference between the Entity Number (13) and Serial Number (8). Each of these identifiers is unique in that as new entities are created, they are assigned numbers. Normally they are just incremented by one, although this is not always the case. The serial number refers to entities created by the model, and the entity number refers to entities created by Arena, including entities created by the model. If you refer to Figure 4-24, you'll see that entities 1, 2, and 3 were created by Arena. They represent events that are scheduled by Arena: the end of the replication, a sealer failure, and a change of capacity for the Rework resource. The first entity created by the model was assigned entity number 5 and serial number 1 (not shown in Figure 4-24). If you want to check this out, you can restart your model and use the *Step* button to step through the first couple of events.

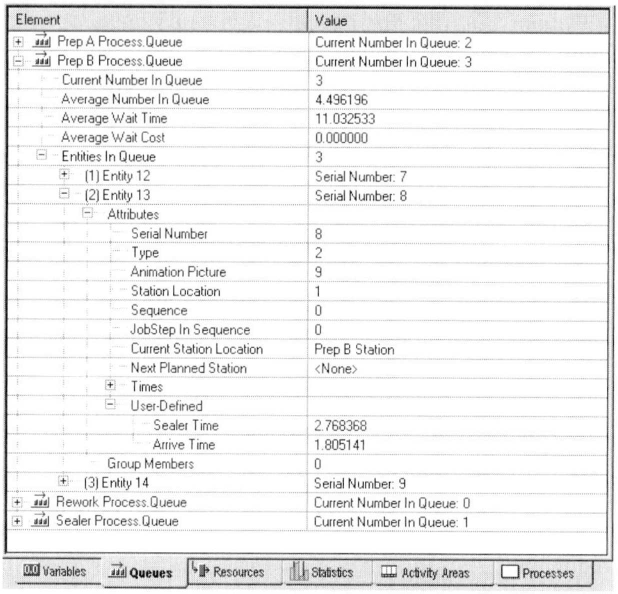

Element	Value
⊞ ▦ Prep A Process.Queue	Current Number In Queue: 2
⊟ ▦ Prep B Process.Queue	Current Number In Queue: 3
Current Number In Queue	3
Average Number In Queue	4.496196
Average Wait Time	11.032533
Average Wait Cost	0.000000
⊟ Entities In Queue	3
⊞ (1) Entity 12	Serial Number: 7
⊟ (2) Entity 13	Serial Number: 8
⊟ Attributes	
Serial Number	8
Type	2
Animation Picture	9
Station Location	1
Sequence	0
JobStep In Sequence	0
Current Station Location	Prep B Station
Next Planned Station	<None>
⊞ Times	
⊟ User-Defined	
Sealer Time	2.768368
Arrive Time	1.805141
Group Members	0
⊞ (3) Entity 14	Serial Number: 9
⊞ ▦ Rework Process.Queue	Current Number In Queue: 0
⊞ ▦ Sealer Process.Queue	Current Number In Queue: 1

| 🔢 Variables | ▦ **Queues** | 🔌 Resources | 📊 Statistics | 🖥 Activity Areas | ☐ Processes |

Figure 4-25. The Queues Window

In either of these views, the values of user-defined and other assignable attributes can be changed, so let's try this feature. First double-click on the value for the Animation Picture, 9, of the entity. Let's change it to 1, which turns out to be an airplane. If you want to know what pictures are available, you'll need to go to the SIMAN Elements view—see Section 7.1.6. If you can see your animation, you'll notice that the airplane is now visible in our Prep B queue. Now double-click on the Sealer Time, 2.768368, and enter a value of 100.

Before we run our model, you might want to look at the other tabs on the Runtime Elements Bar. Most of this information does not require an explanation. You'll find an entry in the Processes tab for each of the Process modules in the model. The Resources tab shows the resources in our model, and the Activity Areas tab shows the model's activity areas. These activity areas were defined by Arena as we constructed our model. Users can also define activity areas, but rather than covering that now, we'll refer you to the Arena online help. Finally, the Statistics tab shows the current values of the statistics being collected by our model.

Let's illustrate some additional features by setting another breakpoint on the sealer queue. Since we changed the sealer process time for entity 13, we should expect to see that queue get rather large once that entity is being processed at the sealer. Open a new breakpoint dialog and select Condition from the drop-down list for the Break On field and enter the expression NQ(Sealer Process.Queue) > 20 in the Condition Expression field. This will cause Arena to stop when the number of entities in the sealer queue exceeds 20.

Now press *Go* and watch the animation. You should see the airplane enter and seize the sealer resource and then see the sealer queue grow. At time 119.7135, the run will pause because of our breakpoint, with 22 entities in the sealer queue. If you check the calendar, you'll see that entity 13 (our airplane) is not scheduled to leave the sealer until time 156.0733, so we would expect the queue to continue to grow for at least the next 35 minutes of our simulation.

Now it's time to introduce you to the Arena command-driven Run Controller. Before we go into the Run Controller, let's define two Arena variables that we'll use: MR and NR. There are many Arena variables that can be used in developing model logic and can be viewed during runtime. The two of interest are:

MR(Resource Name)   –   Resource capacity
NR(Resource Name)   –   Number of busy resource units

Note that these can change during the simulation so their values refer to the status at the moment.

You can think of Arena variables as magic words in that they provide information about the current simulation status. However, they're also reserved words, so you can't redefine their meanings or use them as your own names for anything. We recommend that you take a few minutes to look over the extensive list of Arena variables in the Help system. You might start by looking at the variables summary and then look at the detailed definition if you need more information on a specific variable.

We're now going to use the Run Controller to illustrate just a few of the many commands available. We're not going to explain these commands in any detail; we leave it to you to explore this capability further, and we recommend starting with the Help system.

Open the Run Control Command window (shown in Figure 4-26) by the *Run > Run Control > Command* option or the *Command* button (⬚) on the Run Interaction toolbar.

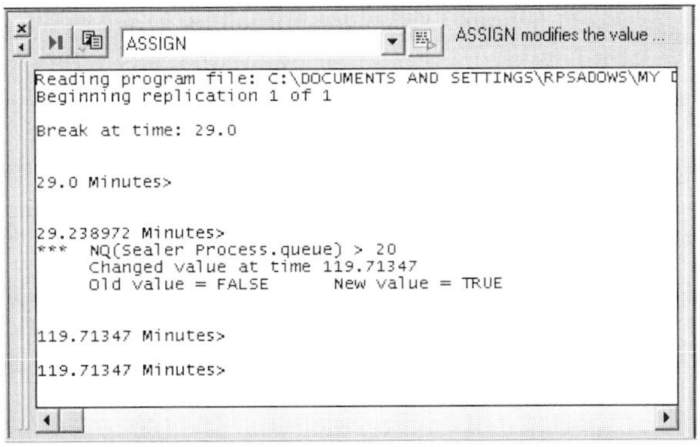

Figure 4-26. The Run Control Command Window

Now let's use the features in the Run Control Command window to increase the capacity of the sealer resource. First we'll check the current values of NR and MR for the Sealer resource. To do this, use the drop-down feature to select the command SHOW (that field currently has the command ASSIGN showing). Then press the *Insert Command* button (🖳) to insert the SHOW command into the window. Now type `MR(Sealer)` and press the *Enter* key. The response will be

MR(Sealer) = 1

This tells you that the Sealer resource currently has a capacity of one. Now repeat the process for NR(Sealer). The response will be

NR(Sealer) = 1

This tells you that one unit of the Sealer resource is currently busy, which should be obvious from the animation since it shows the airplane parked on that resource. Now let's increase the Sealer resource capacity to 3. Use the ASSIGN command to make the following entry

ASSIGN MR(Sealer) = 3

and press the *Enter* key. The Sealer resource now has a capacity of 3. Now press *Go* and watch the animation to see the sealer queue shrink to zero in about 20 minutes of simulation time.

Finally, during a run, you might want to view reports on the status of the model. You can view reports at any time during a run. Simply go to the Project Bar and click on the Reports panel. You can request a report when the model is running or when the model is in Pause mode. Just like the Watch window, you could have an animation running and periodically request a report and never stop the run. Just remember to close the report windows when you're done viewing them.

Even if you don't use the Run Controller, the options available on the Run Interaction toolbar provide the capability to detect model logic errors without your needing to become a SIMAN expert. So that you'll understand how they work, we recommend that you take a simple animated model and practice using these tools. You could wait until you need them, but then not only would you be trying to find an error, but you'd also be trying to learn several new tools!

4.6 Input Analysis: Specifying Model Parameters and Distributions

As you've no doubt noticed, there are a lot of details that you have to specify completely in order to define a working simulation model. Probably you think first of the logical aspects of the model, like what the entities and resources are, how entities enter and maybe leave the model, the resources they need, the paths they follow, and so on. These kinds of activities might be called *structural* modeling since they lay out the fundamental logic of what you want your model to look like and do.

You've also noticed that there are other things in specifying a model that are more numerical or mathematical in nature (and maybe therefore more mundane). For example, in specifying Model 4-2, we declared that interarrival times for Part A were distributed exponentially with a mean of 5 minutes, total processing times for Part B Prep followed a triangular (3, 5, 10) distribution, the "up" times for the sealer were draws from an exponential (120) distribution, and we set up a schedule for the number of rework operators at different times. You also have to make these kinds of specifications, which might be called *quantitative* modeling, and are potentially just as important to the results as are the structural-modeling assumptions.

So where did we get all these numbers and distributions for Model 4-2 (as well as for practically all the other models in this book)? Okay, we admit that we just made them up, after playing around for a while, to get the kinds of results we wanted in order to illustrate various points. We get to do this since we're just writing a book rather than doing any real work, but unfortunately, you won't have this luxury. Rather, you need to observe the real system (if it exists) or use specifications for it (if it doesn't exist), collect data on what corresponds to your input quantitative modeling, and analyze these data to come up with reasonable "models," or representations of how they'll be specified or generated in the simulation. For Model 4-2, this would entail collecting data on actual interarrival times for Part A, processing times for Part B Prep, up times for the sealer, and actual staffing at the rework station (as well as all the other quantitative inputs required to specify the model). You'd then need to take a look at these data and do some kind of analysis on them to specify the corresponding inputs to your model in an accurate, realistic, and valid way.

In this section, we'll describe this process and show you how to use the Arena Input Analyzer (which is a separate application that accompanies and works with Arena) to help you fit probability distributions to data observed on quantities subject to variation.

4.6.1 Deterministic vs. Random Inputs

One fundamental issue in quantitative modeling is whether you're going to model an input quantity as a deterministic (non-random) quantity, or whether you're going to model it as a *random variable* following some probability distribution. Sometimes it's clear that something should be deterministic, like the number of rework operators, though you might want to vary the values from run to run to see what effect they have on performance.

But sometimes it's not so clear, and we can only offer the (obvious) advice that you should do what appears most realistic and valid so far as possible. In Section 4.6.2, we'll talk a little about using your model for what's called *sensitivity analysis* to measure how important a particular input is to your output, which might indicate how much you need to worry about whether it should be modeled as deterministic or random.

You might be tempted to assume away your input's randomness, since this seems simpler and has the advantage that the model's outputs will be non-random. This can be pretty dangerous, though, from the model-validity viewpoint because it's often the randomness itself that leads to important system behavior that you'd certainly want to capture in your simulation model. For instance, in a simple single-server queue, we might assume that all interarrival times are *exactly* 1 minute and that all processing times are *exactly* 59 seconds; both of these figures might agree with *average* values from observed data on arrivals and service. If the model starts empty with the server idle and the first arrival is at time 0, then there will never be a queue since each customer will finish service and leave 1 second before the next customer arrives. However, if the reality is that the interarrival and service times are exponential random variables with respective means of 1 minute and 59 seconds (rather than constant at these values), you get a very different story in terms of queue length (go ahead and build a little Arena model for this, and run it for a long time); in fact, in the long run, it turns out that the average number of customers in the queue is 58.0167, a far cry from 0. Intuitively, what's going on here is that, with the (correct) random model, it sometimes happens that some obnoxious customer has a long service demand, or that several customers arrive at almost the same time; it's precisely these kinds of random occurrences that cause the queue to build up, which never happen in the (incorrect) deterministic model.

4.6.2 Collecting Data

One of the very early steps in planning your simulation project should be to identify what data you'll need to support the model. Finding the data and preparing them for use in your model can be time-consuming, expensive, and often frustrating; and the availability and quality of data can influence the modeling approach you take and the level of detail you capture in the model.

There are many types of data that you might need to collect. Most models require a good bit of information involving time delays: interarrival times, processing times, travel times, operator work schedules, etc. In many cases, you'll also need to estimate probabilities, such as the percentage yield from an operation, the proportions of each type of customer, or the probability that a caller has a touch-tone phone. If you're modeling a system where the physical movement of entities among stations will be represented in the model, the operating parameters and physical layout of the material-handling system will also be needed.

You can go to many sources for data, ranging from electronic databases to interviews of people working in the system to be studied. It seems that when it comes to finding data for a simulation study, it's either "feast or famine," with each presenting its own unique challenges.

If the system you're modeling exists (or is similar to an actual system somewhere else), you may think that your job's easier, since there should be a lot of data available. However, what you're likely to find is that the data you get are not the data you need. For example, it's common to collect processing-time data on machines (that is, a part's time span from arriving at the machine to completing the process), which at first glance might look like a good source for processing times in the simulation model. But, if the observed processing-time data included the queue time or included machine failure times, they might not fit into the model's logic, which explicitly models the queueing logic and the machine failures separately from the machining time.

On the other hand, if you're about to model a brand new system or a significant modification to an existing one, you may find yourself at the other end of the spectrum, with little or no data. In this case, your model is at the mercy of rough approximations from designers, equipment vendors, etc. We'll have a few suggestions (as opposed to solutions) for you in Section 4.6.5.

In either case, as you decide what and how much data to collect, it's important to keep the following helpful hints in mind:

- **Sensitivity analysis:** One often-ignored aspect of performing simulation studies is developing an understanding of what's important and what's not. Sensitivity analysis can be used even very early in a project to assess the impact of changes in data on the model results. If you can't easily obtain good data about some aspect of your system, run the model with a range of values to see if the system's performance changes significantly. If it doesn't, you may not need to invest in collecting data and still can have good confidence in your conclusions. If it does, then you'll either need to find a way to obtain reliable data or your results and recommendations will be coarser.
- **Match model detail with quality of data:** A benefit of developing an early understanding of the quality of your input data is that it can help you to decide how much detail to incorporate in the model logic. Typically, it doesn't make any sense to model carefully the detailed logic of a part of your system for which you have unreliable values of the associated data, unless you think that, at a later time, you'll be able to obtain better data.
- **Cost:** Because it can be expensive to collect and prepare data for use in a model, you may decide to use looser estimates for some data. In making this assessment, sensitivity analysis can be helpful so that you have an idea of the value of the data in affecting your recommendations.
- **Garbage in, garbage out:** Remember that the results and recommendations you present from your simulation study are only as reliable as the model and its inputs. If you can't locate accurate data concerning critical elements of the model, you can't place a great deal of confidence in the accuracy of your conclusions. This doesn't mean that there's no value in performing a simulation study if you can't obtain "good" data. You still can develop tremendous insight into the operation of a complex system, the interactions among its elements, and some level of prediction regarding how it will perform. But take care to articulate the reliability of your predictions based on the quality of the input data.

A final hint that we might offer is that data collection (and some of their analysis) is often identified as the most difficult, costly, time-consuming, and tedious part of a simulation study. This is due in part to various problems you might encounter in collecting and analyzing data, and in part due to the undeniable fact that it's just not as much fun as building the logical model and playing around with it. So, be of good cheer in this activity, and keep reminding yourself that it's an important (if not pleasant or exciting) part of why you're using simulation.

4.6.3 Using Data

If you have historical data (e.g., a record of breakdowns and repair times for a machine), or if you know how part of a system will work (e.g., planned operator schedules), you still face decisions concerning how to incorporate the data into your model. The fundamental choice is whether to use the data directly or whether to fit a probability distribution to the existing data. Which approach you decide to use can be chosen based on both theoretical issues and practical considerations.

From a theoretical standpoint, your collected data represent what's happened in the past, which may or may not be an unbiased prediction of what will happen in the future. If the conditions surrounding the generation of these historical data no longer apply (or if they changed during the time span in which the data were recorded), then the historical data may be biased or may simply be missing some important aspects of the process. For example, if the historical data are from a database of product orders placed over the last 12 months, but four months ago a new product option was introduced, then the order data stored in the preceding eight months are no longer directly useful since they don't contain the new option. The tradeoffs are that if you use the historical data directly, no values other than those recorded can be experienced; but if you sample from a fitted probability distribution, it's possible to get values that aren't possible (e.g., from the tails of unbounded distributions) or to lose important characteristics (e.g., bimodal data, sequential patterns).

Practical considerations can come into play too. You may not have enough historical data to drive a simulation run that's long enough to support your analysis. You may very well need to consider the effect of file access on the speed of your simulation runs as well. Reading a lot of data from a file is typically slower than sampling from a probability distribution, so driving the model with the historical data during a run can slow you down.

Regardless of your choice, Arena supplies built-in tools to take care of the mechanics of using the historical data in your model. If you decide to fit a probability distribution to the data, the Input Analyzer facilitates this process, providing an expression that you can use directly in your model; we'll go into this in Section 4.6.4. If you want to drive the model directly from the historical data, you can bring the values in once to become part of the model's data structure, or you can read the data dynamically during the simulation run, which will be discussed in Section 10.1.

4.6.4   Fitting Input Distributions via the Input Analyzer

If you decide to incorporate your existing data values by fitting a probability distribution to them, you can either select the distribution yourself and use the Input Analyzer to provide numerical estimates of the appropriate parameters, or you can fit a number of distributions to the data and select the most appropriate one. In either case, the Input Analyzer provides you with estimates of the parameter values (based on the data you supply) and a ready-made expression that you can just copy and paste into your model.

When the Input Analyzer fits a distribution to your data, it estimates the distribution's parameters (including any shift or offset that's required to formulate a valid expression) and calculates a number of measures of how good the distribution fits your data. You can use this information to select which distribution you want to use in your model, which we discuss below.

Probability distributions can be thought of as falling into two main types: *theoretical* and *empirical*. Theoretical distributions, such as the exponential and gamma, generate samples based on a mathematical formulation. Empirical distributions simply divide the actual data into groupings and calculate the proportion of values in each group, possibly interpolating between points for more accuracy.

Each type of distribution is further broken into *continuous* and *discrete* types. The continuous theoretical distributions that Arena supports for use in your model are the exponential, triangular, and Weibull distributions mentioned previously, as well as the beta, Erlang, gamma, lognormal, uniform, and normal distributions. These distributions are referred to as continuous distributions because they can return any real-valued quantity (within a range for the bounded types). They're usually used to represent time durations in a simulation model. The Poisson distribution is a discrete distribution; it can return only integer-valued quantities. It's often used to describe the number of events that occur in an interval of time or the distribution of randomly varying batch sizes.

You also can use one of two empirical distributions: the discrete and continuous probability distributions. Each is defined using a series of probability/value pairs representing a histogram of the data values that can be returned. The *discrete empirical distribution* returns only the data values themselves, using the probabilities to choose from among the individual values. It's often used for probabilistically assigning entity types. The *continuous empirical distribution* uses the probabilities and values to return a real-valued quantity. It can be used in place of a theoretical distribution in cases where the data have unusual characteristics or where none of the theoretical distributions provide a good fit.

The Input Analyzer can fit any of the above distributions to your data. However, you must decide whether to use a theoretical or empirical distribution, and unfortunately, there aren't any standard rules for making this choice. Generally, if a histogram of your data (displayed automatically by the Input Analyzer) appears to be fairly uniform or has a single "hump," and if it doesn't have any large gaps where there aren't any values, then you're likely to obtain a good fit from one of the theoretical distributions. However, if there are a number of value groupings that have many observations (multimodal) or there are a number of data points that have a value that's significantly different from the main set of observations, an empirical distribution may provide a better representation of the data; an alternative to an empirical distribution is to divide the data somehow, which we'll describe at the end of this section.

The Input Analyzer is a standard tool that accompanies Arena and is designed specifically to fit distributions to observed data, provide estimates of their parameters, and measure how well they fit the data. There are four steps required to use the Input Analyzer to fit a probability distribution to your data for use as an input to your model:

- Create a text file containing the data values.
- Fit one or more distributions to the data.
- Select which distribution you'd like to use.
- Copy the expression generated by the Input Analyzer into the appropriate field in your Arena model.

To prepare the data file, simply create an ordinary ASCII text file containing the data in free format. Any text editor, word processor, or spreadsheet program can be used. The individual data values must be separated by one or more "white space characters" (blank spaces, tabs, or line feeds). There are no other formatting requirements; in particular, you can have as many data values on a line as you want, and the number of values per line can vary from line to line. When using a word processor or spreadsheet program, be sure to

```
6.1    9.4    8.1    3.2    6.5    7.2    7.8    4.9    3.5    6.6    6.1    5.1    4.9    4.2
6.4    8.1    6.0    8.2    6.8    5.9    5.2
6.5    5.4    5.9    9.3    5.4    6.5    7.4
6.0   12.6    6.8    5.6    5.8    6.2    5.6    6.4    9.5    7.2    5.6    4.7    4.5    7.0
7.7    6.9    5.4    6.3    8.1    4.9    5.3    5.0    4.7    5.7    4.9    5.3    6.4    7.5
4.4    4.9    7.6    3.6    8.3    5.6    6.2    5.0
7.4    5.2    5.0    6.5    8.0    6.2    5.0    4.8    6.2    4.9
7.0    7.7    4.7    5.0    6.0    9.0    5.7    7.1    5.0    5.6
4.9    7.8    7.1    7.1   11.5    5.4    5.2
6.1    6.8    5.4    3.5    7.1    5.7    5.4
5.7    6.1    4.2    8.8    7.4    5.5
5.3    5.9    5.2    6.4    4.5    5.1    5.6    6.1
8.1    8.1    5.1    8.3    7.5    7.6   10.9    6.5    9.0    5.9    6.8    9.0    6.5    6.0
5.8    5.0    6.4    4.7    4.5    6.2    5.2    7.9    5.5    4.9    7.2    4.9    4.5    6.0
6.3    8.3    5.5    7.8    5.4    5.3    6.6    3.6    7.3    5.3    8.9    6.8    7.1    8.7
6.4    3.3    7.0    7.7    6.7    7.6    7.6    7.1    5.6    5.9    4.1    7.5    7.7    5.4
4.8    5.5    8.8    7.2    6.3   10.0    4.3    4.9    5.7    5.1    6.7    6.0    5.6    7.2
7.0    7.8    6.3    6.1    8.4
```

Figure 4-27. Listing of the ASCII File partbprp.dst

save the file in a "text only" format. This eliminates any character or paragraph formatting that otherwise would be included. For example, Figure 4-27 shows the contents of an ASCII file called `partbprp.dst` (the default file extension for data files for the Input Analyzer is *.dst*) containing observations on 187 Part B Prep times; note that the values are separated by blanks or line feeds, there are different numbers of observations per line, and there is no particular order or layout to the data.

To fit a distribution to these data, run the Input Analyzer (e.g., select *Tools > Input Analyzer* from Arena). In the Input Analyzer, load the data file into a data fit window by creating a new window (*File > New* or the ⬜ button) for your fitting session (not for a data file), and then attaching your data file using either *File > Data File > Use Existing* or the 🔾 button. The Input Analyzer displays a histogram of the data in the top part of the window and a summary of the data characteristics in the bottom part, as shown in Figure 4-28.

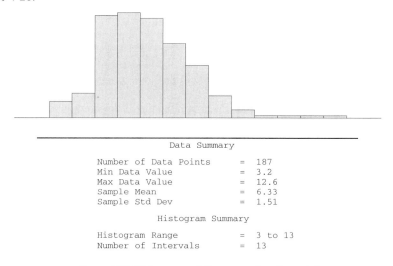

```
                    Data Summary

        Number of Data Points   =   187
        Min Data Value          =   3.2
        Max Data Value          =   12.6
        Sample Mean             =   6.33
        Sample Std Dev          =   1.51

                  Histogram Summary

        Histogram Range         =   3 to 13
        Number of Intervals     =   13
```

Figure 4-28. Histogram and Summary of partbprp.dst

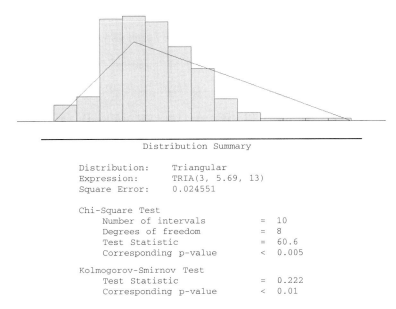

```
                        Distribution Summary

        Distribution:     Triangular
        Expression:       TRIA(3, 5.69, 13)
        Square Error:     0.024551

        Chi-Square Test
            Number of intervals        =   10
            Degrees of freedom         =   8
            Test Statistic             =   60.6
            Corresponding p-value      <   0.005

        Kolmogorov-Smirnov Test
            Test Statistic             =   0.222
            Corresponding p-value      <   0.01
```

Figure 4-29. Fitting a Triangular Distribution to partbprp.dst

You can adjust the relative size of the windows by dragging the splitter bar in the center of the window. Or, to see more of the data summary, you can scroll down through the text. Other options, such as changing the characteristics of the histogram, are described in online Help.

The Input Analyzer's *Fit* menu provides options for fitting individual probability distributions to the data (that is, estimating the required parameters for a given distribution). After you fit a distribution, its density function is drawn on top of the histogram display and a summary of the characteristics of the fit is displayed in the text section of the window. (This information also is written to a plain ASCII text file named DISTRIBUTION.OUT, where DISTRIBUTION indicates the distribution you chose to fit, such as TRIANGLE for triangular.) The exact expression required to represent the data in Arena is also given in the text window. You can transfer this to Arena by selecting *Edit > Copy Expression* in the Input Analyzer, opening the appropriate dialog box in Arena, and pasting the expression (*Ctrl+V*) into the desired field. Figure 4-29 shows this for fitting a triangular distribution to the data in partbprp.dst. Though we'll go into the goodness-of-fit issue below, it's apparent from the plot in Figure 4-29 that the triangular distribution doesn't fit the data particularly well.

If you plan to use a theoretical distribution in your model, you may want to start by selecting *Fit > Fit All* (or the ▲ button on the toolbar). This automatically fits all of the applicable distributions to the data, calculates test statistics for each (discussed below), and displays the distribution that has the minimum square error value (a measure of the quality of the distribution's match to the data). Figure 4-30 shows the results of the Fit All option for our Part B Prep times and indicates that a gamma distribution with $\beta = 0.775$

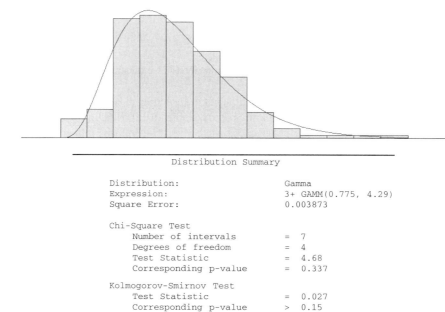

```
─────────────────────────────────────────────
                Distribution Summary

    Distribution:                  Gamma
    Expression:                    3+ GAMM(0.775, 4.29)
    Square Error:                  0.003873

    Chi-Square Test
        Number of intervals      =  7
        Degrees of freedom       =  4
        Test Statistic           =  4.68
        Corresponding p-value    =  0.337

    Kolmogorov-Smirnov Test
        Test Statistic           =  0.027
        Corresponding p-value    >  0.15
```

Figure 4-30. *Fit All Option to* partbprp.dst

and $\alpha = 4.29$, shifted to the right by 3, provides the "best" fit in the sense of minimum square error. Comparing the plot with that in Figure 4-29 indicates that this fitted gamma distribution certainly appears to be a better representation of the data than did the fitted triangular distribution. Other considerations for selecting which theoretical distribution to use in your model are discussed below.

If you want to use a discrete or continuous empirical distribution, use the *Empirical* option from the *Fit* menu. You may first want to adjust the number of histogram cells, which determines how many probability/value pairs will be calculated for the empirical distribution. To do so, select *Options > Parameters > Histogram* and change the number of intervals.

In addition to "eyeballing" the fitted densities on top of the histograms, the Input Analyzer provides three numerical measures of the quality of fit of a distribution to the data to help you decide. The first, and simplest to understand, is the *mean square error*. This is the average of the square error terms for each histogram cell, which are the squares of the differences between the relative frequencies of the observations in a cell and the relative frequency for the fitted probability distribution function over that cell's data range. The larger this square error value, the further away the fitted distribution is from the actual data (and thus the poorer the fit). If you fit all applicable distributions to the data, the Fit All Summary table orders the distributions from smallest to largest square error (select *Window > Fit All Summary*), shown in Figure 4-31 for partbprp.dst. While we see that gamma won the square-error contest, it was

Function	Sq Error
Gamma	0.00387
Weibull	0.00443
Beta	0.00444
Erlang	0.00487
Normal	0.00633
Lognormal	0.00871
Triangular	0.0246
Uniform	0.0773
Exponential	0.0806

Figure 4-31. Fit All Summary for partbprp.dst

followed closely by Weibull, beta, and Erlang, any of which would probably be just as accurate as gamma to use as input to the model.

The other two measures of a distribution's fit to the data are the chi-square and Kolmogorov-Smirnov (K-S) goodness-of-fit hypothesis tests. These are standard statistical hypothesis tests that can be used to assess whether a fitted theoretical distribution is a good fit to the data. The Input Analyzer reports information about the tests in the text window (see the bottoms of Figures 4-29 and 4-30). Of particular interest is the *Corresponding p-value*, which will always fall between 0 and 1.[4] To interpret this, larger *p*-values indicate better fits. Corresponding *p*-values less than about 0.05 indicate that the distribution's not a very good fit. Of course, as with any statistical hypothesis test, a high *p*-value doesn't constitute "proof" of a good fit—just a lack of evidence against the fit.

When it comes down to fitting or choosing a distribution to use, there's no rigorous, universally agreed-upon approach that you can employ to pick the "best" distribution. Different statistical tests (such as the K-S and chi-square) might rank distributions differently, or changes in the preparation of the data (e.g., the number of histogram cells) might reorder how well the distributions fit.

Your first critical decision is whether to use a theoretical distribution or an empirical one. Examining the results of the K-S and chi-square tests can be helpful. If the *p*-values for one or more distributions are fairly high (e.g., 0.10 or greater), then you can use a theoretical distribution and have a fair degree of confidence that you're getting a good representation of the data (unless your sample is quite small, in which case the discriminatory power of goodness-of-fit tests is quite weak). If the *p*-values are low, you may want to use an empirical distribution to do a better job capturing the characteristics of the data.

If you've decided to use a theoretical distribution, there often will be a number of them with very close goodness-of-fit statistics. In this case, other issues may be worth considering for selecting among them:

- First of all, you may want to limit yourself to considering only bounded or unbounded distributions, based on your understanding of how the data will be used in your model. For instance, often the triangular (bounded) and normal

---

[4] More precisely, the *p*-value is the probability of getting a data set that's more inconsistent with the fitted distribution than the data set you actually got, if the fitted distribution is truly "the truth."

(unbounded) distributions will be good fits to data such as process times. Over a long simulation run, they each might provide similar results, but during the run, the normal distribution might periodically return fairly large values that might not practically occur in the real system. A normal distribution with a mean that's close to zero may return an artificially large number of zero-valued samples if the distribution's being used for something that can't be negative, such as a time (Arena will change all negative input values to zero if they occur in a context where negative values don't make sense, like an interarrival or process time); this might be a compelling reason to avoid the normal distribution to represent quantities like process times that cannot be negative. On the other hand, bounding the triangular to obtain a faithful overall representation of the data might exclude a few outlying values that should be captured in the simulation model.

- Another consideration is of a more practical nature; namely, that it's easier to adjust parameters of some distributions than of others. If you're planning to make any changes to your model that include adjusting the parameters of the distribution (e.g., for sensitivity analysis or to analyze different scenarios), you might favor those with more easily understood parameters. For example, in representing interarrival times, Weibull and exponential distributions might provide fits of similar quality, but it's much easier to change an interarrival-time mean by adjusting an exponential mean than by changing the parameters of a Weibull distribution since the mean in the latter case is a complicated function of the parameters.

- If you're concerned about whether you're making the correct choice, run the model with each of your options to see if there's a significant difference in the results (you may have to wait until the model's nearly complete to be able to draw a good conclusion). You can further investigate the factors affecting distribution fits (such as the goodness-of-fit tests) by consulting Arena's online Help or other sources, such as Pegden, Shannon, and Sadowski (1995) or Law and Kelton (2000). Otherwise, select a distribution based on the qualitative and practical issues discussed above.

Before leaving our discussion of the Input Analyzer, we'd like to revisit an issue we mentioned earlier, namely, what to do if your data appear to have multiple peaks or perhaps a few extreme values, sometimes called *outliers*. Either of these situations usually makes it impossible to get a decent fit from any standard theoretical distribution. As we mentioned before, one option is to use an empirical distribution, which is probably the best route unless your sample size is quite small. In the case of outliers, you should certainly go back to your data set and make sure that they're actually correct rather than just being some kind of error, perhaps just a typo; if data values appear to be in error and you can't backtrack to confirm or correct them, then you should probably just remove them.

In the case of either multiple peaks or (correct) outliers, you might want to consider dividing your data set into two (maybe more) subsets before reading them into the Input Analyzer. For example, suppose you have data on machine downtimes and notice from the histogram that there are two distinct peaks; i.e., the data are *bimodal*. Reviewing the original records, you discover that these downtimes resulted from two different situations—breakdowns and scheduled maintenance. Separating the data into two subsets reveals that downtimes resulting from breakdowns tend to be longer than downtimes due to scheduled maintenance, explaining the two peaks. You could then separate the data (before going into the Input Analyzer), fit separate distributions to these data sets, and then modify your model logic to account for both kinds of downtimes.

You can also do a different kind of separation of the data directly in the Input Analyzer, based purely on their range. After loading the data file, select *Options > Parameters > Histogram* to specify cutoffs for the Low Value and High Value, and these cutoffs will then define the bounds of a *subset* of your entire data set. For bimodal data, you'd focus on the left peak by leaving the Low Value alone and specifying the High Value as the point where the histogram bottoms out between the two peaks, and later focus on the right peak by using this cutpoint as the Low Value and using the original High Value for the entire data set; getting the right cutpoint could require some trial and error. Then fit whatever distribution seems promising to each subset (or use the Fit All option) to represent that range of the data. If you've already fit a distribution (or used Fit All) before making Low/High Value cut, the Input Analyzer will immediately re-do the fit(s) and give you the new results for the data subset included in your cut; if you've done Fit All, a different distribution form altogether might come up as "the best." You'd need to repeat this process for each subset of the data. As a practical matter, you probably should limit the number of subsets to two or three since this process can become cumbersome, and it's probably not obvious where the best cutpoints are. To generate draws in your simulation model representing the original entire data set, the idea is to select one of your data subsets randomly, with the selection probabilities corresponding to the relative sizes of the subsets, then generate a value from the distribution you decided on for that subset. For instance, if you started out with a bimodal data set of size 200 and set your cutpoint so that the smallest 120 points represented the left peak and the largest 80 points represented the right peak, you'd generate from the distribution fitted to the left part of the data with probability 0.6 and generate from the distribution fitted to the right part of the data with probability 0.4. This kind of operation isn't exactly provided automatically in a single Arena module, so you'd have to put something together yourself. If the value involved is an activity time like a time delay an entity incurs, one possibility would be to use the Decide module with a Type of 2-way by Chance or N-way by Chance, depending on whether you have two versus more than two subsets to do a "coin flip" to decide from which subset to generate, then Connect to one of several Assign modules to Assign a value to an entity attribute as a draw from the appropriate distribution, and use this attribute downstream for whatever it's supposed to do. A different approach would be to set up an Arena Expression for the entire thing; Expressions are covered in Section 5.2.3.

4.6.5 No Data?

Whether you like it or not, there are times when you just can't get reliable data on what you need for input modeling. This can arise from several situations, like (obviously) the system doesn't exist, data collection is too expensive or disruptive, or maybe you don't have the cooperation or clearance you need. In this case, you'll have to rely on some

Table 4-2. Possible No-Data Distributions

Distribution	Parameters	Characteristics	Example Use
Exponential	Mean	High variance Bounded on left Unbounded on right	Interarrival times Time to machine failure (constant failure rate)
Triangular	Min, Mode, Max	Symmetric or non-symmetric Bounded on both sides	Activity times
Uniform	Min, Max	All values equally likely Bounded on both sides	Little known about process

fairly arbitrary assumptions or guesses, which we dignify as "*ad hoc data.*" We don't pretend to have any great solutions for you here, but have a few suggestions that people have found useful. No matter what you do, though, you really should carry out some kind of sensitivity analysis of the output to these *ad hoc* inputs to have a realistic idea of how much faith to put in your results. You'll either need to pick some deterministic value that you'll use in your study (or run the model a number of times with different values), or you'll want to represent the system characteristic using a probability distribution.

If the values are for something other than a time delay, such as probabilities, operating parameters, or physical layout characteristics, you can either select a constant, deterministic value or, in some cases, use a probability distribution. If you use a deterministic value by entering a constant in the model (e.g., 15% chance of failing inspection), it's a good idea to perform some sensitivity analysis to assess what effect the parameter has on the model's results. If small changes to the value influence the performance of the system, you may want to analyze the system explicitly for a range of values (maybe small, medium, and large) rather than just for your best guess.

If the data represent a time delay, you'll almost surely want to use a probability distribution to capture both the activity's inherent variability as well as your uncertainty about the value itself. Which distribution you'll use will be based on both the nature of the activity and the type of data you have. When you've selected the distribution, then you'll need to supply the proper parameters based on your estimates and your assessment of the variability in the process.

For selecting the distribution in the absence of empirical data, you might first look at the exponential, triangular, normal, and uniform distributions. The parameters for these distributions are fairly easy to understand, and they provide a good range of characteristics for a range of modeling applications, as indicated in Table 4-2.

If the times vary independently (i.e., one value doesn't influence the next one), your estimate of the mean isn't too large, and there's a large amount of variability in the times, the exponential distribution might be a good choice. It's most often used for interarrival times; examples would be customers coming to a restaurant or pick requests from a warehouse.

If the times represent an activity where there's a "most likely" time with some variation around it, the triangular distribution is often used because it can capture processes with small or large degrees of variability and its parameters are fairly easy to understand.

The triangular distribution is defined by minimum, most likely (modal), and maximum values, which is a natural way to estimate the time required for some activity. It has the advantage of allowing a non-symmetric distribution of values around the most likely, which is commonly encountered in real processes. It's also a bounded distribution—no value will be less than the minimum or greater than the maximum—which may or may not be a good representation of the real process.

You may be wondering why we're avoiding what might be the most familiar distribution of all, the normal distribution which is the classical "bell curve" defined by a mean and standard deviation. It returns values that are symmetrically distributed around the mean and is an unbounded distribution, meaning that you could get a very large or very small value once in a while. In cases where negative values can't be used in a model (e.g., the delay time in a process), negative samples from a normal distribution are set by Arena to a value of 0; if the mean of your distribution is close to 0 (e.g., no more than about three or four times the standard deviation from 0), the normal distribution may be inappropriate.

On the other hand, if the mean is positive and quite a bit larger than the standard deviation, there will be only a small chance of getting a negative value, like maybe one in a million. This sounds (and is) small, but remember that in a long simulation run, or one that is replicated many times, *one in a million can really happen*, especially with modern, fast computers; and in that case, Arena would truncate the negative value to zero if its usage in the model can only be positive, perhaps causing an unwanted or extreme action to be taken and possibly even invalidating your results. Figure 4-32 shows a fairly useless little Arena model (Model 4-5) in which entities arrive, spaced exactly an hour apart, an observation from a normal distribution with mean $\mu = 3.0$ and standard deviation $\sigma = 1$ is assigned to the attribute Normal Observation, and the entity goes to one of two Record modules, depending on whether Normal Observation is non-negative or negative, which count the number of non-negative and negative observations obtained out of the total (we'll leave it to you to look through this model on your own). Table 4-3 gives the results for several values of μ (holding σ at 1), and you can see that this is going to occur even if the mean is three or four (or more) standard deviations above zero. Using electronic normal tables, the exact probability of getting a negative value can be computed, and one over this number gives the approximate number of observations (on average) you'd need

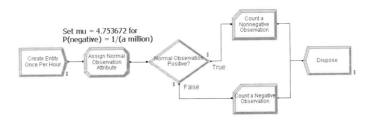

Figure 4-32. Model 4-5 to Count Negative Normal Observations

Table 4-3. *Getting Negative Values from a Normal Distribution with Standard Deviation σ = 1*

Mean μ	Number of Draws	Number of Negative Draws	Exact Probability That an Observation Will Be Negative	"One in This Many" Will Be Negative (on Average)
3.0	One million	1,409	0.001350	741
3.5	One million	246	0.000233	4,298
4.0	One million	37	0.000032	31,560
4.5	One million	3	0.000003	294,048
4.753672	Ten million	7	0.000001	1,000,000

to get a negative value—not too many, considering the speed of generating these (it took about three seconds on a garden-variety 2.1GHz notebook computer to complete each of the million-draw runs, and under 30 seconds to complete the ten-million-draw run). The last value of μ in the table was picked since it yields a probability of exactly one in a million of a getting a negative observation (the first million happened not to produce any negatives, but seven in ten million is close enough to one in a million). Now we respect and admire Gauss as a great mathematician, but we're just not big fans of using the normal as a simulation input distribution to represent logically non-negative things like process times, even though Arena allows it. For data sets that appear to be well fit (or even best fit) by a normal distribution, a different distribution that completely avoids negative values, like Weibull, gamma, lognormal, Erlang, beta, or perhaps empirical, will probably fit almost as well and not expose you to the risk of generating any naughty negative values at all.

Finally, if you really don't know much about the process but can guess what the minimum and maximum values will be, you might use the uniform distribution. It returns all values between a minimum and maximum with equal likelihood.

4.6.6 *Nonstationary Arrival Processes*

This somewhat specialized topic deserves mention on its own since it seems to come up often and can be very important in terms of representing system behavior validly. Many systems subject to external arrivals, like service systems for people, telephone call centers, and manufacturing systems with outside customer demands, experience arrival loads that can vary dramatically over the time frame of the simulation. Examples include a noon rush for burgers, heavy calls to a technical support line in the middle of the afternoon, and strong demand on a manufacturing system during certain seasons. A specific probabilistic model for this, the *nonstationary Poisson process*, is very useful and often provides an accurate way to reflect time-varying arrival patterns. You need to deal with two issues: how to estimate or specify the rate function, and then how to generate the arrival pattern in the simulation.

There are a lot of ways to estimate or specify a rate function from the data, some of which can be pretty complicated. We'll stick to one fairly simple method called the *piecewise-constant* rate function, which seems to work well in many applications. First, identify lengths of time within which the arrival rate appears to be fairly flat; for instance, a call center's arrivals might be fairly constant over half-hour periods but could be quite different in different periods. Count up the numbers of arrivals in each period, and then compute a different rate for each period. For instance, suppose the call center experiences the following numbers of incoming calls for the four 30-minute periods between 8:00 AM and 10:00 AM: 20, 35, 45, and 50. Then the rates, *in units of calls per minute*, for the first four 30-minute periods would be 0.67, 1.17, 1.50, and 1.67.

Once you've estimated the rate function in this way, you need to make sure that Arena follows this pattern in generating the arrivals to your model. The Create module will do this if you select Schedule as the Type for Time Between Arrivals. You then must specify the rate function via the Schedule data module, similarly to what we did in Section 4.2.2 for a Resource Schedule. One caution here: Arena allows you to mix and match whatever time units you want, but you must be careful that the numbers and time units are defined properly. We'll do this in Model 5-2 in Chapter 5. In Chapter 12, we'll have more to say about estimating the rate function, as well as what underlies Arena's generation method for nonstationary arrivals.

4.6.7 Multivariate and Correlated Input Data

Most of the time we assume that all random variables driving a simulation are generated independently of each other from whatever distribution we decide on to represent them. Sometimes, though, this may not be the best assumption, for purely physical reasons. For instance, in Model 4-2 you could imagine that certain parts are "difficult"; maybe you'd detect this by noticing that a large prep time for a specific part tends to be followed by a large sealer time for that part; that is, these two times are positively correlated. Ignoring this correlation could lead to an invalid model and biased results.

There are a number of ways to model situations like this, to estimate the required parameters (including the strength of the correlations), and to generate the required observations on the random variables during the simulation. Some of these methods entail viewing the associated random variables as coordinates of a random *vector* having some joint *multivariate* distribution to be fitted and generated from. You might also be able to specify some kind of formula-based association between related input quantities. Frankly, though, this is a pretty difficult issue in terms of both estimating the behavior and generating it during the simulation. For more on these and related issues, see Law and Kelton (2000) or Devroye (1986).

4.7 Summary and Forecast

If you've read and understood the material in this chapter, you should be getting dangerous in the use of Arena. We encourage you to press other buttons in the modules we've used and even try modules that we've not used. If you get stuck, try the online Help feature, which may not answer your question, but will answer the questions you should be asking. You might also want to try other animation features or provide nicer pictures. At this point, the best advice we can give you is to *use* Arena. Chapters 5-12 will cover most of the modeling capabilities (and some of the statistical-analysis capabilities) of Arena in more depth.

4.8 Exercises

4-1   Travelers arrive at the main entrance door of an airline terminal according to an exponential interarrival-time distribution with mean 1.6 minutes, with the first arrival at time 0. The travel time from the entrance to the check-in is distributed uniformly between 2 and 3 minutes. At the check-in counter, travelers wait in a single line until one of five agents is available to serve them. The check-in time (in minutes) follows a Weibull distribution with parameters $\beta = 7.76$ and $\alpha = 3.91$. Upon completion of their check-in, they are free to travel to their gates. Create a simulation model, with animation (including the travel time from entrance to check-in), of this system. Run the simulation for 16 hours to determine the average time in system, number of passengers completing check-in, and the average length of the check-in queue.

4-2   Develop a model of a simple serial two-process system. Items arrive at the system with a mean time between arrivals of 10 minutes, with the first arrival at time 0. They are immediately sent to Process 1, which has a single resource with a mean service time of 9 minutes. Upon completion, they are sent to Process 2, which is identical to (but independent of) Process 1. Items depart the system upon completion of Process 2. Performance measures of interest are the average numbers in queue at each process and the total time in system of items. Using a replication length of 10,000 minutes, make the following four runs and compare the results (noting that the structure of the model is unchanged, and it's only the input distributions that are changing):

Run 1: exponential interarrival times and exponential service times
Run 2: constant interarrival times and exponential service times
Run 3: exponential interarrival times and constant service times
Run 4: constant interarrival times and constant service times

4-3   Modify the Exercise 4-1 check-in problem by adding agent breaks. The 16 hours are divided into two 8-hour shifts. Agent breaks are staggered, starting at 90 minutes into each shift. Each agent is given one 15-minute break. Agent lunch breaks (30 minutes) are also staggered, starting 3.5 hours into each shift. The agents are rude and, if they're busy when break time comes around, they just leave anyway and make the passenger wait until break time is over before finishing up that passenger. Compare the results of this model to those of the model without agent breaks.

4-4   Two different part types arrive at a facility for processing. Parts of Type 1 arrive with interarrival times following a lognormal distribution with a log mean of 11.5 hours and log standard deviation of 2.0 hours (note that these values are the mean and standard deviation of this lognormal random variable itself); the first arrival is at time 0. These arriving parts wait in a queue designated for only Part Type 1's until a (human) operator is available to process them (there's only one such human operator in the facility) and the processing times follow a triangular distribution with parameters 5, 6, and 8 hours. Parts of Type 2 arrive with interarrival times following an exponential distribution with mean of 15 hours; the first arrival is at time 0. These parts wait in a second queue (designated for Part Type 2's only) until the same lonely (human) operator is available to process them; processing times follow a triangular distribution with parameters 3, 7, and 8 hours. After being processed by the human operator, all parts are sent for processing to an automatic machine not requiring a human operator, which has processing times distributed as triangular with parameters of 4, 6, and 8 hours for both part types; all parts share the same first-come, first-served queue for this automatic machine. Completed parts exit the system. Assume that the times for all part transfers are negligible. Run the simulation for 5,000 hours to determine the average total time in system (sometimes called *cycle time*) for all parts (lumped together regardless of type), and the average number of items in the queues designated for the arriving parts. Animate your model, including use of different pictures for the different part types, and use resources that look different for busy vs. idle.

4-5   During the verification process of the airline check-in system from Exercise 4-3, it was discovered that there were really two types of passengers. The first passenger type arrives according to an exponential interarrival distribution with mean 2.4 minutes and has a service time (in minutes) following a gamma distribution with parameters $\beta = 0.42$ and $\alpha = 14.4$. The second type of passenger arrives according to an exponential distribution with mean 4.4 minutes and has a service time (in minutes) following 3 plus an Erlang distribution with parameters ExpMean = 0.54 and $k = 15$ (i.e., the Expression for the service time is 3 + ERLA(0.54, 15)). A passenger of each type arrives at time 0. Modify the model from Exercise 4-3 to include this new information. Compare the results.

4-6   Parts arrive at a single workstation system according to an exponential interarrival distribution with mean 21 seconds; the first arrival is at time 0. Upon arrival, the parts are initially processed. The processing-time distribution is TRIA(16, 19, 22) seconds. There are several easily identifiable visual characteristics that determine whether a part has a potential quality problem. These parts, about 10% (determined after the initial processing), are sent to a station where they undergo a thorough inspection. The remaining parts are considered good and are sent out of the system. The inspection-time distribution is 95 plus a WEIB(48.5, 4.04) random variable, in seconds. About 14% of these parts fail the inspection and are sent to scrap. The parts that pass the inspection are classified as good and are sent out of the system (so these parts didn't need the thorough inspection, but you know what they say about hindsight). Run the simulation for 10,000 seconds to observe the number of good parts that exit the system, the number of scrapped parts, and the number of parts that received the thorough inspection. Animate your model.

4-7   A proposed production system consists of five serial automatic workstations. The processing times at each workstation are constant: 11, 10, 11, 11, and 12 (all times given in this problem are in minutes). The part interarrival times are UNIF(13, 15). There is an unlimited buffer in front of all workstations, and we will assume that all transfer times are negligible or zero. The unique aspect of this system is that at Workstations 2 through 5 there is a chance that the part will need to be reprocessed by the workstation that precedes it. For example, after completion at Workstation 2, the part can be sent back to the queue in front of Workstation 1. The probability of revisiting a workstation is independent in that the same part could be sent back many times with no change in the probability. At present, it is estimated that this probability, the same for all four workstations, will be between 5% and 10%. Develop the simulation model and make six runs of 10,000 minutes each for probabilities of 5, 6, 7, 8, 9, and 10%. Using the results, construct a plot of the average cycle time (system time) against the probability of a revisit. Also include the maximum cycle time for each run on your plot.

4-8   A production system consists of four serial automatic workstations. The first part arrives at time zero, and then (exactly) every 9.8 minutes thereafter. All transfer times are assumed to be zero and all processing times are constant. There are two types of failures: major and jams. The data for this system are given in the table below (all times are in minutes). Use exponential distributions for the uptimes and uniform distributions for repair times (for instance, repairing jams at Workstation 3 is UNIF(2.8, 4.2)). Run your simulation for 10,000 minutes to determine the percent of time each resource spends in the failure state and the ending status of each workstation queue.

| Workstation | | Major Failure Means | | Jam Means | |
Number	Process Time	Uptimes	Repair	Uptimes	Repair
1	8.5	475	20, 30	47.5	2, 3
2	8.3	570	24, 36	57	2.4, 3.6
3	8.6	665	28, 42	66.5	2.8, 4.2
4	8.6	475	20, 30	47.5	2, 3

4-9   An office that dispenses automotive license plates has divided its customers into categories to level the office workload. Customers arrive and enter one of three lines based on their residence location. Model this arrival activity as three independent arrival streams using an exponential interarrival distribution with mean 10 minutes for each stream, and an arrival at time 0 for each stream. Each customer type is assigned a single, separate clerk to process the application forms and accept payment, with a separate queue for each. The service time is UNIF(8, 10) minutes for all customer types. After completion of this step, all customers are sent to a single, second clerk who checks the forms and issues the plates (this clerk serves all three customer types, who merge into a single first-come, first-served queue for this clerk). The service time for this activity is UNIF(2.66, 3.33) minutes for all customer types. Develop a model of this system and run it for 5,000 minutes; observe the average and maximum time in system for all customer types combined.

A consultant has recommended that the office not differentiate between customers at the first stage and use a single line with three clerks who can process any customer type. Develop a model of this system, run it for 5,000 minutes, and compare the results with those from the first system.

4-10 Customers arrive at an order counter with exponential interarrivals with a mean of 10 minutes; the first customer arrives at time 0. A single clerk accepts and checks their orders and processes payments, taking UNIF(8, 10) minutes. Upon completion of this activity, orders are randomly assigned to one of two available stock persons (each stock person has a 50% chance of getting any individual assignment) who retrieve the orders for the customers, taking UNIF(16, 20) minutes. These stock persons only retrieve orders for customers who have been assigned specifically to them. Upon receiving their orders, the customers depart the system. Develop a model of this system and run the simulation for 5,000 minutes, observing the average and maximum customer time in system.

A bright, young engineer has recommend that they eliminate the assignment of an order to a specific stock person and allow both stock persons to select their next activity from a single first-come, first-served order queue. Develop a model of this system, run it for 5,000 minutes, and compare the results to the first system.

4-11 Using the model from Exercise 4-2, set the interarrival-time distribution to exponential and the process-time distribution for each Process to uniform on the interval $[9 - h, 9 + h]$. Setting the value of h to 1.732, 3.464, and 5.196, compute the (exact) variance of this distribution and make three different runs of 10,000 minutes each and compare the results. Note that the mean of the process time is always 9 and the distribution form is always the same (uniform); the standard deviation (and thus the variance) is the only thing that's changing.

4-12 Using the model from Exercise 4-11, assume the process time has a mean of 9 and a *variance* of 4. Calculate the parameters for the gamma distribution that will give these values. Make a run and compare the results with those from the $h = 3.464$ case of Exercise 4-11. Note that here both the mean *and* the variance are the same—it's only the shape of the distribution that differs.

4-13 Parts arrive at a single machine system according to an exponential interarrival distribution with mean 20 minutes; the first part arrives at time 0. Upon arrival, the parts are processed at a machine. The processing-time distribution is TRIA(11, 16, 18) minutes. The parts are inspected and about 25% are sent back to the same machine to be reprocessed (same processing time). Run the simulation for 20,000 minutes to observe the average and maximum number of times a part is processed, the average number of parts in the machine queue, and the average part cycle time (time from a part's entry to the system to its exit after however many passes through the machine system are required).

4-14 Using the model from Exercise 4-13, make two additional runs with run times of 60,000 minutes and 100,000 minutes and compare the results with those of Exercise 4-13.

4-15 Items arrive from an inventory-picking system according to an exponential interarrival distribution with mean 1.1 (all times are in minutes), with the first arrival at time 0. Upon arrival, the items are packed by one of four identical packers, with a single queue "feeding" all four packers. The packing time is TRIA(2.75, 3.3, 4.0). Packed boxes are then separated by type (20%, international and 80%, domestic), and sent to shipping. There is a single shipper for international packages and two shippers for domestic packages with a single queue feeding the two domestic shippers. The international shipping time is TRIA(2.3, 3.3, 4.8), and the domestic shipping time is TRIA(1.7, 2.0, 2.7). This packing system works three 8-hour shifts, five days a week. All the packers and shippers are given a 15-minute break two hours into their shift, a 30-minute lunch break four hours into their shift, and a second 15-minute break six hours into their shift; use the Wait Schedule Rule. Run the simulation for two weeks (ten working days) to determine the average and maximum number of items or boxes in each of the three queues. Animate your model, including a change in the appearance of entities after they're packed into a box.

4-16 Using the model from Exercise 4-15, change the packer and domestic shipper schedules to stagger the breaks so there are always at least three packers and one domestic shipper working. Start the first 15-minute packer break one hour into the shift, the 30-minute lunch break three hours into the shift, and the second 15-minute break six hours into the shift. Start the first domestic shipper 15-minute break 90 minutes into the shift, the 30-minute lunch break 3.5 hours into the shift, and the second 15-minute break six hours into the shift. Compare the new results to those from Exercise 4-15.

4-17 Using the Input Analyzer, open a new window and generate a new data file (use *File > Data File > Generate New*) containing 50 points for an Erlang distribution with parameters: ExpMean equal to 12, *k* equal to 3, and Offset equal to 5. Once you have the data file, perform a Fit All to find the "best" fit from among the available distributions. Repeat this process for 500, 5,000, and 25,000 data points, using the same Erlang parameters. Compare the results of the Fit All for the four different sample sizes.

4-18 Hungry's Fine Fast Foods is interested in looking at their staffing for the lunch rush, running from 10 AM to 2 PM. People arrive as walk-ins, by car, or on a (roughly) scheduled bus, as follows:

- Walk-ins—one at a time, interarrivals are exponential with mean 3 minutes; the first walk-in occurs EXPO(3) minutes past 10 AM.
- By car—with 1, 2, 3, or 4 people to a car with respective probabilities 0.2, 0.3, 0.3, and 0.2; interarrivals distributed as exponential with mean 5 minutes; the first car arrives EXPO(5) minutes past 10 AM.
- A single bus arrives every day sometime between 11 AM and 1 PM (arrival time distributed uniformly over this period). The number of people on the bus varies from day to day, but it appears to follow a Poisson distribution with a mean of 30 people.

Once people arrive, either alone or in a group from any source, they operate independently regardless of their source. The first stop is with one of the servers at the order/payment counter, where ordering takes TRIA(1, 2, 4) minutes and payment then takes TRIA(1, 2, 3) minutes; these two operations are sequential, first order-taking then payment, by the same server for a given customer. The next stop is to pick up the food ordered, which takes an amount of time distributed uniformly between 30 seconds and 2 minutes. Then each customer goes to the dining room, which has 30 seats (people are willing to sit anywhere, not necessarily with their group), and partakes of the sublime victuals, taking an enjoyable TRIA(10, 20, 30) minutes. After that, the customer walks fulfilled to the door and leaves. Queueing at each of the three "service" stations (order/pay, pickup food, and dining room) is allowed, with FIFO discipline. There is a travel time of EXPO(30) seconds from each station to all but the exit door—entry to order/pay, order/pay to pickup food, and pickup food to dining. After eating, people move more slowly, so the travel time from the dining room to the exit is EXPO(1) minute.

The servers at both order/pay and pickup food have a single break that they "share" on a rotating basis. More specifically, at 10:50, 11:50, 12:50, and 1:50, one server from each station goes on a 10-minute break; if the person due to go on break at a station is busy at break time, he or she finishes serving the customer but still has to be back at the top of the hour (so the break could be a little shorter than 10 minutes).

Staffing is the main issue facing Hungry's. Currently, there are six servers at the order/pay station and two at the pickup food station throughout the 4-hour period. Since they know that the bus arrives sometime during the middle two hours, they're considering a variable staffing plan that, for the first and last hour would have three at order/pay and one at pickup food, and for the middle two hours would have nine at order/pay and three at pickup food (note that the total number of person-hours on the payroll is the same, 32, under either the current staffing plan or the alternate plan, so the payroll cost is the same). What's your advice?

In terms of output, observe the average and maximum length of each queue, the average and maximum time in each queue, and the total number of customers completely served and out the door. Make plots of the queues to get into order/pay, pickup food, and the dining room. Animate all queues, resources, and movements between stations. Pick from a *.plb* picture library a humanoid picture for the entities (different for each arrival source), and make an appropriate change to their appearance after they've finished eating and leave the dining room. Also, while you won't be able to animate the individual servers or seats in the dining room, pick reasonable pictures for them as well.

4-19 In the discussion in Section 4.2.5 of Arena's Instantaneous Utilization vs. Scheduled Utilization output values, we stated that if the Resource has a fixed Capacity (say, $M(t) = c > 0$ for all times t), then Instantaneous Utilization and Scheduled Utilization will be the same. Prove this.

4-20 In the discussion in Section 4.2.5 of Arena's Instantaneous Utilization vs. Scheduled Utilization output values, we stated that neither of the two measures is always larger. Prove this; recall that to prove that a general statement is *not* true you only have to come up with a single example (called a *counterexample*) for which it's not true.

4-21 Modify your solution for Exercise 4-7 to include transfer times between part arrival and the first workstation, between workstations (both going forward and for reprocessing), and between the last workstation and the system exit. Assume all part transfer times are UNIF(2,5). Animate your model to show entity movement and run for 10,000 minutes using a reprocess probability of 8%.

4-22 Management wants to study Terminal 3 at a hub airport with an eventual eye toward improvement. The first step is to model it as it is during the eight hours through the busiest part of a typical weekday. We'll model the check-in and security operations only, i.e., once passengers get through security they're on their way to their gate and out of our model. Passengers arrive one at a time through the front door from curbside ground transportation with interarrival times distributed exponentially with mean 0.5 minute (all times are in minutes unless otherwise noted). Of these passengers, 35% go left to an old-fashioned manual check-in counter, 50% go right to a newfangled automated check-in counter, and the remaining 15% don't need to check in at all and proceed directly from the front door to security (it takes these latter types of passengers between 3 and 5 minutes, uniformly distributed, to make the walk from the front door to the entrance to the security area; the other two passenger types move instantly from their arrival to the manual or automated check-in counter as the case may be). There are two agents at the manual check-in station, fed by a single FIFO queue; manual check-in times follow a triangular distribution between 1 and 5 minutes with a mode of 2 minutes. After manual check-in, passengers walk to the security area, a stroll that takes them between 2 and 6 minutes, uniformly distributed. The automated check-in has two stations (a station consists of a touch-screen kiosk and an employee to take checked bags; view a kiosk-employee pair as a single unified unit, i.e., the kiosk and its employee cannot be separated), fed by a single FIFO queue, and check-in times are triangularly distributed between 0.5 and 1.5 with a mode of 1. After automated check-in, passengers walk to the security area, taking between 1 and 3 minutes, uniformly distributed, to get there (automated check-in passengers are just quicker than manual check-in passengers at everything). All passengers eventually get to the security area, where there are six stations fed by a single FIFO queue; security-check times are triangularly distributed between 1 and 6 with a mode of 2 (this distribution captures all the possibilities there, like x-ray of carry-ons, walking through the metal detector, bag search, body wanding, shoes off, laptop checking, etc.). Once through the security check (everybody passes, though it takes some longer than others to do so), passengers head to their gates and are no longer in our model. Simulate this system for an 8-hour period and look at the average queue lengths, average times in queue, resource utilizations, and average total time in system of passengers (for all passenger types combined). Animate your model, including queues, resources, and passengers walking to security. Put in plots that track the length of each of the three queues over the eight-hour simulation (either three separate plots or three curves in a single plot).

CHAPTER 5

Modeling Detailed Operations

In Chapter 4, we showed you the kinds of modeling you could do with modules mostly from the Basic Process panel. They are relatively high-level, easy-to-use modules that will usually take you a long way toward building a model at a sufficient level of detail. Sometimes it's *all* you'll need.

But sometimes it isn't. As you gain experience in modeling, and as your models become bigger, more complex, and more detailed, you might find that you'd like to be able to control or model things at a lower level, in more detail, or just differently from what the modules of the Basic Process panel offer. Arena doesn't strand you at this level or force you to accept a limited number of "canned" modeling constructs, nor does it force you to learn a programming language or some pseudo-programming syntax to capture complicated system aspects. Rather, it offers a rich and deep hierarchy of several different modeling levels to get the flexibility you might need to model some specific logic appropriately. It's probably a good idea to start with the high-level modules, take them as far as they'll go (maybe that's all the way), and when you need greater flexibility than they provide, go to a lower, more detailed level. This structure allows you to exploit the easy, high-level modeling constructs to the extent possible, yet allows you to drill down when you need to. And because standard Arena modules provide all of this modeling power, you'll be familiar with how to use them. To put them to work, you'll simply need to become familiar with what they do.

This chapter explores some (certainly not all) of the detailed, lower-level modeling constructs available in the Advanced Process and Blocks panels; the latter panel provides the lowest-level model logic where modules correspond to the blocks in the SIMAN simulation language that underlies Arena.

The example we'll use for this is a telephone call center, including technical support, sales, and order-status checking. Section 5.1 describes the initial system and Section 5.2 talks about how to model it using some new Arena modeling concepts. Section 5.3 describes our basic modeling strategy. The model logic and animation are developed in Section 5.4. In Section 5.5, we embellish our model and present several new Arena modeling concepts. We'll then show you how to modify the original model to create the embellished model. We'll also touch on the important topics of nonstationary (time-dependent) arrival processes and a greater level of customization in animation. In Section 5.6, we add more output measures to our enhanced model. We close the chapter in Section 5.7 with an altogether different kind of model, an inventory system, and take this opportunity to illustrate use of one of Arena's lowest and most detailed modeling levels, the Blocks panel that composes the underlying SIMAN simulation language.

After reading this chapter, you should be able to build very detailed and complex models and be able to exploit Arena's rich and deep hierarchy of modeling levels.

5.1   Model 5-1: A Simple Call Center System

Our generic call center system provides a central number in an organization that customers call for technical support, sales information, and order status. Incoming calls arrive with interarrival times being exponentially distributed with a mean of 0.857 minute. This central number feeds 26 trunk lines. If all 26 lines are in use, a caller gets a busy signal; hopefully, the caller will try again later, but for our model, just go away. An answered caller hears a recording describing three options: transfer to technical support, sales information, or order-status inquiry (76%, 16%, and 8%, respectively). The estimated time for this activity is UNIF(0.1, 0.6); all times are in minutes.

If the caller chooses technical support, a second recording requests which of three product types the caller is using, which requires UNIF(0.1, 0.5) minutes. The percentage of requests for product types 1, 2, and 3 are 25%, 34%, and 41%, respectively. If a qualified technical support person is available for the selected product type, the call is automatically routed to that person. If none are currently available, the customer is placed in an electronic queue where he is subjected to annoying rock music until a support person is available. The time for all technical support calls is estimated to be TRIA(3, 6, 18) minutes regardless of the product type. Upon completion of the call, the customer exits the system.

Sales calls are automatically routed to the sales staff. If a salesperson is not available, the caller is treated to soothing new-age space music (after all, we're hoping for a sale). Sales calls are estimated to be TRIA(4, 15, 45)—sales people tend to talk a lot more than technical support people! Upon completion of the call, the happy customer exits the system.

Callers requesting order-status information are automatically handled by the phone system, and there is no limit on the number the system can handle (except that there are only 26 trunk lines, which is itself a limit, since an ongoing order-status call occupies one of these lines). The estimated time for these transactions is TRIA(2, 3, 4) minutes, with 15% of these callers opting to speak to a real person after they have received their order status.

These calls are routed to the sales staff where they wait with a lower priority than sales calls. This means that if an order-status call is in a queue waiting for a salesperson and a new arriving sales call enters, the sales call will be given priority over the order-status call and answered first. These follow-up order-status calls are estimated to last TRIA(3, 5, 10) minutes. These callers then exit the system.

The call center hours are from 8 AM until 6 PM, with a small proportion of the staff on duty until 7 PM. Although the system closes to new calls at 6 PM, all calls that enter the system by that time are answered and served.

Over the course of a day there are eight technical support employees to answer technical support calls. Two are devoted to product type 1 calls; three, to product type 2 calls; and three, to product type 3 calls. There are four sales employees to answer the sales calls and those order-status calls that opt to speak to a real person.

As a point of interest, we'll count the number of customer calls that are not able to get a trunk line and are thus rejected from entering the system (similar to *balking* in queueing systems, though balking usually means a decision on the customer's part not to enter, rather than being rejected as in our model). However, we won't consider *reneging*—customers who get a trunk line initially but later hang up the phone before being served (see Section 9.3 for a discussion of how to model reneging).

Some statistics of interest for these types of systems are: number of customer rejections (busy signals), total time on the line by customer type, time waiting for a real person by customer type, number of calls waiting for service by customer type, and personnel utilization.

5.2   New Modeling Issues

From a simulation viewpoint, this problem is quite different from the ones we covered in Chapters 3 and 4. The most obvious difference is that the previous systems were manufacturing oriented, and this system is of a service nature. Although the original version of SIMAN (the simulation language on which Arena is based) was developed for manufacturing applications, the current Arena capabilities also allow for accurate modeling of service systems. Applications in this area include fast-food restaurants, banks, insurance companies, service centers, and many others. Although these systems have some special characteristics, the basic modeling requirements are largely the same as for manufacturing systems.

Now let's take a look at our call center and explore the new requirements.

5.2.1   *Customer Rejections and Balking*

A call generated by our arrival process is really a customer *trying* to access one of the 26 trunk lines. If all 26 lines are currently in use, a busy signal is received and the customer is rejected and departs the system. The term for this is *balking* if done voluntarily by the customer, though in our call center it is involuntary, so we'll call it a *rejection*; modeling and handling balking and rejection are, however, similar to each other.

Consider a drive-through at a fast-food restaurant that has a single window with room for only five cars to wait for service. The arriving entities would be cars entering a queue to wait to seize a resource called "Window Service." We'd need to set the queue capacity to 5. This would allow one car to be in service and a maximum of five cars to be waiting. If a sixth car attempted to enter the queue, it would balk (or be rejected). You decide as part of your modeling assumptions what happens to these balked/rejected cars or entities. They might be disposed of or we might assume that they would drive around the block and try to re-enter the queue a little later.

There are two methods to model balking/rejection in Arena. The first method is to employ a Queue module from the Blocks panel and define the queue capacity as 0. An arriving entity (an incoming call in our model) enters the zero-capacity queue and immediately attempts to seize one unit of a resource called "Trunk Line." If a unit is available, it's allocated to the call and the call enters the system. If a unit of the resource is not available, the entity attempts to stay in the queue. But since the queue has a capacity of 0, the call would be balked from the queue and disposed of. The second method would have the arriving entity use a Decide module to check whether a unit of the "Trunk Line" resource is available. If a unit is available, the entity is allowed to proceed and seize the resource. If no unit is available, the entity is sent to the balking section of the model. We'll use this second method in our model.

Balking and rejection clearly represent a kind of failure of the system to meet customer needs, so we'll count the number of times this happens in the simulation; smaller is better.

5.2.2 Three-Way Decisions

Once a call is allocated a trunk line and enters the system, we must then determine the call type so we can direct it to the proper part of the system for service. To do this, we need the ability to send entities or calls to *three* different parts of the system based on the given probabilities. The same requirement is true for technical calls since there are three different product types.

We could get out our calculator and dust off our probability concepts and compute the probability of each call type—there are a total of five. We could then define Sequences (see Section 7.1) for each of these call types and route them through the system. Although this might work, you would have to re-compute the probabilities each time you change the distribution of call types, which you might want to do to test the flexibility or robustness of the system.

You may not have been aware of it, but the capability is provided to branch in three or more directions in the same Decide module that we used in the models of Chapter 4.

5.2.3 Variables and Expressions

In many models, we might want to reuse data in several different places. For example, in our call center, there will be several places where we will need to enter the distributions for the time to handle the technical support calls. If we decide to change this value during our experimentation, we'll have to open each dialog that included a call time and change the value. There are other situations where we might want to keep track of the total number of entities in a system or in a portion of the system. In other cases, we may want to use complex expressions throughout the model. For example, we might want to base a processing time on the part type. Arena *Variables* and *Expressions* allow us to fulfill these kinds of needs easily.

The Variables module allows you to define your own global variables and their initial values. Variables can then be referenced in the model by their names. They can also be specified as one- or two-dimensional arrays. The Expressions module allows you to define expressions and their associated values. Similar to variables, expressions are referenced in the model by their names and can also be specified as one- or two-dimensional arrays. Although variables and expressions may appear to be quite similar, they serve distinctly different functions.

User-defined Variables store some real-valued quantity that can be reassigned during the simulation run. For example, we could define a Variable called `Wait Time` with an initial value of 2 and enter the Variable name wherever a wait time was required. We could also define a Variable called `Number in System` with an initial value of 0, add 1 to this variable every time a new part entered the system, and subtract 1 from it every time a part exited the system. For our call center, we'll use two variables. The first will keep track of the number of calls in the system and the second will be used to terminate the generation of incoming calls after ten hours.

User-defined Expressions, on the other hand, don't store a value. Instead, they provide a way of associating a name with some mathematical expression. Whenever the name is referenced in the model, the associated expression is evaluated and its numerical value is returned. Typically, expressions are used to compute values from a distribution or from a complex equation based on the entity's attribute values or even current system variable values. If the mathematical expression is used in only one place in the model, it might be easier to enter it directly where it is required. However, if the expression is used in several places or the form of the expression to be used depends on an entity attribute, a user-defined expression is often better. For our call center, we'll use the Expressions module to define an expression to generate the technical support times.

Variables and Expressions have many other uses that we hope will become obvious as you become more familiar with Arena.

5.2.4 Storages

Storages are an Arena concept that allow the user to animate the presence of entities that fall outside the normal animation features that you have seen to this point. Our previous models all showed an animated entity in a Queue, on a Resource, or moving along a Route. Consider an example where you have an entity arrival that initially enters a delay before it is allowed to proceed to the next module. If the delay represents a travel time from one area of the system to another, you might use the concept of Stations and Routes. This would allow you to animate the movement of an entity through the system, as we did in Model 4-4. But if during the delay the entity is not moving, to animate it we need the Storages concept.

A Storage holds a non-ordered set of entities that are waiting for an event to occur; e.g., the conclusion of a predefined delay time, the arrival of a requested transporter that is en route, a manual removal from the Storage, etc. You might think of a Storage as a variable that we can increment and decrement with the added feature that we can animate the entities that are in a Storage. Placing an entity in a Storage does not prevent the entity from progressing through model logic. For example, we could place an entity in a Storage, and then show it on the animation, as it entered a delay, and then continued through the model logic until we decided to remove it from the Storage at a later point in the model. During the entire time, the entity would be displayed on the animated Storage.

There are two ways to place an entity in a Storage and then remove it. The first uses the Store and Unstore modules found in the Advanced Process panel. The second way is to use a module that has a field to enter a Storage ID. An example is the Delay module from the Blocks panel. We'll show you how to use both methods during our model development.

5.2.5 Terminating or Steady-State

Most (not all) simulations can be classified as either terminating or steady-state. This is primarily an issue of intent or the goal of the study rather than having much to do with internal model logic or construction.

A *terminating* simulation is one in which the model dictates specific starting and stopping conditions as a natural reflection of how the target system actually operates. As the name suggests, the simulation will terminate according to some model-specified rule or condition. For instance, a store opens at 9 AM with no customers present, closes its doors at 9 PM, and then continues operation until all customers are "flushed" out. Another example is a job shop that operates for as long as it takes to produce a "run" of 500 completed assemblies specified by the order. The key notion is that the time frame of the simulation has a well-defined (though possibly unknown at the outset) and natural end, as well as a clearly defined way to start up.

A *steady-state* simulation, on the other hand, is one in which the quantities to be estimated are defined in the long run; that is, over a theoretically infinite time frame. In principle (though usually not in practice), the initial conditions for the simulation don't matter. Of course, a steady-state simulation has to stop at some point, and as you might guess, these runs can get pretty long, so you need to do something to make sure that you're running it long enough. This is an issue we'll take up in Section 7.2. For example, a pediatric emergency room never really stops or restarts, so a steady-state simulation might be appropriate. Sometimes people do a steady-state simulation of a system that actually terminates in order to design for some kind of worst-case or peak-load situation.

We now have to decide which to do for this call center model. Although we'll lead you to believe that the distinction between terminating or non-terminating systems is very clear, that's seldom the case. Some systems appear at first to be one type, but on closer examination, they turn out to be the other. This issue is further complicated by the fact that some systems have elements of both types, and system classification may depend on the types of questions that the analyst needs to address. For example, consider a fast-food restaurant that opens at 11 AM and closes at 11 PM. If we were interested in the daily operational issues of this restaurant, we'd analyze the system as a terminating system. If we were interested only in the operation during the rush that occurs for two hours over lunch, we might assume a stationary arrival process at the peak arrival rate and analyze the system as a steady-state system.

Our call center definitely appears to be a terminating system. The system would appear to start and end empty and idle, so we're going to proceed to analyze our call center system as a terminating system.

5.3 Modeling Approach

In Figure 1-2 of Chapter 1, we briefly discussed the Arena hierarchical structure. This structure freely allows you to combine the modeling constructs from any level into a single simulation model. In Chapters 3 and 4, we were able to develop our models using mostly only the constructs found in the Basic Process panel (yes, we planned it that way), although we did require the Advanced Process panel for our failure and special statistics, and the Advanced Transfer panel for some entity movement.

The general approach that we recommend when you are creating your models is to stay at the highest level possible for as long as you can. However, as soon as you find that these high-level constructs don't allow you to capture the necessary detail, we suggest that you drop down to the next level for some parts of your model rather than sacrifice the accuracy of the simulation model (of course, there are elements of judgment in this kind of decision). You can mix modeling constructs from different levels and panels in the same model. As you become more familiar with the various panels (and modeling levels), you should find that you'll do this naturally. Before we proceed, let's briefly discuss the available panels.

The Basic Process panel provides the highest level of modeling. It's designed to allow you to create high-level models of most systems quickly and easily. Using a combination of the Create, Process, Decide, Assign, Record, Batch, Separate, and Dispose modules allows a great deal of flexibility. In fact, if you look around in these modules, you'll find many additional features we haven't yet discussed. In many cases, these modules alone will provide all the detail required for a simulation project. These modules provide common functions required by almost all models, so it's likely that you'll use them regardless of your intended level of detail.

The Advanced Process panel augments the Basic Process panel by providing additional and more detailed modeling capabilities. For example, the sequence of modules Seize – Delay – Release provides basically the same fundamental modeling logic as a Process module, but could afford more flexibility than the Process module. The handy feature of the Advanced Process panel modules is that you can put them together in almost any combination required for your model. In fact, many experienced modelers start at this level because they feel that the resulting model is more transparent to the user, who may or may not be the modeler.

The Advanced Transfer panel provides the modeling constructs for material-handling activities (like transporters and conveyors) and entity movement in general. Similar to the general modeling capabilities provided by the Advanced Process panel, the Advanced Transfer panel modules give you more flexibility.

The Blocks panel (part of the SIMAN template) provides an even lower level of modeling capability. In fact, it provides the basic functionality that was used to create all of the modules found in the Basic Process, Advanced Process, and Advanced Transfer panels. In addition, it provides many other special-purpose modeling constructs not available in the higher-level modules. Examples would include "while" loops, combined probabilistic and logic branching, and automatic search features. You might note that several of the modules have the same names as those found in the higher-level panels. Although the names are the same, the modules are not. You can distinguish between the two by the color and shape.

The difference between the Blocks panel on the one hand and the Basic Process, Advanced Process, and Advanced Transfer panels on the other hand is easy to explain if you've used SIMAN previously, where you define the model and experiment frames separately, even though you may do this all in Arena. The difference is perhaps best illustrated between the Assign modules on the two panels. When you use the Basic Process panel Assign module, you're given the option of the type of assignment you want to make. If you make an assignment to a new attribute, Arena will automatically define that new attribute and add it to the drop-down lists for attributes everywhere in your model. One reason for staying at the highest level possible is that the Blocks Assign module only allows you to make the assignment—it doesn't define the new attribute. Even with this shortcoming, there are numerous powerful and useful features available only in the Blocks panel.

In addition, the Elements panel hosts the experiment frame modules. This is where, for example, you find the Attributes module to define your new attribute. You'll rarely need these features since they're combined with the modules of the Arena template, but if you need this lowest level for a special modeling feature (you can go to Visual Basic or C, if you're a real glutton for punishment), it's available via the same Arena interface as everything else.

The Blocks and Elements panels also provide modules designed for modeling continuous systems (as opposed to the discrete process we've been looking at). We'll see these in Chapter 11, along with the Flow Process panel specifically designed for this.

Now let's return to our call center system, which *does* require features not found in the Basic Process panel modules. In developing our model, we'll use modules from the Basic Process, Advanced Process, and Blocks panels. In some cases, we'll use lower-level constructs because they are required; in other cases, we'll use them just to illustrate their capabilities. When you model with lower-level constructs, you need to approach the model development in a slightly different fashion. With the higher-level constructs, we tend to group activities and then use the appropriate modules. With the lower-level constructs, we need to concentrate on the actual activities. In a sense, your model becomes a detailed *flowchart* of these activities. Unfortunately, until you're familiar with the logic in the available modules, it's difficult to develop that flowchart.

5.4 Building the Model

At this point, let's divide our model into sections and go directly to their development where we can simultaneously show you the capabilities available. The seven sections, in the order in which they'll be presented, are:

> Section 5.4.1: Create Arrivals and Direct to Service,
> Section 5.4.2: Arrival Cutoff Logic,
> Section 5.4.3: Technical Support Calls,
> Section 5.4.4: Sales Calls,
> Section 5.4.5: Order-Status Calls,
> Section 5.4.6: System Exit and Run Setup, and
> Section 5.4.7: Animation.

As you read through the next seven sections on building the model, you might get the impression that this is exactly how we developed the model. Ideally one would like to be able to plan the construction of the model so it's that easy. In reality, though, you often end up going back to a previously developed section and adding, deleting, or changing modules or data. So as you start to develop more complex models, don't assume that you'll always get it right the first time (we certainly didn't).

5.4.1 *Create Arrivals and Direct to Service*

As you learn more about the capabilities of Arena and develop more complex models, you'll find it necessary to plan your logic in advance of building the model. Otherwise, you may find yourself constantly moving modules around or deleting modules that were erroneously selected. Whatever method you adopt for planning your models will normally come about naturally, but you might want to give it some thought before your

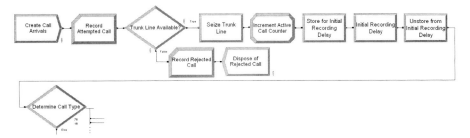

Figure 5-1. Arrivals and Service Selection Process Logic

models become really complicated—and they will! A lot of modelers use pencil (a pen for the overconfident) and paper to sketch out their logic using familiar terminology. Others use a more formal approach and create a logic diagram of sorts using standard flowcharting symbols. Still another approach is to develop a sequential list of activities that need to occur. We've developed such a list for the logic that's required to create and direct arrivals (which happens to look a lot like the flowchart modules in Figure 5-1):

```
Create arriving calls
Record attempted call
        If a trunk line is available - Seize
                Increment call counter
                Delay for recording
                Determine call type
                Direct call type
        Else (all trunk lines are busy)
                Count rejected call
                Dispose of call
```

These types of approaches will help you formalize the modeling steps and reduce the number of errors and model changes. As you become more proficient, you may even progress to the point where you'll start by laying out your basic logic using Arena modules (after all, they're similar to flowchart elements). However, even then we recommend that you lay out your complete logic before you start filling in the details.

The Create module, which creates entities (arrivals) according to our previously defined process, is described in Display 5-1.

Name	Create Call Arrivals
Entity Type	Incoming Call
Time Between Arrivals	
Type	Random (Expo)
Value	0.857
Units	Minutes
Max Arrivals	MaxCalls

Display 5-1. Creating the Call Arrivals

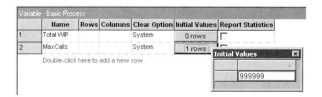

Figure 5-2. The Variable Data Module

We used a very similar Create module in the models in Chapters 3 and 4, with the exception of the entry in the Max Arrivals field. Remember that the system closes to new calls after 6 PM. We'll show you how to use the variable MaxCalls to terminate the call arrivals in the next section.

In most cases where you define a new object (variable, attribute, etc.) in the modules from the Basic Process and Advanced Process panels, Arena will automatically make an entry in the appropriate data module. This always happens when you specify the Type; e.g., variable, attribute, etc. However, in this case, we simply enter the MaxCalls variable without specifying the type. Therefore, we need to define our variable MaxCalls. We do this in the Variable data module found in the Basic Process panel, seen in Figure 5-2. Open this data module by clicking on it and double-clicking to add a new row. Next type in the Name of our new variable, MaxCalls. Then click on the Initial Values cell to open the spreadsheet window where you can enter the initial value. We entered a value of 999999, which means that the arrival process will be terminated after 999999 arriving calls (clearly a ruse that we'll explain below).

As we continue to develop this model, we'll sometimes be thinking ahead to how we would like to animate our model. Let's assume that we want our arriving call to be animated as a black ball. We could assign a picture using an Assign module, but we specified in our Create module that the entity type was Incoming Call so at this point we should click on our Entity data module and select Picture.Black Ball from the drop-down list in the Initial Picture cell.

At this point in the model, we want to record the number of attempted calls using a Record module, as shown in Display 5-2.

We next need to check for an available trunk line. If a line is available, we seize that line; otherwise, the call is rejected from the system. In Section 5.2.1, we discussed two obvious ways (at least to us) to accomplish this. For this model, we'll use a Decide module followed by a Seize module (in the Advanced Process panel) if a line is available.

Name	Record Attempted Call
Type	Count
Value	1
Counter Name	Attempted Calls

Display 5-2. Counting the Attempting Calls

Name	Trunk Line Available?
Type	2-way by Condition
If	Expression
Value	NR(Trunk Line) < MR(Trunk Line)

Display 5-3. Determining if a Trunk Line is Available

The Decide module is used to check whether any units of the `Trunk Line` resource are available. You could use the Expression Builder to develop this condition. The expression `NR(Trunk Line) < MR(Trunk Line)` contains the Arena variables MR and NR. The variable NR is the current number of busy units of the resource, and the variable MR is the current number of units scheduled of the resource, or the total number of units of the resource currently in existence (busy or otherwise). The entries for the Decide module are shown in Display 5-3.

A condition can be any valid expression, but it typically contains some type of comparison. Such conditions may include any of the notations shown in Table 5-1. Logical expressions can also be used.

If a unit of the resource `Trunk Line` is available, we allow the entity to proceed to the Seize module (Display 5-4 from the Advanced Process panel) where we seize that resource. Normally, we would have used a Process module, but in this case, we need to separate the seize part of the Process module from the delay part (you'll see why shortly). Unlike the Process module, the Seize module only includes the queue and seize actions. When you define the resource in the Seize module, that resource will be entered into the Resource data module (Basic Process panel) automatically, but its capacity will default to 1. Since there are 26 trunk lines, you'll need to open the Resource data module and change the capacity to 26.

Table 5-1. The Decide Module Condition Notation

Description	Syntax Options		Description	Syntax Options	
And	.AND.		Or	.OR.	
Greater than	.GT.	>	Greater than or equal to	.GE.	>=
Less than	.LT.	<	Less than or equal to	.LE.	<=
Equal to	.EQ.	==	Not equal to	.NE.	<>

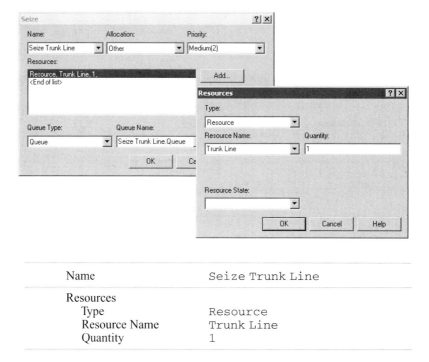

Display 5-4. The Seize Module

Name	Seize Trunk Line
Resources	
Type	Resource
Resource Name	Trunk Line
Quantity	1

After seizing one unit of the resource `Trunk Line`, the entity is sent to the following Assign module (in Display 5-5) where we increment the variable `Total WIP`. This variable will be used to keep track of how many active calls are currently in the system. Later on we'll use this variable to determine when to stop the simulation run.

The entity then enters a Store module (from the Advanced Process panel) where we place the entity in the Storage `Initial Recording Delay Storage`, in Display 5-6. Even though we think of this as placing the entity in the storage, it does not prevent the entity from proceeding in the model logic. We use this Storage so we can show the entity on the animation during the initial recording delay. We'll show you how to add this storage animation in Section 5.4.7.

Name	Increment Active Call Counter
Assignments	
Type	Variable
Variable Name	Total WIP
New Value	Total WIP + 1

Display 5-5. Assign Module to Increment the Active Call Counter

Name	Store for Initial Recording Delay
Type	Storage
Storage Name	Initial Recording Delay Storage

Display 5-6 Placing the Entity in a Storage

Once the entity is in the storage, it enters a Delay module where it incurs a UNIF(0.1,0.6) delay representing the recording and option-selection time, in Display 5-7.

Upon completing the delay, it enters an Unstore module where the entity is removed from the storage where it was placed earlier. You might note that we have defaulted on the Type entry, in Display 5-8. The default option will remove an entity from the last storage that it entered, which was the Initial Recording Delay Storage.

Name	Initial Recording Delay
Delay Time	UNIF(0.1 , 0.6)
Units	Minutes

Display 5-7. The Initial Recording Delay

Name	Unstore from Initial Recording Delay

Display 5-8. Removing the Entity from the Storage

Name	Determine Call Type
Type	N-way by Chance
Percentages	76
	16

Display 5-9. Determining the Call Type

The call then is sent to the Decide module, described in Display 5-9, where the call type is determined according to the probabilities originally stated. We selected the `N-way by Chance` type and specified percentages of `76` and `16`, which represent the technical support and sales calls. The Else exit point will be used to direct the order-status calls (the remaining 8%).

The first branch from the Decide represents a technical support call. Entities taking this branch are sent to an Assign module where we assign the tech support call section of our model. Entities taking the second branch are sent to the sales call section of our model, and the remaining entities, the third branch, are sent to the order-status calls section.

Having dealt with the arriving calls that are successfully allocated a trunk line, we must now deal with those luckless calls that receive a busy signal—the rejected entities. These are sent to a Record module where we count (record) our rejected calls in the counter `Rejected Calls`, Display 5-10.

These entities are finally sent to a Dispose module where they exit the system.

5.4.2   Arrival Cutoff Logic

In our problem description in Section 5.1, we indicated that the system closes to new calls after 6 PM, but all calls that entered the system before that time are answered. This means that we allow calls to enter the system from 8 AM to 6 PM, a total of 600 minutes, but we need to turn off the arrivals after 600 minutes. Normally we would end the simulation run at time 600. However, before we terminate the simulation run, we need to handle any calls that arrived before time 600. We'll show you how to determine when all the calls have been taken care of and how to terminate the run later.

One obvious way to shut off arrivals after 600 minutes would be to insert a Decide module that the arriving calls would enter immediately after the Create module that creates the arriving calls. The Decide module would check to see if the current simulation time was less than 600. If true, the entity would be allowed to continue;

Name	Record Rejected Call
Type	Count
Value	1
Counter Name	Rejected Calls

Display 5-10. The Record Module: Counting the Rejected Calls

Figure 5-3. Arrival Cutoff Logic

otherwise, the entity would be sent to a Dispose module. Although this method would shut off the arrivals (though would not terminate the run), we have chosen to be a little tricky and introduce the concept of a logical entity to cut off our arriving calls.

In the models developed so far, every entity that we have created has represented some physical object, a call in the case of our current model. Logical entities are just like any other entities but they are created and used to perform some logical or control task. In this case, we create a logical entity that will cut off the incoming calls and then be destroyed. The activities to accomplish this are shown here and the flowchart modules are shown in Figure 5-3:

```
Create arrival cutoff entity
Cut off incoming calls
Dispose of entity
```

You might notice that these three modules appear to be totally independent of the rest of our model. They are, except that this section of our model will interact with the rest of the model via a global Variable. We start by creating our control entity with a Create module, in Display 5-11. We have specified a Constant arrival type with a value of 999999. If this was all we entered, we would have an arrival at time 0 and then every 999999 time units. However, we have specified that the first arrival will occur at time 600 (First Creation field) and that there will be only one arrival (Max Arrivals field). As it turns out, we could have selected any Type and any Value or set of values because they are never used.

Name	Create Arrival Cutoff Entity
Entity Type	Arrival Cutoff
Time Between Arrivals	
Type	Constant
Value	999999
Units	Minutes
Max Arrivals	1
First Creation	600

Display 5-11. Creating the Logical Entity

Name	Cut Off Incoming Calls
Assignments	
Type	Variable
Variable Name	MaxCalls
New Value	1

Display 5-12. Assign the Cutoff Variable

The single logical entity created at time 600 enters the following Assign module where we assign the variable MaxCalls a value of 1, in Display 5-12.

By setting the value of the variable MaxCalls to 1, we have stopped the creation of arriving calls. We are assuming that you might be a little confused at this point—you should be. Let's fill in the details so you will understand how this works. At the beginning when we populated the Create module that creates the arriving calls, in Display 5-1, we entered the variable MaxCalls in the Max Arrivals field. We also assigned that Variable an initial value of 999999 in the Variable data module. Because we are at time 600 in our simulation run, we can be assured that we have had more than one arrival, since the First Creation was at time 0. By now (at time 600) setting the value of MaxCalls to 1 we are shutting off our arriving call stream.

Having accomplished its single task in life, our logical entity is now sent to a Dispose module that destroys entities, thus ending our logic-modeling section.

5.4.3 Technical Support Calls

In Section 5.4.1, we used a Decide module to determine the call type. The arriving calls that were determined to be technical support calls would have taken the first branch and been sent to this portion of the model. The activities to service the technical support calls are listed here and the flowchart modules are shown in Figure 5-4:

```
Assign entity type and picture
Delay for recording
Determine call type and direct
Seize technical support resource
Delay for service
Release technical support resource
Send to system exit
```

If a caller selects the technical support option, the entities will be sent to this portion of our model, where we first assign the Entity Type to be Tech Call, and the Entity Picture to be Picture.Red Ball, as in Display 5-13.

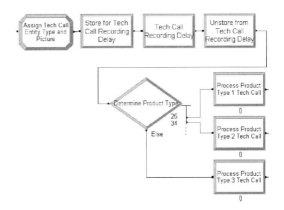

Figure 5-4. Technical Support Call Logic

Here a second recording requests which of the product types the caller is using. In Section 5.4.1, we decided to place an incoming call in a Storage for the first recording so we could show it on our animation. We'll do the same here, so we'll first place that entity in the storage `Tech Call Recording Delay Storage`, as shown below in Display 5-14.

Name	Assign Tech Call Entity Type and Picture
Assignments	
Type	Entity Type
Entity Type	Tech Call
Type	Entity Picture
Entity Picture	Picture.Red Ball

Display 5-13. Assign Tech Call Entity Type and Picture

Name	Store for Tech Call Recording Delay
Type	Storage
Storage Name	Tech Call Recording Delay Storage

Display 5-14. Technical Support Call Storage

Name	Tech Call Recording Delay
Delay Time	UNIF(0.1 , 0.5)
Units	Minutes

Display 5-15. The Technical Support Recording Delay

Once the entity has been placed in the storage, it enters the following Delay module where it incurs a UNIF(0.1,0.5) delay representing the recording and option-selection time, as shown in Display 5-15.

Upon completing the delay, it enters an Unstore module where the entity is removed from the storage where it had been placed, as shown in Display 5-16. Again, the default option will remove the entity from the last storage that it entered.

The call is sent to the Decide module next, as described in Display 5-17, where the product type is determined according to the probabilities originally stated. We selected the N-way by Chance type and specified percentages of 25 and 34, which represent the Type 1 and Type 2 technical support calls. The Else exit point will be used to direct the product Type 3 calls (the remaining 41%).

Each of the three branches from this Decide module connects to a Process module to service that type of technical support call. The entries for each of these three modules are very similar, so we'll only walk you through the first, for product Type 1 technical support, in Display 5-18. We've selected the Seize Delay Release Action and have requested to seize 1 unit of resource Tech 1. As with the Trunk Line resource, you also need to open the Resource data module and change the Capacity from the default (1) to 2. We then delay for a time found in the expression Tech Time, which we have not yet defined. After the delay, the Tech 1 resource is released and the completed call is sent to the exit portion of our model.

Name	Unstore from Tech Call Recording Delay

Display 5-16. Removing the Entity from the Storage

Name	Determine Product Type
Type	N-way by Chance
Percentages	25
	34

Display 5-17. Determining the Tech-Support Product Type

Name	Process Product Type 1 Tech Call
Action	Seize Delay Release
Resources Type	Resource
Resource Name	Tech 1
Delay Type	Expression
Units	Minutes
Expression	Tech Time

Display 5-18. The Technical Support Call Process Module for Product Type 1

Figure 5-5. The Expression Data Module

The expression `Tech Time` is defined using the Expression data module found in the Advanced Process panel. Click on the module, double-click to add a new row, and enter the expression Name, `Tech Time`. Then click on the Expression Values cell, which will open the spreadsheet window, and enter the expression `TRIA(3, 6, 18)`, as shown in Figure 5-5.

The remaining two Process modules in this section accomplish the same task for product Types 2 and 3 that need to seize a `Tech 2` and `Tech 3` resource, respectively. The capacities for both of these resources also need to be set to `3` in the Resource data module.

Next we'll develop the sales calls portion of our model.

5.4.4 Sales Calls

The sales calls from our Decide module in Section 5.4.1 are sent to this portion of our model. The logic for the sales calls is very similar to that described for the technical support calls. However, there are far fewer modules required (actually only two) since we don't have the complexities of multiple product types and different resources. The required logic steps for the sales calls are shown here and the flowchart modules are shown in Figure 5-6:

```
Assign entity type and picture
Seize sales call resource
Delay for service
Release sales call resource
```

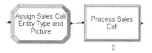

Figure 5-6. Sales Call Logic

An arriving sales call is sent to an Assign module where we assign the Entity Type to `Sales Call`, and the Entity Picture to `Picture.Green Ball`, as in Display 5-19.

It then proceeds to a Process module (Display 5-20) where one unit of the `Sales` resource is seized, it delays for service `TRIA(4, 15, 45)` minutes, and releases its unit of the `Sales` resource.

5.4.5 Order-Status Calls

The order-status calls are handled automatically, so at least initially they don't require a resource other than one unit of `Trunk Line`. The logic steps for order-status calls are:

Name	`Assign Sales Call Entity` `Type and Picture`
Assignments	
Type	`Entity Type`
Entity Type	`Sales Call`
Type	`Entity Picture`
Entity Picture	`Picture.Green Ball`

Display 5-19. Assign Sales Call Entity Type and Picture

Name	`Process Sales Call`
Action	`Seize Delay Release`
Resources	
Type	`Resource`
Resource Name	`Sales`
Delay Type	`Triangular`
Units	`Minutes`
Minimum	`4`
Most likely	`15`
Maximum	`45`

Display 5-20. The Sales Call Process Module

```
Assign entity type and picture
Delay for automated call
If customer wants to speak to a real person
    Set customer priority
    Seize sales person
    Delay for call
    Release sales person
```

The Arena logic is shown in Figure 5-7.

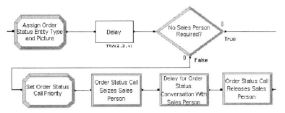

Figure 5-7. The Order-Status Calls Process Logic

The arriving order-status call is sent to an Assign module (Display 5-21) where we assign the entity Type to `Order Status Call`, and the Entity Picture to `Picture.Blue Ball`.

The order-status calls are handled automatically by the phone system, which means that they initially incur a delay like the arriving and the technical support calls. As with the previous two delays, we want to show the call on the animation so we could use a Store–Delay–Unstore module combination like we did for the first two delays. But this time let's take a different approach and use a Delay module from the Blocks panel. By using this module, we can capture the same functionality as we did with the previous combination of modules since it contains the Store/Unstore logic implicitly. The entity will be placed in the defined storage, it will incur a delay, and the entity will be removed from the storage. We made entries for the Label (same as Name), the Duration, and the Storage ID, as shown in Display 5-22.

Name	Assign Order Status Entity Type and Picture
Assignments	
Type	`Entity Type`
Entity Type	`Order Status Call`
Type	`Entity Picture`
Entity Picture	`Picture.Blue Ball`

Display 5-21. Assign Order-Status Call Entity Type and Picture

Label	Order Status Delay
Duration	TRIA(2, 3, 4)
Storage ID	Order Status Delay Storage

Display 5-22. The Delay Module from the Blocks Panel

Note that when you click on the Delay module there is no associated spreadsheet view; this will be the case for any modules from the Blocks or Elements panel. When you use modules from the Blocks panel, you also need to be aware of which Base Time Units you specified in the *Run > Setup > Replication Parameters* dialog and make sure that the units for any time values (e.g., the Duration in our Delay module) are the same. In our model, we've specified minutes, so we are consistent. If you had chosen to use the Delay module from the Advanced Process panel, it would have included an entry for the time units. When we used the Store module to place an entity in a storage, Arena automatically entered the storage name in the Storage data module (Advanced Process panel). That's not the case for the Delay module from the Blocks panel. So now you'll need to click on the Storage data module and enter the storage name, `Order Status Delay Storage`.

Having completed the delay, the entity enters a Decide module, in Display 5-23, where we decide whether it is necessary to connect to a live sales person. If not, the entity is sent to the system exit; otherwise, it needs to seize a sales person to complete the call. We arbitrarily decided that "True" means that it is indeed the case that no sales person is required (hence the module Name and 85 entry for Percent True).

Name	No Sales Person Required?
Type	2-way by Chance
Percentage True	85

Display 5-23. Determining Whether a Sales Person Is Required for an Order-Status Call

In our initial problem description, we indicated that order-status calls are routed to the sales staff where they will wait with a lower priority than sales calls. This means that if an order-status call is in a queue waiting for a sales person and a new sales call enters, the sales call will be given priority over the order-status call and be answered first (though the new sales call would not preempt an order-status call that might already be in progress). At this point, in order to find a solution to our problem, we need to digress and explain how Arena allocates resources to waiting entities.

We hope by now that this allocation process is fairly clear when all entities requesting a resource are resident in the same queue and have the same priority. However, if there are several places within a model (different Queue – Seize combinations) where the same resource could be allocated, some special rules apply. Let's first consider the various circumstances under which a resource could be allocated to an entity:

- an entity requests a resource and the resource is available,
- a resource becomes available and there are entities requesting it in only one of the queues, or
- a resource becomes available and there are entities requesting the resource in more than one queue.

It should be rather obvious what occurs under the first two scenarios, but we'll cover them anyway.

If an entity requests a resource and the resource is available, you guessed it, the resource is allocated to the entity. In our second case, when a resource becomes available and there are entities in only one of the queues, then the resource is allocated to an entity in that queue. Here the determining factor for *which* entity in the queue gets the resource is the queue-ranking rule used to order the entities. Arena provides four ranking options: First In First Out (FIFO), Last In First Out (LIFO), Low Value First, and High Value First. The default, FIFO, ranks the entities in the order that they entered the queue, so the entity that's been in the queue for the longest time would get the resource. LIFO puts the most recent arrival at the front of the queue (like a push/pop stack), so the entity that most recently joined the queue would get the resource. The last two rules rank the queue based on the value of an attribute of entities in it; it's up to you to define an expression for such an attribute. For example, as each entity arrives in the system, you might assign a due date to an attribute of that entity. If you selected Low Value First based on this due-date attribute, you'd have the equivalent of an earliest-due-date queue-ranking rule. As each successive entity arrives in the queue, it is placed in position based on increasing due dates.

The last case, where entities in more than one queue request the resource, is a bit more complicated. Arena first checks the seize priorities; if one of the seize priorities is a smaller number (higher priority) than the rest, the resource is allocated to the first entity in the queue preceding that seize. We could employ this method and return to the Process module we used in our sales call logic and change the Priority selection to High(1). We would then use another Process module to seize the sales resource for our order-status calls and set the Priority selection to Medium (2) or Low (3). This would solve our resource allocation problem. But let's continue our discussion on resource allocation and see if there is another way.

If all priorities are equal, Arena applies a default tie-breaking rule, and the resource is allocated based on the entity that has waited the longest regardless of the queue it's in. Thus, it's essentially a FIFO tie-breaking rule among a merger of the queues involved. This means that if your queues were ranked according to earliest due date, then the entity that met that criterion might not always be allocated the resource. For example, a late job might have just entered a queue, whereas an early job may be at the front of another queue and has been waiting longer.

Arena provides a solution to this potential problem in the form of a *shared queue*. A shared queue is just what its name implies—a single queue that can be shared by two or more seize activities. This allows you to define a single queue where all entities requesting the resource will be placed regardless of where their seize activities are within your model. Arena performs the bookkeeping to ensure that an entity in a shared queue continues in the proper place in the model logic when it's allocated a resource.

In our model, each sales call and a small percentage (15%) of the order-status calls will require a unit of the `Sales` resource. Since these activities are modeled separately, we'll use a shared queue when we attempt to seize a unit of the `Sales` resource. Both types of calls are allocated a unit of the `Sales` resource based on a FIFO rule, or the longest waiting time, so a shared queue is not *required* in our model to assure that the proper call is allocated the resource. However, a shared queue also allows us to collect composite statistics easily on the total number of calls waiting for a `Sales` resource and provides the ability to show all of these waiting calls in the same queue—if we decide to animate that queue.

Since we did not assign the sales calls a priority, we'll take advantage of the fact that for an entity with an unassigned value for an attribute, that value will default to zero. Now we'll use an Assign module to define a new attribute, `Sales Call Priority`, to those order-status calls requiring a unit of the `Sales` resource and set its value to `1`, as in Display 5-24.

The result is that sales calls will have a `Sales Call Priority` with a value of 0 and order-status calls will have a `Sales Call Priority` with a value of 1. Next we'll send the entity to a Seize module. You might ask why we are using a Seize module rather than a Process module. It turns out that if we use a Process module, the queue name defaults to "module name" with a ".queue" extension, which can't be changed. Since we want to use the same queue that was defined for the sales calls seize, we need to use a Seize module instead, which allows us to change the default queue name.

Label	Set Order Status Call Priority
Type	Attribute
Attribute Name	Sales Call Priority
New Value	1

Display 5-24. Assigning the Call Priority

Name	Order Status Call Seizes Sales Person
Resources Type Resource Name Quantity	 Resource Sales 1
Queue Name	Process Sales Call.Queue

Display 5-25. The Order-Status Seize Module

We make this queue a shared queue by first placing and filling out the Seize module and selecting the name of the queue, `Process Sales Call.Queue`, from the drop-down list, as shown in Display 5-25.

Once this is completed, you'll go to the Basic Process panel and open the Queue data module. Your queue will already be in the spreadsheet view, and you'll need to check the Shared box to make it available as a shared queue. Next, click on the Type cell for our queue and select the `Lowest Attribute Value` option. This allows you to select an attribute from the pull-down list in the Attribute Name cell. You should select the Attribute Name `Sales Call Priority`.

Now both the sales call and the order-status call entities will reside in the same queue. Since the sales call entities have a priority of 0, they will always be allocated an available sales resource first. The order-status call entity will only be allocated an available resource if there are no sales call entities currently in the queue. This solves our resource allocation problem.

After seizing a Sales resource, the entity enters a Delay module, as shown in Display 5-26, with a delay of TRIA(3 , 5 , 10) minutes.

It is next sent to a Release module (Advanced Process panel) where we release the sales resource, shown in Display 5-27.

Name	Delay for Order Status Conversation With Sales Person
Delay Time	TRIA(2, 3, 4)
Units	Minutes

Display 5-26. The Order Status Sales Delay

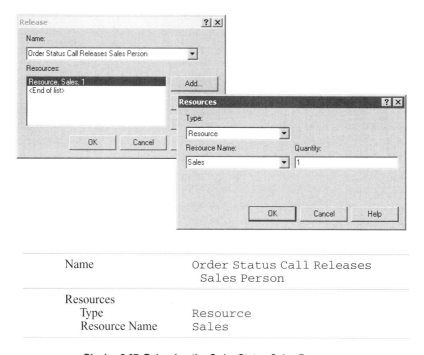

Name	Order Status Call Releases Sales Person
Resources	
Type	`Resource`
Resource Name	`Sales`

Display 5-27. Releasing the Order-Status Sales Resource

After releasing the sales resource, the entity is sent to the system exit.

5.4.6   System Exit and Run Setup

Having completed their mission in life, the calls from the technical support, sales, and order-status sections are all sent to this portion of our model. The logic steps are shown here and the flowchart modules are shown in Figure 5-8:

```
Release Trunk Line
Decrement active call counter
Record completed calls
Dispose of entity
```

The arriving entities enter a Release module where we release the `Trunk Line` resource, in Display 5-28.

Exiting the Release module, they enter an Assign module where we decrement the active call counter, the variable `Total WIP`, shown in Display 5-29.

We then use a Record module to record the completed call in the counter `Completed Calls`, in Display 5-30.

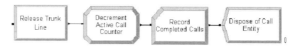

Figure 5-8. The System Exit Logic

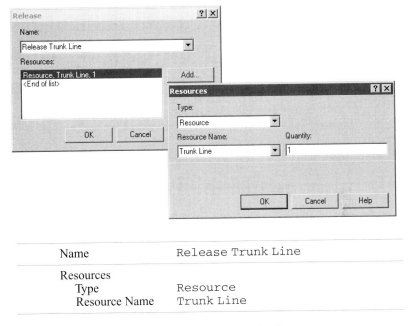

Name	Release Trunk Line
Resources	
Type	Resource
Resource Name	Trunk Line

Display 5-28. Releasing the Trunk Line Resource

The entity is finally sent to a Dispose module where it is destroyed.

Before we can test our model, we need to open the Run Setup dialog and make entries for the Project Title, Analyst Name, and Project Description found under the Project Parameters tab. We also accepted the defaults under the Statistics Collection section, which means that we'll automatically get statistics on the entities, resources, and queues. Now go to the Replication Parameters tab where we requested a single replication with no warm-up period, an infinite replication length, and where we specified the base time units as minutes.

Name	Decrement Active Call Counter
Resources	
Type	Variable
Variable	Total WIP
New Value	Total WIP - 1

Display 5-29. Decrement the Active Call Counter

Name	Record Completed Calls
Counter Name	Completed Calls

Display 5-30. Record the Completed Calls

At this point, you might remember (we are guessing you forgot by this time) that way back in the early part of Section 5.4.1 we told you that we were going to use the variable `Total WIP` to help determine when to stop the simulation. The call center accepts incoming calls only from 8 AM to 6 PM (600 minutes), but any calls that are in the system at 6 PM must be completed. We could have just set an arbitrarily long replication time, but this would have resulted in inaccurate results, specifically on resource utilizations and other time-persistent statistics. There are two conditions that must be true before we can shut down the system and end the simulation run. The first is that the simulation time must be at least 600 minutes. The second condition is that we must have completed all calls. Since the variable `Total WIP` is used to keep track of how many calls are currently in the system, we only need to check whether it is zero, which implies that all calls have been completed. So we'll enter the expression

```
TNOW >= 600.0  &&  Total WIP == 0
```

into the Terminating Condition field; `&&` means "and." Actually, we used the Expression Builder to construct the expression.

We defined this variable specifically for our purposes, although we also could have used the Arena variable `NR(Trunk Line)`. This variable tells us how many trunk lines are currently busy, so when its value is zero, there are no active calls in the system.

By now you should have a fairly good understanding of how to use the modules from the Advanced Process panel, and hopefully, have a better grasp of what the major modules in the Basic Process panel provide. You might have noticed that the logic for the last two sections, sales and order-status calls, have several common components. By defining and assigning a call type and expressions for the time delays, we could have developed one general section of logic that would have handled both call types. (We did consider doing that.) However, this approach can sometimes become quite complicated, and this type of logic is seldom obvious to another modeler (or client) if someone else has to use or modify your model. Unless you can combine significant amounts of model logic and make it relatively transparent, we recommend that you lay out each section of logic separately, even if doing so might be a bit redundant at times. Basically, this is a matter of taste.

That completes our model development, but before we attempt to run our model, let's develop our animation.

5.4.7 Animation

We're not going to get fancy with the animation for our model, but we do want to represent the model activities. We'll animate our recording delays (Storages), resources, resource queues, the number of active calls in each section of the model, and the total number of trunk lines in use. Our final animation is shown in Figure 5-9. This shows the status at approximately 281.6 minutes into the simulation run.

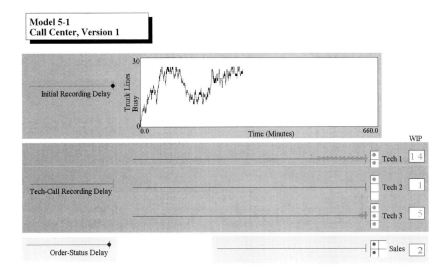

Figure 5-9. The Simple Call Center System Animation

Let's start by placing the storage animations for the recording delays. The storage animation object is located in the Animate Transfer toolbar, which you may need to add to your model. Selecting the *Storage* button will open the Storage dialog shown in Display 5-31. You can then select the storage you want to animate from the Identifier drop-down list. An animated storage contains the same options as an animated queue. We elect to use the defaults, which will result in a line storage. Pressing *OK* places the line storage on your animation. When an entity is in this storage, it will appear on our animation as if it were in a queue.

Having placed the storage for the initial recording delay, we then place the storages for the technical-support call and order-status recording delays.

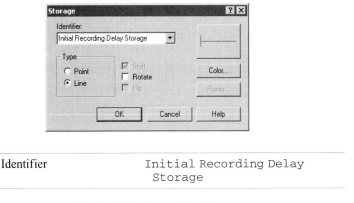

Identifier	Initial Recording Delay Storage

Display 5-31. Storage Animation

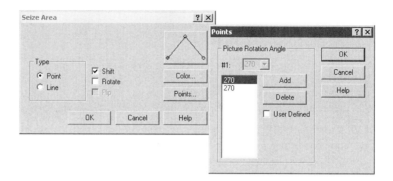

Figure 5-10. Adding Multiple Seize Points

Each time we placed a Process or Seize module in our model, Arena automatically provided a free animated queue. You could simply move these queues to the animation portion of your model, but there is one subtle point to know. These queues are attached to the module that created them. If you move the queue to your animation and then later move the module, the animated queue will be moved as well. This can be rather annoying as you will find yourself creating an initial animation and later moving modules to make your model look pretty or to add logic. There are two ways to overcome this problem. The first is to delete all the queues that are created when a module is placed and then add them back in using the Queue object found on the Animate toolbar. A more convenient way is to click on the queue, press the *Cut* button (or Ctrl-X) to delete it from the animation and then use the *Paste* button (or Ctrl-V) to add it back. We did this for the four queues of interest—the three technical support queues and the sales call queue.

We then added resource animations for our technical support and sales, just like in Model 3-3 (Section 3.5.2), Model 3-5 (Section 3.5.3), Model 4-3 (Section 4.3.3), and Model 4-4 (Section 4.4), as shown in Figures 5-9 and 5-10.

Next we added variables to indicate the total number of calls in our major areas—the three technical support areas and the sales area. We used two different ways to get this information on our animation. First let's look at what we did for the sales area. We added a variable using the Variable button from the Animate toolbar and entered the expression NR(Sales) + NQ(Process Sales Call.Queue). This expression represents the current number of busy Sales resource units plus the number of calls in the queue waiting for a resource.

We used a different approach for the technical support areas. There is a fairly large number of variables that are kept by Arena, so we suggest that you use Arena help to learn more about what Arena variables are available. For our needs, there is an appropriate variable. For every Process module, you place in your model Arena automatically defines a variable NAME.WIP, where NAME is the Name you enter for the module. It also turns out that when you place and populate a Process module, Arena provides that variable as part of the standard animation for that module, along with the queue. If you look closely at your model when it's not running, you'll see a red 0 centered directly under each Process module. For our technical support for product Type 1, that turns out to be the variable `Process Product Type 1 Tech Call.WIP`. You can place a new variable using the Variable button and enter the name—it's not on the drop-down list—or you can cut and paste the variable supplied by the Process module. We actually used the copy and paste functions in creating our animation. Arena will not let you animate multiple queues or resources with the same name, but it will allow it for variables. We used this approach to add the three WIP variables to our technical support portion of the animation.

The last thing that we added to our animation was the total number of trunk lines in use. We could have used the Arena variable `NR(Trunk Lines)`, which would have given us the current number of calls in the system. Instead, we decided to add a plot, of the same variable, which gives us a good view of how the demand on our system changes over the simulation run. To complete our animation, we added some labels and background color.

If you run your model, you might notice by looking at our plot of the number of busy trunk lines that the system is either full or close to full most of the time; the "flat top" on the plot at height 26 confirms our arrival logic that refuses incoming calls when all 26 trunk lines are already taken. Looking at the output from the run, we see that there were 735 attempted calls with 644 of them completed and 91 rejected or balked. You might also notice that the simulation run length was just over 641 minutes. The last 41 minutes were the time it took to complete the calls that were in the system at 6 PM; note that the trunk-lines plot declines to zero at the end of the simulation, confirming our termination logic described above.

That completes Model 5-1.

5.5 Model 5-2: The Enhanced Call Center System

In developing models for real systems, you want to keep your model as simple as possible and yet you want that model to capture the systems behavior accurately. Our ultimate goal is to use our model to find the most cost effective way to increase our service level or customer satisfaction. We would like to consider the impact of such changes as increasing the number of trunk lines or changing our staffing levels. Although our current model would allow us to capture these changes, it's lacking in detail. So let's revisit our problem description and determine where we need to add more detail.

5.5.1 The New Problem Description

Let's start with a more detailed examination of call arrivals. The call-arrival rate to this system actually varies over the course of the day according to a nonstationary Poisson process (see Section 12.3), which is typical of these types of systems. So we might collect data expressed in calls per hour for each 30-minute period during which the system is open. These call-arrival rates are given in Table 5-2.

Table 5-2. Call Arrival Rates (Calls Per Hour)

Time	Rate	Time	Rate	Time	Rate	Time	Rate
8:00 - 8:30	20	10:30 - 11:00	75	1:00 - 1:30	110	3:30 - 4:00	90
8:30 - 9:00	35	11:00 - 11:30	75	1:30 - 2:00	95	4:00 - 4:30	70
9:00 - 9:30	45	11:30 - 12:00	90	2:00 - 2:30	105	4:30 - 5:00	65
9:30 - 10:00	50	12:00 - 12:30	95	2:30 - 3:00	90	5:00 - 5:30	45
10:00 - 10:30	70	12:30 - 1:00	105	3:00 - 3:30	85	5:30 - 6:00	30

Next let's take a look at our staffing. Although our initial model assumed constant staffing levels for our sales and technical-support areas, the staffing level actually varies over the day. It turns out that there are six sales people with the staggered daily schedules summarized as (number of people @ time period in minutes): 1@60, 3@60, 4@90, 5@60, 6@60, 5@90, 6@90,5@30, 3@60, and 2@60.

Our technical support employees work an eight-hour day with 30 minutes off for lunch (lunch is not included in the eight hours). There are 11 technical support people whose work schedules are shown in Table 5-3. Charity and Noah are only qualified to handle calls for product Type 1; Tierney, Aidan, and Emma are only qualified to handle calls for product Type 2; Shelley, Jenny, and Christie are only qualified to handle calls for product Type 3. Molly is qualified to handle product Types 1 and 3, and Anna and Sammy are qualified to handle calls for all three product types.

The last bit of detail that we omitted from our initial model was that four percent of technical calls require further investigation after completion of the phone call. The questions raised by these callers are forwarded to another technical group, outside the boundaries of our model, that prepares a response. The time to prepare these responses is

Table 5-3. Technical Support Schedules

Name	Product Lines	Time Period (30 minutes)																					
		1	2	3	4	5	6	7	8	9	10	11	12	13	14	15	16	17	18	19	20	21	22
Charity	1	•	•	•	•	•	•	•		•	•	•	•	•	•	•	•	•					
Noah	1						•	•	•	•	•	•		•	•	•	•	•	•	•	•	•	•
Molly	1, 3			•	•	•	•	•	•		•	•	•	•	•	•	•	•	•	•			
Anna	1, 2, 3						•	•	•	•	•		•	•	•	•	•	•	•	•	•	•	
Sammy	1, 2, 3				•	•	•	•	•	•	•		•	•	•	•	•	•	•	•	•		
Tierney	2	•	•	•	•	•	•	•		•	•	•	•	•	•	•	•	•					
Aidan	2						•	•	•	•		•	•	•	•	•	•	•	•	•	•	•	•
Emma	2				•	•	•	•	•		•	•	•	•	•	•	•	•	•	•	•		
Shelley	3	•	•	•	•	•	•	•	•		•	•	•	•	•	•	•	•					
Jenny	3						•	•	•	•		•	•	•	•	•	•	•	•	•	•	•	•
Christie	3				•	•	•	•	•		•	•	•	•	•	•	•	•	•	•	•		

estimated to be EXPO(60) minutes. The resulting response is sent back to the same technical support person who answered the original call. This person then calls the customer, which takes TRIA(2, 4, 9) minutes. These returned calls require one of the 26 trunk lines and receive priority over incoming calls. If a returned call is not completed on the same day the original call was received, it's carried over to the next day.

If we are going to consider changing our staffing levels to increase customer satisfaction, we might also want more detail on when the system is congested. So let's add counters to our model that we'll use to count the number of rejected calls during each hour of operation.

5.5.2 New Concepts

Our enhanced model requires two new concepts. The first is the arrival process whose rate varies over time. The second concept shows that when a call attempts to seize a unit of a technical staff resource, it is really trying to seize a resource from a pool or set of resources qualified to handle the call. Let's discuss each of these new concepts before we build our model.

We'll start with the arrival process, which has a rate that varies over time. This type of arrival process is fairly typical of service systems and requires a different approach. Arrivals at many systems are modeled as a *stationary Poisson process* in which arrivals occur one at a time, are independent of one another, and the average rate is constant over time. For those of you who are not big fans of probability, this implies that we have exponential interarrival times with a fixed mean. You may not have realized it, but this is the process we used to model arrivals in our previous models (with the exception of Model 4-5, which was contrived to illustrate a particular point). There was a slight variation of this used for the Part B arrivals in Models 4-1 through 4-4, where we assumed that an arrival was a batch of four; therefore, our arrivals still occurred one *batch* at a time according to a stationary Poisson process.

For this model, the mean arrival rate is a function of time. These types of arrivals are usually modeled as a *nonstationary Poisson process*. An obvious, but incorrect, modeling approach would be to enter for the Time Between Arrivals in a Create module an exponential distribution with a user-defined variable as a mean Value, then change this Value based on the rate for the current time period. For our example, we'd change this Value every 30 minutes. This would provide an approximate solution if the rate change between the periods was rather small. But if the rate change is large, this method can give very misleading (and wrong) results. The easiest way to illustrate the potential problem is to consider an extreme example. Let's say we have only two periods, each 30 minutes long. The rate for the first period is 3 (average arrivals per hour), or an interarrival-time mean of 20 minutes, and the rate for the second period is 60, or an interarrival-time mean of 1 minute. Let's suppose that the last arrival in the first time period occurred at time 29 minutes. We'd generate the next arrival using an interarrival-time mean Value of 20 minutes. Using an exponential distribution with a mean of 20 could easily[1] return a value more than 31 for the time to the next arrival. This would result in no arrivals during the second period, when in fact there should be an expected value of 30 arrivals[2]. In general, using this simplistic method causes an incorrect decrease in the number of arrivals when going from one period to the next with an increase in the rate, or a decrease in the interarrival time. Going from one period to the next with a decrease in the rate will incorrectly increase the number of arrivals in the second period.

Nevertheless, it's important to be able to model and generate such arrival processes correctly since they seem to arise all the time, and ignoring the nonstationarity can create serious model-validity errors since the peaks and troughs can have significant impact on system performance. Fortunately, Arena has a built-in ability to generate nonstationary Poisson arrivals (and to do so correctly) in the Create module. The underlying method used is described in Section 12.3.

The second new concept is the need to model an entity arriving at a location or station and selecting from one of several similar (but not quite identical) objects. The most common situation is the selection of an available resource from a pool of resources. Let's assume you have three operators: Sean, Lynn, and Michael. Any one of these operators can perform the required task, and you would like to select any of the three, as long as one is currently available. The Sets data module in the Basic Process panel provides the basis for this functionality. Arena *sets* are groups of objects of the same type that can be referenced by a common name (the *set name*) and a *set index*. The objects that make up the set are referred to as *members* of the set. Members of a particular set must all be the same type of object, such as resources, queues, pictures, etc. You can collect almost any type of Arena objects into a set, depending on your modeling requirements. An object can also reside in more than one set. Let's assume in our `Operators` set that Lynn is also qualified as a setup person. Therefore, we might define a second resource set called `Setup` as Lynn and Doug (Doug's not an operator). Now, if an operator is required, we'd select from the set called `Operators`; if a setup person is required, we would select from the set called `Setup`. Lynn might be chosen via either case because she's a member of both sets. You can have as many sets as you want with as much or as little overlap as required.

For our call center, we'll need to use sets to model the technical support staff correctly. We also need to consider how to model the returned technical support calls. These are unique in that they must be returned by the same tech-support person who handled the original call, so we must have a way to track who handled the original call. We'll do this

---

[1] With probability $e^{-31/20} = 0.21$, to be (almost) exact. Actually this figure is the *conditional* probability of no arrivals in the second period, *given* that there were arrivals in the first period and that the last of these was at time 29. This is not quite what we want, though; we want the *unconditional* probability of seeing no arrivals in the second period. It's possible to work this out, but it's complicated. However, it's easy to see that a lower bound on this probability is given by the probability that the first arrival after time 0, generated as exponential with mean 20 minutes, occurs after time 60—this is one way (not the only way) to have no arrivals in the second period, and has probability $e^{-60/20} = e^{-3} = 0.0498$. Thus, the incorrect method would give us at least a 5% chance of having no arrivals in the second period. Now, go back to the text, read the next sentence, and see the next footnote.

[2] The probability of no arrivals in the second period should be $e^{-60(1/2)} = 0.000000000000093576$.

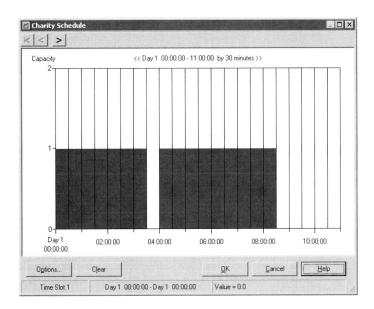

Figure 5-11. The Graphical Schedule Editor: The Charity Schedule

by storing the set index of the specific resource allocated from the selected technical support staff, so we'll know which individual needs to return the call, if necessary.

5.5.3 Defining the Data

In our first model, we had a total of five resources (one for the 26 trunk lines, one for the sales staff, and three for the technical support staff). In our new model, we'll have a total of 13 resources (one for the 26 trunk lines, one for the sales staff, and the remaining 11 for the individual technical support staff). All but the trunk lines follow a schedule (we discussed schedules in Section 4.2.2). We defined a separate schedule for each resource; however, as several of the technical support staff follow the same schedule (e.g., Charity, Tierney, and Shelley), you could simply reuse a single schedule (though doing so would make your model less flexible in terms of possible future changes). In developing the schedules for the technical support staff, we used the Graphical Schedule Editor, set the number of time slots to 22, and the maximum capacity on the *y*-axis of the graph to two (we could have made it one) in the Options dialog. The Charity schedule is shown in Figure 5-11.

If you're building this model along with us, be sure that each schedule covers the entire 660 minutes of each day, from 8 AM to 7 PM (22 half-hour time periods), even if you have to fill in many zeroes at the beginning or end (as we do at the end for Charity). Although the schedule shown in Figure 5-11 may look like it covers the entire 660 minutes, it really only covers the first 17 time periods (through 8:30 as shown in Figure 5-11). The graphical schedule editor does not assume that each schedule covers one day (they can be of any length). In this case, the schedule would start over in time period 18, which is only 8.5 hours into the 11-hour day. We could have checked the box under the

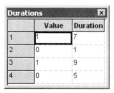

Figure 5-12. The Durations Spreadsheet Window

Options button to "Remain at capacity when at the end of the schedule," and set that capacity to zero. This would have resulted in the correct schedule for the first day. The problem in using this option is that if we decide to have multiple replications and we chose not to initialize the system at the start of each day (in the Run Setup dialog), this would forever result in a zero capacity after the first replication. Thus, we would like to make sure that the last five half-hour time periods explicitly have a capacity of zero. You can do this by right-clicking on the Durations cell of the schedule and selecting Edit via Dialog or Edit via Spreadsheet. Selecting Edit via Spreadsheet opens the Durations window in Figure 5-12, which allows you to double-click to add a new row and then enter the Value, Duration combination of 0, 5. This represents five 30-minute periods with a capacity of zero, thus explicitly filling out the entire 11 hours of the day.

Selecting Edit via Dialog opens the Durations dialog, which allows you to enter the same values as shown in Display 5-32. If you reopen the Graphical Schedule Editor after adding these data, the figure will look the same as before. A word of caution—make sure you exit the editor without saving the data or the added data (0, 5) will be deleted.

We also developed a schedule for the sales staff and the arrival process. You develop the arrival process schedule just like you developed the resource schedules, with the exception that you select `Arrival` in the Type cell, rather than `Capacity`.

The next step is to define our 13 resources in the Resource data module. We selected the `Ignore` option for the Schedule type, even though it would appear that the `Wait` option is the most likely option for this type of operation. If a tech support person is on the phone when a break is scheduled to occur, that person would typically complete the call and then take the full break (the `Wait` option). This would work fine if we run only one replication or we run multiple replications and initialized the system at the start of each replication. Unfortunately, that's not always the case. If we chose the Wait option and a resource was delayed for lunch by 10 minutes, the rest of the day's schedule would be pushed out by 10 minutes. Although this is not a problem, that 10-minute delay would show up in the next day's schedule. All delays would be cumulative over time. That could be a problem!

We've also included some costing data (Busy/Hour and Idle/Hour) for our technical-support and sales staff. The final spreadsheet view is shown in Figure 5-13.

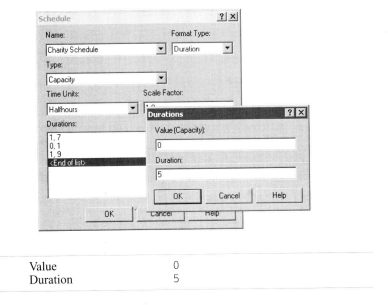

| Value | 0 |
| Duration | 5 |

Display 5-32. The Durations Dialog

Having defined our resources, we can now define our sets using the Set data module found in the Basic Process panel. We first define three resource sets. They are:

```
Product 1: Charity, Noah, Molly, Anna, Sammy
Product 2: Tierney, Aidan, Emma, Anna, Sammy
Product 3: Shelley, Jenny, Christie, Molly, Anna, Sammy
```

The contents of these sets correspond to the technical staff qualified to handle the calls for each product type. Note that Molly, Anna, and Sammy are included in more than one set. Also note that we've consistently listed these three versatile staff members at the end of each set for a deliberate reason. For instance, take the Product 1 set, which consists of Charity, Noah, Molly, Anna, and Sammy, in that order. If possible, we'd like to allocate

	Name	Type	Capacity	Schedule Name	Schedule Rule	Busy / Hour	Idle / Hour	Per Use	StateSet Name	Failures	Report Statistics
1	Sales	Based on Schedule	Sales Schedule	Sales Schedule	Wait	20	20	0.0		0 rows	✓
2	Trunk Line	Fixed Capacity	26	26	Wait	0.0	0.0	0.0		0 rows	✓
3	Charity	Based on Schedule	Charity Schedule	Charity Schedule	Ignore	18	18	0.0		0 rows	✓
4	Noah	Based on Schedule	Noah Schedule	Noah Schedule	Ignore	18	18	0.0		0 rows	✓
5	Molly	Based on Schedule	Molly Schedule	Molly Schedule	Ignore	20	20	0.0		0 rows	✓
6	Anna	Based on Schedule	Anna Schedule	Anna Schedule	Ignore	22	22	0.0		0 rows	✓
7	Sammy	Based on Schedule	Sammy Schedule	Sammy Schedule	Ignore	22	22	0.0		0 rows	✓
8	Tierney	Based on Schedule	Tierney Schedule	Tierney Schedule	Ignore	18	18	0.0		0 rows	✓
9	Aidan	Based on Schedule	Aidan Schedule	Aidan Schedule	Ignore	18	18	0.0		0 rows	✓
10	Emma	Based on Schedule	Emma Schedule	Emma Schedule	Ignore	18	18	0.0		0 rows	✓
11	Shelley	Based on Schedule	Shelley Schedule	Shelley Schedule	Ignore	18	18	0.0		0 rows	✓
12	Jenny	Based on Schedule	Jenny Schedule	Jenny Schedule	Ignore	18	18	0.0		0 rows	✓
13	Christie	Based on Schedule	Christie Schedule	Christie Schedule	Ignore	18	18	0.0		0 rows	✓

Double-click here to add a new row.

Figure 5-13. The Resources Spreadsheet Window

Figure 5-14. The Members Spreadsheet Window for the Product 1 Set

Charity or Noah to an incoming call first because they can only handle product Type 1 calls. We'll use the Preferred Order selection rule which allocates resources based on their order in the set, starting with the first (Charity). In our model, this will try to keep Molly, Anna, and Sammy available since they can handle other product types as well.

The Set module allows us to form sets of resources, counters, tallies, entity types, and entity pictures. You can also form sets of any other Arena objects by using the Advanced Set data module found in the Advanced Process panel. Form a set by clicking on the Set module and then double-clicking in the spreadsheet view to add a new entry. We'll walk you through the details for creating the first resource set. Enter Product 1 in the Name cell, and select Resource from the pull-down in the Type cell. To define the members of the set, click on "0 Rows" in the Members column to open a spreadsheet for listing the members, and then double-click to open a new blank row. Since we have already defined our resources, you can use the pull-down option to select the Resource Name for each member. To add additional members to the set, just double-click where indicated to add more new rows. The completed member spreadsheet view for the Product 1 set is shown in Figure 5-14. You'll need to repeat this for the Product 2 and Product 3 sets.

We also defined a Counter set (called Rejected Calls) with ten members, which we'll use to keep track of the number of rejected calls (busy signals) for each hour time period. For this set, we first defined the ten counters using the Statistic data module (in the Advanced Process panel). We assigned the statistic names as Period 01 through Period 10 and the corresponding Counter Names as Period 01 Rejected Calls through Period 10 Rejected Calls (the leading 0s for periods 1 through 9 are so things will be alphabetized properly in the output reports). We also selected Replicate for the Initialize Option cell, which will cause the counters to be cleared whenever the other statistics are cleared, as specified in the *Run > Setup* dialog. We could have also selected the Yes option; if the No option is specified and multiple replications are performed, then the value of the counter at the end of a replication will be retained as the initial value at the beginning of the next replication so that the reported values will be cumulative as the replications are made. The first five counter entries are shown in Figure 5-15.

The completed Set data module is shown in Figure 5-16, which includes three Resource sets and one Counter set.

	Name	Type	Counter Name	Limit	Initialization Option
1	Period 01	Counter	Period 01 Rejected Calls		Replicate
2	Period 02	Counter	Period 02 Rejected Calls		Replicate
3	Period 03	Counter	Period 03 Rejected Calls		Replicate
4	Period 04	Counter	Period 04 Rejected Calls		Replicate
5	Period 05	Counter	Period 05 Rejected Calls		Replicate

Figure 5-15. The Statistic Spreadsheet Window for the Counters

	Name	Type	Members
1	Product 1	Resource	5 rows
2	Product 2	Resource	5 rows
3	Product 3	Resource	6 rows
4	Rejected Calls	Counter	10 rows

Figure 5-16. The Set Spreadsheet Window

5.5.4 Modifying the Model

We'll modify Model 5-1 that we created earlier to develop our new Model 5-2. We've already covered almost all of the needed concepts contained in these modules, so we won't bore you with the details of each one. Instead, we'll focus on the new modules and concepts.

Let's start with the call arrival section. We made our first change to the Create module that creates our incoming calls. In the Time Between Arrivals section, we selected Schedule from the Type drop-down list. The arrival schedule, Arrival Schedule, that we created above was entered in the Schedule Name field. Since the Create module will follow our schedule and automatically stop creating new arrivals 11 hours into the simulation, we can delete the three modules that we had in our arrival cutoff section of Model 5-1. We also deleted the Assign module, Increment Active Call Counter, used to increment the variable Total WIP. We used this variable in the expression that we entered in the Terminating Condition portion in the *Run > Setup* dialog for Model 5-1. We replaced that variable with the variable NR (Trunk Lines). We did this for two reasons. First, we really didn't need the first variable, so we made our model simpler by deleting it from our new model. The second reason is that using the variable Total WIP in the new logic required to handle the returned calls in our technical support area causes some modeling problems. We would have had to decrement the variable when the returned call was sent to the back office and then increment it when the response was returned to technical support. So we took the easy way out since that variable was really not needed and eliminated it from our model. The last change required for this first section of our model was to the Record module used to record the number of rejected calls. As mentioned earlier, we decided to keep track of the number of rejected calls in each one-hour time period. To accomplish this, we defined 10

Name	Record Rejected Call for Current Hour Period
Type	Count
Value	1
Record into Set	Check
Counter Set Name	Rejected Calls
Set Index	AINT((TNOW/60) + 1)

Display 5-33. Counting the Rejected Calls by One-Hour Period

counters in the Statistic data module and formed a set called Rejected Calls containing these counters. We altered the Record module by checking the Record into Set box and entered the set name. We then developed an expression AINT((TNOW/60)+1) that would yield an integer value that represents the current period based on the simulation run time, in Display 5-33; to understand how this expression works, note that TNOW is the internal Arena simulation-clock value (in minutes), AINT is a built-in Arena function that truncates its real-number argument toward zero, and "play computer" for some minute values of TNOW.

Now let's proceed to the changes we made to the Tech Support Calls section of our model. The first five modules (up to and including the Determine Product Type Decide module) remain the same as in Model 5-1. We'll walk you through the changes required for the product Type 1 technical support and refer you to the file Model 05-02.doe for the differences for the other two types of technical support calls.

Directly following the Determine Product Type Decide module, we inserted an Assign module where we made three assignments. We first assigned a new entity type and entity picture so we could tell the difference between the calls for different product types in both the animation and the reports. We also defined a new attribute called Tech Call Type, which is assigned an integer value of 1, 2 or 3, depending on the product type. Soon you'll see why we need this new attribute.

The following Process module is the same except for the information in the Resources section. Since we replaced the single resource, Tech 1, in Model 5-1 with resource set Product 1, we needed to modify this section. We changed the Type to Set and entered Product 1 for the Set Name. We selected Preferred Order for the Selection Rule and defined a new attribute, Tech Agent Index, for the Save Attribute field. When we defined our resource sets, we consistently listed the most versatile staff members at the end of each set. Recall that this set consists of Charity, Noah, Molly, Anna, and Sammy, in that order; notice that we placed Molly, Anna, and Sammy as the last resources in the set. Thus, if possible, we'd like to allocate Charity or Noah to an incoming call first because they can only handle product Type 1 calls. The Preferred Order selection rule allocates resources based on their order in the set, if there is a choice among multiple idle resources, starting with the first (Charity). In our model, this will try to keep Molly, Anna, and Sammy available; remember that they can handle other call types as well. We've also stored the index of the resource allocated in the Save Attribute Tech Agent Index because we'll need to know which person took the call if further

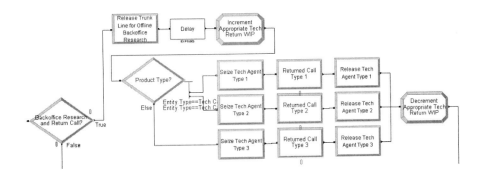

Figure 5-17. The Returned Technical Calls Section

investigation and a return call are required. This completes the changes to this section of our model, but as long as we are dealing with the technical support calls, we'll develop a new section for our model to handle the returned technical calls.

The logic for this section is shown here and the flowchart modules are shown in Figure 5-17:

```
If customer call requires further investigation
     Release trunk line resource
     Delay for back office research
     Increment Tech Return WIP variable
     Determine product type and direct
     Seize technical support resource
     Seize trunk line resource
     Delay for return-call service
     Release trunk line resource
     Release technical support resource
     Decrement Tech Return WIP variable
Send to system exit
```

The completed tech call entity is sent to a Decide module named "Backoffice Research and Return Call?" where 96% of the calls are considered complete (we set this up as the False branch) and are directed to the same system exit we used in Model 5-1. The remaining 4% (the True branch) are sent to Release module where the Trunk Line is released since the call is going offline now. The following Delay module holds up the entity for EXPO(60) minutes, which is the time required for further investigation, conducted by undefined resources outside the boundaries of our model (any queueing out there is rolled into the EXPO(60) time). If a call enters this delay toward the end of a day and cannot be finished during that day, it would, in reality, be carried over into the next day. Since our model runs for just one day, such calls are simply not finished within the run. We also defined and entered the Storage ID Backoffice Research Storage so we could show these entities in our animation. After the delay, the entity is sent to an Assign module where we increment the one-dimensionally arrayed variable Tech Return WIP using the value of the attribute Tech Call Type as the index. You can define this variable in the Assign module, but you first need to open the Variable data module and enter a value of 3 in the Rows cell, which defines the array size. We next sense the product type with an N-way by Condition test in the "Product Type?" Decide module and direct the return call to the proper branch.

The return-call entity must seize both a trunk line (outgoing calls need a trunk line as do incoming calls), and the appropriate tech support person, being the one who took the original incoming call. We'll describe things for product Type 1; the other two are similar. In the following Seize module, the return tech support call for product Type 1 seizes from the appropriate Resource Set of tech support people. Queueing is possible, and we use a separate queue (and animation) for this, and specify the Seize priority as High (1), as outgoing return calls take priority over incoming calls. Here we seize only the tech support person and not the trunk line (we'll do that in the following module). This return call must go to the same tech support person who took the original customer call, so we use the Specific Member selection rule and specify the Attribute Tech Agent Index in the Set Index field to ensure this.

Having already seized the right Tech 1 agent, the return Tech 1 call now seizes a Trunk Line in the following Process module, with High priority over Medium-priority incoming calls. The processing delay is based on the expression Returned Tech Time and when completed, the trunk line is released. Queueing could occur in this module, if no trunk is immediately available. During that time, the animation is still accurate since the entity picture shows in the Resource animation's Seize Area, so there is no separate queue animation for this. When the call is completed, the specified tech support person is released in the following Release module.

In our first attempt at developing this section, we used a single Process module with a high priority to seize both the tech support person and the trunk line at the same time. When we ran our simulation and watched the animation, we realized that these return calls were waiting a long time before they were processed. It turns out that when you take this modeling approach both resources must be available at the same time for the entity to proceed. If only one of the specified resources is available and the system is busy, it would most likely be allocated to another waiting entity. Since these return calls have a high priority, we decided to seize the resources separately. We realized that in the real system the returned call would most likely be given to the specified tech support person, who would try to return the call as soon as he or she was available. Of course in doing so, he or she would have to first seize an available trunk line. So we modified our model logic by separating the two seize actions as previously described.

Upon completion of the return call, the entity enters an Assign module where the arrayed variable Tech Return WIP is decremented. The completed call is now ready to be sent to the system exit. In every other case where a completed call was sent to the system exit, it still had the trunk line resource, so the first module was a Release module to release that resource. These entities have already released the trunk line resource so we need to send the entities to the Record module following that Release module.

If you look in the Statistic data module (Advanced Process panel), you'll see the ten Counter-type entries we described earlier. We also set up four statistics of the Time-Persistent type to request results (time average, minimum, and maximum) on the total number of calls in backoffice research for a callback (tracking variable `NSTO(Backoffice Research Storage)`) and for the total numbers of each tech-support product type in the system (tracking expressions `Tech 1 Total Online WIP`, `Tech 2 Total Online WIP`, and `Tech 3 Total Online WIP`).

Our new model does not require any changes to the Sales Calls or Order-Status Calls of Model 5-1 so no further changes are required.

Now let's review the modifications to our animation. The three animated storages for the recording delays, the Trunk Lines Busy plot, the variable for the Sales WIP, and the queues for the calls waiting for tech support and sales remain the same. We first deleted the Tech 1, Tech 2, and Tech 3 resources from our animation. Then we changed the variable names for the three WIP displays to match our new model. For example, we changed the WIP variable for product Type 1 from `Process Product Type 1 Tech Call.WIP` to `Tech 1 Total Online WIP`, which is an expression we defined in the Expression data module to be `Process Product Type 1 Tech Call.WIP + Tech Return WIP(1)` so is at all times the total number o tech calls for product Type 1 that are "open," either as an incoming call or in the back office for a return call. We also rearranged our new animation to allow for a Back Office Research area and added a WIP variable, `NSTO(Backoffice Research Storage)`; `NSTO` is a built-in Arena function that returns the number of entities in the Storage given in its argument.

We also added three new queues to the tech support area to shown the return calls waiting for service. This queue for the Tech 1 area is `Seize Tech Agent Type 1.Queue`. We then added a storage animation to show the return calls that we had under investigation for the new back office research area, `Backoffice Research Storage`.

The final change was to add the resources for our new tech support personnel. We selected Charity from the resource list and opened the Arena office picture library, `Office.plb`. There we found a picture of a person sitting at a desk with an idle phone and a second picture with the person talking on the phone. We copied the first picture to represent our Idle picture and the second to be the Busy picture. For our Inactive state, we altered the first picture to remove the person so the desk was unattended. We then closed the window and placed our resource. After sizing our resource, we used the Text option to label it. Then we made a copy of these two objects and changed the identifiers to represent Noah. We repeated this for Tierney, but changed the color of the shirt in the Idle and Busy pictures. Our plan was to use different shirt colors to represent the different capabilities of the support staff.

The final animated simulation is similar to Figure 5-18, which was taken at approximately time 340.7. In the Category Overview report, we see from the User Specified section's Counters that most reject calls occur in hours 5–8 (corresponding to noon to 4:00 PM), so this might be a time period to look at adding staff (though we've made only one replication here so such a conclusion is statistically suspect); we'll take this up in a more organized and statistically reliable way in Chapter 6.

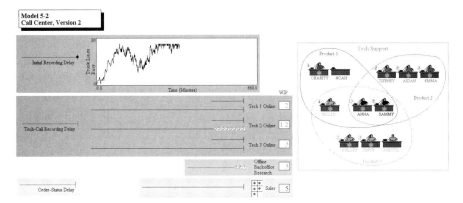

Figure 5-18. Model 5-2 Animation

5.6   Model 5-3: The Enhanced Call Center with More Output Performance Measures

While Model 5-2 produces plenty of output performance measures, it doesn't produce an overall economic figure of merit on which we might easily make comparisons across different configurations. A lot of simulation studies focus on cost reduction (or ideally, cost minimization), so we'll create an overall cost measure as the primary output. At the same time, we'll add some options to the model to set the stage for comparison of alternatives and for optimum-seeking in Chapter 6. To get a little better statistical precision, we'll also make five replications, representing a work week and will focus on weekly costs.

There are two basic areas in which costs appear: (1) staffing and resource costs, which are quite tangible and easily measured, and (2) costs due to poor customer service, which are less tangible and can be difficult to quantify. We'll consider these two kinds of costs separately.

First, let's look at staffing and resource costs, some of which we defined in Model 5-2 but didn't use. In the Resource module of Model 5-2, we defined a cost of $20/hour for sales staff and $18-$22/hour for tech support staff, depending on their level of training and flexibility; these costs were incurred whenever the staff were present according to their schedules, regardless of whether they were busy or idle. Using the staffing information described in Section 5.5 (and defined in the Schedule data module of Model 5-2), the weekly cost for the current staff is:

> Sales staff altogether: 45 hours/day × $20/hour × 5 days/week = $4,500/week
> Tech support staff:
> > 8 people × (8 hours/day)/person × $18/hour × 5 days/week = $5,760/week
> > 1 person × (8 hours/day)/person × $20/hour × 5 days/week = $800/week
> > 2 people × (8 hours/day)/person × $22/hour × 5 days/week = $1,760/week

Adding, we get a current weekly payroll cost of $12,820.

To try to improve service, we now generalize the model to allow for additional staff. Looking at the hourly counts of the busy-signal balks from Model 5-2, it seems that the most severe staffing shortfalls are between noon and 4:00 PM, so we'll create the ability in the model to add both sales and tech support during that four-hour period, which corresponds to the four hour-long time periods 5 through 8.

To add an easily controlled variable number of sales staff, we define a variable New Sales to be the number of additional sales staff we hire for this four-hour period each day, at a cost of $17/hour. Each additional sales staff person will thus work 20 hours/week, so the additional cost is $17/hour × 20 hours/week = $340/week for each new sales staff person. To place the new sales staff in the model, we edit the Sales Schedule line in the Schedule data module. Since we'll use a Variable (New Sales) as part of the schedule, we cannot use the Graphical Schedule Editor and must access this schedule via its dialog (or spreadsheet); right-click in the Sales Schedule line and then select Edit via Dialog. Display 5-34 illustrates where the edits are made. Of course, if we set New Sales to 0, we have the same sales-staff configuration as in the base model.

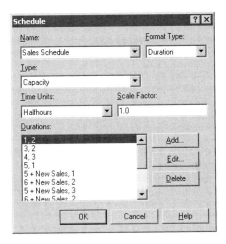

Name	Sales Schedule
Type	Capacity
Time Units	Halfhours
Value (Capacity), Duration	1,2
Value (Capacity), Duration	3,2
Value (Capacity), Duration	4,3
Value (Capacity), Duration	5,1
Value (Capacity), Duration	5 + New Sales, 1
Value (Capacity), Duration	6 + New Sales, 2
Value (Capacity), Duration	5 + New Sales, 3
Value (Capacity), Duration	6 + New Sales, 2
Value (Capacity), Duration	6,1
Value (Capacity), Duration	5,1
Value (Capacity), Duration	3,2
Value (Capacity), Duration	2,2

Display 5-34. The Changed Sales Schedule Dialog for Model 5-3

We follow a similar strategy to add tech support people during this 12:00 PM to 4:00 PM period. We defined new Variables New Tech 1, New Tech 2, and New Tech 3 to be the number of additional tech support staff added who are qualified only on Product Types 1, 2, and 3, respectively, and New Tech All to be the number of tech support people added who are qualified on all three product types. Naturally, the new Type 1 techs are all named Larry, the new Type 2 techs are all named Moe, the new Type 3 techs are all named Curly, and the new All-type techs are all named Hermann. Four new entries in the Resource data module are added to create these resources and define hourly rates for them ($16 for each Larry, Moe, and Curly, and $18 for each Hermann). To express their qualifications, we add Larry to the Product 1 set in the Set module, Moe to the Product 2 set, Curly to the Product 3 set, and Hermann to all three of these sets. We created schedule entries for all four of these new resources. The Larry schedule is in Display 5-35 and the other three are similar; like the sales schedule, these can be edited only via the dialog (or spreadsheet) and not by the Graphical Schedule Editor since they involve variables. As for costs, we incur $16 \times 4 \times 5 = \$320$/week for every Larry, Moe, and Curly and $17 \times 4 \times 5 = \$360$/week for every Hermann.

The final way to alter the resource mix is to consider changing the number of trunk lines, which was previously assumed to be fixed at 26. The Trunk Line resource is defined in the Resource module as a simple Fixed Capacity resource, so we'll just alter the entry there if we want to change the number of trunk lines. To build a cost into this, we incur a flat $98/week for each trunk line, a deal that includes all the local and long-distance usage of that line.

Name	Larry Schedule
Type	Capacity
Time Units	Halfhours
Value (Capacity), Duration	0,8
Value (Capacity), Duration	New Tech 1,8
Value (Capacity), Duration	0,6

Display 5-35. The Larry Schedule Dialog for Model 5-3

Name	New Res Cost
Expression Value	New Sales*340 +(New Tech 1+New Tech 2+New Tech 3)*320 +New Tech All*360 +98*MR(Trunk Line)

Display 5-36. The New Res Cost Expression for Model 5-3

To put all of the above resource costs together, we define an Expression called New Res Cost. Display 5-36 shows the entry for it in the Expression data module. Everything involved in this Expression was discussed above, except for MR(Trunk Line), which uses the built-in Arena function MR to give the number of units of the Resource in the argument, in this case, Trunk Line. Note that this Expression does not depend on what happens during the simulation and is used only at the end in the Statistic module to help produce the final output performance measures.

Now let's turn to the other category of costs to the system, those incurred by making customers wait on hold. Clearly, these trade off against the resource and staffing costs; we can use our model to explore this trade-off, and maybe even to try to optimize it. In practice, these kinds of customer-dissatisfaction costs are difficult to quantify, so we'll have to rely on customer surveys and techniques like regression analysis to come up with the numbers.

We assume that most people are by now hardened enough to expect some waiting time on hold when dealing with a call center, but at some point, people will start getting mad and the system will start incurring a cost. For tech calls, this point is 3 minutes; for sales calls, it's 1 minute; and for order-status calls, it's 2 minutes. Beyond this tolerance point for each call type, the system will incur a cost of 36.8 cents/minute for tech calls that continue on hold, 81.8 cents/minute for sales calls, and 34.6 cents/minute for order-status calls. For each call type, we accumulate in a variable the "excess" waiting times in queue (i.e., the waiting time in queue beyond the tolerance cutoff) of all completed calls via a new Assign module through which entities pass after their calls are done.

Display 5-37 shows this new Assign module for tech support calls, and the other two are similar (we put bright orange boxes behind these new Assign modules so you can easily spot them in the model). Note that ENTITY.WAITTIME is a built-in Arena attribute that automatically accumulates the total of all times in queue for an entity as it goes along, as well as any other delay times that are allocated to "Wait." In our model, there are no such delay allocations, and there are no upstream queue delays other than that on hold while waiting for the call to be answered, so this value indeed contains what we want in this case. (Computation of ENTITY.WAITTIME happens only if the Costing box is checked in the *Run > Setup > Project Parameters* dialog, so we need to make sure it is checked.) At the end of the simulation, the variable Excess Tech Wait Time will be the total number of minutes beyond the 3-minute tolerance that completed tech calls endured during the one-day run, and over the five replications, these will be averaged to produce, in the final output, an average *daily* excess tech wait time. If we multiply this by the per-minute cost of 36.8 cents, we would get the cost for the excess tech wait time for

Name	Assign Excess Tech Wait Time
Assignments	
Type	Variable
Variable Name	Excess Tech Wait Time
New Value	Excess Tech Wait Time
	+ MAX(ENTITY.WAITTIME − 3, 0)

Display 5-37. The New Assign Module to Accumulate Excess Tech Waiting Times for Model 5-3

a *day*, but we're costing things out on a *weekly* basis, so we need to multiply instead by 36.8 cents × 5 = $1.84 to get the *weekly* cost. The costs for the other two types of calls are computed similarly (the factors by which we need to multiply the excess wait times to get weekly costs are 81.8 cents × 5 = $4.09 for sales calls, and 34.6 cents × 5 = $1.73 for order-status calls). While we are making five one-day replications, so it could be thought of as a work week, the model constructed in this way will produce weekly cost estimates regardless of the number of replications made (we'll use this model in Chapter 6 with far more than five replications, to improve output precision).

Putting all these various costs together into a single overall cost measure is fairly simple. Given all the above discussion and setup, the overall `Total Cost` output performance measure is

```
New Res Cost
+ Excess Sales Wait Time * 4.09
+ Excess Status Wait Time * 1.73
+ Excess Tech Wait Time * 1.84
+ 12820,
```

which we enter in the Statistic data module (in the Advanced Process panel); the Type of this Statistic is chosen to be `Output`, meaning that it is a quantity that's computed only at the end of the simulation and we just want to see its value in the reports. Since we're defining this output ourselves, rather than having Arena compute it internally, it will appear under User Specified → Output in the Category Overview report.

At this point, you're probably wondering about those poor unfortunate rejected souls who called only to get immediately brushed off by a busy signal. They didn't even have the opportunity to get mad after their tolerance time on hold and start charging cost against the system (even though they're *really* mad). We could have estimated a (*big*) cost for each busy signal and built that into our cost structure, but instead we decided just to compute the percent of incoming calls that receive a busy signal and produce that as a separate output. Instead of viewing this as part of the performance-measure objective of the model, we'll view it later as a requirement (kind of like a constraint, except it's on the output rather than on the input) that no more than 5% of incoming calls get a busy signal; any model configuration not meeting this requirement will be regarded as unacceptable no matter how low its `Total Cost` output might be.

To compute the percent of incoming calls that get busy signals, we need to know the total number of attempted calls and rejected calls. We already have a Record module that counts the number of attempted calls (the counter `Attempted Calls`). We also count the number of rejected calls, but we are currently recording these rejected calls on a per-hour basis. We could just add up the values from these ten counters, but we took the easy way out and added another Record module in the call-arrival section of the model logic (with a bright orange box behind it) that will give us that total (counter `Total Rejected Calls`). To compute the output value `Percent Rejected`, we added a line to the Statistic module with that name and the expression `100*NC(Total Rejected Calls)/NC(Attempted Calls)`; NC is the built-in Arena function that returns the value of the counter named in its argument.

Because we've requested multiple (five) replications, we need to tell Arena what to do between replications. There are four possible options.

Option 1: Initialize System (yes), Initialize Statistics (yes)

This will result in five statistically independent and identical replications and reports, each starting with an empty system at time 0 and each running for up to 660 minutes. The random-number generator (see Section 12.1) just keeps on going between replications, making them independent and identically distributed (IID). Possible returned technical support calls that are carried over to the next day are lost.

Option 2: Initialize System (yes), Initialize Statistics (no)

This will result in five independent replications, each starting with an empty system at time 0 and each running for up to 660 minutes, with the reports being cumulative. Thus, Report 2 would include the statistics for the first two replications, Report 3 would contain the statistics for the first three replications, etc. The random-number generator behaves as in Option 1.

Option 3: Initialize System (no), Initialize Statistics (yes)

This will result in five runs, the first starting at time 0, the second at the time the first completed (remember that we are using a stopping rule to terminate a replication), the third at the time the second completed, etc. Since the system is not initialized between replications, the time continues to advance, and any technical support calls that were not returned will be carried over to the next day. The reports will contain only the statistics for a single replication or day, rather than being cumulative.

Option 4: Initialize System (no), Initialize Statistics (no)

This will result in five runs, the first starting at time 0, the second at the time the first completed, the third at the time the second completed, etc. Since the system is not initialized between replications, the time continues to advance, and any technical support calls that were not returned will be carried over to the next day. The reports will be cumulative.

Ideally we would like to select either Option 3 or 4 since we'd like to carry all non-returned technical calls into the next day. However, this would require us to change our stopping rule and let each replication run for 660 minutes. So for now let's just accept the defaults—Option 1.

As you've gone through the discussion in this section, you've probably thought of other ways we could have accomplished many of the things we did. It would have been possible, for instance, to exploit Arena's built-in costing capabilities more fully rather than doing a lot of it on our own and to have parameterized the model differently with only the "primitive" input values like wage rates and number of employees. Or we could have let Arena do some of the calculations internally that we did externally on our own little calculators. That's all true, but it also would have taken more time and effort for us to do so. This is a typical kind of judgment call you'll have to make when building a model— trading off your model-building time against model generality (and maybe elegance), usually being guided by who's going to use the model, and for what. The configuration of this model will be useful in the next chapter, where we show you several of Arena's capabilities in the important activity of statistical analysis of simulation output data.

We ran the model in what might be called the "base case," with no additional employees or trunk lines (i.e., New Sales, New Tech 1, New Tech 2, New Tech 3, and New Tech All were all set to 0, and the number of trunk lines was set to 26). The Total Cost came out to be $22,500.07 for the week, and 12.9% of incoming calls got a busy signal. Since this percent of rejected calls is above our 5% target, we made another run where we increased each of the six input resource "control" variables by 3 (i.e., New Sales, New Tech 1, New Tech 2, New Tech 3, and New Tech All were all set to 3, and the number of trunk lines was set to 29); this produced a Total Cost of $23,668.69 with only 1.6049% of incoming calls getting a busy signal; evidently, the cost of hiring the extra staff and adding three trunk lines is not a wise investment looking purely at Total Cost. But it does reduce the percent of balked calls to an acceptable level. However, you should have an uneasy feeling about this "analysis" (we do) since each result came from only five replications. We don't know if the results we're seeing are just random bounce, or if we could confidently say that the second configuration is better, or how we'd confidently choose the best from among several different scenarios, or what might be the best scenario of all. These are exactly the issues we take up in Chapter 6, and we'll use Model 5-3 to look into them.

5.7 Model 5-4: An (s, S) Inventory Simulation

We close this chapter by considering a completely different kind of system, an inventory, to indicate how such operations might be modeled and to illustrate the breadth of simulation in general and of Arena in particular (it's not just for queueing[3]-type models). We'll use modules from the Blocks and Elements panels only, primarily to demonstrate their use and make you aware that they're there if you need them (so, in effect, we're using the SIMAN simulation language); in Exercise 5-17, we ask you to recreate this model using the higher-level modeling constructs of the Basic Process and Advanced Process panels. This model is essentially the same as the one in Section 1.5 of Law and Kelton (2000).

5.7.1 System Description

Widgets by Bucky, a multi-national holding company, carries inventory of one kind of item (of course, they're called widgets). Widgets are indivisible, so the inventory level must always be an integer, which we'll denote as $I(t)$ where t is time (in days) past the beginning of the simulation. Initially, 60 widgets are on hand: $I(0) = 60$.

---

[3] As pointed out by the late Carl M. Harris, "queueing" is evidently the only word in English with five or more consecutive vowels; see Gass and Gross (2000).

Customers arrive with interarrival times distributed as exponential with mean 0.1 day (it's a round-the-clock operation), with the first arrival occurring not at time zero but after one of these interarrival times past time zero. Customers demand 1, 2, 3, or 4 widgets with respective probabilities 0.167, 0.333, 0.333, and 0.167. If a customer's demand can be met out of on-hand inventory, the customer gets the full demand and goes away happy. But if the on-hand inventory is less than the customer's demand, the customer gets whatever is on hand (which might be nothing), and the rest of the demand is *backlogged* and the customer gets it later when inventory will have been sufficiently replenished; this is kept track of by allowing the inventory level $I(t)$ to go negative, which makes no sense physically but is a convenient accounting artifice. Customers with backlogged items are infinitely patient and never cancel their orders. If the inventory level is already negative (i.e., we're already in backlog) and more customers arrive with demands, it just goes more negative. We don't keep track of specifically which widgets arriving in the future will satisfy which backlogged customers (they're also infinitely polite, so it doesn't matter).

At the beginning of each day (including at time zero, the beginning of day 1), Bucky "takes inventory" to decide whether to place an order with the widget supplier at that time. If the inventory level (be it positive or negative) is (strictly) less than a constant s (we'll use $s = 20$), Bucky orders "up to" another constant S (we'll use $S = 40$). What this means is that he orders a quantity of widgets so that, if they were to arrive instantly, the inventory level would pop up to exactly S. So if t is an integer and thus $I(t)$ is the inventory level at the beginning of a day (could be positive, negative, or zero) and $I(t) < s$, Bucky orders $S - I(t)$ items; if $I(t) \geq s$, Bucky does nothing, lets the day go by, and checks again at the beginning of the next day, that is, at time $t + 1$. Due to the form of this review/replenishment policy, systems like this are often called *(s, S) inventory models*.

However, an order placed at the beginning of a day does *not* arrive instantly, but rather sometime during the last half of that day, after a *delivery lag* (a.k.a. *lead time*) distributed uniformly between 0.5 and 1 day. So when the order arrives, the inventory level will pop up by an amount equal to the original order quantity but, if there were any demands since the order was placed, it will pop up to something less than S when the order is finally delivered. Note that the relative timings of the inventory evaluations and delivery lags are such that there can never be more than one order on the way, since an order placed at the beginning of a day will arrive, at the very latest, just before the end of that day, which is the beginning of the next day, the first opportunity to place another order; see Exercise 5-18 for modeling implications of this particular numerical situation.

Bucky is interested in the average total operating cost per day of this system over 120 days, which will be the sum of three components:

- *Average ordering cost per day.* Every time an order is placed, there's a cost incurred of $32 regardless of the order quantity, plus $3 per item ordered; if no order is placed there's no ordering cost, not even the fixed cost of $32. The $3 is not the (wholesale) price of a widget, but rather the administrative operational cost to Bucky of *ordering* a widget (we're not considering prices at all in this model). At the end of the 120-day simulation, the total of all the ordering costs accrued is divided by 120 to get the average ordering cost per day.

■ *Average holding cost per day*. Whenever there are items actually physically in inventory (i.e., $I(t) > 0$), a holding cost is incurred of $1 per widget per day. The total holding cost is thus

$$\int_0^{120} 1 \times \max(I(t),0)dt$$

(think about it), and the average holding cost per day is this total divided by the length of the simulation, 120 days.

■ *Average shortage cost per day*. Whenever we're in backlog (i.e., $I(t) < 0$), a shortage cost of $5 per widget per day is incurred, a harsher penalty than holding positive inventory. The total shortage cost is thus

$$\int_0^{120} 5 \times \max(-I(t),0)dt$$

(think about it a little harder), and the average shortage cost per day is this total divided by the simulation length.

Note that for periods when we have neither backlog nor items physically in inventory (i.e., $I(t) = 0$) there is neither shortage nor holding cost — cost-accountant nirvana. Also, you might notice that we're not accounting for the wholesale or retail price of the widgets anywhere; in this model, we assume that these prices are fixed and induce this demand, which will happen regardless, so the revenues and profits are fixed and it's only the operating cost that we can try to affect.

One final fine point before we build our (rather simple) simulation. Inventory evaluations occur at the beginning of each day—i.e., when the clock is an integer, and any ordering cost is incurred at that time. It happens that the run is supposed to end at an integer time (120), so normally there would be an inventory evaluation then, and possibly an order placed that would not arrive until after the end of the world, so we'd never get it but we'd have to pay the ordering cost. So we should prevent an inventory evaluation from happening at time 120, which we'll do by actually stopping the run at time 119.9999 (a lazy fudge on our part, but we'll ask you to clean up after us in Exercise 5-19).

5.7.2 Simulation Model

As we mentioned earlier, we'll build this model using only modules from the Blocks and Elements panels. It would have been a lot easier to do this with the higher-level (and less ancient) modules from the Basic Process and Advanced Process panels, but we leave that joy to you as Exercise 5-17.

Figure 5-19 shows the completed model, including the animation at the top right. The modules from the Blocks and Elements panels at the bottom are divided into three sections, as indicated by the outline boxes behind the modules.

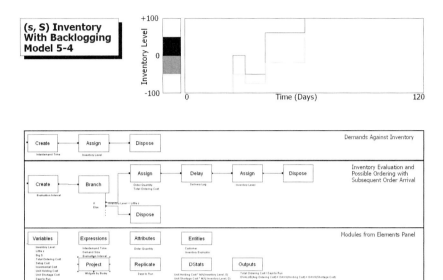

Figure 5-19. The Completed Inventory Model

Let's start with the data structure for the model. The modules in the bottom section of Figure 5-19 are from the Elements panel, so are themselves called *elements*. Note that they're not connected to anything, since they define various objects for the whole model. As we go through these, you'll already be familiar with many of the terms, since many of the same constructs and functions in the higher levels of Arena are available here as well.

The Variables element, shown in its completed form with the entry for Inventory Level visible in Figure 5-20, defines and optionally initializes Arena variables (much like the Variable module from the Basic Process panel). If there's no initialization for a variable, it defaults to zero. We'll let you poke through the entries in this module on your own, which are defined as follows:

> Inventory Level: At any time in the simulation, this is the inventory level (positive, zero, or negative), and is initialized to 60 here. This variable is the function $I(t)$.
>
> Little s: This is s, initialized to 20.
>
> Big S: This is S, initialized to 40.
>
> Total Ordering Cost: A statistical-accumulator variable to which all ordering costs are added; not initialized (so is implicitly initialized to 0).
>
> Setup Cost: The fixed cost of ordering, initialized to 32.
>
> Incremental Cost: The variable (per-widget) ordering cost, initialized to 3.
>
> Unit Holding Cost: The cost of holding one widget in inventory for one day, initialized to 1.

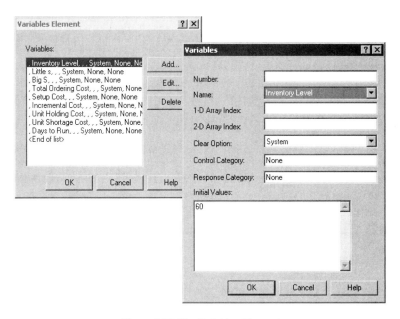

Figure 5-20. The Variables Element

`Unit Shortage Cost`: The cost of having one widget in backlog for one day, initialized to 5.

`Days to Run`: The length of the simulation (in days), initialized to `119.9999` per our admitted fudge, as described above.

We defined four Expressions via the Expressions element, in Figure 5-21 with the entry for `Demand Size` visible. These Expressions define the probability distributions for the quantities indicated by their names and should be self-explanatory as you look through them. We decided to define `Evaluation Interval` as an Expression rather than a variable, even though it's just a constant (1), for reasons of model generality should we want to try out some kind of random evaluation interval in the future (but, as seen in Exercise 6-12, there's also an advantage to defining it as a Variable).

The Attributes element has only one entry, used to declare that `Order Quantity` is an attribute to be attached to entities. The Entities element declares the two types of entities we'll be using, `Customer` and `Inventory Evaluator`. The Project element allows you to specify a Title, Analyst Name, and other documentation, as well as control some of the reporting, similar to the *Run > Setup > Project Parameters* option we've used before. There's not much to see in these modules, so we'll skip the figures and let you take a look at them on your own.

Figure 5-21. The Expressions Element

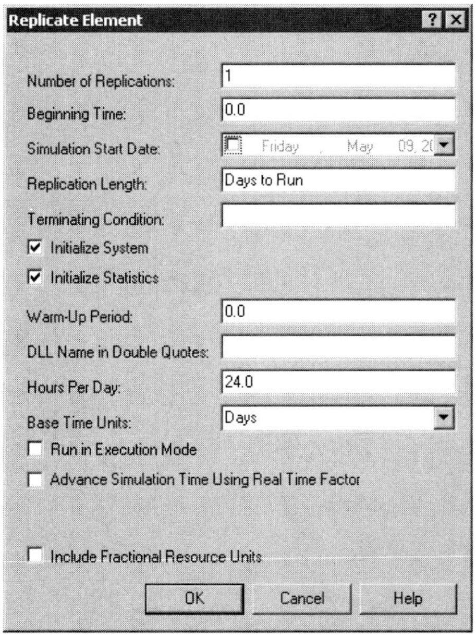

Figure 5-22. The Replicate Element

The Replicate element, shown in Figure 5-22, basically replicates what's available via *Run > Setup > Replication Parameters*. We have only two non-default entries. First, we changed the Base Time Units to Days, since all of our input time measurements are in days and the blocks from the Blocks panel have no provision for specifying input time units, assuming that they're all in the Base Time Units. Second, we specified the Replication Length to be Days to Run, the Variable we initialized to 119.9999, our cheap fudge to avoid the useless inventory evaluation at time 120.

The DStats element, in Figure 5-23 with the Holding Cost entry visible, does what Statistic module entries of type Time-Persistent do, in this case, saying that we want to accumulate the time-persistent values for the holding and shortage costs. The partly hidden SIMAN Expression in Figure 5-23 is

```
Unit Holding Cost * MX(Inventory Level, 0)
```

and is the instantaneous holding cost to be charged whenever Inventory Level is positive (recall that MX is the built-in Arena function that returns the maximum of its arguments). Similarly, the other entry in the DStats element, with Name Shortage Cost, accumulates the SIMAN Expression

```
Unit Shortage Cost * MX(-Inventory Level, 0)
```

for the shortage cost. What we want are the time-averages of these values, which we'll get next.

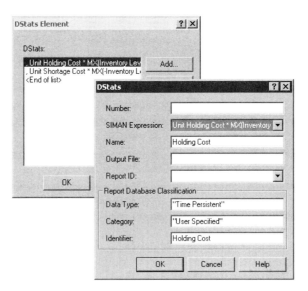

Figure 5-23. The DStats Element

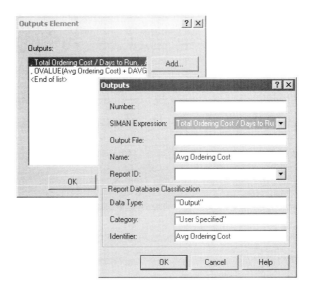

Figure 5-24. The Outputs Element

The Outputs element of Figure 5-24 does what Statistic module entries of type Output do, and here we need to do two things. First (and visible in Figure 5-24), we'll get the average ordering cost per day by dividing `Total Ordering Cost` by `Days to Run`, the length of the run, and storing this in the output name `Avg Ordering Cost`. Second, we add up all three components of the overall output performance measure into what we call `Avg Total Cost`, via the expression

```
OVALUE(Avg Ordering Cost) + DAVG(Holding Cost) + DAVG(Shortage Cost)
```

The Arena function OVALUE returns the last (most recent) value of its argument, which in this case is what we defined in the preceding line of this module (we could have avoided this by entering `Total Ordering Cost / Days to Run` here instead, but doing it the way we did produces a separate output for the average-ordering-cost component of average total cost, which might be of interest on its own). And the Arena function DAVG returns the time-persistent average of its argument, so is used twice here to get the average daily holding and shortage costs (also produced separately in the reports).

Now let's turn to the logic modules from the Blocks panel, which are themselves called *blocks*. The top group of blocks in Figure 5-19 represents customers arriving, making demands against the inventory, and leaving.

The Create block, shown in Figure 5-25, required three non-default entries. For both First Creation and Interval, we entered `Interdemand Time`, defined in the Expressions element as `EXPO(0.1)`. The Entity Type here we defined as `Customer`.

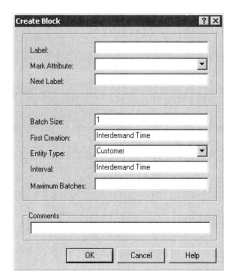

Figure 5-25. The Create Block for Customer Arrivals

The next Assign block, in Figure 5-26, decrements the customer's demand from the `Inventory Level`. The Expression `Demand Size` was defined in the Expressions module as a discrete random variable to return demands of 1, 2, 3, or 4 with the appropriate probabilities.

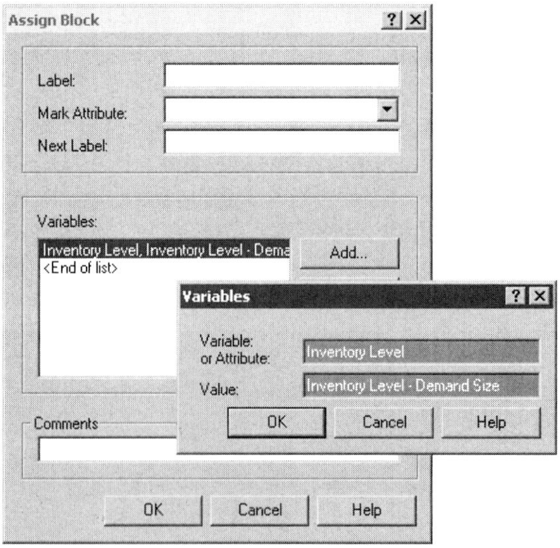

Figure 5-26. The Assign Block for Customer Demands Against Inventory

The Customer entity then goes away via the Dispose block, where the only thing we did was clear the box to Record Entity Statistics (we don't care about them in this model).

The center group of blocks in Figure 5-19 represent the periodic inventory evaluation and decision about ordering. If an order is placed, we wait for it to arrive and then increment the inventory level; if no order is placed, there's nothing to do.

This logic starts with a Create block, in Figure 5-27, to insert into the model what you can think of as a widget (not bean) counter who will count the widgets, decide whether to place an order, and then, if an order is placed, wait around for it to be delivered, and then put the widgets on the shelf. We want the First Creation to be at time 0 since our system calls for an inventory evaluation at the beginning of the run. The Entity Type is Inventory Evaluator, and the Interval of time between successive creations is the Expression Evaluation Interval, which was specified to be the constant 1 in the Expressions element.

Once created, the Inventory Evaluator entity proceeds to the Branch block, in Figure 5-28, which does some of the same things as the Basic Process panel's Decide module, with which you're already familiar. In this case, we want to decide whether to place an order at this time.

First we Add a branch of type If with the condition Inventory Level < Little s, which, if true, indicates that we want to place an order now (the dialog box for this branch is visible in Figure 5-28). The other branch is of type Else and is the only other possibility, indicating that no order is to be placed right now. The exit points from the Branch block correspond to truth of the corresponding branches; in this case, the first exit point is followed if we are to place an order (and leads to the Assign block to the right of and just above the Branch block), and the second exit point corresponds to the Else

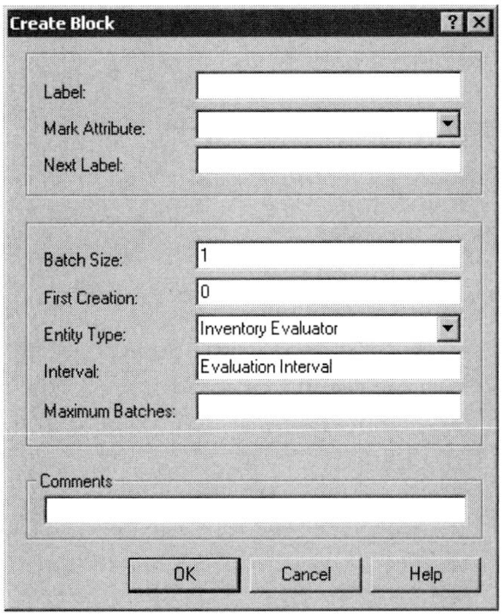

Figure 5-27. The Create Block for Inventory Evaluator

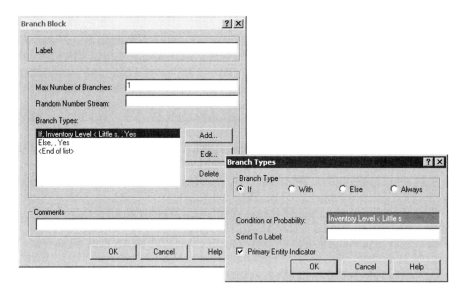

Figure 5-28. The Branch Block

branch and means that nothing is to be done (so the entity immediately goes to the Dispose block to the right of and just below the Branch block). An important part of the Branch block is the Max Number of Branches field, which defaults to infinity (and which we set instead to 1); in general, the Branch block evaluates each branch in the list in sequence, and sends the incoming entity (or a duplicate of it) out through each branch that evaluates to True, until Max Number of Branches is used up. This is a powerful capability but one that can lead to errors if the entries in the block are not made and coordinated with care.

If we need to place an order, the `Inventory Evaluator` entity next goes to the following Assign block, seen in Figure 5-29, which first computes the `Order Quantity` (an attribute attached to the entity) as `Big S - Inventory Level`. The next assignment in this block incurs the ordering cost for this order by replacing the Variable `Total Ordering Cost` by the expression

```
Total Ordering Cost + Setup Cost + Incremental Cost * Order Quantity
```

Note that it was important to do the assignments here in this order, since the result of the first is used in the second.

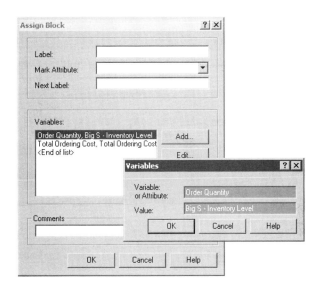

Figure 5-29. The Assign Block to Place an Order

Now it's time to wait for the order to arrive, which is accomplished by sending the `Inventory Evaluator` entity to the Delay block in Figure 5-30, where it simply sits for `Delivery Lag` time units (remember, only the Base Time Units are available in blocks). Recall that `Delivery Lag` was defined in the Expressions element to be `UNIF(0.5, 1.0)`.

After the delivery lag, the `Inventory Evaluator` goes to the next Assign block, in Figure 5-31, where it increments `Inventory Level` by its attribute `Order Quantity`.

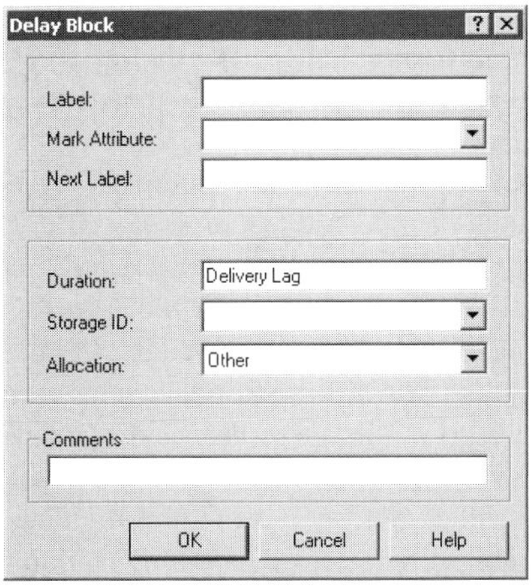

Figure 5-30. The Delay Block for the Delivery Lag

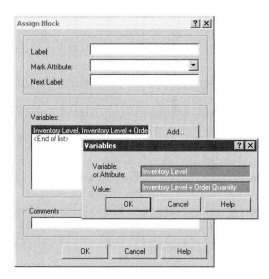

Figure 5-31. The Assign Block for the Order Arrival

After the order is delivered, this `Inventory Evaluator`'s job is done, so it goes to the final Dispose block (don't be sad, though, because another one of these entities will be created soon).

Now let's add a little (just a little) animation. Our Plot dialog box is shown in Figure 5-32, which graphs the `Inventory Level` over time. We'd like to have different colors depending on whether the `Inventory Level` is positive (in the black) or negative (in the red), which we accomplish by plotting two separate curves. To show positive inventory levels, we'll plot the expression `MX(Inventory Level, 0)` in black, and for negative inventory levels, we'll plot `MN(Inventory Level, 0)`, which will be a negative value, in red (MN is Arena's built-in function to return the minimum of its arguments). Note that when 0 "wins" in either of these plots, we'll get a flat line at zero, which maybe isn't so bad since it will visually delineate where zero is (in a playful, multicolor fashion).

It might also be fun to see the inventory itself in some way (no, we're not going to animate bins of little widgets or ghosts of them). For this, we installed a pair of Level animations (sometimes called *thermometer animations*) just to the left of the plot. The top animation (the black one, seen in Figure 5-33) plots the number of widgets physically in the inventory, which is expressed by `MX(Inventory Level, 0)`, and we'll plot it so that it goes up as the inventory becomes more positive. The bottom animation (the red one, for which we didn't bother making a figure here) plots the number of widgets in backlog, which is `MX(-Inventory Level, 0)`; when we're in backlog, this is a positive number, but we'd like to plot it diving lower from a zero level as we go deeper into backlog, so we selected the Fill Direction to be Down.

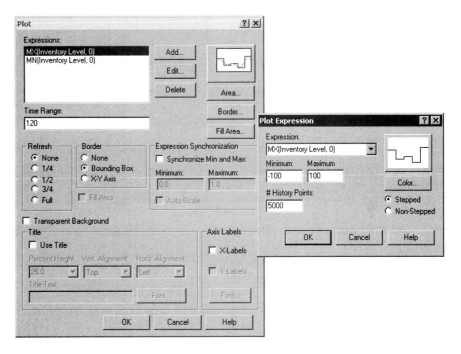

Figure 5-32. The Plot Dialog Box

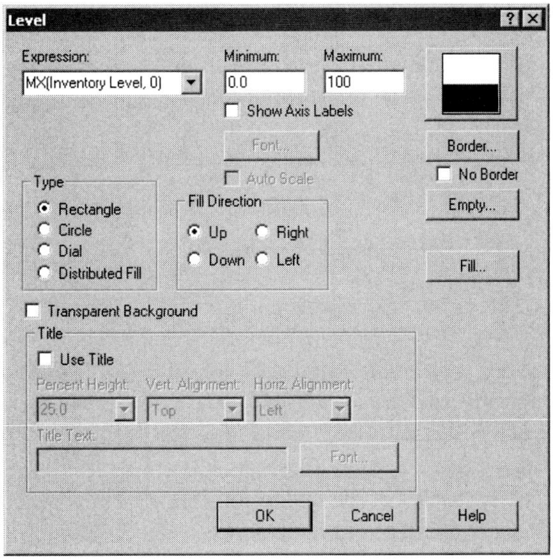

Figure 5-33. The Level Animation Dialog Box

After adding a few labels and such, we're ready to go. Watch the plot and the thermometers to see the inventory level drop as demands occur, then pop back up when orders arrive. The average daily costs (rounded to the nearest penny) were 9.37 for holding, 17.03 for shortage, and 100.39 for ordering, for a total of 126.79. Whether $(s, S) = (20, 40)$ is the best policy is a good question and one that we'll ask you to take up (in a statistically valid way) in Exercises 6-10 and 6-11 at the end of Chapter 6.

5.8   Summary and Forecast

This chapter has gone into some depth on the detailed lower-level modeling capabilities, as well as correspondingly detailed topics like fine-tuned animation. While we've mentioned how you can access and blend in the SIMAN simulation language, we've by no means covered it; see Pegden, Shannon, and Sadowski (1995) for the complete treatment of SIMAN. At this point, you should be armed with a formidable arsenal of modeling tools to allow you to attack many systems, choosing constructs from various levels as appropriate. In several of the following chapters, we'll continue to expand on Arena's modeling capabilities.

5.9   Exercises

5-1   Develop a model of the problem we described in Chapter 2 and modeled as Model 3-1, but this time only using modules from the Advanced Process panel to replace the Process module. Use the Plot and Variable features from the Animate toolbar to complete your model. Run it for 20 minutes and compare your results to what we got earlier.

5-2   Parts arrive at a two-machine system according to an exponential interarrival distribution with mean 20 minutes. Upon arrival, the parts are sent to Machine 1 and processed. The processing-time distribution is TRIA(4.5, 9.3, 11) minutes. The parts are then processed at Machine 2 with a processing-time distribution as TRIA(16.4, 19.1, 21.8) minutes. The parts from Machine 2 are directed back to Machine 1 to be processed a second time (same processing time). The completed parts then exit the system. Run the simulation for a single replication of 20,000 minutes to observe the average number in the machine queues and the average part cycle time.

5-3   Stacks of paper arrive at a trimming process with interarrival times of EXPO(10); all times are in minutes. There are two trimmers, a primary and a secondary. All arrivals are sent to the primary trimmer. If the queue in front of the primary trimmer is shorter than five, the stack of paper enters that queue to wait to be trimmed by the primary trimmer, an operation of duration TRIA(9, 12, 15). If there are already five stacks in the primary queue, the stack is balked to the secondary trimmer (which has an infinite queue capacity) for trimming, of duration TRIA(17, 19, 21). After the primary trimmer has trimmed 25 stacks, it must be shut down for cleaning, which lasts EXPO(30). During this time, the stacks in the queue for the primary trimmer wait for it to become available. Animate and run your simulation for 5,000 minutes. Collect statistics, by trimmer, for cycle time, resource utilization, number in queue, and time in queue. So far as possible, use modules from the Advanced Process panel.

5-4   Trucks arrive with EXPO(9) interarrival times (all times are in minutes) to an unload area that has three docks. The unload times are TRIA(25, 28, 30), TRIA(23, 26, 28), and TRIA(22, 25, 27) for docks 1, 2, and 3, respectively. If there is an empty dock, the truck proceeds immediately to that dock. Assume zero travel times to all docks. If there is more than one empty dock, the truck places preference on the higher-numbered dock (3, 2, 1). If all the docks are busy, it chooses the dock with the minimum number of trucks waiting. If there is a tie, it places preference on the lowest numbered dock (1, 2, 3). Develop a simulation model with modules from the Advanced Process panel, using required modules from the Basic Process panel to implement the selection logic. Run your model for 20,000 minutes and collect statistics on dock utilization, number in queue, time in queue, and the time in system.

5-5   Kits of ceiling fans arrive at an assembly system with TRIA(2, 5, 10) interarrival times (all times are in minutes). There are four assembly operators and the kits are automatically sent to the first available operator for assembly. The fan assembly time is operator-dependent as given below.

Operator	Assembly Time
1	TRIA(15, 18, 20)
2	TRIA(16, 19 ,22)
3	TRIA(16, 20, 24)
4	TRIA(17, 20, 23)

Upon completion of the assembly process, the fans are inspected with approximately 7% being found defective. A defective fan is sent back for repair to the same operator who assembled it. These defective fans have priority over incoming kits. Since the fan needs to be disassembled and then reassembled, the repair time is assumed to be 30% greater than the normal assembly time. Run your model for 20,000 minutes and collect statistics on operator utilization and the time in system.

5-6   The quality-control staff for the fan-assembly area of Exercise 5-5, has decided that if a fan is rejected a second time it should be rejected from the system and sent to a different area for rework. Make the necessary changes to the model and run the simulation for 20,000 minutes and compare the results (based on just one replication of each model). Also keep track of the number of fans rejected from the system.

5-7   Develop a model of a three-workstation serial production line with high reject rates: 7% after each workstation. Parts rejected after the first workstation are sent to scrap. Parts rejected after the second workstation are returned to the first workstation where they are reworked, which requires a fresh "draw" from the processing-time distribution but increased by 50% from the distribution of the original operation. (This penalty factor of 1.5 applies only at Workstation 1 and not at Workstation 2 when the part returns to it.) Parts rejected at the third workstation are returned to the second workstation where they are reworked, with a 50% penalty there (but not on its revisit to Workstation 3). The operation times are TRIA(6, 9, 12), TRIA(5, 8.5, 13), and TRIA(6.5, 8.9, 12.5) for workstations 1, 2, and 3 respectively (all times are in minutes). Part interarrival times to the system are UNIF(6,14). Run the model for 20,000 minutes, collecting statistics on the number in queue at each workstation, the number of scrapped parts, workstation utilizations, and average and maximum cycle times for parts that are not rejected at any workstation and for parts that are rejected at least once. Also, collect statistics on the number of times a rejected part was rejected.

5-8   In order to decrease the part cycle time in Exercise 5-7, a new priority scheme is being considered. The queue priority is based on the total number of times a part has been rejected, regardless of where it was rejected, with the more rejections already convicted against a part, the further back it is in the queue. Is there a difference in cycle times for this new priority scheme? HINT: Use the Queue data module to use the reject count as a queue-ranking scheme.

5-9   Parts arrive at a machine shop with EXPO(25) interarrival times (all times are in minutes). The shop has two machines, and arriving parts are assigned to one of the machines by flipping a (fair) coin. Except for the processing times, both machines operate in the same fashion. When a part enters a machine area, it requires operator attention to set up the part on the machine (there is only one operator in the shop). After the part is set up, the machine can process it without the operator. Upon completion of the processing, the operator is once again required to remove the part. After completion, the parts exit the system (parts have to go to only one machine). The same operator does all setups and part removals, with priority given to the machine waiting the longest for an operator. The times are (parameters are for triangular distributions):

Machine Number	Setup Time	Process Time	Removal Time
1	8, 11, 16	20, 23, 26	7, 9, 12
2	6, 8, 14	11, 15, 20	4, 6, 8

The run length is 25,000 minutes. Observe statistics on machine utilization, operator utilization, cycle times for parts separated by which machine they used, overall cycle times (i.e., not separated by machine used), and the time that each machine spends waiting for operator attention (both at setup and removal). Animate the process using storages for the setup, process, and removal activities.

5-10 A small warehouse provides work-in-process storage for a manufacturing facility that produces four different part types. The part-type percentages and inventory costs per part are:

Part Type	Inventory Cost Percentage	Per Part
1	20	$5.50
2	30	$6.50
3	30	$8.00
4	20	$10.50

The interpretation of "inventory cost per part" is as follows. Each part in inventory contributes an amount from the last column of the above table to the total cost (value) of inventory being held at the moment. For instance, if the current inventory is three units of Part 1, none of Part 2, five of Part 3, and one of Part 4, then the current inventory cost is

$$3 \times \$5.50 + 0 \times \$6.50 + 5 \times \$8.00 + 1 \times \$10.50 = \$67.00.$$

As parts arrive and depart, as described below, this inventory cost will rise and fall.

Parts arrive with TRIA(1.5, 2.0, 2.8) interarrival times (all times are in minutes). Two cranes store and retrieve parts with a travel time of UNIF(1.2, 2.9), each way. Requests for part removal follow the same pattern as for arrivals. If no part is available, the request is not filled. All part requests are given priority over part storages, and priority is given to retrieving based on the highest part cost.

For part arrivals, increment the inventory cost upon arrival, and increment the total number of parts in inventory after the part is stored. For part requests, decrement the total number of parts in inventory as soon as you know there is a part to retrieve, and decrement the inventory cost after the part is retrieved.

Run your model for 5,000 minutes starting with four of each part type in the warehouse. Collect statistics on the crane utilization, the average inventory cost, the average number of each part type in the warehouse, and the number of unfilled requests due to having no parts of the requested type.

HINTS: Use index variables for part inventory and per-part cost. (Note that you need to use the "Other" option in the Assign module when assigning to index variables.) Use the Discrete distribution to determine the part type and the Statistic data module to collect some of the required statistics.

5-11 A medium-sized airport has a limited number of international flights that arrive and require immigration and customs. The airport would like to examine the customs staffing and establish a policy on the number of passengers who should have bags searched and the staffing of the customs facility. Arriving passengers must first pass through immigration (immigration is outside the boundaries of this model). They then claim their bags and proceed to customs. The interarrival times to customs are distributed as EXPO(0.2); all times are in minutes. The current plan is to have two customs agents dedicated to passengers who will not have their bags searched, with service times distributed as EXPO(0.55). A new airport systems analyst has developed a probabilistic method to decide which customers will have their bags searched. The decision is made when the passengers are about to enter the normal customs queue. The decision process is as follows: a number is first generated from a Poisson distribution with a mean of 7.0. This number is increased by 1, to avoid getting a zero, and a count is started. When the count reaches the generated number, that unlucky passenger is sent to a second line to have his or her bags searched. A new search number is generated and the process starts over. A single agent is dedicated to these passengers, with service times distributed as EXPO(3). The number of passengers who arrive on these large planes is uniformly distributed between 240 and 350, and the simulation is to run until all passengers on the plane have been completely processed. Develop a simulation of the proposed system and make 20 replications, observing statistics on the system time by passenger type (searched vs. not searched), the number of passengers, and agent utilizations.

5-12 A state driver's license exam center would like to examine its operation for potential improvement. Arriving customers enter the building and take a number to determine their place in line for the written exam, which is self administered by one of five electronic testers. The testing times are distributed as EXPO(8); all times are in minutes. Thirteen percent of the customers fail the test (it's a hard test with lots of questions). These customers are given a booklet on the state driving rules for further study and leave the system (on foot). The customers who pass the test select one of two photo booths where their picture is taken and the new license is issued. The photo booth times are distributed TRIA(2.5, 3.6, 4.3). The photo booths have separate lines, and the customers enter the line with the fewest number of customers waiting in queue, ignoring whether anyone is in service; if there is a tie, they enter the nearest booth, Booth 1. Note that this set of rules could lead to what might appear to be irrational customer behavior in the case that neither booth has a queue (i.e., the lengths of both queues are zero), Booth 1 is busy, and Booth 2 is idle—a customer coming into the photo area would choose to queue up for Booth 1 (via the tie-breaking rule, since the queue lengths are tied at zero) rather than go right into service at Booth 2—but hey, they can't see into the photo booths! These customers then leave the system (driving), proudly clutching their new licenses. The center is open for arriving customers eight hours a day, although the services are continued for an additional hour to accommodate the remaining customers. The customer arrival pattern varies over the day and is summarized below:

Hour	Arrivals per Hour	Hour	Arrivals per Hour
1	22	5	35
2	35	6	43
3	40	7	29
4	31	8	22

Run your simulation for ten days, keeping statistics on the average number of test failures per day, electronic-tester and photo-booth utilization (not utilization for the testing resource overall, but separate utilizations for each photo booth), average number in queue, and average customer system time for those customers passing the written exam. Animate all the electronic-test booths and the photo booths.

5-13 An office of a state license bureau has two types of arrivals. Individuals interested in purchasing new plates are characterized to have interarrival times distributed as EXPO(6.8) and service times as TRIA(8.7, 13.7, 15.2); all times are in minutes. Individuals who want to renew or apply for a new driver's license have interarrival times distributed as EXPO(8.7) and service times as TRIA(16.7, 20.5, 29.2). The office has two lines, one for each customer type. The office has five clerks: two devoted to plates (Mary and Kathy), two devoted to licenses (Sue and Jean), and the team leader (Neil) who can serve both customer types. Neil will serve the customer who has been waiting the longest. Assume that all clerks are available all the time for the eight-hour day. Note that when entities from the front of multiple FIFO queues (corresponding to multiple Process modules) try to seize the same Resource, the logic to select which entity "wins" is as though all the queues were merged together into a single FIFO queue. Observe the system or cycle time for both customer types.

5-14 The office described in Exercise 5-13 is considering cross-training Kathy so she can serve both customer types. Modify the model to represent this, and see what effect this has on system time by customer.

5-15 Modify the model from Exercise 5-14 to include 30-minute lunch breaks for each clerk. Start the first lunch break 180 minutes into the day. Lunch breaks should follow one after the other covering a 150-minute time span during the middle of the day. The breaks should be given in the following order: Mary, Sue, Neil, Kathy, and Jean. What impact does this have on system time by customer?

5-16 Modify the probability-board model from Exercise 3-10 so that the bounce-right probabilities for all the pegs can be changed at once by changing the value of just a single variable. Run it with the bounce-right probabilities set to 0.25 and compare with the results of the wind-blown version of the model from Exercise 3-10.

5-17 Recreate Model 5-4 (the inventory model) without using anything from the Blocks or Elements panels, and using only modules from the Basic Process and Advanced Process panels.

5-18 In Model 5-4, the relative timings of the inventory-evaluation interval and the delivery lag were such that at no time could there be more than one order outstanding. What if the numbers were different so that there could be multiple orders on the way at a given time? Would Model 5-4 still work? (Note that in Model 5-4 we represented the order quantity, if any, by an attribute of the inventory-evaluator entity; what if that order quantity had been represented instead by a variable?)

5-19 In Model 5-4, remove the "fudge factor" of ending at time 119.9999 rather than the correct 120. Run the simulation to exactly time 120, but add logic to prevent a useless inventory evaluation at time 120; what's the effect on the output?

5-20 Generalize Model 5-4 to have two additional types of items (doodads and kontraptions), as well as widgets. The customers arrive in the same pattern as before, but now each customer will have a demand for doodads and kontraptions, as well as for widgets. Widget demands are as before, doodad demands are POIS(1.9), and kontraption demands are POIS(2.3); assume that a customer's demand for an item is independent of his or her demands for the other two items. There's still one inventory evaluator, who still arrives at the beginning of each day, but now has to look at all three inventories and order according to separate (s, S) policies for each of the three inventories. For widgets, $(s, S) = (20, 40)$ as before; for doodads, $(s, S) = (15, 35)$; and for kontraptions, $(s, S) = (25, 45)$. Delivery lags for widgets are UNIF(0.5, 1.0) as before; for doodads, it's UNIF(0.4, 0.8); and for kontraptions, it's UNIF(0.8, 1.7). Ordering costs (both setup and incremental), holding, and shortage costs for doodads and kontraptions are the same as for widgets. Run the simulation for the same length of time as Model 5-4 (i.e., it's okay to fudge the ending point to avoid useless inventory evaluations at time 120), and get the total daily cost, as well as separate holding and shortage costs for each type of item.

5-21 In Exercise 5-20, suppose that the suppliers for the three items merge and offer a deal to eliminate multiple setup costs on a given day's orders—that is, if Bucky orders any items of any type at the beginning of a day, he only has to pay the $32 setup cost once for that day, not a separate $32 for each type of item he orders. (If no order is placed for anything, there's still no setup cost.) Modify your model for Exercise 5-20 to do this. What kind of incentives do you think this alternate cost structure might place on Bucky in terms of picking better values of s and S for each item (see Exercises 6-13 and 6-14)?

5-22 In the machine-repair model of Exercise 3-14, suppose it costs $80 in lost productivity for each hour that each machine is broken down and $19/hour to employ each repair technician (the technicians are paid this hourly wage regardless of whether they are busy or idle). Modify the model to collect these cost data, and report the average total cost per hour. Also, show the machines in the "up" state in a separate area of the animation, make separate animation pictures for up vs. down machines, and keep track of the number of machines up (and plot it). Make the number of machines in the shop a Variable. If you want to minimize total average cost, is two the right number of repair technicians? Experiment (though see Exercise 6-21 for a statistically valid way to experiment).

CHAPTER 6

Statistical Analysis of Output from Terminating Simulations

As we warned you back in Section 2.6 with the hand simulation, there's an issue of randomness and thus statistical analysis when you build a model with any random (i.e., distribution- or probability-driven) inputs, as has been the case with all our models so far. In this section, we'll show you how to collect the appropriate data from the simulation and then statistically analyze them from the reports you're already getting, using Model 5-3 for the third version of the call-center model as created in Chapter 5. We'll also show you how to do more sophisticated statistical analyses with the help of the Output Analyzer (to compare two alternative versions of your model, called *scenarios*), the Process Analyzer (to run several scenarios conveniently and perhaps select the best or gauge the effects of inputs on outputs), and OptQuest for Arena (which "takes over" running your model in a quest for a configuration of input controls that optimizes a selected output response).

In Section 6.1, we'll talk about the time frame of simulations, which affects the kind of analysis you do. Then in Sections 6.2 and 6.3 we'll describe the basic strategy for data collection and statistical analysis in the context of just a single variant (the *base case* of Model 5-3, with the number of trunk lines still at 26 and no additional people hired for hours 5 through 8). We'll make some simple changes to the model's input parameters in Section 6.4 and use the Arena Output Analyzer to see if it makes a (statistically significant) difference. In Section 6.5, we'll introduce several more model variants and use the Arena Process Analyzer to run them in an efficient and organized way, as well as sort out which of them is probably best. Finally, in Section 6.6, we'll use OptQuest for Arena to search intelligently and efficiently among the bewildering number of possible input-parameter combinations for a model configuration that appears to be optimal in some sense. Throughout, we'll be illustrating methods that result in a reliable and precise statistical analysis, which promote informed and sound decisions.

In the past, which is fortunately now gone, a lot of people pretty much ignored these kinds of issues, and that's a real shame. By just running your model once, and then trying out a few haphazardly chosen scenarios (and running them only once), you just have no idea how valid or precise or general your results or conclusions might be. Sometimes the truth about validity and precision and generality can be ugly, and thus dangerous if it's unknown, since you'd run a real risk of getting poor estimates and making bad decisions. As you'll see, it doesn't take much work at all on your part to do these things right; your computer might have to work hard, but it spends most of its time loafing anyway and besides, unlike you, it's cheap (compared to the importance of the decisions you'll make based on your simulation's results). You've worked hard to build your model, so now it's time to let the model (and your computer) work hard for you to find out how it really behaves, and then you can confidently transport that knowledge into solid decisions.

6.1   Time Frame of Simulations

As in Section 5.2.5, most (not all) simulations can be classified as either terminating or steady state. For many models, this is primarily an issue of intent or the goal of the study, rather than having much to do with internal model logic or construction.

A *terminating* simulation is one in which the model dictates specific starting and stopping conditions as a natural reflection of how the target system actually operates. As the name suggests, the simulation will terminate according to some model-specified rule or condition. For instance, a store opens at 9 AM with no customers present, closes its doors at 9 PM, but continues operation for a little while longer until all customers are "flushed" out. Another example is a job shop that operates for as long as it takes to produce a "run" of 500 completed assemblies specified by the order. The key notion is that the time frame of the simulation has a well-defined (though possibly unknown-at-the-outset) and natural end, as well as a clearly defined way to start up.

A *steady-state* simulation, on the other hand, is one in which the quantities to be estimated are defined in the long run; i.e., over a theoretically infinite time frame. In principle (though usually not in practice), the initial conditions for the simulation don't matter. Of course, a steady-state simulation has to stop at some point, and as you might guess, these runs can get pretty long; you need to do something to make sure that you're running it long enough, an issue we'll take up in Section 7.2. For example, an emergency room never really stops or restarts, so a steady-state simulation might be appropriate. Sometimes people do a steady-state simulation of a system that actually terminates in order to design for some kind of worst-case or peak-load situation.

In this chapter, we'll stick to a terminating statistical analysis of the third version of the call-center model from Chapter 5 in its Model 5-3 incarnation, viewing it as a terminating simulation starting and stopping as described in Chapter 5. Steady-state analyses have to be done differently (we'll do this in Section 7.2 using the model we'll develop in Section 7.1).

6.2   Strategy for Data Collection and Analysis

With a terminating simulation, it's conceptually simple to collect the appropriate data for statistical analysis—just make some number n of independent replications.[1]

To do this, just open the *Run > Setup > Replication Parameters* dialog box and enter the value of n you want for the Number of Replications. Be sure that the boxes under Initialize Between Replications are both checked (the default) to cause both the system-state variables and the statistical accumulators to be cleared at the end of each replication. There are reasons to leave one or both of these boxes unchecked, but to get true, statistically independent and identically distributed (IID) replications for terminating analysis, you need to make sure that both boxes are checked. These changes will cause

---

[1] While conceptually simple, this could still imply a lot of run time for big or highly variable models.

Table 6-1. Total Cost and Percent Rejected from Ten Replications of Model 6-1 (Same as Model 5-3)

Replication	Total Cost ($)	Percent Rejected
1	21,281.24	12.6836
2	20,612.12	11.6059
3	20,023.67	9.2958
4	25,834.40	17.6084
5	24,748.90	13.3240
6	19,667.52	13.0201
7	19,565.40	11.0803
8	23,145.32	12.2655
9	19,931.75	9.6403
10	20,667.84	12.7830

the simulation to be replicated n times, with each replication starting afresh (fresh system state and fresh statistical accumulators) and using separate basic random numbers[2] to drive the simulation. For each replication, a separate section is generated in the Category by Replication report containing the results on that replication (such by-replication results are not in the Category Overview report).

For instance, we created Model 6-1 to be the same as the base case of Model 5-3 (the number of trunk lines still at 26 and no additional people hired for hours 5 through 8), except we made $n = 10$ replications to obtain the values in Table 6-1 for the Total Cost and Percent Rejected performance measures. It's important to remember that each of these values is the result over an entire simulation run and that each is an "observation" (or "realization") of a random variable representing the Total Cost and Percent Rejected over a "random" replication with these starting and stopping conditions.

How did we know ahead of time that $n = 10$ was the appropriate number of replications to make? We didn't. And we really still don't since we haven't done any analysis on these data. This is typical since you don't know up front how much variation you have in your output. We'll have more to say below about picking (or guessing) an appropriate value for the number of replications. By the way, when cranking out replications like this for statistical analysis, you might want to turn off the animation altogether to move things along, as we did for Model 6-1: choose *Run > Run Control* and select (check) *Batch Run (No Animation)*; to get the animation back later, you'll need to go back and clear this same entry.

You're probably not going to want to copy out all the values for all the performance measures of interest over all the replications and then type or paste them into some statistical package or spreadsheet or (gasp!) calculator for analysis. Fortunately, Arena

---

[2] Actually, each replication just keeps advancing within the same random-number "streams" being used; see Chapter 12 for more on how random-number generators work and can be controlled.

internally keeps track of all the results in the reports for all the replications. If you're making more than one replication, the Category Overview report will give you the average of each result over the replications, together with the half width of a (nominal) 95% confidence interval on the expected value of the output result; we'll discuss further what this all means in Section 6.3.

You can also have Arena save to binary ".*dat*" files (later to be fed into the Arena Output Analyzer, a separate application we'll discuss in Section 6.4) whatever you want from the summary of each replication. You do this in the Statistic data module, by specifying file names in the Output File column on the right of each row that has Output as the selection in its Type column. This file will then contain the kind of data we listed in Table 6-1.

6.3   Confidence Intervals for Terminating Systems

Just as we did for the hand simulation back in Section 2.6.2 (and using the same formulas given there), we can summarize our output data across the $n = 10$ replications reported in Table 6-1. We give the sample mean, sample standard deviation, half width of a 95% confidence interval, and both the minimum and maximum of the summary output values across the replications, in Table 6-2.

Arena will automatically produce the information in Table 6-2 (except for the sample standard deviation, but the half width contains essentially the same information) in the Category Overview report if you call for more than a single replication in *Run > Setup > Replication Parameters.* Figure 6-1 shows the relevant part of the Category Overview report for Model 6-1 after we ran it for ten replications; except for a little bit of round-off error, this agrees with the information in Table 6-2 that we computed by hand.

If you want to control the conditions and reporting in some way, such as specifying the confidence level to something other than 95%, producing the results in particular groupings or orderings, or if you want to get the graphical displays of the confidence intervals, minima, and maxima, you can save the summary data to a *.dat* file in the Statistic data module as described earlier and then use the Output Analyzer (see Section 6.4). You can also get graphical displays of confidence intervals as one of the capabilities of the Arena Process Analyzer (see Section 6.5).

Table 6-2. Statistical Analysis from Ten Replications of Model 6-1 (Same as Model 5-3)

	Total Cost ($)	Percent Rejected
Sample Mean	21,547.82	12.33
Sample Standard Deviation	2,243.38	2.31
95% Confidence Interval Half Width	1,604.82	1.66
Minimum Summary Output Value	19,565.40	9.30
Maximum Summary Output Value	25,834.40	17.61

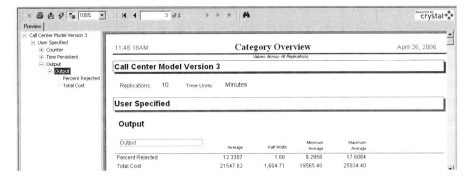

Figure 6-1. Results from Ten Replications of Model 6-1 (Same as Model 5-3)

It's probably obvious that the way to reduce the half width of the confidence interval on expected `Total Cost` or on `Percent Rejected` (or on anything, for that matter) is to increase the sample size *n*. But by how much? If you have a certain "smallness" in mind that you'd like (or could tolerate) for the half width, you can easily get an idea, but not an exact answer, of how big *n* will have to be to achieve this goal. Suppose you have an initial set of replications from which you compute a sample average and standard deviation, and then a confidence interval whose half width is disappointingly large. For instance, from our initial ten replications above in the case of `Total Cost`, we got a sample mean of $\overline{X} = \$21,547.82$, a sample standard deviation of $s = \$2,243.38$, and the half width of the 95% confidence interval turned out to be

$$t_{n-1,1-\alpha/2}\frac{s}{\sqrt{n}} = 2.262\frac{2,243.38}{\sqrt{10}} = \$1,604.71$$

(up to round-off), which represents a 7.4% error in the point estimate $21,547.82. If you want to achieve a specific half-width *h*, presumably smaller than the one you got from your initial set of replications, try setting *h* equal to the half-width formula above and solve for *n*:

$$n = t_{n-1,1-\alpha/2}^{2}\frac{s^{2}}{h^{2}}.$$

The difficulty with this is that it isn't really solved for *n* since the right-hand side still depends on *n* (via the degrees of freedom in the *t* distribution critical value and, though the notation doesn't show it, via the sample standard deviation *s*, which depends not only on *n* but also on the data obtained from the initial set of *n* replications). However, to get at least a rough approximation to the sample size required, you could replace the *t* distribution critical value in the formula above with the standard normal critical value (they're close for *n* more than about 30), and pretend that the current estimate *s* will be about the same when you compute it from the larger sample. This leads to the following as an approximate required sample size to achieve a confidence interval with half width equal to a pre-specified desired value *h*:

$$n \cong z^2_{1-\alpha/2} \frac{s^2}{h^2}$$

where s is the sample standard deviation from an initial set of n replications (which you'd have to make before doing this). An easier but slightly different approximation is (we'll leave the algebra to you)

$$n \cong n_0 \frac{h_0^2}{h^2}$$

where n_0 is the number of initial replications you have and h_0 is the half width you got from them. In the `Total Cost` example above, to reduce the half width from $h_0 = \$1{,}604.71$ to, say, $h = \$500.00$, we'd thus need a total of something like

$$n \cong 1.96^2 \frac{2{,}243.38^2}{500^2} = 77.3 \text{ (first approximation)}$$

or

$$n \cong 10 \frac{1{,}604.71^2}{500^2} = 103.0 \text{ (second approximation)}$$

(round up) replications instead of the ten we originally made. The second approximation will always be bigger since it uses $t_{n_0-1,1-\alpha/2}$ rather than $z_{1-\alpha/2}$. Note the depressing quadratic growth of the number of replications as h shrinks (i.e., we demand more precision)—to reduce the half width to half its initial value, you need about four times as many replications. While this might seem unfair (to do twice as well you have to work four times as hard), the intuition is that as you add more and more replications, each additional replication carries less and less percentage increase in your accumulating storehouse of knowledge.

Since this model runs very quickly without the animation, we decided to be conservative (statistically, not necessarily politically) and make 110 replications. The only change required to Model 6-1 was in *Run > Setup > Replication Parameters* where we changed the `10` in Number of Replications to `110`; we'll call the result Model 6-2. This produced a 95% confidence interval of

$$22241.71 \pm 413.52,$$

or in other words, [21828.12, 22655.16] (we'll drop the "$" units from here on). Note that this meets our goal of getting the half width down to 500 or less, and indeed it's well under 500, so maybe we overshot a little on the number of replications; on the other hand, it could have happened that the half width would still be above 500, since these are indeed only approximations and are not exact. In Section 12.5, we'll discuss sequential sampling, which for us will be ways to trick Arena into continuing to run until it produces confidence intervals that really are of the desired precision.

By the way, from this 110-replication run of Model 6-2, the approximate 95% confidence interval on the expected percentage of incoming calls that are rejected was 11.6536 ± 0.53. This seems like a pretty tight confidence interval, with half width of about half a percentage point (on a dimensionless scale that's always between 0 and 100 for a percent). However, if you wanted even tighter precision, you could repeat the above approximations for the required number of replications and then use the *maximum* of this and the 110 we got for the `Total Cost` output performance measure so that you would meet your precision needs on *both* outputs.

It's important to understand just what a confidence interval is (and what it isn't). Take the `Total Cost` output measure as an example. Each replication produces a `Total Cost` value for that replication, and due to random inputs, these values vary across replications. The average of the 110 values, you'll agree, is probably a "better" indicator of what to expect from a "typical" run than is any one of the individual values. Also, it's intuitive that the more replications you make, the "better" this average will be. The expected average, which is usually denoted by some kind of notation like μ, can be thought of as the `Total Cost` averaged over an infinite number of replications; as such, μ will have no uncertainty associated with it.

Unfortunately, mere mortals like us can't wait for an infinite number of replications, so we have to make do with a finite-sample estimate like the 22,241.71 from our 110 replications. The confidence interval centered around 22,241.71 can be thought of as a "random" interval (the next set of 110 replications will give you a different interval) that has approximately a 95% (in this case) chance of containing or "covering" μ in the sense that if we made a lot of sets of 110 replications and made such an interval from each set, about 95% of these intervals would cover μ. Thus, a confidence interval gives you both a "point" estimate (22,241.71) of μ, as well as an idea of how precise this estimate is.

A confidence interval is not an interval in which, for example, 95% of the `Total Cost` measures from replications will fall. Such an interval, called a *prediction interval*, is useful as well and can basically be derived from the same data. One clear difference between these two types of intervals is that a confidence interval will shrink to a single point as n increases, but a prediction interval won't since it needs to allow for variation in (future) replications.

A final word about confidence intervals concerns our hedging the confidence-level statement ("approximately" 95%, etc.). The standard methods for doing this, which Arena uses, assumes that the basic data, such as the 110 observations on `Total Cost` across the replications, are IID (that's satisfied for us) and normally distributed (that's not satisfied). Basically, use of the t distribution in the confidence-interval formula requires normality of the data. So what's the effect on the actual (as opposed to stated) confidence level of violation of this normality assumption? We can firmly and absolutely state that it depends on several things, including the "true" distribution as well as the number, n, of replications. The *central limit theorem*, one of the cornerstones of statistical theory, reassures us that we'll be pretty much okay (in terms of actual confidence being close to stated confidence) if n is "large." But how large? The answer to this is fairly imponderable and depends on how closely the distribution of the data resembles a normal

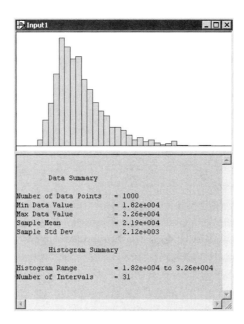

Figure 6-2. Histogram of 1,000 Total Cost Values

distribution, particularly in terms of symmetry. This can be at least qualitatively checked out by making a histogram of the data values, as we do in Figure 6-2 for $n = 1,000$ replications rather than the original 110 (the histogram with only 110 points didn't illustrate much). To make this plot, we modified Model 6-2 into Model 6-3 by first changing the Number of Replications from 110 to 1000 in *Run > Setup > Replication Parameters* and also arranged to save the 1000 individual Total Cost values to a file called Total Cost.dat, which we named in the Output File column in the row for Total Cost in the Statistic data module. We then ran the Arena Output Analyzer (we'll have more to say about this separate piece of software in Section 6.4) to read Total Cost.dat, exported it to a plain ASCII text file (*File > Data File > Export* in the Output Analyzer), rearranged it and streamlined it somewhat in a spreadsheet, saved it as a plain ASCII text file, and then brought it into the Arena Input Analyzer (see Section 4.6.4), even though it's not "input" data for a simulation model.

Though admittedly just eyeballing the data, we see that the shape of the histogram is not too far from the familiar "bell" curve of the normal density function (though it appears to be a little bit skewed out to the right), suggesting that we're probably okay in terms of using the standard confidence-interval method even for a relatively small value of *n*. For this model, which took about 40 seconds to run all 1,000 replications on a 2.13 GHz machine, we had the luxury of doing these 1,000 replications to check for normality; with a big, industrial-strength model, you won't be able to do anything of the sort, so how do you check this out in practice? Or is it just an article of faith? Though far from being a statement of general truth, a lot of simulation experience (backed up by some theory in the form of more general versions of the central limit theorem) indicates that if the value you're getting out of each individual replication is a sum or average of something, as opposed to an extreme value, use of standard normal-theory statistical-inference methods is fairly safe. Our `Total Cost` measure is of the "sum" form, perhaps explaining its approximate normality in Figure 6-2, and justifying use of Arena's normal-theory-based statistical methods.

6.4   Comparing Two Scenarios

In most simulation studies, people eventually become interested in comparing different versions, or *alternatives*, or *scenarios*, of some general model. What makes the scenarios differ from each other could be anything from a simple parameter change to fundamentally different logic. In any case, you need to take care to apply the appropriate statistical methods to the output from the scenarios to ensure that valid conclusions are drawn. In this section, we'll restrict ourselves to the case of just two scenarios; in Section 6.5, we'll allow for more.

We'll begin by modifying Model 6-3 into what we'll call Model 6-4, by simply knocking the Number of Replications in *Run > Setup > Replication Parameters* back down to 110; we left in the entry in the Statistic data module, Output File Column, `Total Cost` row to save the `Total Costs` from each replication to the file `Total Cost.dat` (more on this below). Let's consider two versions. The first, which we'll call the *base case*, is the same as Model 5-3 in Chapter 5, with exactly the same input parameters and run conditions (except that we're doing 110 rather than five replications), i.e., we still have 26 trunk lines and have added no Larrys (Variable `New Tech 1` is 0), no Moes (`New Tech 2` is 0), no Curlys (`New Tech 3` is 0), no Hermanns (`New Tech All` is 0), or sales staff (`New Sales` is 0) during hours 5–8. The alternative scenario, what we'll call the *more-resource scenario* is just the base case except that we'll add three trunk lines (in the `Trunk Line` row of the Resource data module change the Capacity from 26 to 29), and hire three each of Larry, Moe, Curly, Hermann, and sales staff (in the Variable data module initialize each of `New Tech 1`, `New Tech 2`, `New Tech 3`, `New Tech All`, and `New Sales` to 3 rather than 0). By adding resources in this way, we'll certainly increase salary cost, but will also reduce waiting and thus excess-wait costs. So maybe it's not so clear what the overall cost impact of this change will be, much less what its magnitude might be. And while these changes should reduce the percent of incoming calls that are rejected, it's hard to say by how much. Sounds like a job for (you guessed it) simulation.

We made a run (just one replication) of both the base case and the more-resources case, and got `Total Cost` outputs of $21,281.24 and $23,306.75, respectively; `Percent Rejected` was 12.6836 and 0.2833, respectively. On the surface, this looks mixed for the more-resources scenario since `Total Cost` seems to have increased somewhat while `Percent Rejected` decreased, apparently substantially. But if you were conscious when you read Section 6.3 (or Section 2.6.3), you'll know better than to fall into this trap of concluding much of anything from just one run with no statistical analysis of the output. So we'll do this comparison in a way that will allow us to make a statistically valid conclusion.

Before doing the comparison the right way, we'll do it in a reasonable-but-not-quite-right way. Focusing on the `Total Cost` output measure, we saw in Section 6.3 that, from 110 replications of the base-case scenario, a 95% confidence interval on the expected `Total Cost` is

$$22241.71 \pm 413.52 \text{ or } [21828.12, 22655.16].$$

Rerunning the model in the more-resources configuration, also with 110 replications, the confidence interval becomes

$$24560.12 \pm 315.93, \text{ or } [24244.19, 24876.05].$$

These two intervals do not overlap at all, suggesting that the expected `Total Costs` are significantly different. Turning to the `Percent Rejected` output measure, the base-case confidence interval was 11.6536 ± 0.53, while the more-resources interval was 1.6396 ± 0.28; these two intervals do not overlap, suggesting a statistically significant difference on this measure. However, looking at whether confidence intervals do or don't overlap is not quite the right procedure; to do the comparison the right way, we'll use the Arena Output Analyzer discussed next.

As we've mentioned before, the Output Analyzer is a separate application that runs apart from Arena itself, but operates on output files created by Arena through the Statistic data module (the *.dat* files we've mentioned before). While some of the things the Output Analyzer does are also done by Arena itself (like forming confidence intervals on expected output performance measures), the Output Analyzer has additional capabilities, and statistical comparison of two scenarios is among them.

To save the values of `Total Cost` for each replication to a *.dat* file for the Output Analyzer, we entered `Total Cost.dat` in the Output File column for the `Total Cost` row in the Statistic data module, and similarly for `Percent Rejected`. After the simulation has been run, these files will contain the 110 values of these output performance measures over the 110 replications; the file format is binary, however, and it can be read only by the Output Analyzer so will not be readable from applications like word processors or spreadsheets. Since we'll want to save the `Total Cost` and `Percent Rejected` values over our 110 replications of both the base-case and more-resource scenarios, and since we'll have two such files for each output performance measure, we either need to change their file names in the Arena model before each run, or rename them out in the operating system after each run to avoid overwriting them on subsequent runs; we took the latter route and appended "`- Base Case`" or "`- More Resources`" to the file names. Once you've made your runs of 110 replications each of the two scenarios, start the Output Analyzer (it's probably in the same folder that contains Arena, unless you did something weird during your installation).

In the Output Analyzer, select *File > New* (or *Ctrl+N* or 🗋) to open a new data group. This is not a *.dat* file containing data, but is rather a list of *.dat* files that you select as the ones in which you have current interest; this basically screens out the possibly many other *.dat* files present and, while not required, simplifies things. You can save this data group file (the default file name extension is *.dgr*) after you populate it with *.dat* file names so that next time you can just open it rather than selecting the *.dat* files all over again; we saved this data group as `Model 06-04.dgr`.

Use the *Add* button in the data group window to select (open) the files `Total Cost - Base Case.dat` and `Total Cost - More Resources.dat` to make them available for analysis; then do the same for the two `Percent Rejected`.*dat* files. The Output Analyzer can make the comparison we want via *Analyze > Compare Means*. Display 6-1 fills in the information for the Compare Means function (the graphical part of Display 6-1 also shows the data group as the upper-left window behind the others). Note that, in the Compare Means dialog, we cleared the check box for Scale Display; leaving this checked would put the graphical displays for the two comparisons on the same scale, which could make sense if the measurement units were the same for the two comparisons, but in our case, one is in dollars and the other is in percents, so scaling them together would not make sense.

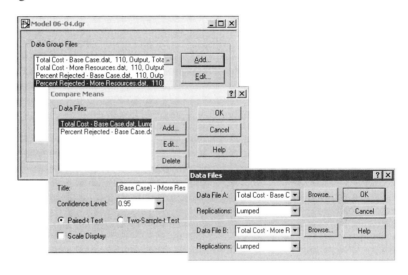

Title	(Base Case) - (More Resources)
Scale Display	*clear*
Add	
Data File A	Total Cost - Base Case.dat
Replication	Lumped
Add	
Data File B	Total Cost - More Resources.dat
Replication	Lumped
Add	
Data File A	Percent Rejected - Base Case.dat
Replication	Lumped
Add	
Data File B	Percent Rejected - More Resources.dat
Replication	Lumped

Display 6-1. Using the Output Analyzer's Compare Means Facility

We first Add the data files, one pair for `Total Cost` and a second pair for `Percent Rejected`, requiring specification of those from both scenarios (called A and B in the dialog box, both Lumped so that all 110 replications for each scenario are "lumped" together for our analysis), then maybe fill in a Title and accept or change the Confidence Level for the comparison. The option button group for Paired-t Test (the default) vs. Two-Sample-t Test refers to an issue of random number allocation and statistical independence, which we'll take up in Section 12.4.1; the Paired-t approach is somewhat more general and will be the one to use unless we've taken deliberate steps to make the scenarios statistically independent (which we didn't), or if we try to improve precision by allocating the random numbers carefully (which we haven't done here). As noted above, we cleared the check box for Scale Display since in this case the units are different for the two comparisons.

The results are in Figure 6-3. The Output Analyzer does the subtraction of means in the direction A – B; since we called the base case A and the more-resources scenario B, and more resources tended to increase `Total Cost` but decrease `Percent Rejected`, the signs of the differences are as expected. To see whether these are statistically significant differences (that is, whether these differences are too far away from zero to be reasonably explained by random noise), the Output Analyzer gives you 95% confidence intervals on the expected differences; since the intervals both miss zero, we conclude that there are statistically significant differences on both output performance measures, and that increasing resources in this way significantly increases `Total Cost` and significantly decreases `Percent Rejected` (the results of the equivalent two-sided hypothesis test for zero-expected difference are also shown in the lower part of the window). Confidence intervals are really a better way to express all this

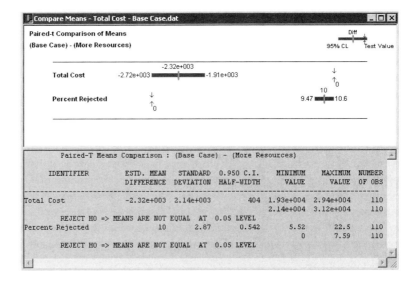

Figure 6-3. Confidence Interval and Hypothesis Test on the Expected Difference Between Total Costs and Percents Rejected

than are hypothesis tests, since they contain the "do-not-reject" or "reject" conclusions from the tests (the interval contains or misses zero), but also quantify the magnitude of any statistically significant differences. If you did this comparison because you're considering implementing this change, you now have an idea of what the increase in `Total Cost` will be and what the decrease in `Percent Rejected` will be if you increase the resources in this way.

6.5 Evaluating Many Scenarios with the Process Analyzer (PAN)

In Section 6.4, we defined two different alternative system configurations in terms of six input parameters (capacity of the `Trunk Line` resource, and the variables `New Tech 1`, `New Tech 2`, `New Tech 3`, `New Tech All`, and `New Sales`), ran the model in both configurations, and carried out a statistical analysis to be confident that we were drawing the correct conclusions about the difference between the two scenarios in terms of two particular output performance measures (`Total Cost` and `Percent Rejected`). What if you had more (maybe a lot more) than two scenarios that you wanted to run and compare? You'd face two issues:

(1) Managing the practical mechanics of making the model changes for all the different scenarios, which could be tedious and laborious if you have a lot of scenarios, and defining each one involves a lot of parameter changes in the model. And don't forget about changing the name of any saved output files, either by editing the model or by changing the file names out in the operating system.

(2) Evaluating the results in a statistically valid way to sort out which scenarios differ significantly from which others relative to which output performance measures, and which ones might be significantly better than the others, or even the best among all of those you considered.

Arena comes with another separate application called the *Process Analyzer* (or PAN for short) that greatly eases your burden on issue (1) above and also provides support for issue (2) to help you evaluate your results in a statistically valid way and make sound decisions. PAN operates on Arena program files, which normally have .*p* as the file name extension; after you run a model, the .*p* file will be there, or you can create it from Arena without running it by checking your model (*Run > Check Model* or the *F4* key or the ✓ button). You start PAN as you would any application, for instance by navigating to it via the Windows *Start* button (by default it's in the same place as Arena itself), or from within Arena via *Tools > Process Analyzer*. Either way, PAN will run on its own, in a window separate from Arena (if you have Arena running).

A *scenario* for PAN is a combination of a program (.*p*) file that exists somewhere on your system, a set of values for input *controls* for that program file that you select, a set of output *responses* for that program file that you also choose, as well as a descriptive Name for the scenario. A PAN *project* is a collection of such scenarios and can be saved in a PAN file (.*pan* extension) for later use. The program (.*p*) files defined for different scenarios can in fact be the same program file or can result from different Arena model (.*doe*) files, depending on what you want to run and compare. As we'll see, you select the controls from among your models' Variables and Resource capacities, and you select the responses from your models' outputs (as produced by Arena) and your own Variables. Once you have your scenarios defined, PAN will execute the ones you select (you can select all of them if you want), with the controls set at the values you define for each scenario and deliver the response results in a table together with a record of the control values for each scenario. This is equivalent to your going in and editing your model for the control values for each scenario and then running them all individually from within Arena; of course, doing this via PAN is a lot easier and faster and also supports a valid statistical comparison between the results. To make effective use of PAN, you should think ahead as you build your model to what input parameters you'll want to change to define different scenarios, and be sure to set them up as Arena Variables (or Resource capacities) in your model so that they can be selected to become controls in a PAN scenario.

After starting PAN, either create a new PAN project via *File > New* (or *Ctrl+N* or □), or open a previously saved project file (*.pan* extension) via *File > Open* (or *Ctrl+O* or 📂); at present, PAN can have only one project window open at a time, but you can have multiple instances of PAN running simultaneously. To add a new scenario line to the project, double-click where indicated to bring up the Scenario Properties dialog box where you can give the scenario a Name (we chose `Base Case` for this scenario) and Tool Tip Text, and associate an existing program (*.p*) file with this scenario (use the *Browse* button to navigate to the *.p* file you want). The model we'll use for our PAN experiments will be Model 6-5, which is the same as Model 6-4 except that we removed the Output File entries `Total Cost.dat` and `Percent Rejected.dat` in the Statistic data module since PAN will itself keep track of these data and our writing them out to our own file would only waste time; we're still making 110 replications for each scenario. We checked this model to produce `Model 06-05.p`, and we selected that for our Program File here. If you want to edit these things later, just right-click in the line for the scenario and select Scenario Properties. To select the controls for this scenario, right-click in its line (except at the left edge) or in the top Scenario Properties label, or use the *Insert > Controls* menu option, and select Insert Control to bring up a dialog box containing an expandable tree of possible controls from among the Resources, System variables (number and length of replications), and User-Specified variables; clicking the plus sign in any tree expands it to allow you to select any entry by double-clicking on it, causing it to appear as a Control in the scenario row. For our experiments with this model, we selected six Controls, one from Resources and the other five from among the User Specified variables:

- Under Resources, the capacity of `Trunk Line`; select Integer for the Data Type in the drop-down list and then click *OK*.
- Under User Specified, each of `New Tech 1`, `New Tech 2`, `New Tech 3`, `New Tech All`, and `New Sales`, selecting Integer for the Data Type in each case.

Once you've selected all the controls you want, right-click again in the row or at the top and select Insert Response to select your response(s) in the same way; we selected two Responses from the User-Specified group, `Total Cost` (selecting two Decimal

Figure 6-4. The PAN Project Experiment 06-05.pan *After Defining
the First (Base-Case) Scenario*

Places) and `Percent Rejected` (one Decimal Place). The values of the controls show up as the ones they have in the model as it was defined, but (here's the power of PAN) you can change them for execution of this scenario by simply editing their values in this scenario row in the PAN project window; for the base-case scenario, we left the control values as they were in the original model (26 and the five 0s). The values for the responses are blank since we haven't run the scenario yet. Figure 6-4 shows the status of part of the PAN project window after we defined this scenario and saved the project as `Experiment 06-05.pan` (we chose "Experiment" since this really is a whole experiment going on, which could even be a statistically designed one though it isn't here—see Section 12.6).

You could repeat the above process for additional scenarios. However, if they are similar to one that you've already defined, you can duplicate it (right-click in its scenario number on the left and select Duplicate Scenario(s)), and then just make edits in the duplicated scenario.

To define a sensible set of scenarios for this model, suppose that you received an order from above giving you $1300/week more to spend on additional resources, but that you have to devote the entire $1300 to just a single kind of resource. To which of the six "expandable" resources should you allocate the new money? Given the weekly costs of the resources, you could get 13 more trunk lines ($98 each), or 4 more of any one of the additional single-product tech-support people ($320 each), or 3 of the additional all-product tech-support people ($360 each), or 4 additional sales people ($340 each). Figure 6-5 shows the PAN project filled out with the six possible scenarios, plus the base-case scenario at the top.

To run some or all of the scenarios, select their rows by clicking in the left-most column of each row (*Ctrl+Click* to extend the selection or *Shift+Click* to select a contiguous set of lines). Then select *Run > Go* (or the *F5* function key or the ▶ button). Figure 6-6 shows the results of running all seven scenarios; the Responses column gives the average over the 110 replications of each scenario. Looking at the `Total Cost` column, it appears that beefing up the sales-call staff would be best; looking at the `Percent Rejected` column, it appears that adding more trunk lines would be best.

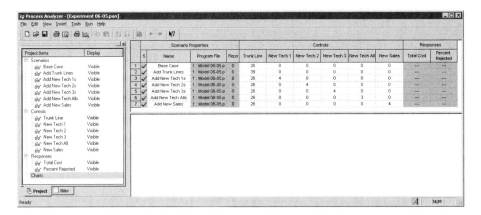

Figure 6-5. The PAN Project Experiment 06-05.pan *After Defining All Scenarios*

While the above conclusions are based on more than just one replication of each scenario (remember that we set the number of replications in the model file to 110 per scenario, so this is what PAN runs since we didn't override it via Num Reps in the System Controls), it would be far better to have some idea of their statistical validity. To investigate this on the `Total Cost` measure, select that column (click in the Responses label above the `Total Cost` cell at the top to select this column of results), and then select *Insert > Chart* (or) or right-click in this column and select Insert Chart. There are a lot of options here, so we'll just guide you through one of them and let you explore the others on your own, with the help of Help. Under Chart Type, select Box and Whisker, click Next, make sure `Total Cost` is selected under Use these Responses, and accept the defaults in the Next (third) window. In the last (fourth) window, check the Identify Best Scenarios box, then select "Smaller is Better" and enter `0` for the Error Tolerance (overriding what's there by default). Then click "Show best

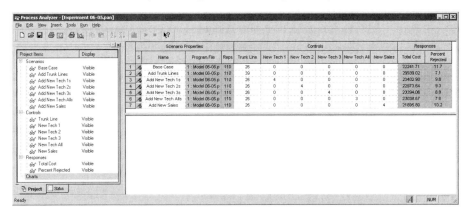

Figure 6-6. Results of Running the PAN Project for All Scenarios

scenarios" and note what appears in the window below that button; click Finish to get the plot. Finally, repeat all of the above for the `Percent Rejected` column to produce a second Box and Whisker plot for those values.

In the bottom part of the PAN window, you should see something like what's in Figure 6-7; you might want to resize the charts to make them bigger, by selecting them and dragging the corners around. If you right-click on a chart, select Chart Options, and then the Data tab, you can see, for each scenario, numerical results beyond the means that are in the table, such as the minimum and maximum across the replications, the half width of 95% confidence intervals, and the low and high values that are the bounds of the confidence intervals. In the chart itself, the vertical box plots indicate 95% confidence intervals on expected values for each alternative scenario, and those colored red are determined to be significantly better (smaller) than those colored blue (we've circled the red boxes so you can see what's what in this crummy monochrome book). More precisely, what this means is that the "red" scenarios form a subset of scenarios that is 95% certain to contain the true best scenario in terms of the true (and unknown) expected value of that response. If we'd chosen an Error Tolerance of greater than 0, the "red" subset will contain the best scenario or one within the Error Tolerance of the best, with 95% confidence. Keep in mind that these conclusions are based on the number of replications you specified for each scenario (in our case, we specified 110 replications

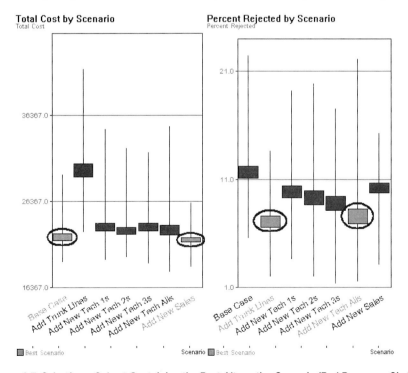

Figure 6-7. Selecting a Subset Containing the Best Alternative Scenario (Red Boxes are Circled)

for each scenario, but we could have specified different numbers of replications for different scenarios); if you want to narrow the selected (red) subset, you could increase the number of replications for the scenarios. You could also narrow the selected (red) subset by specifying the Error Tolerance to be a positive number that represents an amount small enough that you don't care if the selected scenario(s) are actually inferior to the true best one by at most this amount; so a positive Error Tolerance may reduce the number of selected scenarios at the risk of being off by a little bit. This selection procedure is firmly based on sound statistical theory developed by Nelson, Swann, Goldsman, and Song (2001).

So depending on whether you regard `Total Cost` or `Percent Rejected` as the more important response, you now have an indication of which among these six (seven, if you include the base case) scenarios would likely be best. It's hard to think of a sensible way to combine these two responses into one on which to optimize. Wouldn't it be nice if, for instance, we could seek to minimize `Total Cost` while holding `Percent Rejected` to no more than something like five? Press on (to Section 6.6), gentle reader.

6.6   Searching for an Optimal Scenario with OptQuest

The scenarios evaluated in Section 6.5 were just a few of the myriad of possibilities that we might explore in a quest to minimize `Total Cost`, subject to the requirement that the `Percent Rejected` response turn out to be no more than five. Suppose you're now set free to explore all the possibilities you'd like in terms of the `Trunk Line` capacity (but you're contractually obligated to keep at least the 26 you already have, and the wiring closet has room for at most 50), as well as `New Tech 1`, `New Tech 2`, `New Tech 3`, `New Tech All`, and `New Sales` (but you have room for at most 15 new people in all of the tech-support and sales categories combined). Since you have six input control variables to play with, you can think of this as wandering around in part of six-dimensional space looking for a six-vector (of non-negative integers) that minimizes `Total Cost` while holding `Percent Rejected` to five or less. This is a lot of possibilities (1,356,600, actually), and considering them all exhaustively is not practical (with the 110 replications per scenario that we've been using, it would run around the clock for over two months on a 2.13 GHz machine).

Where do you start? This is a hard question, but a reasonable idea would be the best point you know about so far. None of the scenarios in the PAN experiment of Figure 6-6 meets the requirement that `Percent Rejected` ≤ 5 (though a couple are close, and we could start with such an infeasible point though we'd need to move away from it eventually, of course, as in Exercise 6-24). However, the more-resources case from Section 6.4 (the `Trunk Line` resource has a capacity of 29, and `New Tech 1`, `New Tech 2`, `New Tech 3`, `New Tech All`, and `New Sales` are all set to 3) does meet the `Percent Rejected` ≤ 5 requirement, so we'll start there, even though its `Total Cost` is rather high in comparison to other scenarios with which we've experimented, and we'll seek to reduce it while obeying the `Percent Rejected` ≤ 5 requirement.

But where do you go from here? This is not an easy question to answer either, to say the least. A lot of people have been doing research into such topics for many years and have developed a variety of methods to address this problem. Arena comes with a package called OptQuest®, from OptTek Systems Inc., that uses heuristics known as *tabu search* and *scatter search* (and other methods) to move around intelligently in the input-control space and try to converge quickly and reliably to an optimal point. OptQuest is, in a way, similar to PAN in that it "takes over" the execution of your Arena model; the difference is that rather than relying on you to specify which specific scenarios to simulate, OptQuest decides on its own which ones to consider in an iterative fashion that hopefully leads to something close to an optimal combination of input control values. For detailed documentation on how OptQuest works with Arena, see *Help > Product Manuals > Arena OptQuest User's Guide*.

To run OptQuest for Arena, first make active the Arena window for the model of interest. We'll basically use Model 6-5, except that OptQuest requires that a finite Run Length be specified in *Run > Setup > Replication Parameters*, so we entered 1000 hours, which will never be reached since we retain our model-sensitive stopping rule that will almost certainly stop the replication well before time 1000 hours. We called the result Model 6-6. Then select *Tools > OptQuest for Arena* to bring up the OptQuest application window, then New Optimization.

OptQuest takes a moment to look through your model for potential Controls (including capacities of resources and Arena Variables you've defined) and Responses (including the built-in Arena costing outputs, and all outputs you've defined), and presents them in an expandable/contractible tree on the left. These are the input parameters that you can choose to allow OptQuest to vary as it searches; in mathematical-programming or optimization parlance, these are the *choice variables* or *decision variables*, and we seek a combination that will *optimize* (minimize or maximize) an *objective* that will be chosen below from the simulation's outputs. Start by clicking on the word Controls in the tree on the left (or the menu option *View > Controls*), to bring up a list of all Resources and User Specified variables, and check the boxes in the Included column for `Trunk Line`, `New Sales`, `New Tech 1`, `New Tech 2`, `New Tech 3`, and `New Tech All`. In each of these Included Control lines, double-click to open a window where you specify the appropriate Lower Bound, Suggested Value (i.e., starting values as discussed above), Upper Bound, and maybe a brief Description; also check the Type as appropriate, which will be Integer for all of our controls. You can then click on the Included column heading and its arrow to collect the controls you've selected at the top or bottom. Note that, since the total of all new staff hired must be no more than 15, certainly each one must be no more than 15, establishing those Upper (High) Bounds. Figure 6-8 shows the OptQuest window with the completed Controls section visible.

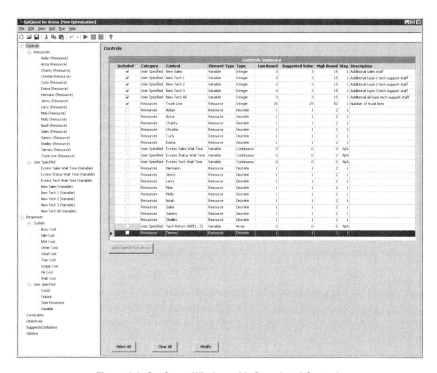

Figure 6-8. OptQuest Window with Completed Controls

Next, click on Responses in the tree on the left (or the menu option *View > Responses*) to select the outputs of interest. Here, simply check the boxes in the Included column for `Total Cost` and `Percent Rejected`, then maybe click the Included header to collect them at the top. The result is in Figure 6-9, this time without the tree on the left.

To specify the constraints, click on Constraints in the tree on the left (or *View > Constraints*) to bring up that window. Click *Add* at the bottom, and then successively click each of the first five Controls, then a + sign from the calculator pad at the right, and finally `<= 15` to specify this constraint in the field at the bottom on the inputs; maybe add a Name and Description, then click *OK*. Then click *Add* at the bottom again, and select `Percent Rejected` under Responses, then `<= 5` to specify this constraint on an output; note that this "constraint" is not really like a constraint in mathematical programming, as it's on a model output (that's different from the objective) rather than on an input choice or decision variable. The completed Constraints window is in Figure 6-10.

To tell OptQuest what it is you want to optimize, click Objectives in the tree on the left (or *View > Objectives*), then *Add*, then click `Total Cost` under the Responses, then the Minimize radio button (and a Name and Description if you like). The result is in Figure 6-11.

Responses

Responses Summary				
Included ▽	Category	Data Type	Response	Response Type
☑	User Specified	Output	Percent Rejected	Output Value
☑	User Specified	Output	Total Cost	Output Value
☐	System	NVA Cost	All Entities.NVACost	Output Value
☐	System	Other Cost	All Entities.OtherCost	Output Value
☐	System	Total Cost	All Entities.TotalCost	Output Value
☐	System	Tran Cost	All Entities.TranCost	Output Value
☐	System	VA Cost	All Entities.VACost	Output Value
☐	System	Wait Cost	All Entities.WaitCost	Output Value
☐	System	Busy Cost	All Resources.BusyCost	Output Value
☐	System	Idle Cost	All Resources.IdleCost	Output Value
☐	System	Total Cost	All Resources.TotalCost	Output Value
☐	System	Usage Cost	All Resources.UsageCost	Output Value
☐	System	Total Cost	System.TotalCost	Output Value
☐	User Specified	Count	Attempted Calls	Counter Value
☐	User Specified	Time Persistent	Backoffice Research WIP	DStat Average
☐	User Specified	Count	Completed Calls	Counter Value
☐	User Specified	Variable	Excess Sales Wait Time	Variable Value
☐	User Specified	Variable	Excess Status Wait Time	Variable Value
☐	User Specified	Variable	Excess Tech Wait Time	Variable Value
☐	User Specified	Variable	New Sales	Variable Value
☐	User Specified	Variable	New Tech 1	Variable Value
☐	User Specified	Variable	New Tech 2	Variable Value
☐	User Specified	Variable	New Tech 3	Variable Value
☐	User Specified	Variable	New Tech All	Variable Value
☐	User Specified	Count	Period 01 Rejected Calls	Counter Value
☐	User Specified	Count	Period 02 Rejected Calls	Counter Value
☐	User Specified	Count	Period 03 Rejected Calls	Counter Value
☐	User Specified	Count	Period 04 Rejected Calls	Counter Value
☐	User Specified	Count	Period 05 Rejected Calls	Counter Value
☐	User Specified	Count	Period 06 Rejected Calls	Counter Value
☐	User Specified	Count	Period 07 Rejected Calls	Counter Value
☐	User Specified	Count	Period 08 Rejected Calls	Counter Value
☐	User Specified	Count	Period 09 Rejected Calls	Counter Value
☐	User Specified	Count	Period 10 Rejected Calls	Counter Value
☐	User Specified	Time Persistent	Tech 1 Total Online WIP Stat	DStat Average
☐	User Specified	Time Persistent	Tech 2 Total Online WIP Stat	DStat Average
☐	User Specified	Time Persistent	Tech 3 Total Online WIP Stat	DStat Average
☐	User Specified	Variable	Tech Return WIP[1..3]	Variable Value
☐	User Specified	Count	Total Rejected Calls	Counter Value

Figure 6-9. Completed Responses for OptQuest

Constraints

Constraints Summary				
Included	Name	Type	Description	Expression
☑	Desk Space	Linear	Available space for new staff	[New Sales] + [New Tech 1] + [New Tech 2] + [New Tech 3] + [New Tech All] <= 15
☑	Rejected Calls	NonLinear	Limit on percent of calls rejected	[Percent Rejected] <= 5

Figure 6-10. Completed Constraints for OptQuest

Figure 6-11. Completed Constraints for OptQuest

The Options entry on the tree to the left (or *View > Options*) provides several ways to control how OptQuest decides to stop searching, a tolerance for regarding two objective values as equal, control on the replications per scenario, and a place for a complete log of what OptQuest considers during its run. Since we know that this model has considerable variation across replications, under Replications per simulation, specify a minimum of 10 and a maximum of 110 replications. Otherwise we'll accept the default options, and in particular the Automatic Stop option, which means that OptQuest will stop looking when it has not seen significant improvement for 100 different scenarios in a row.

From the Options window, you could click the *Optimize* button to start OptQuest, or alternatively click the *Start Optimization* (▶) button on the toolbar, or select *Run > Start Optimization* (or the F5 key). This causes OptQuest to simulate your model at the starting point you chose, and then decide on its own how best to search through the feasible region of six-dimensional space for a combination of input controls that minimizes `Total Cost` while satisfying both constraints. As it does so, you can see its progress by selecting Optimization in the tree on the left, to see the best objective value so far, the current objective value, and how many scenarios (what OptQuest calls "simulations") have been run. You can also see the best and current controls, and that the constraints are being satisfied during the search (or occasionally violated). The graph at the bottom plots the best (smallest for us) objective found so far, as a function of how many "simulations" (scenarios) have thus far been tried. The final view of this information is in Figure 6-12, which displays the best set of controls found and the associated objective function. This graph is typical, with large improvements (reductions) realized early (the low-hanging-fruit phenomenon), followed by harder-fought improvements and eventually a tapering off. OptQuest evaluated 429 scenarios (0.03% of the 1,356,600 possible), finding the best one among them as the 329th scenario it considered (and then no improvement for 100 more scenarios, its default stopping rule). The best `Total Cost` was 20368.83, some 15% lower than the starting point; the way to achieve that would be to stick with 26 trunk lines, hire two new sales staffers, and one each of type-1, type-3, and all-type tech-support people. You can check the log file OptQuest creates (see the Options window for its name and location on your system) for the complete history of its journey through six space. Also, by clicking on Best Solutions in the tree on the left you get a window giving not just the best solutions, but the runners-up, in order, so you can see other good scenarios in case the best one is problematic for some external reasons (e.g., office politics). You can save all this to a *.opt* file, to be opened later from within OptQuest, via *File > Save* or 🖫 (we saved this as the file `Optimum Seeking 06-06.opt`).

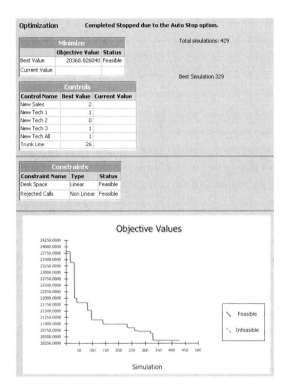

Figure 6-12. Final Optimization Window for OptQuest

Now OptQuest cannot absolutely guarantee that it's finding the exact optimum in every case; after all, this is a very challenging problem, not only since there are so many possibilities to examine, but also since the objective (the simulation-model output performance measure to be optimized) cannot even be measured with certainty. However, in most cases, OptQuest will do a far better job than we could by just messing around and trying things, so in general has considerable value. For background on what OptQuest is doing, see Glover, Kelly, and Laguna (1999), for example.

6.7   Summary and Forecast

You should now have a pretty good sense of what you and your computer need to do in order to use your terminating simulation models effectively to form accurate and precise estimates of expected results, draw valid conclusions about how different alternatives (or scenarios) of your model might differ in terms of key output performance measures, and to search efficiently for a model configuration that's optimal in some sense. This takes you a long way toward becoming an effective user of your models (not just an effective modeler, though that's certainly important too). In Chapter 7, we'll go through some more modeling ideas and then at the end revisit output analysis if your interest is in long-run (or steady-state) results, in which case you have to do things a little differently from what we've described in this chapter. Chapter 12 discusses further statistical issues in simulation, including how the underlying randomness is generated, how to reduce output variability (without just more computing), getting Arena to decide itself how much to run to give you acceptably precise results, and a little more on designing simulation experiments.

6.8   Exercises

6-1   In Exercise 5-1, statistically compare your results to what we got earlier; do this comparison informally by making 95% confidence intervals based on 50 replications from both models and see if they overlap. As performance measures, use the average waiting time in queue, average queue length, and server utilization.

6-2   Using the model from Exercise 5-2, change the processing time for the second pass on Machine 1 to TRIA(6.7, 9.1, 13.6). Run the simulation for 20,000 minutes and compare the results with respect to the same three output performance measures. Do an "informal" statistical comparison by making 20 replications of both versions of the model, forming 95% confidence intervals, and looking at the relation between the confidence intervals.

6-3   In Exercise 5-3, about how many replications *would* be required to bring the half width of a 95% confidence interval for the expected average cycle time for both trimmers down to one minute? (You don't need to *do* this, just do something to estimate the approximate number of replications that *would* be required.)

6-4   The facility of Exercise 5-4 is in a congested urban area where land is expensive and you've been asked to decide how much space should be planned for the trucks in queue to unload; address this question (being mindful of statistical issues).

6-5   In Exercise 5-5, suppose you could hire one more person and that person could be a second (identical) operator at any of the four locations. Which one should it be? Use PAN with five replications per scenario and select the best in terms of lowest average time in system. In retrospect, is your choice surprising? How critical is it to get this choice right?

Now suppose you could hire up to five more people and allocate them any way you want to the four existing operators, including allocating all five of the new people to a single operator. What's the best thing to do? Use OptQuest to search for the optimal allocation of this maximum of five more people, with the objective of minimizing the average time in system.

6-6   In Exercise 4-1, suppose that 7% of arriving customers are classified just after their arrival as being frequent fliers. Run your simulation for five replications (five days). Assume all times are the same as before, and observe statistics on the average time in system by customer type (frequent fliers and non-frequent fliers) under each of the following scenarios:

   (*a*)   Assign four of the current agents to serve non-frequent fliers only and the fifth to serve frequent fliers only.

(b) Change your model from part (a) so that the frequent-flier agent can serve regular customers when there are no frequent fliers waiting in queue.

(c) Change your model so that any agent can serve any customer, but priority is always given to frequent fliers.

Which of the three scenarios above do you think is "best" for the frequent fliers? How about for the 93% of customers who are non-frequent fliers? Viewing this as a terminating simulation, address these comparison questions in a statistically valid way.

6-7 In Exercise 5-13, make 30 replications and compute a 95% confidence interval on the expected system or cycle time for both customer types.

6-8 In Exercise 5-14, make 30 replications, and estimate the expected difference between this and the model of Exercise 5-13 (based on system time by customer). Be sure to use an appropriate formal statistical technique.

6-9 In Exercise 5-15, make 30 replications and estimate the expected difference between this model and the one in Exercise 5-14 (based on system time by customer), using an appropriate formal statistical technique.

6-10 For the inventory system of Model 5-4, set up a PAN comparison to investigate the average total cost resulting from each of the following (s, S) pairs:

(20, 40)	(20, 60)	(20, 80)	(20, 100)
	(40, 60)	(40, 80)	(40, 100)
		(60, 80)	(60, 100)
			(80, 100)

Run 50 replications of each case, and pick the best (lowest-cost) among these in a statistically valid way. This comparison is from Section 1.5.5 of Law and Kelton (2000).

6-11 For the inventory system of Model 5-4, use OptQuest to look for the best (total-cost-minimizing) setting for (s, S). Let s run between 1 and 99 (by ones) and let S run between 2 and 100 (by ones); remember that s and S must be integers and that we must have $s < S$. Make your own judgments about run time, number of replications, etc.

6-12 In Exercise 6-11, investigate whether taking inventory at the beginning of each day is necessarily the best policy by adding the inventory-evaluation interval (currently set at one day) to the optimum-seeking variable mix, and allow OptQuest to vary it (continuously) between a half day and five days, as well as simultaneously allowing s and S to vary as in Exercise 6-11. If you use `Model 05-04.doe` as a base, you'll need to change `Evaluation Interval` from an Expression to a Variable so OptQuest can get at it.

6-13 Apply OptQuest on all six values of (s, S) in Exercise 5-20. Let each s range between 1 and 99 (by ones), and let each S range between 2 and 100 (by ones); of course, s and S must be integers and for each of the three items in inventory we must have $s < S$.

6-14 Apply OptQuest on all six values of (s, S) in Exercise 5-21, with the same ranges as in Exercise 6-13. Compare your results with those from Exercise 6-13 and explain this in terms of the incentives offered by the supplier in this different cost structure.

6-15 In Exercise 4-22, suppose you could hire one more person, and that person could be assigned either to manual check-in, automated check-in, or security. Where would this additional person be best used? As an overall criterion of "goodness" of a system, use the average total time in system of all types of passengers combined. Set up a PAN run and decide about the number of replications needed to arrive at a reliable conclusion.

6-16 In Exercise 6-15, suppose you could hire a total of five more people (rather than just one more) to be allocated in any way to the existing staff at manual check-in, automated check-in, or security. You can't reduce staffing anywhere from the levels in Exercise 4-22. Using average total time in system of all types of passengers combined, what would be the best deployment of these five additional people? Should you even use all five of them? Set up an OptQuest run and decide about run time and numbers of replications.

6-17 In Exercise 4-22 (with staffing fixed at the original levels), the airline noticed that a lot of the people who opt for the manual check-in really don't need the extra services there and could have used the automated check-in. Instead of the original 35% and 50% that go to manual and automated, respectively, suppose that through more effective "encouragement" from people with red blazers and loud voices, it could be made 10% and 75% instead. (There probably are 10% of passengers who genuinely need the extra services of manual check-in.) How would this shift in passenger attitudes (or maybe bravery) affect average total time in system of passengers? Base your comparison on 200 replications of the model in each configuration.

6-18 Exercise 2-8 described a change in Model 3-1. Carry out a statistically valid study to measure the effect of this change on the average waiting time in queue and on the server utilization. Make your own decisions about numbers of replications in order to arrive at meaningful and reasonably precise conclusions.

6-19 Recall Models 3-2, 3-3, 3-4, and 3-5 from Chapter 3, comparing specialized serial vs. generalized parallel operations in environments of both high-variance and low-variance (no-variance, actually) service times. It *appeared* in Table 3-1 that the generalized parallel arrangement provided better service, though the improvement was much stronger in the high-variance service-time environment. Choose relevant output performance measures and make valid statistical comparisons to see if what *appears* to be true in Table 3-1 (from just a single run) is statistically valid.

6-20 Recall Exercise 3-13, which modified Model 3-3 to add 18% onto the individual-task times in the generalized integrated-work parallel configuration of Model 3-3. Choose relevant output performance measures and make a valid statistical comparison to see if the model of Exercise 3-13 differs significantly from the original specialized serial configuration of Model 3-2. In other words, if you had to endure this 18% increase in task-processing times, should you still move from the specialized-work serial setup to the generalized-work parallel setup?

6-21 For the enhanced machine-repair model of Exercise 5-22, do a valid statistical analysis to defend your choice of the number of repair technicians to hire in order to minimize total average cost. Make your own decision on number of replications, and modify it if necessary.

6-22 Use Arena to simulate the newsvendor problem of Section 2.7.1. Consider just the case $q = 140$, and run for 30 days to get an average daily profit and 95% confidence interval, as well as the proportion of days in which a loss is incurred.

6-23 Generalize your Arena model from Exercise 6-22 to consider all five values of q in Section 2.7.1, using the same daily demand realization for each value of q.

6-24 Rerun the OptQuest search in Section 6.6, except starting with the base-case model (26 trunk lines and no new employees). Also, rerun the search starting with scenario 6 from the PAN experiment in Figure 6-6 (which is almost, but not quite, feasible with respect to `Percent Rejected`). Comment on the effect of different starting points in optimization searches.

CHAPTER 7

Intermediate Modeling and Steady-State Statistical Analysis

Many of the essential elements of modeling with Arena were covered in Chapters 3, 4, and 5, including the basic use of some of the Basic Process and Advanced Process panel modules, controlling the flow of entities, Resource Schedules and States, Sets, Variables, Expressions, Stations, Transfers, and enhancing the animation. In this chapter, we'll expand on several concepts that allow you to do more detailed modeling. As before, we'll illustrate things concretely by means of a fairly elaborate example. We'll first start by introducing a new example as described in Section 7.1; expressing it in Arena requires the concept of entity-dependent Sequences, discussed in Section 7.1.1. Then in Section 7.1.2, we take up the general issue of how to go about modeling a system, the level of detail appropriate for a project, and the need to pay attention to data requirements and availability. The data portion required for the model is built in Section 7.1.3 and the logical model in Section 7.1.4. In Section 7.1.5, we develop an animation, including discussion of importing existing CAD drawings for the layout. We conclude this portion with a discussion on verifying that the representation of a model in Arena really does what you want, in Section 7.1.6.

Continuing our theme of viewing all aspects of simulation projects throughout a study, we resume the topic of statistical analysis of the output data in Section 7.2, but this time it's for steady-state simulations, using the model from Section 7.1.

By the time you read and digest the material in this chapter, you'll have a pretty good idea of how to model things in considerable detail. You'll also be in a position to draw statistically valid conclusions about the performance of systems as they operate in the long run.

7.1 Model 7-1: A Small Manufacturing System

A layout for our small manufacturing system is shown in Figure 7-1. The system to be modeled consists of part arrivals, four manufacturing cells, and part departures. Cells 1, 2, and 4 each have a single machine; Cell 3 has two machines. The two machines at Cell 3 are not identical; one of these machines is a newer model that can process parts in 80% of the time required by the older machine. The system produces three part types, each visiting a different sequence of stations. The part steps and process times (in minutes) are given in Table 7-1. All process times are triangularly distributed; the process times given in Table 7-1 at Cell 3 are for the older (slower) machine.

The interarrival times between successive part arrivals (all types combined) are exponentially distributed with a mean of 13 minutes; the first part arrives at time 0. The distribution by type is 26%, Part 1; 48%, Part 2; and 26%, Part 3. Parts enter from the

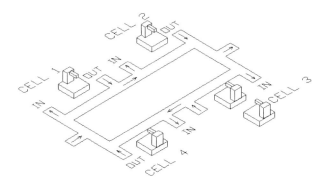

Figure 7-1. The Small Manufacturing System Layout

left, exit at the right, and move only in a clockwise direction through the system. For now, we'll also assume that the time to move between any pair of cells is two minutes, regardless of the distance (we'll fix this up later). We want to collect statistics on resource utilization, time and number in queue, as well as cycle time (time in system, from entry to exit) by part type. Initially, we'll run our simulation for 32 hours.

7.1.1   New Arena Concepts

There are several characteristics of this problem that require new Arena concepts. The first characteristic is that there are three part types that follow different process plans through the system. In our previous models, we simply sent all entities through the same sequence of stations. For this type of system, we need a process plan with an automatic routing capability.

The second characteristic is that the two machines in Cell 3 are not identical—the newer machine can process parts faster than the old machine. Here we need to be able to distinguish between these two machines.

The third characteristic is the nature of the flow of entities through the system. In our previous models, the flow of entities through the system was accomplished using the direct Connect or direct Route option. When you use the Connect option, an entity is sent immediately to the next module, according to the connection, with no time advance in

Table 7-1. Part Routings and Process Times

Part Type	Cell/Time	Cell/Time	Cell/Time	Cell/Time	Cell/Time
1	1	2	3	4	
	6, 8, 10	5, 8, 10	15, 20, 25	8, 12, 16	
2	1	2	4	2	3
	11, 13, 15	4, 6, 8	15, 18, 21	6, 9, 12	27, 33, 39
3	2	1	3		
	7, 9, 11	7, 10, 13	18, 23, 28		

the simulation. If we used the Connect option in this new model, we would have to include a number of Decide modules to direct the parts to the correct next station in the sequence followed by Route modules both to model the two-minute transfer time and to enable animation of the part movements. As you might expect, there is an Arena concept, *sequences*, that allows easy modeling of the flow of entities through the system while allowing for the part transfer time and enabling its animation.

Many systems are characterized by entities that follow predefined, but different paths through the system. Most manufacturing systems have part or process plans that specify a list of operations each part type must complete before exiting the system. Many service systems have similar types of requirements. For example, a model of passenger traffic in an airport may require different paths through the airport depending on whether the passengers check baggage or have only carry-on pieces, as well as whether the passengers have domestic or international flights.

Arena can send entities through a system automatically according to a predefined *sequence* of station visitations. The Sequence data module, on the Advanced Transfer panel, allows us to define an ordered list of Stations that can include assignments of attributes or variables at each station. To direct an entity to follow this pattern of station visitations, we assign the sequence to the entity (using a built-in sequence attribute, described below) and use the "By Sequence" option in the Route module when we transfer the entity to its next destination.

As the entity makes its way along its sequence, Arena will do all the necessary book-keeping to keep track of where the entity is and where it will go next. This is accomplished through the use of three special, automatically defined Arena attributes: Entity.Station (or M), Entity.Sequence (or NS), and Entity.JobStep (or IS). Each entity has these three attributes, with the default values for newly created entities being 0 for each. The Entity.Station attribute contains the current station location of the entity or the station to which the entity is currently being transferred. The Entity.Sequence attribute contains the sequence the entity will follow, if any; you need to assign this to each entity that will be transferring via a sequence. The Entity.Jobstep attribute specifies the entity's position within the sequence, so normally starts at 0 and then is incremented by 1 as the entity moves to each new station in its sequence.

We first define and name the list of stations to be visited for each type of entity (by part type, in our example) using the Sequence data module in the Advanced Transfer panel. Then, when we cause a new part to arrive into the system, we associate a specific Sequence with that entity by assigning the name of the sequence to the entity's Entity.Sequence attribute, NS.[1] When the entity is ready to transfer to the next station in its sequence, we select the By Sequence option in the Destination Type field of the module we're using to transfer the entity to the next station. At this point during the run, Arena first increments the Entity.Jobstep attribute (IS) by 1. Then it retrieves the destination station from the Sequence based on the current values for the Entity.Sequence

---

[1] Arena uses M, NS, and IS internally as names for these attributes, but provides the aliases *Entity.Station*, *Entity.Sequence*, and *Entity.Jobstep* in the pull-down list.

and Entity.Jobstep attributes. Any optional assignments are made (as defined in the Sequence) and the entity's Station attribute (Entity.Station) is set to the destination station. Finally, Arena transfers the entity to that station.

Typically, an entity will follow a sequence through to completion and will then exit the model. However, this is not a requirement. The Entity.Jobstep attribute is incremented only when the entity is transferred using the By Sequence option. You can temporarily suspend transfer via the sequence, transfer the entity directly to another station, and then re-enter the sequence later. This might be useful if some of the parts are required to be reworked at some point in the process. Upon completion of the rework, they might re-enter the normal sequence.

You can also reassign the sequence attributes at any time. For example, you might handle a part failure by assigning a new Entity.Sequence attribute and resetting the Entity.Jobstep attribute to 0. This new Sequence could transfer the part through a series of stations in a rework area. You can also back up or jump forward in the sequence by decreasing or increasing the Entity.Jobstep attribute. However, caution is advised as you must be sure to reset the Entity.Jobstep attribute correctly, remembering that Arena will first increment it, then look up the destination station in the sequence.

As indicated earlier, attribute and variable assignments can also be made at each step in a sequence. For example, you could change the entity picture or assign a process time to a user-defined attribute. For our small manufacturing model, we'll use this option to define some of our processing times that are part- and station-specific.

7.1.2   The Modeling Approach

The modeling approach to use for a specific simulation model will often depend on the system's complexity and the nature of the available data. In simple models, it's usually obvious what modules you'll require and the order in which you'll place them. But in more complex models, you'll often need to take a considerable amount of care developing the proper approach. As you learn more about Arena, you'll find that there are often a number of ways to model a system or a portion of a system. You will often hear experienced modelers say that there's not just a single, correct way to model a system. There are, however, plenty of wrong ways if they fail to capture the required system detail correctly.

The design of complex models is often driven by the data requirements of the model and what real data are available. Experienced modelers will often spend a great deal of time determining how they'll enter, store, and use their data, and then let this design determine which modeling constructs are required. As the data requirements become more demanding, this approach is often the only one that will allow the development of an accurate model in a short period of time. This is particularly true of simulation models of supply-chain systems, warehousing, distribution networks, and service networks. For example, a typical warehouse can have hundreds of thousands of uniquely different items, called SKUs (Stock-Keeping Units). Each SKU may require data characterizing its location in the warehouse, its size, and its weight, as well as reorder or restocking data for this SKU. In addition to specifying data for the contents of the warehouse, you also have customer order data and information on the types of storage devices or equipment that hold the SKUs. If your model requires the ability to change SKU locations, storage devices, restocking options, and so forth, during your experimentation, the data structure you use is critical. Although the models we'll develop in this book are not that complicated, it's always advisable to consider your data requirements before you start your modeling.

For our small manufacturing system, the data structure will, to a limited extent, affect our model design. We will use Sequences to control the flow of parts through the system, and the optional assignment feature in the Sequence data module to enter attributes for part process times for all but Cell 1. We'll use an Expression to define part process times for Cell 1. The part transfer time and the 80% factor for the time required by the new machine in Cell 3 will exploit the Variables concept. Although we don't normally think of Sets as part of our data design, their use can affect the modeling method. In this model, we'll use Sets, combined with a user-defined index, to ensure that we associate the correct sequence and picture with each part type.

First, we'll enter the data modules as discussed earlier. We'll then enter the main model portion, which will require several new modules. Next, we'll animate the model using a CAD or other drawing as a starting point. Finally, we'll discuss briefly the concept of model verification. By now you should be fairly familiar with opening and filling Arena dialog boxes, so we will not dwell on the mundane details of how to enter the information in Arena. In the case of modules and concepts that were introduced in previous chapters, we'll only indicate the data that must be entered. To see the "big picture" of where we're heading, you may want to peek ahead at Figure 7-5, which shows the complete model.

7.1.3 The Data Modules

We start by editing the Sequence data module from the Advanced Transfer panel. Double-click to add a new row and enter the name of the first sequence, Part 1 Process Plan. Having entered the sequence name, you next need to enter the process Steps, which are lists of Arena stations. For example, the Part 1 Process Plan requires you to enter the following Arena stations: Cell 1, Cell 2, Cell 3, Cell 4, and Exit System. We have given the Step Names as Part 1 Step 1 through Part 1 Step 5. The most common error in entering sequences is to forget to enter the last step, which is typically where the entity exits the system. If you forget this step, you'll get an Arena run-time error when the first entity completes its process route and Arena is not told where to send it next. As you define the Sequences, remember that after you've entered a station name once, you can subsequently pick it from station drop-down lists elsewhere in your model. You'll also need to assign attribute values for the part process times for Cells 2, 3, and 4. Recall that we'll define an Expression for the part process times at Cell 1, so we won't need to make an attribute assignment for process times there.

Display 7-1 shows the procedure for Sequence Part 1 Process Plan, Step Part 1 Step 2, and Assignment of Process Time. Using the data in Table 7-1, it should be fairly straightforward to enter the remaining sequence steps. Soon we'll show you how to reference these sequences in the model logic.

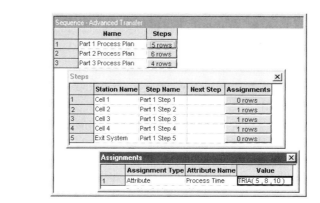

Sequence

	Name	Steps
1	Part 1 Process Plan	5 rows
2	Part 2 Process Plan	6 rows
3	Part 3 Process Plan	4 rows

Steps

	Station Name	Step Name	Next Step	Assignments
1	Cell 1	Part 1 Step 1		0 rows
2	Cell 2	Part 1 Step 2		1 rows
3	Cell 3	Part 1 Step 3		1 rows
4	Cell 4	Part 1 Step 4		1 rows
5	Exit System	Part 1 Step 5		0 rows

Assignments

	Assignment Type	Attribute Name	Value
1	Attribute	Process Time	TRIA(5 , 8 , 10)

Name	Part 1 Process Plan
Steps	
Station Name	Cell 2
Step Name	Part 1 Step 2
Assignments	
Assignment Type	Attribute
Attribute Name	Process Time
Value	TRIA(5, 8, 10)

Display 7-1. The Sequence Data Module

Next, we'll define the expression for the part process times at Cell 1, using the Expression data module in the Advanced Process panel. The expression we want will be called `Cell 1 Times`, and it will contain the part-process-time distributions for the three parts at Cell 1. We could just as easily have entered these in the previous Sequences module, but we chose to use an expression so you can see several different ways to assign the processing times. We have three different parts that use Cell 1, so we need an arrayed expression with three rows, one for each part type. Display 7-2 shows the data for this module.

Next, we'll use the Variable data module from the Basic Process panel to define the machine-speed factor for Cell 3 and the transfer time. Prior to defining our `Factor` variable, we made the following observation and assumption. The part-process times for Cell 3, entered in the Sequence data module, are for the old machine. We'll assume that

Expression	
Name	Cell 1 Times
Rows	3
Expression Values	
Expression Value	TRIA(6, 8, 10)
Expression Value	TRIA(11, 13, 15)
Expression Value	TRIA(7, 10, 13)

Display 7-2. The Expression for Cell 1 Part Process Times

Name	Factor
Rows	2
Initial Values	
Initial Value	0.8
Initial Value	1.0
Name	Transfer Time
Initial Values	
Initial Value	2

Display 7-3. The Factor and Transfer Time Variables

the new machine will be referenced as 1 and the old machine will be referenced as 2. Thus, the first factor value is 0.8 (for the new machine), and the second factor is 1.0 (for the old machine). The transfer-time value (which doesn't need to be an array) is simply entered as 2. This allows us to change this value in a single place at a later time. If we thought we might have wanted to change this to a distribution, we would have entered it as an Expression instead of a Variable. Display 7-3 shows the required entries.

We'll use the Set data module from the Basic Process panel to form sets for our Cell 3 machines, part pictures, and entity types. The first is a Resource set, Cell 3 Machines, containing two members: Cell 3 New and Cell 3 Old. Our Entity Picture set is named Part Pictures and it contains three members: Picture.Part 1, Picture.Part 2, and Picture.Part 3. Finally, our Entity Type set is named Entity Types and contains three members: Part 1, Part 2, and Part 3.

As long as we're creating sets, let's add one more for our part sequences. If you attempt to use the Set module, you'll quickly find that the types available are Resource, Counter, Tally, Entity Type, and Entity Picture. You resolve this problem by using the Advanced Set data module found in the Advanced Process panel. This module lists three types: Queue, Storage, and Other. The Other type is a catch-all option that allows you to form sets of almost any similar Arena objects. We'll use this option to enter our set named Part Sequences, which contains three members: Part 1 Process Plan, Part 2 Process Plan, and Part 3 Process Plan.

Before we start placing logic modules, let's open the *Run > Setup* dialog box and set our Replication Length to 32 hours and our Base Time Units as Minutes. We also selected *Edit > Entity Pictures* to open the Entity Picture Placement window, where we created three different pictures—Picture.Part 1, Picture.Part 2, and Picture.Part 3. In our example, we copied blue, red, and green balls; renamed them; and placed a text object with 1, 2, and 3, respectively, to denote the three different part types. Having defined all of the data modules, we're now ready to place and fill out the modules for the main model to define the system's logical characteristics.

7.1.4 The Logic Modules

The main portion of the model's operation will consist of logic modules to represent part arrivals, cells, and part departures. The part arrivals will be modeled using the four modules shown in Figure 7-2. The Create module uses a Random(Expo) distribution with a mean of 13 minutes to generate the arriving parts.

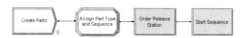

Figure 7-2. The Part Arrival Modules

At this point, we have not yet associated a sequence with each arriving entity. We make this association in the Assign module as shown in Display 7-4. These assignments serve two purposes: they determine which part type has arrived, and they define an index, `Part Index`, for our sets that will allow us to associate the proper sequence, entity type, and picture with each arrival. We first determine the part index, or part type, with a discrete distribution. The distribution allows us to generate certain values with given probabilities. In our example, these values are the integers 1, 2, and 3 with probabilities 26%, 48%, and 26%, respectively. You enter these values in pairs—*cumulative* probability and value. The cumulative probability for the last value (in our case, 3) should be 1.0. In general, the values need not be integers, but can take on any values, including negative numbers. The part index values of 1, 2, and 3 allow us not only to refer to the part type, but in this case, they also allow us to index into the previously defined set `Part Sequences` so that the proper sequence will be associated with the part. To do so, we assign the proper sequence to the automatic Arena Entity.Sequence attribute by using the `Part Index` attribute as an index into the `Part Sequences` set.

We also need to associate the proper entity type and picture for the newly arrived part. We do this in the last two assignments by using the `Part Index` attribute to index into the relevant sets. Recall that we created the `Entity Types` set (consisting of `Part 1`, `Part 2`, and `Part 3`). We also defined three pictures (`Picture.Part 1`, `Picture.Part 2`, and `Picture.Part 3`) and grouped them into a set called `Part Pictures`. You can think of these sets as an arrayed variable (one-dimensional) with

Name	Assign Part Type and Sequence
Assignments	
Type	Attribute
Attribute Name	Part Index
New Value	DISC(0.26,1 , 0.74,2 , 1.0,3)
Type	Attribute
Attribute Name	Entity.Sequence
New Value	Part Sequences(Part Index)
Type	Attribute
Attribute Name	Entity.Type
New Value	Entity Types(Part Index)
Type	Attribute
Attribute Name	Entity.Picture
New Value	Part Pictures(Part Index)

Display 7-4. The Assign Module: Assigning Part Attributes

the index being the `Part Index` attribute we just assigned. This is true in this case because we have a one-to-one relationship between the part type (`Part Index`) and the sequence, picture, and entity type. An index of 1 implies `Part 1` follows the first sequence, etc. (Be careful in future models because this may not always be true. It is only true in this case because we defined our data structure so we would have this one-to-one relationship.)

In this case, the completed Assign module determines the part type, and then assigns the proper sequence, entity type, and picture. Note that it was essential to define the value of `Part Index` first in this Assign module, which performs multiple assignments in the order listed, since the value of `Part Index` determined in the first step is used in the later assignments. We're now ready to send our part on its way to the first station in its sequence.

We'll accomplish this with the Route module from the Advanced Transfer panel. Before we do this, we need to tell Arena where our entity, or part, currently resides (its current station location). If you've been building your own model along with us (and paying attention), you might have noticed the attribute named `Entity.Station` on the drop-down list in the Assign module that we just completed. (You can go back and look now if you like.) You might be tempted to add an assignment that would define the current location of our part. Unfortunately if you use this approach, Arena will return an error when you try to run your model; instead, we use a Station module as described next.

Our completed model will have six stations: `Order Release`, `Cell 1`, `Cell 2`, `Cell 3`, `Cell 4`, and `Exit System`. The last five stations were defined when we filled in the information for our part sequences. The first station, `Order Release`, will be defined when we send the entity through a Station module (Advanced Transfer panel), which will define the station and tell Arena that the current entity is at that location. Our completed Station module is shown in Display 7-5.

We're finally ready to send our part to the first station in its sequence with the Route module (Advanced Transfer panel). The Route module transfers an entity to a specified station, or the next station in the station visitation sequence defined for the entity. A Route Time to transfer to the next station may be defined. In our model, the previously defined variable `Transfer Time` is entered for the Route Time; see Display 7-6. We selected the `By Sequence` option as the Destination Type. This causes the Station Name field to disappear, and when we run the model, Arena will route the arriving entities according to the sequences we defined and attached to entities after they were created.

Name	Order Release Station
Station Type	Station
Station Name	Order Release

Display 7-5. The Station Module

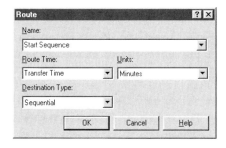

Name	Start Sequence
Route Time	Transfer Time
Units	Minutes
Destination Type	By Sequence

Display 7-6. The Route Module

Figure 7-3. Cell 1 Logic Modules

Now that we have the arriving parts being routed according to their assigned part sequences, we need to develop the logic for our four cells. The logic for all four cells is essentially the same. A part arrives to the cell (at a station), queues for a machine, is processed by the machine, and is sent to its next step in the part sequence. All four of these cells can be modeled easily using the Station – Process – Route module sequence shown in Figure 7-3 (for Cell 1).

The Station module provides the location to which a part can be sent. In our model, we are using sequences for all part transfers, so the part being transferred would get the next location from its sequence. The entries for the Station module for Cell 1 are shown in Display 7-7.

A part arriving at the Cell 1 station is sent to the following Process module (using a direct connect). For the Expression for the delay time, we've entered the previously defined arrayed expression `Cell 1 Times` using the `Part Index` attribute to reference the appropriate part-processing time in the arrayed expression. This expression will generate a sample from the triangular distribution with the parameters we defined earlier. The remaining entries are shown in Display 7-8.

Name	Cell 1 Station
Station Type	Station
Station Name	Cell 1

Display 7-7. The Cell 1 Station Module

Name	Cell 1 Process
Action	Seize Delay Release
Resources	
Type	Resource
Resource Name	Cell 1 Machine
Quantity	1
Delay Type	Expression
Units	Minutes
Expression	Cell 1 Times(Part Index)

Display 7-8. Cell 1 Process Module

Upon completion of the processing at Cell 1, the entity is directed to the Route module, described in Display 7-9, where it is routed to its next step in the part sequence. Except for the Name, this Route module is identical to the Route module we used to start our sequence at the Order Release station in Display 7-6.

The remaining three cells are very similar to Cell 1, so we'll skip the details. To create their logic, we copied the three modules for Cell 1 three times and then edited the required data. For each of the new Station and Route modules, we simply changed all occurrences of Cell 1 to Cell 2, Cell 3, or Cell 4. We made the same edits to the three additional Process modules and changed the delay Expression for Cells 2 and 4 to Process Time. Recall that in the Sequence module we defined the part-processing times for Cells 2, 3, and 4 by assigning them to the entity attribute Process Time. When the part was routed to one of these cells, Arena automatically assigned this value so it could be used in the module.

At this point, you should be aware that we've somewhat arbitrarily used an Expression for the Cell 1 part-processing times, and attribute assignments in the sequences for the remaining cells. We actually used this approach to illustrate that there are often several different ways to structure data and access them in a simulation model. We could just as easily have incorporated the part-processing times at Cell 1 into the sequences, along with the rest of the times. We also could have used Expressions for these times at Cells 3 and 4. However, it would have been difficult to use an Expression for these times at Cell 2, because Part 2 visits that cell twice and the processing times are different, as shown in Table 7-1. Thus, we would have had to include in our model the ability to know whether the part was on its first or second visit to Cell 2 and define our expression to recognize that. It might be an interesting exercise, but from the modeler's point of view, why not just define these values in the sequences? You should also recognize that this sample model is

Name	Route from Cell 1
Route Time	Transfer Time
Units	Minutes
Destination Type	By Sequence

Display 7-9. Cell 1 Route Module

Name	Cell 3 Process
Action	Seize Delay Release
Resources	
Type	Set
Set Name	Cell 3 Machines
Quantity	1
Save Attribute	Machine Index
Delay Type	Expression
Units	Minutes
Expression	Process Time * Factor(Machine Index)

Display 7-10. Cell 3 Process Module

relatively small in terms of the number of cells and parts. In practice, it would not be uncommon to have 30 to 50 machines with hundreds of different part types. If you undertake a problem of that size, we strongly recommend that you take great care in designing the data structure since it may make the difference between a success or failure of the simulation project.

The Process module for Cell 3 is slightly different because we have two different machines, a new one and an old one, that process parts at different rates. If the machines were identical, we could have used a single resource and entered a capacity of 2. We made note of this earlier and grouped these two machines into a Set called Cell 3 Machines. Now we need to use this set at Cell 3. Display 7-10 shows the data entries required to complete the Process module for this cell.

In the Resource section, we select Set from the drop-down list for our Resource Type. This allows us to select from our specified Cell 3 Machines set for the Set Name field. You can also Seize a Specific Member of a set; that member can be specified as an entity attribute. For our selection of Resource Set, we've accepted the default selection rule, Cyclical, which causes the entity to attempt to select the first available resource beginning with the successor of the last resource selected. In our case, Arena will try to select our two resources alternately; however, if only one resource is currently available, it will be selected. Obviously, these rules are used only if more than one resource is available for selection. The Random rule would cause a random (equiprobable) selection from among those available, and the Preferred Order rule would select the first (lowest-numbered) available resource in the set. Had we selected this option, Arena would have always used the new machine, if available, because it's the first resource in the set. The remaining rules would apply only if one or more of our resources had a capacity greater than 1.

The Save Attribute option allows us to save the index, which is a reference to the selected set member, in a user-specified attribute. In our case, we will save this value in the Attribute Machine Index. If the new machine is selected, this attribute will be assigned a value of 1, and if the old machine is selected, the attribute will be assigned a value of 2. This numbering is based on the order in which we entered our resources when we defined the set. The Expression for the delay time uses our attribute Process

Figure 7-4. The Exit System Logic Modules

Time, assigned by the sequence, multiplied by our variable Factor (Machine Index). Recall that our process times are for the old machine and that the new machine can process parts in 80% of that time. Although it's probably obvious by now how this expression works, we'll illustrate it. If the first resource in the set (the new machine) is selected, Machine Index will be assigned a value of 1 and the variable Factor (Machine Index) will use this attribute to take on a value of 0.8. If the second machine (the old machine) is selected, the Machine Index is set to 2, the Factor (Machine Index) variable will take on a value of 1.0, and the original process time will be used. Although this method may seem overly complicated for this example, it is used to illustrate the tremendous flexibility that Arena provides. An alternate method, which would not require the variable, would have used the following logical expression:

```
Process Time * (((Machine Index == 1)*0.8) + (Machine Index == 2))
```

or

```
Process Time * (1 - ((Machine Index == 1)*0.2)).
```

We leave it to you to figure out how these expressions work (hint: "a==b" evaluates to 1 if a equals b and to 0 otherwise).

Having completely defined all our data, the part arrival process, and the four cells, we're left only with defining the parts' exit from the system. The two modules that accomplish this are shown in Figure 7-4.

As before, we'll use the Station module to define the location of our Exit System Station. The Dispose module is used to destroy the completed part. Our completed model (although it is not completely animated) appears in Figure 7-5.

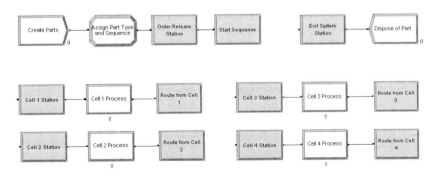

Figure 7-5. The Completed Model

At this point, we could run our model, but it would be difficult to tell if our sequences are working properly. So before we start any analysis, we'll first develop our animation to help us determine whether the model is working correctly. Let's also assume that we'll have to present this model to higher-level management, so we want to develop a somewhat fancier animation.

7.1.5 Animation

We could develop our animation in the same manner as we did in Chapter 5—create our own entity pictures, resource symbols, and background based on the picture that was presented in Figure 7-1. However, a picture might already exist that reflects an accurate representation of the system. In fact, it might even exist as a CAD file somewhere in your organization. For instance, the drawing presented in Figure 7-1 was developed using the AutoCAD® package from AutoDesk. Arena supports integration of files from CAD programs into its workspaces. Files saved in the DXF format defined by AutoCAD can be imported directly into Arena. Files generated from other CAD or drawing programs (e.g., Microsoft® Visio®) should transfer into Arena as long as they adhere to the AutoCAD standard DXF format.

If the original CAD drawing is 2-D, you only need to save it in the DXF format and then import that file directly into Arena. Most CAD objects (polygons, etc.) will be represented as the same or similar objects in Arena. If the drawing is 3-D, you must first convert the file to 2-D. Colors are lost during this conversion to 2-D, but they may be added again in AutoCAD or in Arena. This conversion also transforms all objects to lines, so the imported drawing can only be manipulated in Arena as individual lines, or the lines may be grouped as objects. We'll assume that you have a DXF file and refer you to online help (the Importing DXF Files topic) for details on creating the DXF file or converting a 3-D drawing. One word of advice: a converted file is imported into Arena as white lines so if your current background is white, it may appear that the file was not imported. Simply change the Window Background Color (▨ ▾ on the Draw toolbar) or select all the lines and change the line color (▨ ▾).

For our small manufacturing system, we started with a 3-D drawing and converted it to 2-D. We then saved the 2-D file in the DXF format (`Model 07-01.dxf`). A DXF file is imported into your current model file by selecting *File > DXF Import*. Select the file, and when you move your pointer to the model window, it will appear as cross hairs. Using the pointer, draw a box to contain the entire drawing to be imported. If the imported drawing is not the correct size, you can select the entire drawing and resize it as required. You can now use this drawing as the start of your animation.

For our animation, you should delete all the lettering and arrows. Then we'll move the Cell 1 queue animation object from the Process module (`Cell 1 Process`) to its approximate position on the drawing.

Now position your pointer near the upper-left corner of a machine and drag your pointer so the resulting bounding outline encloses the entire drawing. Use the *Copy* button to copy the drawing of the machine to the clipboard. Now we'll open the Resource Picture Placement window by clicking the *Resource* button (⬚▸) from the Animate toolbar. Double-click the idle resource symbol and replace the current picture with a copy of the contents of the clipboard. Delete the base of the machine and draw two boxes on top of the representation for the top part of the machine. This is necessary as our drawing consists of all lines, and we want to add a fill color to the machine. Now delete all the lines from the original copy and fill the boxes with your favorite color. Copy this new symbol to your library and then copy that symbol to the Busy picture. Save your library, exit the resource window, and finally place the resource. You now have a resource picture that will not change, but we could go back later and add more animation effects.

For the next step, draw a box the same size as the base of the machine and then delete the entire drawing of the machine. Fill this new box with a different color and then place the resource symbol on top of this box (you may have to resize the resource symbol to be the correct size). Now move the seize point so that it is positioned at about the center of our new machine. Now make successive copies of our new resource and the machine base, and then change the names. If you follow this procedure, note that you will need to flip the resources for Cells 3 and 4.

To complete this phase of the animation, you need to move and resize the remaining queues. Once this is complete, you can run your animation and see your handiwork. You won't see parts moving about the system (there are no routes yet), but you will see the parts in the queues and at the machines. If you look closely at one of the machines while the animation is running, you'll notice that the parts sit right on top of the entire machine. This display is not what we want. Ideally, the part should sit on top of the base, but under the top part of our machine (the resource). Representing this is not a problem. Select a resource (you can do this in edit mode or by temporarily pausing the run) and use the Bring to Front feature with *Arrange > Bring to Front* or the button on the Arrange toolbar (🖫). Now when you run the animation, you'll get the desired effect. We're now ready to animate our part movement.

In our previous animations, we basically had only one path through the system so adding the routes was fairly straightforward. In our small manufacturing system, there are multiple paths through the system, so you must be careful to add routes for each travel possibility. For example, a part leaving Cell 2 can go to Cell 1, 3, or 4. The stations need to be positioned first; add station animation objects using the *Station* button on the Animate Transfer toolbar. Next, place the routes. If you somehow neglect to place a route, the simulation will still (correctly) send the entity to the correct station with a transfer time of 2; however, that entity movement will not appear in your animation. Also be aware that routes can be used in both directions. For example, let's assume that you added the route from Cell 1 to Cell 2, but missed the route from 2 to 1. When a Part 2 completes its processing at Cell 2, Arena looks to route that part to Cell 1, routing from 2 to 1. If the route were missing, it would look for the route from 1 to 2 and use that. Thus, you would see that part moving from the entrance of Cell 2 to the Exit of Cell 1 in a counterclockwise direction (an animation mistake for this model).

If you run and watch your newly animated model, you should notice that occasionally parts will run over or pass each other. This is due to a combination of the data supplied and the manner in which we animated the model. Remember that all part transfers were assumed to be two minutes regardless of the distance to be traveled. Arena sets the speed of a routed entity based on the transfer time and the physical length of the route on the animation. In our model, some of these routes are very short (from Cell 1 to Cell 2) and some are very long (from Cell 2 to Cell 1). Thus, the entities being transferred will be moving at quite different speeds relative to each other. If this was important, we could request or collect better transfer times and incorporate these into our model. The easiest way would be to delete the variable `Transfer Time` and assign these new transfer times to a new attribute in the Sequences module. If your transfer times and drawing are accurate, the entities should now all move at the same speed. The only potential problem is that a part may enter the main aisle at the same time another part is passing by, resulting in one of the parts overlapping the other until their paths diverge. This is a more difficult problem to resolve, and it may not be worth the bother. If the only concern is for purposes of presentation, watch the animation and find a long period of time when this does not occur—show only this period of time during your presentation. The alternative is to use material-handling constructs, which we'll do in Chapter 8.

After adding a little annotation, the final animation should look something like Figure 7-6 (this is a view of the system at time 541.28). At this point in your animation, you might want to check to see if your model is running correctly or at least the way you intended it. We'll take up this topic in our next section.

7.1.6 Verification

Verification is the process of ensuring that the Arena model behaves in the way it was intended according to the modeling assumptions made. This is actually very easy compared to model *validation*, which is the process of ensuring that the model behaves the same as the real system. We'll discuss both of these aspects in more detail in Chapter 13. Here we'll only briefly introduce the topic of model verification.

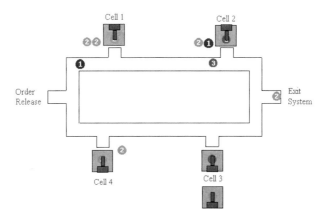

Figure 7-6. The Animated Small Manufacturing System: Model 7-1

Verification deals with both obvious problems as well as the not-so-obvious. For example, if you had tried to run your model and Arena returned an error message that indicated that you had neglected to define one of the machine resources at Cell 3, the model is obviously not working the way you intended. Actually, we would normally call this the process of debugging! However, since we assume that you'll not make these kinds of errors (or, if you do, you can figure out yourself how to fix them), we'll deal with the not-so-obvious problems.

Verification is fairly easy when you're developing small classroom-size problems like the ones in this book. When you start developing more realistically sized models, you'll find that it's a much more difficult process, and you may never be 100% sure on very, very large models.

One easy verification method is to allow only a single entity to enter the system and follow that entity to be sure that the model logic and data are correct. You could use the *Step* button (▶︎) found on the Standard toolbar to control the model execution and step the entity through the system. For this model, you could set the Max Arrivals field in the Create module to 1. To control the entity type, you could replace the discrete distribution in the Assign module that determines the part type with the specific part type you want to observe. This would allow you to check each of the part sequences. Another common method is to replace some or all model data with constants. Using deterministic data allows you to predict system behavior exactly.

If you're going to use your model to make real decisions, you should also check to see how the model behaves under extreme conditions. For example, introduce only one part type or increase/decrease the part interarrival or service times. If your model is going to have problems, they will most likely show up during these kinds of stressed-out runs. Also, try to make effective use of your animation—it can often reveal problems when viewed by a person familiar with the operation of the system being modeled.

It's often a good idea, and a good test, to make long runs for different data and observe the summary results for potential problems. One skill that can be of great use during this process is that of *performance estimation*. A long, long time ago, before the invention of calculators and certainly before personal computers, engineers carried around sticks called *slide rules* (often protected by leather cases and proudly attached to engineers' belts). These were used to calculate the answers to the complicated problems given to them by their professors or bosses. These devices achieved this magic feat by adding or subtracting logs (not tree logs, but logarithms). Although they worked rather well (but not as well, as easily, or as fast as a calculator), they only returned (actually you read it off the stick) a sequence of digits, say 3642. It was up to the engineer to figure out where to put the decimal point. Therefore, engineers had to become very good at developing rough estimates of the answers to problems before they used the sticks. For example, if the engineer estimated the answer to be about 400, then the answer was 364.2. However, if the estimate was about 900, there was a problem. At that point, it was necessary to determine if the problem was in the estimation process or the slide-rule process. We suspect that at this point you're asking two questions: 1) why the long irrelevant story, and 2) did we really ever use such sticks? Well, the answers are: 1) to illustrate a point, and 2) all of the authors did!

So how do you use this great performance-estimation skill? You define a set of conditions for the simulation, estimate what will result, make a run, and look at the summary data to see if you were correct. If you were, feel good and try it for a different set of conditions. If you weren't correct (or at least in the ballpark), find out why not. It may be due to bad estimation, a lack of understanding of the system, or a faulty model. Sometimes not-so-obvious (but real) results are created by surprising interactions in the model. In general, you should thoroughly exercise your simulation models and be comfortable with the results before you use them to make decisions. And it's best to do this early in your project—and often.

Back in Section 2.8 when we mentioned verification, we suggested that you verify your code. Your response could have been, "What code?" Well, there *is* code, and we'll show you a little in Figures 7-7 and 7-8. But you still might be asking, "Why code?" To explain the reason for this code requires a little bit of background (yes, another history lesson). The formation of Systems Modeling (the company that originally developed Arena) and the initial release of the simulation language SIMAN® (on which Arena is based and to which you can gain access through Arena) occurred in 1982. Personal computers were just starting to hit the market, and SIMAN was designed to run on these new types of machines. In fact, SIMAN required only 64 Kbytes of memory, which was a lot in those days. There was no animation capability (Cinema, the accompanying animation tool, was released in 1985), and you created models using a text editor, just like using a programming language. A complete model required the development of two files, a *model file* and an *experiment file*. The model file, often referred to as the *.mod* file, contained the model logic. The experiment file, referred to as the *.exp* file, defined the experimental conditions. It required the user to list all stations, attributes, resources, etc., that were used in the model. The creation of these files required that the user follow a rather exacting syntax for each type of model or experiment construct. In other words, you had to start certain statements only in column 10; you had to follow certain entries with the required comma, semicolon, or colon; all resources and attributes were referenced only by number; and only a limited set of key words could be used. (Many of your professors, or maybe their professors, learned simulation this way.)

Since 1982, SIMAN has been enhanced continually and still provides the basis for an Arena simulation model. When you run an Arena simulation, Arena examines each option you selected in your modules and the data that you supplied and then creates SIMAN *.mod* and *.exp* files. These are the files that are used to run your simulations. The implication is that all the modules found in the Basic Process, Advanced Process, and Advanced Transfer panels are based on the constructs found in the SIMAN simulation language. The idea is to provide a simulation tool (Arena) that is easy to use, yet one that is still based on the powerful and flexible SIMAN language. So you see, it is still possible, and sometimes even desirable, to look at the SIMAN code. In fact, you can even write out and edit these files. However, it is only possible to go down (from Arena to SIMAN code) and not possible to go back up (from SIMAN code to Arena). As you become more proficient with Arena, you might occasionally want to look at this code to be assured that the model is doing exactly what you want it to—verification.

```
;
;    Model statements for module:  Process 3
;
10$     ASSIGN:     Cell 3 Process.NumberIn=Cell 3 Process.NumberIn + 1:
                    Cell 3 Process.WIP=Cell 3 Process.WIP+1;
138$    QUEUE,      Cell 3 Process.Queue;
137$    SEIZE,      2,VA:
                    SELECT(Cell 3 Machines,CYC,Machine Index),1:NEXT(136$);
136$    DELAY:      Process Time * Factor( Machine Index ),,VA;
135$    RELEASE:    Cell 3 Machines(Machine Index),1;
183$    ASSIGN:     Cell 3 Process.NumberOut=Cell 3 Process.NumberOut + 1:
                    Cell 3 Process.WIP=Cell 3 Process.WIP-1:NEXT(11$);
```

Figure 7-7. SIMAN Model File for the Process Module

We should point out that you can still create your models in Arena using the base SIMAN language. If you build a model using only the modules found in the Blocks and Elements panels, you are basically using the SIMAN language. Recall that we used several modules from the Blocks panel when we constructed our inventory model in Section 5.7.

You can view the SIMAN code for our small manufacturing model by using the *Run > SIMAN > View* option. Selecting this option will generate both files, each in a separate window. Figure 7-7 shows a small portion of the *.mod* file, the code written out for the Process module used at Cell 3. The SIMAN language is rather descriptive, so it is possible to follow the general logic. An entity that arrives at this module increments some internal counters, enters a queue, `Cell 3 Process.Queue`, waits to seize a resource from the set `Cell 3 Machines`, delays for the process time (adjusted by our factor), releases the resource, decrements the internal counters, and exits the module.

Figure 7-8 shows a portion of the *.exp* file that defines our three attributes and the queues and resources used in our model.

We won't go into a detailed explanation of the statements in these files; the purpose of this exercise is merely to make you aware of their existence. For a more comprehensive explanation, refer to Pegden, Shannon, and Sadowski (1995).

```
ATTRIBUTES:  Machine Index:
             Part Index:
             Process Time;
QUEUES:      Cell 1 Process.Queue,FIFO,,AUTOSTATS(Yes,,):
             Cell 2 Process.Queue,FIFO,,AUTOSTATS(Yes,,):
             Cell 3 Process.Queue,FIFO,,AUTOSTATS(Yes,,):
             Cell 4 Process.Queue,FIFO,,AUTOSTATS(Yes,,);
RESOURCES:   Cell 2 Machine,Capacity(1),,,COST(0.0,0.0,0.0),
               CATEGORY(Resources),,AUTOSTATS(Yes,,):
             Cell 1 Machine,Capacity(1),,,COST(0.0,0.0,0.0),
               CATEGORY(Resources),,AUTOSTATS(Yes,,):
             Cell 3 Old,Capacity(1),,,COST(0.0,0.0,0.0),
               CATEGORY(Resources),,AUTOSTATS(Yes,,):
             Cell 3 New,Capacity(1),,,COST(0.0,0.0,0.0),
               CATEGORY(Resources),,AUTOSTATS(Yes,,):
             Cell 4 Machine,Capacity(1),,,COST(0.0,0.0,0.0),
               CATEGORY(Resources),,AUTOSTATS(Yes,,);
```

Figure 7-8. SIMAN Experiment File for the Process Module

If you're familiar with SIMAN or you would just like to learn more about how this process works, we might suggest that you place a few modules, enter some data, and write out the *.mod* and *.exp* files. Then edit the modules by selecting different options and write out the new files. Look for the differences between the two *.mod* files. Doing so should give you a fair amount of insight as to how this process works.

7.2 Statistical Analysis of Output from Steady-State Simulations

In Chapter 6, we described the difference between terminating and steady-state simulations and indicated how you can use Arena's reports, PAN, and the Output Analyzer to do statistical analyses in the terminating case. In this section, we'll show you how to do statistical inference on steady-state simulations.

Before proceeding, we should encourage you to be *sure* that a steady-state simulation is appropriate for what you want to do. Often people simply *assume* that a long-run, steady-state simulation is the thing to do, which might in fact be true. But if the starting and stopping conditions are part of the essence of your model, a terminating analysis is probably more appropriate; if so, you should just proceed as in Chapter 6. The reason for avoiding steady-state simulation is that, as you're about to see, it's a *lot* harder to carry out anything approaching a valid statistical analysis than in the terminating case if you want anything beyond Arena's standard 95% confidence intervals on mean performance measures; so if you don't need to get into this, you shouldn't.

One more caution before we wade into this material: As you can imagine, the run lengths for steady-state simulations need to be pretty long. Because of this, there are more opportunities for Arena to sequence its internal operations a little differently, causing the random-number stream (see Chapter 12) to be used differently. This doesn't make your model in any sense "wrong" or inaccurate, but it can affect the numerical results, especially for models that have a lot of statistical variability inherent in them. So if you follow along on your computer and run our models, there's a chance that you're going to get numerical answers that differ from what we report here. Don't panic over this since it is, in a sense, to be expected. If anything, this just amplifies the need for some kind of statistical analysis of simulation output data since variability can come not only from "nature" in the model's properties, but also from internal computational issues.

In Section 7.2.1, we'll discuss determination of model warm-up and run length. Section 7.2.2 describes the truncated-replication strategy for analysis, which is by far the simplest approach (and, in some ways, the best). A different approach called batching is described in Section 7.2.3. A brief summary is given in Section 7.2.4, and Section 7.2.5 mentions some other issues in steady-state analysis.

7.2.1 Warm-Up and Run Length

As you might have noticed, our examples have been characterized by a model that's initially in an *empty-and-idle* state. This means that the model starts out empty of entities and all resources are idle. In a terminating system, this might be the way things actually start out, so everything is fine. But in a steady-state simulation, initial conditions aren't supposed to matter, and the run is supposed to go on forever.

Actually, though, even in a steady-state simulation, you have to initialize and stop the run somehow. And since you're doing a simulation in the first place, it's a pretty safe bet that you don't know much about the "typical" system state in steady state or how "long" a run is long enough. So you're probably going to wind up initializing in a state that's pretty weird from the viewpoint of steady state and just trying some (long) run lengths. If you're initializing empty and idle in a simulation where things eventually become congested, your output data for some period of time after initialization will tend to understate the eventual congestion; that is, your data will be *biased* low.

To remedy this, you might try to initialize in a state that's "better" than empty and idle. This would mean placing, at time 0, some number of entities around the model and starting things out that way. While it's possible to do this in your model, it's pretty inconvenient. More problematic is that you'd generally have no idea how many entities to place around at time 0; this is, after all, one of the questions the simulation is supposed to answer.

Another way of dealing with initialization bias is just to run the model for so long that whatever bias may have been there at the beginning is overwhelmed by the amount of later data. This can work in some models if the biasing effects of the initial conditions wear off quickly.

However, what people usually do is initialize empty and idle, realizing that this is unrepresentative of steady state, but then let the model *warm up* for a while until it appears that the effects of the artificial initial conditions have worn off. At that time, you can clear the statistical accumulators (but not the system state) and start afresh, gathering statistics for analysis from then on. In effect, this *is* initializing in a state other than empty and idle, but you let the model decide how many entities to have around when you start (afresh) to watch your performance measures. The run length should still be long, but maybe not as long as you'd need to overwhelm the initial bias by sheer arithmetic.

It's very easy to specify an initial Warm-up Period in Arena. Just open the *Run > Setup > Replication Parameters* dialog box and fill in a value (be sure to verify the Time Units). Every replication of your model still runs starting as it did before, but at the end of the Warm-up Period, all statistical accumulators are cleared and your reports (and any Outputs-type saved data from the Statistic module of results across the replications) reflect only what happened after the warm-up period ends. In this way, you can "decontaminate" your data from the biasing initial conditions.

The hard part is knowing how long the warm-up period should be. Probably the most practical idea is just to make some plots of key outputs from within a run, and eyeball when they appear to stabilize. To illustrate this, we took Model 7-1 from Section 7.1 and made the following modifications, calling the result Model 7-2:

- To establish a single overall output performance measure, we made an entry in the Statistic module to compute the total work in process (WIP) of all three parts combined. The Name and Report Label are both `Total WIP`, and the Type is `Time-Persistent`. To enter the Expression we want to track over time, we right-clicked in that field and selected Build Expression, clicked down the tree via Basic Process Variables → Entity → Number in Process, selected `Part 1` as the Entity Type, and got `EntitiesWIP(Part 1)` for the Current

Figure 7-9. The Completed Total WIP Entry in the Statistic Module

Expression, which is part of what we want. Typing a + after that and selecting `Part 2` for the Entity Type, another +, then `Type 3` for the Entity Type finally gives us `EntitiesWIP(Part 1) + EntitiesWIP(Part 2) + EntitiesWIP(Part 3)`, which is the total WIP. This will create an entry labeled `Total WIP` in the Category Overview (under User Specified) giving the time-average and maximum of the total number of parts in process.

- However, we also want to track the "history" of the `Total WIP` curve during the simulation rather than just getting the post-run summary statistics since we want to "eyeball" when this curve appears to have stabilized in order to specify a reasonable Warm-up Period. You could place an animated Plot in your model, as we've done before; however, this will disappear as soon as you hit the *End* button, and will also be subject to perhaps-large variation, clouding your judgment about the stabilization point. We need to save the curve history and make more permanent plots, and we also need to plot the curve for several replications to mitigate the noise problem. To do so we made a file-name entry, `Total WIP History.dat`, in the Output File field of the `Total WIP` entry in the Statistic module, which will save the required information into that file, which can be read into the Arena Output Analyzer (see Section 6.4) and plotted after the run is complete. Depending on how long your run is and how many replications you want, this file can get pretty big since you're asking it to hold a lot of detailed, within-run information (our run, described below, resulted in a file of about 176 KB). The complete entry in the Statistic module is in Figure 7-9.
- Since we aren't interested in animation at this point, we accessed the *Run > Run Control* option and checked *Batch Run (No Animation)* to speed things up. We also cleared all the boxes under Statistics Collection in *Run > Setup > Project Parameters* to increase speed further. To get a feel for the variation, we specified `10` for the Number of Replications in *Run > Setup > Replication Parameters*; since we're now interested in long-run steady-state performance, we increased the Replication Length from 32 Hours (1.33 Days) to 5 Days. We freely admit that these are pretty much arbitrary settings, and we settled on them after some trial and error. For the record, it took less than two seconds to run all this on a 2.13 GHz notebook computer.

To make a plot of WIP vs. time in the Output Analyzer (see Section 6.4 for basics on the Output Analyzer), we created a new data group and added the file `Total WIP History.dat` to it. We then selected Plot (⟨icon⟩ or *Graph > Plot*), Added the *.dat* file (selecting All in the Replications field of the Data File dialog box), typed in a Title, and changed the Axis Labels; the resulting dialog boxes are shown in Figure 7-10.

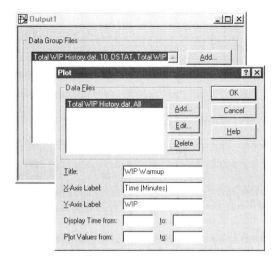

Figure 7-10. The Output Analyzer's Plot Dialog Box

Figure 7-11 shows the resulting plot of WIP across the simulations, with the curves for all ten replications superimposed. From this plot, it seems clear that as far as the WIP output is concerned, the run length of five days (7,200 minutes, in the Base Time Units used by the plot) is enough for the model to have settled out, and also it seems fairly clear that the model isn't "exploding" with entities, as would be the case if processing couldn't keep up with arrivals. As for a Warm-up Period, the effect of the empty-and-idle initial conditions for the first several hundred minutes on WIP is evident at the left end of each curve and is reasonably consistent across replications, but it looks like things settle down after about 2,000 minutes; rounding up a little to be conservative, we'll select 2 Days (2,880 minutes) as our Warm-up Period.

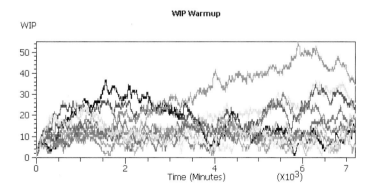

Figure 7-11. Within-Run WIP Plots for Model 7-2

If we'd made the runs (and plot) for a longer time period, or if the model warm-up occurred more quickly, it could be difficult to see the appropriate Warm-up Period on the left end of the curves. If so, you could "zoom in" on just the first part of each replication via the "Display Time from . . . to . . ." area in the Plot dialog box.

In models where you're interested in multiple output performance measures, you'd need to make a plot like Figure 7-11 for each output. There could be disagreement between the output measures on warm-up rate, in which case, the safe move is to take the maximum of the individual warm-ups as the one to use overall.

7.2.2 Truncated Replications

If you can identify a warm-up period, and if this period is reasonably short relative to the run lengths you plan to make, then things are fairly simple for steady-state statistical analysis: just make IID (that's independent and identically distributed, in case you've forgotten) replications, as was done for terminating simulations in Chapter 6, except that you also specify a Warm-up Period for each replication in *Run > Setup > Replication Parameters*. With these appropriate warm-up and run-length values specified, you just proceed to make independent replications and carry out your statistical analyses as in Chapter 6 with warmed-up independent replications so that you're computing steady-state rather than terminating quantities. Life is good.

This idea applies to comparing or selecting alternatives (Sections 6.4 and 6.5) and optimum seeking (Section 6.6), as well as to single-system analysis. However, there could be different warm-up and run-length periods needed for different alternatives; you don't really have the opportunity to control this across the different alternatives that OptQuest might decide to consider, so you should probably run your model ahead of time for a range of possible scenarios and specify the warm-up to be (at least) the maximum of those you experienced.

For the five-day replications of Model 7-2, we entered in *Run > Setup > Replication Parameters* the 2-Day Warm-up Period we determined in Section 7.3.1 and again asked for ten replications. Since we no longer need to save the within-run history, we deleted the Output File entry in the Statistic module. We called the resulting model Model 7-3 (note our practice of saving successive changes to models under different names so we don't cover up our tracks in case we need to backtrack at some point). The result was a 95% confidence interval for expected average WIP of 16.39 ± 6.51; from Model 7-2 with no warm-up, we got 15.35 ± 4.42, illustrating the downward-biasing effect of the initial low-congestion period. To see this difference more clearly, since these confidence intervals are pretty wide, we made 100 replications (rather than ten) and got 15.45 ± 1.21 for Model 7-3 and 14.42 ± 0.88 for Model 7-2. Note that the confidence intervals, based on the same number of replications, are wider for Model 7-3 than for Model 7-2; this is because each replication of Model 7-3 uses data from only the last three days whereas each replication of Model 7-2 uses the full five days, giving it lower variability (at the price of harmful initialization bias, which we feel is a desirable tradeoff in favor of truncating as in Model 7-3).

If more precision is desired in the form of narrower confidence intervals, you could achieve it by simulating "some more." Now, however, you have a choice as to whether to make more replications with this run length and warm-up or keep the same number of replications and extend the run length. (Presumably the original warm-up would still be adequate.) It's probably simplest to stick with the same run length and just make more replications, which is the same thing as increasing the statistical sample size. There is something to be said, though, for extending the run lengths and keeping the same number of replications; the increased precision with this strategy comes not from increasing the "sample size" (statistically, degrees of freedom) but rather from decreasing the variability of each within-run average since it's being taken over a longer run. Furthermore, by making the runs longer, you're even more sure that you're running long enough to "get to" steady state.

If you can identify an appropriate run length and warm-up period for your model, and if the warm-up period is not too long, then the truncated-replication strategy is quite appealing. It's fairly simple, relative to other steady-state analysis strategies, and gives you truly independent observations (the results from each truncated replication), which is a big advantage in doing statistical analysis. This goes not only for simply making confidence intervals as we've done here, but also for comparing alternatives as well as other statistical goals.

7.2.3 *Batching in a Single Run*

Some models take a long time to warm-up to steady state, and since each replication would have to pass through this long warm-up phase, the truncated-replication strategy of Section 7.2.2 can become inefficient. In this case, it might be better to make just one *really* long run and thus have to "pay" the warm-up only once. We modified Model 7-3 to make a single replication of length 50 days including a (single) warm-up of two days (we call this Model 7-4); this is the same simulation "effort" as making the ten replications of length five days each, and it took about the same amount of computer time. Since we want to plot the within-run WIP data, we reinstated a file name entry in the Output File field in the Statistic data module, calling it `Total WIP History One Long Run.dat` (since it's a long run, we thought it deserved a long file name too). Figure 7-12 plots Total WIP across this run. (For now, ignore the thick vertical bars we drew in the plot.)

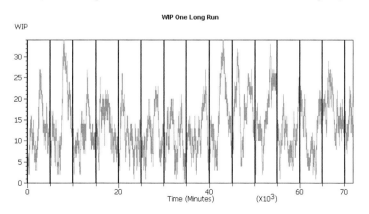

Figure 7-12. Total WIP Over a Single Run of 50 Days

The difficulty now is that we have only one "replication" on each performance measure from which to do our analysis, and it's not clear how we're going to compute a variance estimate, which is the basic quantity needed to do statistical inference. Viewing each individual observation or time-persistent value within a run as a separate "data point" would allow us to do the arithmetic to compute a within-run sample variance, but doing so is extremely dangerous since the possibly heavy correlation (see Section C.2.4 in Appendix C) of points within a run will cause this estimate to be severely biased with respect to the true variance of an observation. Somehow, we need to take this correlation into account or else "manufacture" some kind of "independent observations" from this single long run in order to get a decent estimate of the variance.

There are several different ways to proceed with statistical analysis from a single long run. One relatively simple idea (that also appears to work about as well as other more complicated methods) is to try to manufacture almost-uncorrelated "observations" by breaking the run into a few large *batches* of many individual points, compute the averages of the points within each batch, and treat them as the basic IID observations on which to do statistical analysis (starting with variance estimation). These *batch means* then take the place of the means from within the multiple replications in the truncated-replication approach—we've replaced the replications by batches. In the WIP plot of Figure 7-12, the thick vertical dividing lines illustrate the idea; we'd take the time-average WIP level between the lines as the basic "data" for statistical analysis. In order to obtain an unbiased variance estimate, we need the batch means to be nearly uncorrelated, which is why we need to make the batches big; there will still be some heavy correlation of individual points near either side of a batch boundary, but these points will be a small minority within their own large batches, which we hope will render the batch means nearly uncorrelated. In any case, Arena will try to make the batches big enough to look uncorrelated, and will let you know if your run was too short for it to manufacture batch means that "look" (in the sense of a statistical test) nearly uncorrelated, in which case, you'd have to make your run longer.

As we just hinted, Arena automatically attempts to compute 95% confidence intervals via batch means for the means of all output statistics and gives you the results in the Half Width column next to the Average column in the report for each replication. If you're making just one replication, as we've been discussing, this is your batch-means confidence interval. On the other hand, if you've made several replications, you see this in the Category by Replication report for each replication; the Half Width in the Category Overview report is across replications, as discussed in Sections 6.3 and 7.2.2. We say "attempts to compute" since internal checks are done to see if your replication is long enough to produce enough data for a valid batch-means confidence interval on an output statistic; if not, you get only a message to this effect, without a half-width value (on the theory that a wrong answer is worse than no answer at all) for this statistic. If you've specified a Warm-up Period, data collected during this period are not used in the confidence-interval calculation. To understand how this procedure works, think of Arena as batching "on the fly" (that is, observing the output data during your run and throwing them into batches as your simulation goes along).

So what does "enough data" mean? There are two levels of checks to be passed, the first of which is just to get started. For a Tally statistic, Arena demands a minimum of 320 observations. For a time-persistent statistic, you must have had at least five units of simulated time during which there were at least 320 changes in the level of the discrete-change variable. Admittedly, these are somewhat arbitrary values and conditions, but we need to get started somehow, and the more definitive statistical test, which must also be passed, is done at the end of this procedure. If your run is not long enough to meet this first check, Arena reports "(Insufficient)" in the Half-Width column for this variable. Just making your replication longer should eventually produce enough data to meet these getting-started requirements.

If you have enough data to get started, Arena then begins batching by forming 20 batch means for each Tally and time-persistent statistic. For Tally statistics, each batch mean will be the average of 16 consecutive observations; for time-persistent statistics, each will be the time average over 0.25 base time unit. As your run progresses, you will eventually accumulate enough data for another batch of these same "sizes," which is then formed as batch number 21. If you continue your simulation, you'll get batches 22, 23, and so on, until you reach 40 batches. At this point, Arena will re-batch these 40 batches by averaging batches one and two into a new (and bigger) batch one, batches three and four into a new (bigger) batch two, etc., so that you'll then be back to 20 batches, but each will be twice as "big." As your simulation proceeds, Arena will continue to form new batches (21, 22, and so on), each of this larger size, until again 40 batches are formed, when re-batching back down to 20 is once again performed. Thus, when you're done, you'll have between 20 and 39 complete batches of some size. Unless you're really lucky, you'll also have some data left over at the end in a partial batch that won't be used in the confidence-interval computation. The reason for this re-batching into larger batches stems from an analysis by Schmeiser (1982), who demonstrated that there's no advantage to the other option, of continuing to gather more and more batches of a fixed size to reduce the half width, since the larger batches will have inherently lower variance, compensating for having fewer of them. On the other hand, having batches that are too small, even if you have a lot of them, is dangerous since they're more likely to produce correlated batch means, and thus an invalid confidence interval.

The final check is to see if the batches are big enough to support the assumption of independence between the batch means. Arena tests for this using a statistical hypothesis test due to Fishman (1978). If this test is passed, you'll get the Half Width for this output variable in your report. If not, you'll get "(Correlated)" indicating that your process is evidently too heavily autocorrelated for your run length to produce nearly uncorrelated batches; again, lengthening your run should resolve this problem, though—depending on the situation, you may have to lengthen it a lot.

Returning to Model 7-4, after making the one 50-day run and deleting the first two days as a warm-up period, Arena produced in the Category Overview report a 95% batch-means confidence interval of 13.6394 ± 1.38366 on expected average WIP. The reason this batch-means confidence interval shows up here in the Category Overview report is that we've only made a single replication.

In most cases, these automatic batch-means confidence intervals will be enough for you to understand how precise your averages are, and they're certainly easy (requiring no work at all on your part). But there are a few considerations to bear in mind. First, don't forget that these are relevant only for steady-state simulations; if you have a terminating simulation (Chapter 6), you should be making independent replications and doing your analysis on them, in which case Arena reports cross-replication confidence-interval half-widths as in Chapter 6, and does no within-run batching. Secondly, you still ought to take a look at the Warm-up Period for your model, as in Section 7.2.1, since the automatic batch-means confidence intervals don't do anything to correct for initialization bias if you don't specify a Warm-up Period. Finally, you can check the value of the automatic batch-means half width as it is computed during your run, via the Arena variables THALF(Tally ID) for Tally statistics and DHALF(Dstat ID) for time-persistent (a.k.a. Dstat) statistics; one reason to be interested in this is that you could use one of these for a critical output measure in the Terminating Condition field of your Simulate module to run your model long enough for it to become small (i.e., precise) enough to suit you; see Section 12.5 for more on this and related ideas.

We should mention that, if you really want, you can decide on your own batch sizes, compute and save the batch means, then use them in statistical procedures like confidence-interval formation and comparison of two alternatives (see Sections 6.3 and 6.4). Briefly, the way this works is that you save your within-run history of observations just as we did to make our warm-up-determination plots, read this file into the Output Analyzer, and use its *Analyze > Batch/Truncate Obs'ns* capability to do the batching and averaging, saving the batch means that you then treat as we did the cross-replication means in Sections 6.3 and 6.4. When batching, the Output Analyzer performs the same statistical test for uncorrelated batches, and will let you know if the batch size you selected is too small. Some reasons to go through this are if you want something other than a 95% confidence level, if you want to make the most efficient use of your data and minimize waste at the end, or if you want to do a statistical comparison between two alternatives based on their steady-state performance. However, it's certainly a lot more work.

7.2.4   *What To Do?*

We've shown you how to attack the same problem by a couple of different methods and hinted that there are a lot more methods out there. So which one should you use? As usual, the answer isn't obvious (we warned you that steady-state output analysis is difficult). Sometimes there are tradeoffs between scientific results and conceptual (and practical) simplicity, and that's certainly the case here.

In our opinion (and we don't want to peddle this as anything more than opinion), we might suggest the following list, in decreasing order of appeal:

1. Try to get out of doing a steady-state simulation altogether by convincing yourself (or your patron) that the appropriate modeling assumptions really entail specific starting and stopping conditions. Go to Chapter 6 (and don't come back here).

2. If your model is such that the warm-up is relatively short, probably the simplest and most direct approach is truncated replication. This has obvious intuitive

appeal, is easy to do (once you've made some plots and identified a reasonable Warm-up Period), and basically winds up proceeding just like statistical analysis for terminating simulations, except for the warm-up. It also allows you to take advantage of the more sophisticated analysis capabilities in PAN and OptQuest.

3. If you find that your model warms up slowly, then you might consider batch means, with a single warm-up at the beginning of your single really long run. You could either accept Arena's automatic batch-means confidence intervals or handcraft your own. You cannot use the statistical methods in PAN or OptQuest with the batch-means approach, however (part of the reason this is last in our preference list).

7.2.5   Other Methods and Goals for Steady-State Statistical Analysis

We've described two different strategies (truncated replications and batch means) for doing steady-state statistical analysis; both of these methods are available in Arena. This has been an area that's received a lot of attention among researchers, so there are a number of other strategies for this difficult problem: econometric time-series modeling, spectral analysis from electrical engineering, regenerative models from stochastic processes, standardized time series, as well as variations on batch means like separating or weighting the batches. If you're interested in exploring these ideas, you might consult Chapter 9 of Law and Kelton (2000), a survey paper like Sargent, Kang, and Goldsman (1992), or peruse a recent volume of the annual *Proceedings of the Winter Simulation Conference*, where there are usually tutorials, surveys, and papers covering the latest developments on these subjects.

7.3   Summary and Forecast

Now you should have a very good set of skills for carrying out fairly detailed modeling and have an understanding of (and know what to do about) issues like verification and steady-state statistical analysis. In the following chapter, we'll expand on this to show you how to model complicated and realistic material-handling operations. In the chapters beyond, you'll drill down deeper into Arena's modeling and analysis capabilities to exploit its powerful and flexible hierarchical structure.

7.4   Exercises

7-1   In Exercise 4-7, is the run long enough to generate a batch-means-based confidence interval for the steady-state expected average cycle time? Why or why not?

7-2   Modify your solution for Exercise 5-2 to include transfer times between part arrival and the first machine, between machines, and between the last Machine 1 and the system exit. Assume all part transfer times are UNIF(1.5,3.2). Animate your model to show part transfers with the part entity picture changing when it departs from Machine 2. Run for 20,000 minutes. To the extent possible, indicate the batch-means-based confidence intervals on expected steady-state performance measures from this run.

7-3   Using the model from Exercise 7-2, change the processing time for the second pass on Machine 1 to TRIA(6.7, 9.1, 13.6) using Sequences to control the flow of parts through the system and the assignment of process times at each machine. Run the simulation for 20,000 minutes. To the extent possible, indicate the batch-means-based confidence intervals on expected steady-state performance measures from this run.

7-4   A part arrives every ten minutes to a system having three workstations (A, B, and C), where each workstation has a single machine; the first part arrives at time 0. There are four part types, each with equal probability of arriving. The process plans for the four part types are given below. The entries for the process times are the parameters for a triangular distribution (in minutes).

Part Type	Workstation/ Process Time	Workstation/ Process Time	Workstation/ Process Time
Part 1	A 5.5,9.5,13.5	C 8.5,14.1,19.7	
Part 2	A 8.9,13.5,18.1	B 9,15,21	C 4.3,8.5,12.7
Part 3	A 8.4,12,15.6	B 5.3,9.5,13.7	
Part 4	B 9.2,12.6,16.0	C 8.6,11.4,14.2	

Assume that the transfer time between arrival and the first station, between all stations, and between the last station and the system exit is three minutes. Use the Sequence feature to direct the parts through the system and to assign the processing times at each station. Use the Sets feature to collect cycle times (total times in system) for each of the part types separately. Animate your model (including the part transfers) and run the simulation for 10,000 minutes.

7-5   Modify your solution for Exercise 7-4 to use the Expressions feature for determining the processing times (rather than assigning them in the Sequence data module). Run for 10,000 minutes and compare the results to those from Exercise 7-4. Are the results different? If so, why?

7-6   Modify your solution for Exercise 7-5 so that all parts follow the same path through the system: Workstation A – Workstation B – Workstation C. If a part does not require processing at a workstation, it must still wait in queue, but incurs a zero processing-time delay. Compare the results to those obtained from Exercises 7-4 and 7-5.

7-7   Three types of customers arrive at a small airport: check baggage (30%), purchase tickets (15%), and carry-on (55%). The interarrival-time distribution for all customers combined is EXPO(1.3); all times are in minutes and the first arrival is at time 0. The bag checkers go directly to the check-bag counter to check their bags—the time for which is distributed TRIA(2, 4, 5)—proceed to X-ray, and then go to the gate. The ticket buyers travel directly to the ticket counter to purchase their tickets—the time for which is distributed EXPO(7)—proceed to X-ray, and then go to the gate. The carry-ons travel directly to the X-ray, then to the gate counter to get a boarding pass—the time for which is distributed TRIA(1, 1.5, 3). All three counters are staffed all the time with one agent each. The X-ray time is EXPO(1). All travel times are EXPO(2), except for the carry-on time to the X-ray, which is EXPO(3). Run your model for 920 minutes, and collect statistics on resource utilization, queues, and system time from entrance to gate for all customers combined.

7-8 Parts arrive at a four-machine system according to an exponential interarrival distribution with mean 10 minutes. The four machines are all different and there's just one of each. There are five part types with the arrival percentages and process plans given below. The entries for the process times are the parameters for a triangular distribution (in minutes).

Part Type	%	Machine/ Process Time	Machine/ Process Time	Machine/ Process Time	Machine/ Process Time
1	12	1	2	3	4
		10.5,11.9,13.2	7.1,8.5,9.8	6.7,8.8,10.1	6,8.9,10.3
2	14	1	3	2	
		7.3,8.6,10.1	5.4,7.2,11.3	9.6,11.4,15.3	
3	31	2	4	1	3
		8.7,9.9,12	8.6,10.3,12.8	10.3,12.4,14.8	8.4,9.7,11
4	24	3	4	3	2
		7.9,9.4,10.9	7.6,8.9,10.3	6.5,8.3,9.7	6.7,7.8,9.4
5	19	2	1	4	
		5.6,7.1,8.8	8.1,9.4,11.7	9.1,10.7,12.8	

The transfer time between arrival and the first machine, between all machines, and between the last machine and the system exit follows a triangular distribution with parameters 8, 10, 12 (minutes). Collect system cycle time (total time in system) and machine utilizations. Animate your model (including part transfers) and run the simulation for 10,000 minutes. If the run is long enough, give batch-means-based confidence intervals on the steady-state expected values of the results.

7-9 Modify your solution for Exercise 7-8 to include the travel times that are move-specific. The travel times are given below as the parameters for a triangular distribution (in minutes). Compare your results. Is this a statistically reliable comparison?

From/To	Machine 1	Machine 2	Machine 3	Machine 4	Exit System
Enter System	7,11,19	7,11,16	8,12,19		
Machine 1		9,13,20	10,13,18	7,9,13	
Machine 2	8,10,15		7,12,18	7,9,12	8,9,14
Machine 3		9,13,20		9,14,21	6,8,11
Machine 4	11,13,17		10,13,21		6,10,12

7-10 Modify your solution to Exercise 4-21 to use sequences to control the flow of parts through the system. (HINT: Reset the value of Entity.Jobstep or IS.)

7-11 Modify Model 7-1 to account for acquiring a new customer, in addition to the one supplying the existing three part types. This new customer will supply two new types of parts—call them Type 4 and Type 5. The arrival process is in addition to and independent of that for the original three part types and has exponential interarrival times with mean 15 minutes; the first arrival is at time 0. When the parts arrive, assign 40% of the new parts to be Type 4 and the rest to be Type 5. Here are the process plans and mean processing times (in minutes) for the new part types:

Part Type	Cell/ Mean Proc. Time	Cell/ Mean Proc. Time	Cell/ Mean Proc. Time	Cell/ Mean Proc. Time
4	1	3	2	4
	6.1	5.2	1.3	2.4
5	2	3	4	1
	3.5	4.1	3.2	2.0

While people feel comfortable with these mean processing times, there's not very good information on their distributions, so you're asked just to assume that the distributions are uniform with the indicated mean but plus or minus 0.2 minute in each case. For example, if a mean of 6.1 is shown, assume that the distribution is uniform between 5.9 and 6.3. Make all necessary changes to the model, including the modules, animation pictures (create new entity pictures for the new part types), and anything else required. Be sure that the animation works properly, including the clockwise-only movement of all entities and the new part types.

(**a**) Clearly, adding this new customer is going to clog the system compared to what it was. Using the time-average total number of parts in all four queues combined, how bad does it get compared to the original system? Just make one run of each alternative (you'll do a better job on statistical analysis in the next part of this exercise).

(**b**) In an effort to alleviate the added congestion introduced by the new customer, you've been given a budget to buy one new machine to be placed either in Cell 1, 2, or 4 (not in Cell 3). Where would you recommend placing it in terms of the time-average total number of parts in all four queues combined? Assume that the new machine will work at the same rate as the machine it joins in whatever cell. You might want to refer to Section 6.5 and specify your own output statistic to use PAN. View this as a terminating simulation and make 50 replications with the goal to select the best placement of the new machine; include for comparison a base-case scenario in which no machines are added anywhere.

7-12 Modify your solution to Exercise 5-2 to use Sequences to control the flow of parts through the system. Also add a transfer time between arrival and the first machine, between both machines, and between the last machine and the system exit that follows a triangular distribution with parameters 7, 9, 14 (minutes). View this as a terminating simulation, and make ten replications of this model as well as the one from Exercise 5-2, using PAN to compare the results for average total time in system and the average lengths of each queue.

7-13 Modify your solution to Exercise 7-12 to account for a 20% increase in processing time when the part returns to the first machine for its last operation. View this as a terminating simulation, and make 50 replications of this model as well as the one from Exercise 7-12, using PAN to compare the results for the same output performance measures as in Exercise 7-12.

CHAPTER 8

Entity Transfer

Up to now, we've considered two different cases with regard to how long it takes entities to move from one place to another. We've had them move from module to module with no travel time via Connections. In other models, we've moved them by Routing between stations with some transit-time delay. In both cases, the entities proceed uninhibited, as though they all had their own feet and there was enough room in the transitways for as many of them at a time as wanted to be moving.

Of course, things aren't always so nice. There could be a limit on the concurrent number of entities in transit, such as a communications system where the entities are packets of information and the bandwidth is limited to a certain number of packets in transmission at a time. There could also be situations in which something like a forklift or a person needs to come pick up an entity and then transport it. There are also different kinds of conveyors where entities can be viewed as competing for space on the belt or line. We'll explore some of these modeling ideas and capabilities in this chapter. It is often important to model such *entity transfer* accurately since studies have shown that delays and inefficiencies in operations might be caused more by the need just to move things around rather than in actually doing the work.

Section 8.1 discusses in more detail the different kinds of entity movement and transfers and how they can be modeled. In Section 8.2, we'll indicate how you can use the Arena modeling tools you already have to represent a constraint on the number of entities that can be in motion at a time (though all entities still have their own feet). Transport devices like forklifts, pushcarts, and (of course) people are taken up in Section 8.3. Modeling conveyors of different types is described in Section 8.4.

After reading this chapter and considering the examples, you'll be able to model a rich variety of entity movement and transfer that can add validity to your model and realism to your animations.

8.1 Types of Entity Transfers

To transfer entities between modules, we initially used the Connect option (Chapter 3) to transfer entities directly between modules with no time delay. In Chapter 4, we introduced the concept of a Route that allows you to transfer entities between stations with a time delay in the transfer. We first showed how to use Routes for entity transfer to a specific station; then we generalized this concept in Chapter 7 by using Sequences.

Although this gives us the ability to model most situations, we sometimes find it necessary to limit or constrain the number of transfers occurring at any point in time. For example, in modeling a communications network, the links have a limited capacity. Thus, we must have a method to limit the number of simultaneous messages that are being transferred by each network link or for the entire network. The solution is rather simple;

we think of the network links as resources with a capacity equal to the maximum number of simultaneous messages allowed. If the capacity depends on the size of the messages, then we define the resource capacity in terms of this size and require each message to seize the required number of units of this resource, determined by its size, before it can be transferred. Let's call this type of entity transfer *resource constrained*, and we'll discuss it in more detail in Section 8.2.

Using a resource to constrain the number of simultaneous transfers may work fine for a communications network, but it doesn't allow us to model accurately an entire class or category of entity transfers generally referred to as *material handling*. The modeling requirements for different material-handling systems can vary greatly, and the number of different types of material-handling devices is enormous. In fact, there is an entire handbook devoted to this subject (see Kulwiec, 1985). However, it's possible to divide these devices into two general categories based on their modeling requirements.

- The first category constrains the number of simultaneous transfers based on the number of individual material-handling devices available. Material-handling devices that fall into this category are carts, hand trucks, fork trucks, AGVs, people, and so on. However, there is an additional requirement in that each of these devices has a physical location. If a transfer is required, we may first have to move the device to the location where the requesting entity resides before we can perform the actual transfer. From a modeling standpoint, you might think of these as *moveable resources*, referred to in Arena as *Transporters*.
- The second category constrains the ability to start a transfer based on space availability. It also requires that we limit the total number of simultaneous transfers between two locations, but this limit is typically based on the space requirement. Material-handling devices that fall into this category include conveyors, overhead trolleys, power-and-free systems, tow lines, and more. An escalator is a familiar example of this type of material-handling device. If a transfer is required, we first have to obtain the required amount of available or unoccupied space on the device, at the location where we are waiting, before we can initiate our transfer. These devices require a significantly different modeling capability, referred to in Arena as *Conveyors*.

The Arena Transporter and Conveyor constructs allow us to model almost any type of material-handling system easily. However, there are a few material-handling devices that have very unique requirements that can create a modeling challenge. Gantry or bridge cranes are classic examples of such devices. A single crane is easily modeled with the transporter constructs. If you have more than one crane on a single track, the method used to control how the cranes move is critical to modeling these devices accurately. Unfortunately, almost all systems that have multiple cranes are controlled by the crane operators who generally reside in the cabs located on the cranes. These operators can see and talk to one another, and their decisions are not necessarily predictable. In this case, it is easy to model the cranes; the difficult part is how to incorporate the human logic that prevents the cranes from colliding or gridlocking.

Thus, we've defined three types of constrained entity transfers: resource constrained, transporters, and conveyors. We'll first briefly discuss resource-constrained transfers in Section 8.2, then introduce transporters and conveyors by using them in our small manufacturing system from Chapter 7 (Model 7-1) in Sections 8.3 and 8.4.

8.2 Model 8-1: The Small Manufacturing System with Resource-Constrained Transfers

In our initial model (Model 7-1) of the small manufacturing system, we assumed that all transfer times were two minutes. If these transfer times depend on the availability of material-handling devices, the actual times may be quite different from each other during operation of the system. Because of this, the earlier model might give us reasonable estimates of the maximum potential system capacity; however, it would most likely not provide very accurate estimates on the part cycle times. The simplest method is to include resource-constrained transfers. So let's assume that our transfer capacity is two, with the same two-minute transfer time in every case. This means that a maximum of two transfers can occur at the same time. If other entities are ready for transfer, they will have to wait until one of the ongoing transfers is complete.

Using a resource to constrain the number of simultaneous entity transfers is a relatively easy addition to a model. We need to define what we think of as a new kind of transferring resource, seize one unit of that resource before we initiate our route to leave a location, and release the resource when we arrive at the next station or location. This is an ordinary Arena resource, but we're thinking of it differently to model the transfers.

There are two different ways we could add this logic to our existing model. The most obvious (at least to us) is to insert a Seize module from the Advanced Process panel just prior to each of the existing Route modules (there are five in the current model). Each Seize module would request one unit of the transfer resource; e.g., `Transfer`. We would also need to change the capacity of this new resource to 2 in the Resource data module. There is one additional concept that you should consider while making these modifications, "Should each Seize module have a separate queue?" If you use a single *shared queue* (see Section 5.4.5 and Model 5-1) for all of our newly inserted Seize modules, then the resource would be allocated based on a FIFO rule. If you specified a different queue at each Seize module, you would be able to apply priorities to the selection process. For example, you might want to give the new arriving entities top priority. This would allow the newly arriving parts to be sent to their first operation as soon as possible, in the hope that it would allow the processing of the part to start with a resulting reduction in the part cycle time. (If you spend much time thinking about this, you might conclude that this is faulty logic.) If all priorities are the same, you end up with the same FIFO rule as with the shared-queue approach. So it would appear that the two approaches are the same. This is not quite true if you plan to show the waiting parts in your animation (which we do). By using separate queues (the default) prior to each route, we can show the waiting parts easily on our animation.

Having seized the transfer resource prior to each part transfer via a route, we now need to release that resource once the part arrives at its destination. We accomplish this by inserting a Release module after each Station module in the model. (The single exception is `Order Release Station`, which is used only to tell Arena the location of the newly arrived parts.) In each of these modules, we simply release the `Transfer` resource.

The second method is to replace the Route modules with Leave modules (from the Advanced Transfer panel) and the Station modules with Enter modules (also from the Advanced Transfer panel). The Leave module allows us to seize the resource and also to route the part in the same module. Similarly, the Enter module allows us to combine the features of the Station and Release modules. We'll choose this second method because it allows us to introduce two new modules, and it facilitates the conversion of the model to include other transfer features.

The Leave module allows you to seize a transferring resource before leaving the station. Which resource to seize can be defined as a particular resource, a resource set, a specific member of a set, or a resource based on the evaluation of an expression. You can also specify a Priority in the Transfer Out logic that would allow you to control which entity would be given the transferring resource if more than one entity was waiting. The smaller the number, the higher the priority. For our system, you will only have to select the `Seize Resource` option and enter a transferring resource name. We'll also have to enter the additional routing information.

Let's start by replacing the `Start Sequence` Route module with the Leave module shown in Display 8-1. We'll use the same module name, `Start Sequence`, select the Transfer Out logic as `Seize Resource`, and enter the Resource Name as `Transfer`. Finally, we select `Route` as the Connect Type, enter `Transfer Time` for the Move Time, select `Minutes`, and select `By Sequence` as the Station Type. We've defaulted on the remaining boxes. Note that this gives us our individual queues for parts waiting for the `Transfer` resource. Notice also that with the selection of the Transfer Out resource logic, a queue (`Start Sequence.Queue`) was added to the module when the dialog box was accepted. At this point, you might want to move this queue to the proper position on your animation.

Now you can replace the remaining four Route modules with Leave modules. With the exception of the Name (we used the same name that was on the replaced Route module), the data entries are the same as shown in Display 8-1.

The Enter module allows you to release the transferring resource and provides the option of including an unload delay time. Let's start with the replacement of the `Cell 1 Station` Station module with an Enter module, as in Display 8-2. Delete the existing Station module and add the new Enter module. We'll use the same Name, `Cell 1 Station`, and select `Cell 1` from the drop-down list for the station Name. In the logic section, you need only select `Release Resource` for the Transfer In entry and select the `Transfer` resource for the Resource Name to be released.

Replace the remaining four Station modules with Enter modules (but not the `Order Release Station`). As before, the data entries are the same as shown in Display 8-2, with the exception of the Name (we used the same name that was in the replaced Station module).

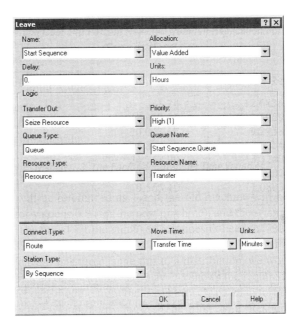

Name	Start Sequence
Logic	
Transfer Out	Seize Resource
Resource Name	Transfer
Connect Type	Route
Move Time	Transfer Time
Units	Minutes
Station Type	By Sequence

Display 8-1. The Leave Module for Resources

Provided that you have already moved the new wait queues to the animation, we are almost ready to run our model. Before we do that, you should be aware that we have not yet defined the capacity of our constraining resource, which determines the maximum number of parts that could be in transit at one time. Although we have defined the existence of this resource, Arena defaults to a capacity of 1 for it. You'll need to open the Resource data module (Basic Process panel) and change the capacity of the Transfer resource to 2. The resulting model looks identical to Model 7-1 because our replacements were one-for-one and we used the same names. We used the same animation, with the addition of the five new wait queues. The positions of these queues are shown in Figure 8-1.

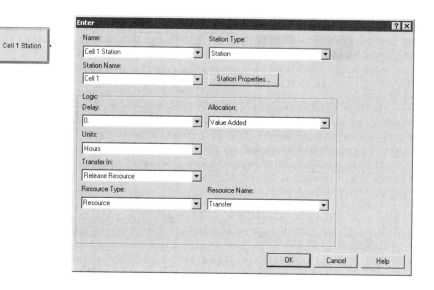

Name	Cell 1 Station
Station Name	Cell 1
Logic	
Transfer In	Release Resource
Resource Name	Transfer

Display 8-2. The Enter Module for Resources

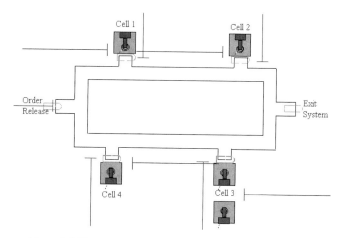

Figure 8-1. The Revised Small Manufacturing System Animation

If you run this model and watch the animation, you'll quickly observe that there is seldom a part waiting for transfer. If you have doubts about the correctness of your model, change the capacity of the `Transfer` resource to 1 and watch the animation. If you compare the results of this model (with two `Transfer` resources) to the results of Model 7-1, you'll see a slight increase in the part total time in system due to time spent waiting in a queue for the `Transfer` resource. The difference is between 3 and 9 minutes, depending on part type.

8.3   The Small Manufacturing System with Transporters

Let's now assume that all material transfers are accomplished with some type of Transporter such as carts, hand trucks, or fork trucks (we'll call them carts). Let's further assume that there are two of these carts, each moving at 50 feet per minute, whether loaded or empty. Each cart can transport only one part at a time, and there are 0.25-minute load and unload times at the start and end of each transport. We'll provide the distances between stations after we've added the carts to our model.

There are two types of Arena Transporters: Free-Path and Guided. *Free-Path Transporters* can move freely through the system without encountering delays due to congestion. The time to travel from one point to another depends only on the transporter velocity and the distance to be traveled. *Guided Transporters* are restricted to moving within a predefined network. Their travel times depend on the vehicles' speeds, the network paths they follow, and potential congestion along those paths. The most common type of guided vehicle is an *automated guided vehicle* (AGV). The carts for our system fall into the free-path category.

The transfer of a part with a transporter requires three activities: Request a transporter, Transport the part, and Free the transporter. The key words are Request Transporter, Transport, and Free Transporter. The *Request Transporter* activity, which is analogous to seizing a resource, allocates an available transporter to the requesting entity and moves the allocated transporter to the location of the entity, if it's not already there. The *Transport* activity causes the transporter to move the entity to the destination station. This is analogous to a route, but in the case of a Transport, the transporter and entity move together. The *Free Transporter* activity frees the transporter for the next request, much like the action of releasing a resource.

If there are multiple transporters in the system, we face two issues regarding their assignment to entities. First, we might have the situation during the run where an entity requests a transporter and more than one is available. In this case, a *Transporter Selection Rule* dictates which one of the transporter units will fulfill the request. In most modeled systems, the *Smallest Distance* rule makes sense—allocate the transporter unit that's closest to the requesting entity. Other rules include Preferred Order, Cyclical, Random, Specific Member, and Largest Distance (though it takes a creative mind to imagine a case where it would be sensible). The second issue concerning transporter allocation arises when a transporter is freed and there are multiple entities waiting that have requested one. In this case, Arena applies a priority, allocating the transporter to the waiting entity that has the highest priority (lowest priority number). If there's a priority tie among entities, the transporter will be allocated to the closest waiting entity.

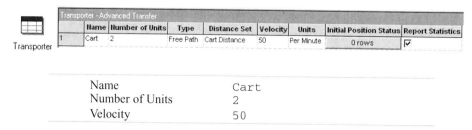

Display 8-3. The Transporter Data Module

8.3.1 Model 8-2: The Modified Model 8-1 for Transporters

To represent the carts in our small manufacturing system model, let's start by defining the carts. We do this with the Transporter data module found on the Advanced Transfer panel. If you're building the model along with us, we suggest that you start with a copy of Model 8-1, which we just completed. In this case, we need only enter the transporter Name, Number of Units, Type (Free Path vs. Guided), the Distance Set to which the transporter will refer when moving and calculating time to move (discussed below), Velocity and time Units for this velocity, and information on the transporters' Initial Position Status (Display 8-3). We'll accept the default for the Distance Set name (we'll return to this concept later), time Units, Initial Position of the carts, and the Report Statistics.

We need to take care when defining the Velocity that we enter a quantity that's appropriate for the time and distance units we're using in the model. We defaulted our time Units to Per Minute, the same units as for our stated velocity, but we'll need to make sure that our distances are specified in feet.

Having defined the transporter, we can now develop the picture that represents the cart. This is done in almost the same way as it was for an entity or resource. Clicking on the *Transporter* button (▶▶), found on the *Animate Transfer* toolbar, opens the Transporter Picture Placement window. Let's replace the default picture with a box (we used a 5-pixel line width) that is white, with a green line when the cart is idle and blue line when it's busy. When the cart is carrying a part, Arena also needs to know where on the cart's busy picture to position the part. This placement is similar to a seize point for a resource. In this case, it's called a *Ride Point* and can be found under the Object menu of the Arena Picture Editor window when you're editing the busy picture. Selecting *Object > Ride Point* changes your pointer to cross hairs. Move this pointer to the center of the cart and click. The ride point, which appears as ⊗ on the busy cart, results in the entity being positioned so its reference point is aligned with this ride point. When you've finished creating your pictures, accept the changes to close the Transporter Picture Placement window. Your pointer becomes cross hairs that you should position somewhere near your animation and click. This action places the picture of your idle cart in the model window. You needn't worry about where you place the transporter as it is only placed in the model window in case you need to re-edit it later. During a run, this picture will be hidden, and replicas of it will move across the animation for each individual transporter unit.

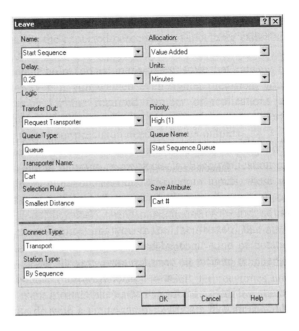

Name	Start Sequence
Delay	0.25
Units	Minutes
Logic	
Transfer Out	Request Transporter
Transporter Name	Cart
Selection Rule	Smallest Distance
Save Attribute	Cart #
Connect Type	Transport
Station Type	By Sequence

Display 8-4. The Leave Module for Transporters

To request a cart in the model logic, we'll use the same modules that we did for seizing a resource—the Leave modules. Display 8-4 shows the entries for the Start Sequence Leave module. The other Leave module entries are identical, except for their names. We've entered a 0.25-minute delay to account for the load activity and selected the Request Transporter option from the lists for the Transfer Out type. The Selection Rule entry determines which transporter will be allocated if more than one is currently free: Cyclic, Random, Preferred Order, Specific Member, Largest Distance, or Smallest Distance. The Cyclic rule attempts to cycle through the transporters, thus leveling their utilizations. The Preferred Order rule attempts always to select the available transporter with the lowest number. We've chosen the Smallest Distance rule, which results in an allocation of the cart closest to the requesting entity. A new Attribute, Cart #, is defined and used to save the number of the cart that was allocated (more on this later). We also select the Transport option for the Connect Type. Note that when you make this selection, the Move Time field disappears. The actual move time will be calculated from the distance traveled and the transporter velocity.

The Leave module performs four activities: allocate a transporter, move the empty transporter to the location of the requesting part, delay for the load time, and transport the part to its destination. There are alternate ways to model these activities. For example, we could have used a Request – Delay – Transport module combination to perform the same activities. The Request module (Advanced Transfer panel) performs the first two; the Delay module, the load activity; and the Transport module (Advanced Transfer panel), the last activity. Although this requires three modules, it does provide you with some additional modeling flexibility. You can specify different travel velocities for both moving the empty transporter and transporting the part to its destination location. In fact, you could specify the transporting velocity as an expression of an attribute of the part; for example, part weight. You could also replace the Delay module with a Process module, which could require the availability of an operator or material handler (modeled as a resource) to load the part. Finally, you could replace the Request module with an Allocate – Move module combination. The Allocate module (Advanced Transfer panel) allocates a transporter, while the Move module (Advanced Transfer panel) moves the empty transporter to the requesting part location. Again, this would allow you to insert additional logic between these two activities if such logic were required to model your system accurately.

When the part arrives at its next location, it needs to free the cart so it can be used by other parts. To free the cart, we'll use the same modules that we did for releasing the resource in the previous model—the Enter modules. Display 8-5 shows the entries for the Enter Cell 1 Station module. The other Enter module entries are identical, except for their names. We've entered the unload Delay in minutes and selected the Free Transporter option for Transfer In. In our model, we also entered the Transporter Name, Cart, and name of the attribute where we saved the transporter Unit Number, Cart #. We could have left both of these last two fields empty. Arena keeps track of which cart was allocated to the entity and frees that cart. However, if you enter the Transporter Name and there is more than one transporter, you also need to enter the Unit Number. If you only entered the Transporter Name, Arena would always try to free the first transporter.

As with the Leave module, the Enter module performs multiple activities: defining the station, delaying for the unload time, and freeing the transporter. Again there are alternate ways to model these functions. We could have used a Station – Delay – Free module combination to perform the same activities. The Station module defines the station; the Delay module, the unload activity; and the Free module (Advanced Transfer panel) frees the transporter. Separating these three activities would allow you to insert additional model logic, if required for your system. For example, you could replace the Delay module with a Process module, which could require the availability of an operator or material handler (modeled as a resource) to unload the part.

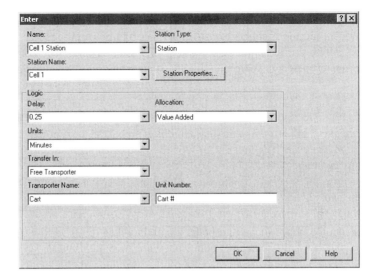

Name	Cell 1 Station
Station Name	Cell 1
Logic	
Delay	0.25
Units	Minutes
Transfer In	Free Transporter
Transporter Name	Cart
Unit Number	Cart #

Display 8-5. The Enter Module for Transporters

So far we've defined the carts, and we've changed the model logic to Request, Transport, and Free the carts to model the logic required to transfer the parts through the system. The actual travel time depends on the transporter velocity, which is defined with the carts, and the distance of the transfer. When we defined the carts (Display 8-3), we accepted a default reference to a Distance Set with the (default) name Cart.Distance. We must now provide these distances so that Arena can reference them whenever a request or transport occurs. Let's start by considering only the moves from one station to the next station made by the parts as they make their way through the system (see Table 7-1). This information is provided in Table 8-1. These table entries include a pair of stations, or locations, and the distance (in feet) between those stations. For example, the first entry is for the move from Order Release to Cell 1, which is a distance of 37 feet. Blank entries in Table 8-1 are for from-to pairs that don't occur for part movements (see Table 7-1).

Table 8-1. The Part Transfer Distances

From	To Cell 1	Cell 2	Cell 3	Cell 4	Exit System
Order Release	37	74			
Cell 1		45	92		
Cell 2	139		55	147	
Cell 3				45	155
Cell 4		92			118

We enter our distances using the Distance data module found on the Advanced Transfer panel. The 11 entries are shown in Display 8-6. You really only need to type the Distance since the first two entries can be selected from the lists. You should also note that a direction is implied, *from* Order Release *to* Cell 1 in the case of the first row in Display 8-6. As was the case with the Route module, we can define the distance from Cell 1 to Order Release if the distance or path is different (but no entity ever takes that route in this model, so it's unnecessary).

Now let's take a look at the animation component for these distances. We already defined the transporter picture, but we need to add our distances to the animation. If you're building this model along with us, now would be a good time to delete all the route paths (but leave the stations), before we add our distances. Add your distances by using the *Distance* button (⌀) found in the Animate Transfer toolbar. Click on this button and add the distances much as we added the route paths in Model 4-4. When you transport a part, the cart (and the part) will follow these lines (or distances) on the animation. When you click on the *Distance* button, a dialog box appears that allows us to select From and To stations, and modify the path characteristics, as in Display 8-7. In this case, we specified that this distance animation was From the station, Order Release, To the station Cell 1, and requested that neither the cart nor the part riding on the cart be rotated or flipped as they move along the path. These features are not necessary as our pictures are all squares and circles and, for easy identification, we placed numbers on the parts (we'll leave it as a challenge to you to figure out which numbers go with which part type).

Distance

Stations				
	Beginning Station	Ending Station	Distance	
1	Order Release	Cell 1	37	
2	Order Release	Cell 2	74	
3	Cell 1	Cell 2	45	
4	Cell 1	Cell 3	92	
5	Cell 2	Cell 1	139	
6	Cell 2	Cell 3	55	
7	Cell 2	Cell 4	147	
8	Cell 3	Cell 4	45	
9	Cell 3	Exit System	155	
10	Cell 4	Cell 2	92	
11	Cell 4	Exit System	118	

Display 8-6. The Part Transfer Distances

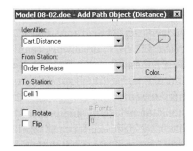

Display 8-7. The Distance Dialog Box

As you add the 11 distances, you might also want to consider activating the *Grid* and *Snap to Grid* commands from the *View* toolbar. We suggest that you add these distances in the order in which they appear in the spreadsheet for the Distance data module. This will avoid confusion and reduce the possibility of missing one. If you find that your distances are not where you want them, they are easy to edit. Click on the distance line to select it. Pay close attention to the shape of the pointer during this process. When you highlight the line, the handles (points) will appear on the line. When you move your pointer directly over a point, it will change to cross hairs; click and hold to drag the point. If the pointer is still an arrow and you click and hold, all the interior points on the line will be moved. If you accidentally do this, don't forget that you can use the *Undo* button. If you find you need additional points, or you have too many, simply double-click on the distance line to open the dialog box and change the number of points. When you add or subtract distance-line points, they're always added or subtracted at the destination-station end of the line.

If you ran the model and watched the animation now, you would see the transporter moving parts through the system, but once the transporter was freed, it would disappear from the animation until it started moving again. This is because we have not told Arena where the transporters should be when they are idle. We do this by adding parking areas to our animation. Clicking the *Parking* button ( ) from the *Animate Transfer* toolbar will open the Parking dialog box shown in Display 8-8. Accepting the dialog will change your pointer to cross hairs. Position your pointer near the lower left-hand corner of a Station on the animation and click. As you move your pointer, you should see a bounding outline stretching from the station to your current pointer position. If you don't have this, click again until you're successful. Now position your pointer where you want the transporter to sit when it becomes idle, and click; move your pointer to a position where a second idle transporter would sit and double-click. This should exit you from the parking area activity and revert your pointer to its normal arrow. If you accidentally create too many (or too few) parking areas, you can double-click one of the parking areas you added to reopen the Parking Area dialog box. Use the *Points* button to edit the number of points or parking areas. We want two because it's possible to have as many as two carts at the same location at the same time. Repeat this action for each station in your model.

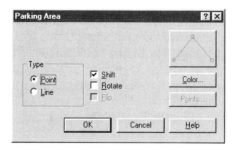

Display 8-8. The Parking Area Dialog Box

The final positioning for the Order Release and Cell 1 stations should look something like Figure 8-2.

You can run your model and animation at this point, but if you do, very quickly the execution will pause and the Arena Error/Warning window will appear. A warning will be displayed telling you that a distance between a pair of stations has not been specified, forcing Arena to assume a distance of 0, and it will recommend that you fix the problem. (Arena makes the rash assumption that you forgot to enter the value.) You can close this window and continue the run, but the message will reappear. The problem is most likely caused by a cart being freed at the `Exit System`, `Cell 3`, or `Cell 4` locations when the cart is being requested at the `Order Release` location. Arena attempts to transfer the cart to `Order Release` and fails to find a distance; thus, we have a problem. In fact, there are more problems than just this. If a cart is freed at `Cell 1` or `Cell 2` and is requested at `Order Release`, it will travel backward rather than clockwise as we desire. To avoid this, we'll add distances for all possible loaded and empty transfers that might occur.

In general, if you have *n* locations, there are $n(n - 1)$ possible distances. You can roughly think of this as an *n* by *n* matrix with 0's along the diagonal. This assumes that all moves are possible and that the distance between any two locations could depend on the direction of travel. For our example, we have six locations, or 30 possible distances

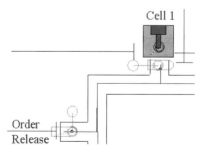

Figure 8-2. Place the Distance and Parking Animation Objects

Table 8-2. Possible Cart Moves, Including Empty Moves

		To					
		Order Release	Cell 1	Cell 2	Cell 3	Cell 4	Exit System
From	Order Release		37	74			
	Cell 1	155		45	92	129	
	Cell 2	118	139		55	147	
	Cell 3	71	92	129		45	155
	Cell 4	34	55	92	139		118
	Exit System	100	121	158	37	74	

(in our model, distance *does* depend on direction of travel). If the distances are not direction-dependent, you only have half this number of distances. Also, there are often many moves that will never occur. In our example, the empty cart may be required to move from the `Exit System` location to the `Order Release` location, but never in the opposite direction. (Think about it!) The data contained in Table 8-1 were for *loaded* cart moves only. If we enumerate all possible *empty* cart moves and eliminate any duplicates from the first set, we have the additional distances given in Table 8-2. The shaded entries are from the earlier distances in Table 8-1. Altogether, there are 25 possible moves.

If the number of distances becomes excessive, we suggest that you consider switching to guided transporters, which use a network rather than individual pairs of distances. For a review of the concepts of guided transporters, we refer you to Chapter 5 of the *Arena User's Guide* (found via *Help > Product Manuals*) or Chapter 9, "Advanced Manufacturing Features," of Pegden, Shannon, and Sadowski (1995). You could also search in the Arena online Help system for topics related to guided transporters, as well as look at the example in `Smarts195.doe` in the Smarts folder, which is inside the Arena 10.0 folder.

The final animation will look very similar to our first animation, but when you run the model, you'll see the carts moving around and the parts moving with the carts. It's still possible for one cart to appear on top of another since we're using free-path transporters. However, with only two carts in the system, it should occur much less frequently than in the previous model that used unconstrained routes. A view of the running animation at approximately time 643 is shown in Figure 8-3.

8.3.2 Model 8-3: Refining the Animation for Transporters
If you ran your model (or ours) and watched the animation closely, you might have noticed that the parts had a tendency to disappear while they were waiting to be picked up by a cart. This was only temporary as they suddenly reappeared on the cart just before it pulled away from the pickup point. If you noticed this, you might have questioned the correctness of the model. It turns out that the model is accurate, and this disappearing act is due only to a flaw in the animation. When a new part arrives, or a part completes its processing at a cell, it enters a request queue to wait its turn to request a cart. Once a cart is allocated, the entity is no longer in the request queue during the cart's travel to the

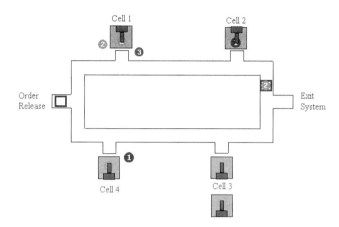

Figure 8-3. The Small Manufacturing System with Transporters

entity location. (We don't need to go into detail about where the entity really is at this time.) The simplest way to animate this activity is with the Storage feature, used in Models 5-2 and 5-3. An Arena storage is a place for the entity to reside while it's waiting for the transporter to move to its location so that it will show on the animation. Each time an entity enters a storage, a storage variable is incremented by 1, and when the entity exits the storage, this variable is decremented by 1. This also allows us to obtain statistics on the number in the storage.

Unfortunately, the storage option is not available with the modules found in the Advanced Transfer panel. However, it is available if we use modules from the Blocks panel (the spreadsheet view is not available with the modules from the Blocks or Elements panels). We'll briefly show you how to make the necessary changes to our model. Let's start with Model 8-2 and save it as Model 8-3. The first thing you should do is delete all five Leave modules from the model and replace them with the Queue – Request – Delay – Transport module combination shown in Figure 8-4. (The modules shown in Figure 8-4 replace the `Start Sequence` Leave module.) All four of these modules were taken from the Blocks panel.

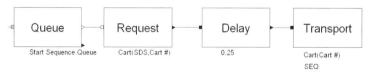

Figure 8-4. The Queue-Request-Delay-Transport Modules: Blocks Panel

Display 8-9. The Storage Data Module

Next you'll need to select the Queue data module (Basic Process panel) and add the five queues where parts wait for a cart: `Start Sequence.Queue`, `Route from Cell 1.Queue`, `Route from Cell 2.Queue`, `Route from Cell 3.Queue`, and `Route from Cell 4.Queue`. At this point, you might be thinking, "We already added these queues!" You're correct, but they were added in the Leave modules that we just deleted. Once you've completed that task, you'll need to add ten storages to the Storage data module (Advanced Process panel), as shown in Display 8-9.

As was the case with the missing queues, our attribute `Cart #` had also been defined in the deleted Leave modules. So let's start by adding this attribute to our `Assign Part Type and Sequence` Assign module. The additional assignment is shown in Display 8-10. We've used this Assign module only to define the attribute so the value assignment has no meaning. We could have also used the Attributes module from the Elements panel to add this attribute.

Finally, we can edit our four-module combination shown in Figure 8-4. We'll show you the data entries for the arriving parts and let you poke through the model to figure out the entries for the four cells. Let's start with the Queue data module where we only need to enter the queue name, shown in Display 8-11. This is the queue in which parts will wait until they have been allocated a cart. We could have omitted this queue and Arena would have used an internal queue to hold the waiting entities. This would have resulted in an accurate model, but we would not have been able to animate the queue or review statistics on parts waiting for transfer.

Type	Attribute
Name	Cart #
Value	0

Display 8-10. Assign Module to Define the Cart # Attribute

Queue ID	Start Sequence.Queue

Display 8-11. Defining Start Sequence.Queue: The Queue Data Module

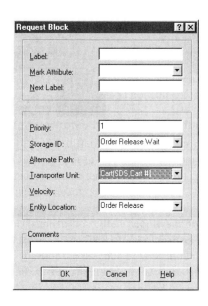

Storage ID	`Order Release Wait`
Transporter Unit	`Cart(SDS,Cart #)`
Entity Location	`Order Release`

Display 8-12. The Request Block Module

The Queue Block module is followed by a Request Block module. This module allocates an idle Cart and moves it to the part location. It also allows us to specify a storage location, which we can use to animate the part during the time the cart is moved to its location. The data for the Storage ID and Entity Location entries can be selected from the list in Display 8-12. The Transporter Unit entry requires further explanation. First, you might have noticed that the box is shaded. This indicates that this is a required field for this module. There are three different forms that can be used to specify the transporter. The first form is *Transporter ID(Number)*, where *Transporter ID* is the transporter name; in this case, `Cart`. The *Number* is the unit number of the transporter being requested. In our example, there are two carts, so the unit number would be 1 or 2. Thus, we are, in effect, requesting a specific cart if we use this form. You should also be aware that if you omit the *Number*, Arena defaults it to 1. The second form is *Transporter ID(TSR)*, where *TSR* is the transporter selection rule. For our example, this would be the shortest distance rule, SDS (which stands for Smallest Distance to Station). When using the Request Block module, you must use the Arena designation for these rules. The last form is *Transporter ID(TSR,AttributeID)*, where *AttributeID* is the name of the attribute where Arena should store the unit number of the transporter selected. In our example, this would be the attribute `Cart #`. Thus, our entry is `Cart(SDS,Cart #)`.

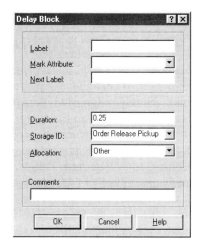

Duration	0.25
Storage ID	Order Release Pickup

Display 8-13. The Delay Block Module

When the Cart arrives at the pickup station, the part will exit the Request Block module and enter the Delay Block module, as illustrated in Display 8-13. We have entered the load delay, 0.25, and a second Storage ID, Order Release Pickup. We'll discuss why we added this second storage when we modify the animation.

Once the part has been loaded onto the Cart, it is sent to the following Transport Block module, seen in Display 8-14. There we entered the cart to be used, Cart (Cart #), and the destination, SEQ (the shortened Arena name for Sequence). In this example, we've intentionally kept track of the Cart that was allocated so we could make sure that the correct cart was freed upon arrival at the destination station.

The remaining modules were filled with the same entries as these, except for the Entity Location and the Storage ID values (which are specific to the location).

Now let's modify our animation. You should first check to see if your wait queues have disappeared (ours did). The reason they disappeared is that they were part of the Leave modules that we replaced. We had simply moved the queues that came with our Leave modules to the proper place on the animation. Had you used the *Queue* button (🔜) found on the *Animate* toolbar to add your queues, they would still be there. If not, go ahead and use the *Queue* button to add the five wait queues.

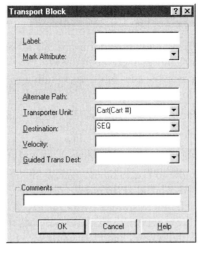

| Transporter Unit | Cart(Cart #) |
| Destination | SEQ |

Display 8-14. The Transport Block Module

Now we need to add our ten storages. We'll use the *Storage* button (⊞) from the *Animate Transfer* toolbar to make these additions. We placed the wait queues so there was room between the pickup point and the queue for us to place the pickup storages. We need two positions for each pickup, because there might be as many as two carts being moved to a location at the same time. We then placed the pickup storage directly over the parking position where the cart will sit during its loading. We used only one storage for the pickup position, even though it's possible to have two parts being picked up at the same station at the same time. If this were the case, the second cart and part would sit on top of the first one. If you want to see exactly how we did this, we suggest you take a look at the file `Model 08-03.doe`. If you run your (or our) simulation, you'll see the part initially in the wait queue. As soon as a cart becomes available and is allocated to the part, the part picture will move to the pickup storage and remain there until the cart arrives at the station. At that time, the part will appear on the transporter and remain there until the load delay is complete. The part will then move off with the cart.

There is a simpler way to create an animation where the parts don't disappear during the move and delay activity. This method would use a Store – Request – Delay – Unstore – Transport module combination in place of the Leave modules. The Request and Transport modules are from the Advanced Transfer panel, while the Store and Unstore modules are found on the Advanced Process panel. These last two modules allow you to increment and decrement storages. If you used this method, you would not animate the wait queue. All parts would be placed into a single storage while they waited for a cart to be allocated, for the cart to be moved to the station, and for the delay for the load time. We've chosen not to provide the details, but feel free to try it on your own.

Before you run your model, you should use the *Run > Setup* command and request that the transporter statistics be collected (the Project Parameters tab). If you watch the animation, you'd rarely see a queue of parts waiting for the carts, so you would not expect any major differences between this run and the original run (Model 8-1). The carts are in use only about 60% of the time. Again, we caution you against assuming there are no differences between these two systems. The run time is relatively short, and we may still be in a transient startup state (if we're interested in steady-state performance), not to mention the effects of statistical variation that can cloud the comparison.

If we change the cart velocity to 25 (half the original speed) and run the model, we'll see that there is not a huge difference in the results between the runs. If you think about this, it actually makes sense. The average distance traveled per cart movement is less than 100 feet, and when the carts become busier, they're less likely to travel empty (sometimes called *deadheading*). Remember that if a cart is freed and there are several requests, the cart will take the closest part. So the average part movement time is only about 2 minutes. Also, there was a load and unload time of 0.25 minute, which remains unchanged (remember the slide rule?). We might suggest that you experiment with this model by changing the cart velocity, the load and unload times, the transporter selection rule, and the number of carts to see how the system performs. Of course, you ought to make an appropriate number of replications and perform the correct statistical comparisons, as described in Chapter 6.

8.4 Conveyors

Having incorporated carts into our system for material movement, let's now replace the carts with a conveyor system. We'll keep the system fairly simple and concentrate on the modeling techniques. Let's assume we want a loop conveyor that will follow the main path of the aisle in the same clockwise direction we required for the transporters. There will be conveyor entrance and exit points for parts at each of the six locations in the system. The conveyor will move at a speed of 20 feet per minute. The travel distances, in feet, are given in Figure 8-5. We'll also assume that there is still a requirement for the 0.25-minute load and unload activity. Let's further assume that each part is 4 feet per side, and we want 6 feet of conveyor space during transit to provide clearance on the corners and to avoid any possible damage.

Arena conveyors operate on the concept that each entity to be conveyed must wait until sufficient space on the conveyor is available at its location before it can gain access to the conveyor. An Arena conveyor consists of cells of equal length that are constantly moving. When an entity tries to get on the conveyor, it must wait until a defined number of unoccupied consecutive cells are available at that point. Again, it may help to think in terms of a narrow escalator in an airport, with each step corresponding to a cell and different people requiring different numbers of steps. For example, a traveler with several bags may require two or three steps, whereas a person with no bags may require only one step. When you look at an escalator, the cell size is rather obvious. However, if you consider a belt or roller conveyor, the cell size is not at all obvious.

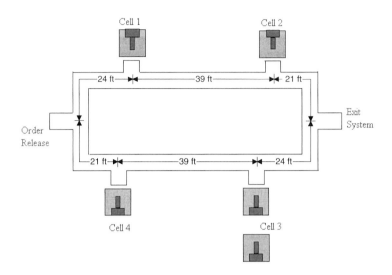

Figure 8-5. Conveyor Lengths

In order to make the conveyor features as flexible as possible, Arena allows you to define the cell size; that is, to divide the conveyor length into a series of consecutive, equal-sized cells. Each cell can hold no more than one entity. This creates a rather interesting dilemma because you would like the cells to be as small as possible to obtain the greatest modeling accuracy, yet you would also like the cells to be as large as possible to obtain the greatest computational efficiency. Let's use a simple example to illustrate this dilemma. You have a conveyor that is 100 feet long and you want to convey parts that are 2 feet long. Because we've expressed our lengths in feet, your first response might be to set your cell size to 1 foot. This means that your conveyor has 100 cells and each part requires two cells for conveyance. We could have just as easily set the cell size at 2 feet (50 cells), or 1 inch (1200 cells). With today's computers, why should we worry about whether we have 50 or 1200 cells? There should be a negligible impact on the computation speed of the model, right? However, we've seen models that have included over five *miles* of conveyors, and there is certainly a concern about the speed of the model. The difference between a cell size of 1 inch or 100 feet would have a significant impact on the time to run the model.

So why not always make your cells as large as possible? Well, when an entity attempts to gain access to a conveyor, it's only allowed on the conveyor when the end of the previous cell, or the start of the next cell, is lined up with the entity location. In our escalator analogy, this corresponds to waiting for the next step to appear at the *load* area. Consider the situation where an entity has just arrived at an empty conveyor and tries to get on. You have specified a cell size of 100 feet, and the end of the last cell has just passed that location, say by 1/2 inch. That entity would have to wait for the end of the current cell to arrive, in 99 feet 11 1/2 inches. If you had specified the cell size at 1 inch, the wait would only have been for 1/2 inch of conveyor space to pass by.

The basic message is that you need to consider the impact of cell size with respect to the specific application you're modeling. If the conveyor is not highly utilized, or the potential slight delay in timing has no impact on the system's performance, use a large cell size. If any of these are critical to the system performance, use a small cell size. There are two constraints to consider when making this decision. The entity size, expressed in number of cells, must be an integer so you cannot have an entity requiring 1.5 cells; you would have to use one or two cells as the entity size. Also, the conveyor segments (the length of a conveyor section from one location to another) must consist of an integral number of cells. So, if your conveyor were 100 feet long, you could not use a cell size of three.

Before we start adding conveyors to our model, let's go over a few more concepts that we hope will be helpful when you try to build your own models with conveyors. As with Resources and Transporters, Conveyors have several key words: Access, Convey, and Exit. To transfer an entity with Arena conveyors, you must first *Access* space on the conveyor, then *Convey* the entity to its destination, and finally *Exit* the conveyor (to free up the space). For representing the physical layout, an Arena conveyor consists of a series of segments that are linked together to form the entire conveyor. Each segment starts and ends at an Arena station. You can only link these segments to form a line or loop conveyor. Thus, a single conveyor cannot have a diverge point that results in a conveyor splitting into two or more downstream lines, or a converge point where two or more upstream lines join. However, you can define multiple conveyors for these types of systems. For example, in a diverging system, you would convey the entity to the diverge point, Access space on the appropriate next conveyor, Exit the current conveyor, and Convey the entity to its next destination. For our small manufacturing system, we'll use a single loop conveyor.

Arena has two types of Conveyors: *Nonaccumulating* and *Accumulating*. Both types of conveyors travel in only a single direction, and you can't reverse them. These conveyors function very much as their name would imply. Examples of a non-accumulating conveyor would be a bucket or belt conveyor or our escalator. The spacing between entities traveling on these types of conveyors never changes, unless one entity exits and re-accesses space. Because of this constraint, nonaccumulating conveyors operate in a unique manner. When an entity accesses space on this type of conveyor, the entire conveyor actually stops moving or is disengaged. When the entity is conveyed, the conveyor is re-engaged while the entity travels to its destination. The conveyor is stopped when the entity reaches its destination, the entity exits, and the conveyor is then restarted. If there is no elapsed time between the Access and Convey, or when the entity reaches its destination and then exits, then it's as if the conveyor was never stopped. However, if there is a delay, such as for a load or unload activity of positive duration (which is the case in our system), you'll see the conveyor temporarily stop while the load or unload occurs.

Accumulating conveyors differ in that they never stop moving. If an entity is stopped on an accumulating conveyor, all other entities on that conveyor will continue on their way. However, the stopped entity blocks any other entities from arriving at that location so that the arriving entities accumulate behind the blocking entity. When the blocking entity exits the conveyor or conveys on its way, the accumulated entities are also free to move on to their destinations. However, depending on the spacing requirements specified in the model, they may not all start moving at the same time. You might think of cars accumulated or backed up on a freeway. When the cars are stopped, they tend to be fairly close together (mostly to prevent some inconsiderate driver from sneaking in ahead of them). When the blockage is removed, the cars start moving one at a time in order to allow for more space between them. We'll describe these data requirements later in this chapter.

8.4.1   Model 8-4: The Small Manufacturing System with Nonaccumulating Conveyors

We're now ready to incorporate nonaccumulating conveyors into our system. If you're building this model along with us (after all, that was the idea), you might want to consider making another change. After including conveyors and running our model, we're likely to find that the load and unload times are so small relative to the other times that it's hard to see if the conveyor is working properly. (Actually, we've already run the model and know that this will happen.) Also, you may want to do some experimentation to see what impact these activities have on system performance. So we suggest that you define two new Variables, `Load Time` and `Unload Time`, and set the initial values for both to `0.25` minute. Then replace the current constant times with the proper variable name; this will allow you to make global changes for these times in one place.

Let's start by taking Model 8-1 and again deleting the route paths. Define the conveyor using the Conveyor data module from the Advanced Transfer panel as in Display 8-15.

Two of these entries, Cell Size and Max Cells Occupied, require some discussion. Based on the results from the previous model, it's fairly clear that there is not a lot of traffic in our system. Also, if there's a slight delay before a part can Access space on the conveyor, the only impact would be to increase the part cycle time slightly. So we decided to make the conveyor cells as large as possible. We chose a cell size of 3 feet, because this will result in an integer number of cells for each of the conveyor lengths (actually, we cheated a little bit on these lengths to make this work). Since we require 6 feet of space for each part, there is a maximum of two cells occupied by any part. This

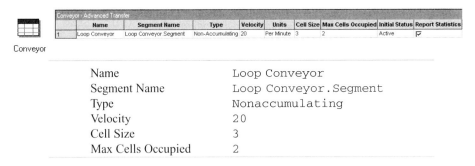

Display 8-15. The Conveyor Module

information is required by Arena to assure that it has sufficient storage space to keep track of an entity when it arrives at the end of the conveyor. Since our conveyor is a loop and has no end, it really has no impact for this model. But, in general, you need to enter the largest number of cells that any entity can occupy during transit.

Incorporating conveyors into the model is very similar to including transporters, so we won't go into a lot of detail. We need to change every Leave and Enter module, much like we did for transporters. The entries for the `Start Sequence` Leave module are shown in Display 8-16. We've included only the changes that were required starting with Model 8-1. You might note that each part requires two cells (with a size of 3 feet each) to access the conveyor, and that we've used the variable `Load Time` that we suggested earlier.

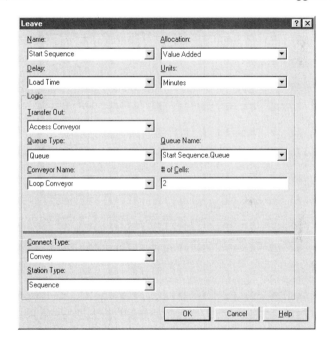

Name	Start Sequence
Delay	Load Time
Units	Minutes
Logic	
Transfer Out	Access Conveyor
Conveyor Name	Loop Conveyor
# of Cells	2
Connect Type	Convey

Display 8-16. The Leave Module for Conveyors

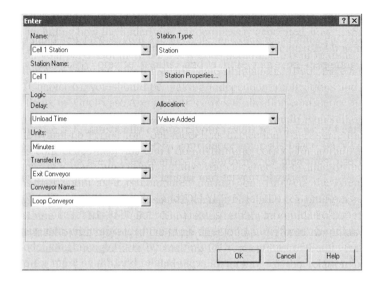

	Name	Cell 1 Station
	Station Type	Station
	Station Name	Cell 1
Logic		
	Delay	Unload Time
	Units	Minutes
	Transfer In	Exit Conveyor
	Conveyor Name	Loop Conveyor

Display 8-17. The Enter Module for Conveyors

The entries for the `Cell 1 Station` Enter module are shown in Display 8-17.

We're now ready to place the conveyor segments on our animation. As was the case with the Distances used in our last model, we'll first need to define the segment data required for both the model and animation. We define these segments using the Segment data module found on the Advanced Transfer panel. Display 8-18 shows the first entry for the main spreadsheet view of the Segment data module. We arbitrarily started our `Loop Conveyor` at the `Order Release` station.

	Name	Beginning Station	Next Stations
1	Loop Conveyor.Segment	Order Release	6 rows

Name	Loop Conveyor.Segment
Beginning Station	Order Release

Display 8-18. The Segment Data Module

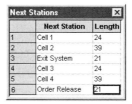

Display 8-19. The Segment Data Module

Display 8-19 shows the spreadsheet data entries for the six segments. Notice that the Length is expressed in terms of the actual length, not the number of cells, for that segment. These lengths are taken directly from the values provided in Figure 8-5.

We're now ready to add our segments to the animation. You'll first need to move each station symbol to the center of the main loop directly in front of where the part will enter or exit the conveyor. To add the segments, we use the *Segment* button (⌨) found on the *Animate Transfer* toolbar. We add segments much like we added distances, except in this model, we need to add only six segments. The Segment dialog box from `Order Release` to `Cell 1` is shown in Display 8-20, where we've simply accepted the defaults.

For this model, you need to place only six segments, as shown in Figure 8-6. As you're placing your segments, you may find it necessary to reposition your station symbols in order to get the segments to stay in the center of the loop (we did!).

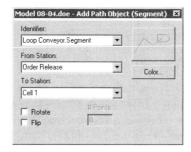

Display 8-20. The Segment Dialog Box

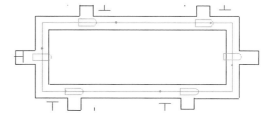

Figure 8-6. The Conveyor Segments

You're now almost ready to run your animated model. You should first use the *Run > Setup* command and request that the conveyor statistics be collected (the Project Parameters tab). If you run the model to completion, you'll notice that two new statistics for the conveyor are included in the report. The first statistic tells us that the conveyor was blocked, or stopped, about 17.65% of the time. Although this may initially appear to be a large value, you must realize that each time a part leaves or enters a station, it has a load or unload delay. During this time, the conveyor is stopped or blocked. If you consider the number of moves and the number of parts that have gone through the system, this value seems plausible. If you're still not convinced, the numbers are all there for you to approximate what this value should be.

The second statistic tells us that the conveyor was utilized only about 6.6% of the time. This is not the amount of time that a part was on the conveyor, but the average amount of space occupied by parts on the conveyor. This one is easy to check out. You can determine the total length (168 feet) of the conveyor using the lengths provided in Figure 8-5. Knowing that each cell is 3 feet, we can determine the number of cells (56). Each part requires two cells, meaning that the conveyor can only hold a maximum of 28 parts at any point in time. Thus, if we multiply the 28-part conveyor capacity by the average utilization (0.066), we know that on the average there were about 1.85 parts on the conveyor. If you watch the animation, this would appear to be correct.

If you compare the system times for the parts to those from Model 8-1, you'll find that they are slightly higher here. Given the conveyor blocking and speed, this also appears to be reasonable.

If you can't see the conveyor stop during the load and unload activity (even if you set the animation speed factor at its lowest setting), change these variables to 2 or 3 minutes. The stopping should be quite obvious with these new, higher values.

There are also several alternate ways to change variable values during a simulation. If you want to make such changes frequently during a simulation run, you might want to consider using the Arena interface for Visual Basic® for Applications (VBA) (see Chapter 10). The second method is to use the Arena command-driven Run Controller. Begin running your model, and after it has run for a while (we waited until about time 445), use the *Run > Pause* command or the *Pause* button (**II**) on the *Standard* toolbar to suspend the execution of the model temporarily. Then use *Run > Run Control > Command* or the *Command* button (⊡) on the *Run Interaction* toolbar to open the command window. This text window will have the current simulation time displayed, followed by ">" as a prompt. Arena is ready for you to enter your commands. We used the SHOW command to view the current value of the variable Load Time and then the ASSIGN command to change that value—see Figure 8-7. We then repeated these two steps for the Unload Time. Now close this window and use the *Run > Go* menu option or the *Go* button (▶) on the *Standard* toolbar to run the simulation with the new variable values from the current time.

If you watch the animation, in a very short period of time, you'll see that the conveyor movement is quite jerky, and it quickly fills up with over ten parts. You can stop the simulation run at any time to change these values. It's worth noting that the changes you make in the Run Controller are temporary. They aren't saved in the model, and the next time you run your model, it will use the original values for these variables.

```
458.39347 Minutes>
SIMAN Run Controller.

458.39347 Minutes>SHOW Load Time
    LOAD TIME =          0.25

458.39347 Minutes>ASSIGN Load Time = 2

458.39347 Minutes>SHOW Unload Time
    UNLOAD TIME =          0.25

458.39347 Minutes>ASSIGN Unload Time = 2

458.39347 Minutes>
```

Figure 8-7. Changing Variables with the Run Controller

Although viewing the animation and making these types of changes as the model is running are excellent ways to verify your model, you should not change the model conditions during a run when you finally are using the model for the purposes of evaluation. This can give you very misleading performance values on your results and will be essentially impossible to replicate.

8.4.2   Model 8-5: The Small Manufacturing System with Accumulating Conveyors

Changing our conveyor model so the conveyor is accumulating is very easy. We started by using the *File > Save As* command to save a copy of Model 8-4 as Model 8-5. We need to make only two minor changes in the Conveyor data module. Change the conveyor Type to `Accumulating`, and enter an Accumulation Length of 4, as in Display 8-21. Adding the Accumulation Length allows the accumulated parts to require only 4 feet of space on the conveyor. Note that this value, which applies only when an entity is stopped on the conveyor, does not need to be an integer number of cells. When the blockage is removed, the parts will automatically re-space to the 6 feet, or two cells, required for transit on the conveyor.

Having made these changes, run your model and you should note that very little accumulation occurs. You can confirm this by looking at the conveyor statistics on the summary report. The average accumulation is less than 1 foot and the average utilization is less than 6%. To increase the amount of accumulation, which is a good idea for verification, simply change the load and unload times to 4 or 5 minutes. The effect should be quite visible.

You can see that these results are very similar to those from the nonaccumulating system except for the part system times, which are much higher. We strongly suspect that this is due to the short run time and the resulting statistical variation for the model. We'll leave it to the interested reader to confirm (or refute) this suspicion.

Type	Accumulating
Accumulation Length	4

Display 8-21. The Accumulating Conveyor Data Module Dialog Box

8.5 Summary and Forecast

Entity movement is an important part of most simulation models. This chapter has illustrated several of Arena's facilities for modeling such movement under different circumstances to allow for valid modeling.

This is our last general "tutorial" chapter on modeling with Arena. It has gone into some depth on the detailed lower-level modeling capabilities, as well as correspondingly detailed topics like debugging and fine-tuned animation. While we've mentioned how you can access and blend in the SIMAN simulation language, we've by no means covered it; see Pegden, Shannon, and Sadowski (1995) for the complete treatment of SIMAN. At this point, you should be armed with a formidable arsenal of modeling tools to allow you to attack many systems, choosing constructs from various levels as appropriate.

There are, however, more modeling constructs and, frankly, "tricks" that you might find handy. These we take up next in Chapter 9.

8.6 Exercises

8-1   Change your model for Exercise 7-4 to include fork trucks to transport the parts between stations. Assume that there are two fork trucks that each travel at 85 feet per minute. Loading or unloading a part by the fork truck requires 0.25 minute. The distance between stations is given (in feet) in the following table; note that the distances are, in general, directional:

		To				
		Arrive	WS A	WS B	WS C	Exit System
From	Arrive	0	100	100	200	300
	WS A	100	0	150	100	225
	WS B	100	150	0	100	200
	WS C	250	100	100	0	100
	Exit	350	250	225	100	0

Run your simulation for 100,000 minutes (you may want to turn off the animation via the *Run > Run Control > Batch Run (No Animation)* command after confirming that things are working properly). Assume that fork trucks remain at the station where they unloaded the last part if no other request is pending. If both fork trucks are available, assume that the closest one is selected.

8-2   Change your model for Exercise 7-4 to use nonaccumulating conveyors to transfer the parts between stations. Assume that there is a single conveyor that starts at the arrive area and continues to the exit area: Arrive – WS A – WS B – WS C – Exit. Assume that the distances between all adjacent stations on the conveyor are 100 feet. Further assume that the cells of the conveyor are 2 feet and that each part requires 4 feet of conveyor space. Load and unload times are each 0.25 minute. The conveyor speed is 20 feet per minute.

8-3 Change your model for Exercise 5-2 to use a fork truck (45 feet/minute) for transportation of parts in the system. Assume that the parts arrive at an incoming dock and exit at a second dock. Assume that the distance between the incoming dock and Machine 2 is 200 feet; all other distances are 100 feet. Animate your solution.

8-4 Using the model from Exercise 8-3, set the number of transporters to four and make three runs using transporter selection rules of Smallest Distance, Largest Distance, and Cyclical. Compare the results using average Cycle Time.

8-5 Modify Model 4-3 to include the use of a single truck to transfer parts from the two prep areas to the sealer. Assume that the distance between any pair of the three stations is 100 feet and that the truck travels at a rate of 75 feet per minute. Animate your solution.

8-6 Modify Model 4-3 to include the use of two conveyors to transfer parts from the two prep areas to the sealer. Both conveyors are 100 feet long and are made up of 20 cells of 5 feet each. The conveyor velocity is 30 feet per minute. Animate your solution.

8-7 A prototype of a new airport bag-screening system is currently being designed, as in the figure below. Bags arrive to the system with interarrival times of EXPO(0.25) (all times are in minutes), and are loaded on a load conveyor (75 feet long) and conveyed to an initial scan area. At the initial scan area, the bags are dropped into a chute to wait for the initial scan, which is of duration TRIA(0.1, 0.25, 0.35). Based on the results of the scan, accepted bags (79%) are loaded on an out conveyor (40 feet long) and conveyed to exit the system, whereas rejected bags are loaded on a secondary conveyor (20 feet long) and conveyed to a secondary scan area. At the secondary scan area, the bags are dropped into a chute to wait for a visual inspection, which takes TRIA(0.6,1.2,1.4). After the inspection, they are sent through a secondary scanner that is 10 feet long. This scan takes 1.2 minutes and only one bag can be in the scanner at a time. At the end of the secondary scanner, the bags are examined by another inspector, taking EXPO(0.4). Based on the results of this secondary inspection, 10% of the bags are sent to a reject area for further inspection, accepted bags are loaded on an out 2 conveyor (20 feet long) and conveyed to exit the system. All conveyors are nonaccumulating with a velocity of 40 feet per minute. Each bag requires one foot of space on each of the conveyors. Develop a model of this system and run it for 2,000 minutes. Use a resource-constrained route to model the second scanner.

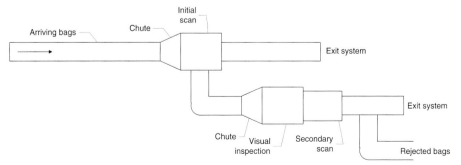

The current design calls for a chute capacity of 20 at the initial scan area and a total capacity of 25 at the secondary scan area, not including the bag undergoing the visual inspection. What percent of the time would you expect this capacity be exceeded? (Work this exercise with infinite chute capacities, and just watch the percent of time the chute capacities contemplated above would be exceeded if they were put into effect.) HINT: Add two time-persistent statistics to collect this information using logical expressions that evaluate to zero if the number of bags is less than the capacity, and to one if the capacity is exceeded.

8-8 Develop a model of a cross-dock system that groups and transfers material for further shipment. This facility has five incoming docks and three outgoing docks. Trucks arrive at each of the incoming docks with loads of material on pallets. The interarrival time is UNIF(30, 60) between truck arrivals on each incoming dock (all times are in minutes). Each truck will have a number of pallets drawn from a UNIF(10, 22) distribution (round to the nearest integer) that need to be transferred to one of the outgoing docks. Assume an equal probability of any incoming pallet going to any of the three outgoing docks. When trucks arrive, an automatic unloading device unloads the pallets at the incoming dock. Assume that this requires no time, so the pallets are immediately ready for transfer. The pallets are transferred by one of five fork trucks to the outgoing docks, which are located on the other side of the building. The distance between any incoming dock and any outgoing dock is 75 feet, and the fork trucks travel at 75 feet per minute. The time for a fork truck to load a pallet is UNIF(0.3, 0.4) and to unload a pallet is UNIF(0.2, 0.4). The performance measure of interest is the average time a pallet spends in the system. (HINT: If you place all fork trucks at the outgoing docks for their initial position, you don't have to add distances between the incoming docks.) Set the run length to be 720 minutes and regard this as a terminating simulation. Develop a model for each of the following two cases, and animate your models:

(*a*) Assume the priority is such that you want to transfer the pallets that have been waiting the longest. (HINT: Enter the name of the attribute with the arrival time in the priority box of the Transfer Out dialog box.)

(*b*) Because we want to be assured that an incoming truck can always unload, modify your model from part (*a*) above so that the pick-up priority is placed on the incoming dock with the most pallets. (HINT: Develop an expression using NQ for the priority field.)

Which of the above two options would appear to be "best"? Be sure to justify your conclusions with a valid statistical analysis. Furthermore, for each option, study the effect of having more fork trucks (up to seven); again, be aware of the gap in credibility that results from an inadequate statistical analysis.

CHAPTER 9

A Sampler of Further Modeling Issues and Techniques

In Chapters 4-8, we gave you a reasonably comprehensive tour of how to model different kinds of systems by means of a sequence of progressively more complicated examples. We chose these examples with several goals in mind, including reality and importance (in our experience) of the application, illustration of various modeling issues, and indication of how you can get Arena to represent things the way you want—in many cases, fairly easily and quickly. Armed with these skills, you'll be able to build a rich variety of valid and effective simulation models.

But no reasonable set of digestible examples could possibly fathom all the nooks and crannies of the kinds of modeling issues (and, yes, tricks) that people sometimes need to consider, much less all of the features of Arena. And don't worry, we're not going to attempt that in this chapter either. But we would like to point out some of what we consider to be the more important additional modeling issues and techniques (and tricks) and tell you how to get Arena to perform them for you.

We'll do this by constructing more examples, but these will be more focused toward specific modeling techniques and Arena features, so will be smaller in scope. In Section 9.1, we'll refine the conveyor models we developed in Chapter 8; in Section 9.2, we'll discuss a few more modeling refinements to the transporters from Chapter 8. In service systems, especially those involving humans standing around in line, there is often consideration given to customer *reneging* (in other words, jumping out of line at some point); this is taken up in Section 9.3, along with the concept of balking. Section 9.4 goes into methods (beyond the queues you've already seen) for holding entities at some point, as well as batching them together with the possibility of taking the batch apart later. In Section 9.5, we'll discuss how to represent a *tightly coupled* system in which entities have to be allocated resources downstream from their present position before they can move on; this is called *overlapping resources* from the entity viewpoint. Finally, Section 9.6 briefly mentions a few other specific topics, including guided transporters, parallel queues, and the possibility of complex decision logic and looping.

This chapter is structured differently from the earlier ones in that the sections are not necessarily meant to be read through in sequence. Rather, it is intended to provide a sampler of modeling techniques and Arena features that we've found useful in a variety of applied projects.

9.1 Modeling Conveyors Using the Advanced Transfer Panel

In this section, we indicate some refinements to the basic conveyor models described in Chapter 8.

9.1.1   Model 9-1: Finite Buffers at Stations

In Chapter 8, we introduced you to Arena conveyors. In Section 8.4.1, we developed a model, Model 8-4, for our small manufacturing system using nonaccumulating conveyors as the method for transferring parts within our system. In developing that model, we assumed that there was an infinite buffer in front of each cell for the storage of parts waiting to be processed. This assumption allowed us to use the conveyor capabilities found in the Enter and Leave modules. We did, however, need to add the Conveyor and Segment data modules from the Advanced Transfer panel to define our conveyor.

Now let's modify that assumption and assume that there is limited space at Cells 1 and 2 for the storage of unprocessed parts. In fact, let's assume that there is only room for one unprocessed part at each cell. For this type of model, we need to define what happens to a part that arrives at Cell 1 or 2 and finds that there is already a part occupying the limited buffer space. Assuming that we could determine a way to limit the buffer using the Enter module (it can't be done), it would be tempting simply to let the arriving part wait until the part already in the buffer is moved to the machine in that cell. Of course, this could cause a significant logjam at these cells. Not only would parts not be able to enter the cell, but parts on their way to other cells would queue up behind them creating yet another problem. It turns out that processed parts trying to leave the cell need to access space on the conveyor before they can be conveyed to their next destination station. Yes, the space they're trying to access is the same space occupied by the part waiting to enter the cell. This would create what's sometimes called a *deadlock* or *gridlock*.

So let's use the following strategy for parts arriving at Cell 1 or 2. If there is not a part currently waiting in the buffer for the cell, the arriving part is allowed to enter the buffer. Otherwise, the arriving part is conveyed around the loop conveyor back to the same point to try a second (or third, or fourth, etc.) time. To do this, we need to alter our model so we can better control when the part exits the conveyor. The Advanced Transfer panel provides five new modules for conveyors (Access, Convey, Exit, Start, and Stop) that allow us to model conveyor activities at a more detailed level. The Exit module causes an entity to exit a conveyor, releasing the conveyor cell(s) it occupied. This is essentially what happens when you select the Exit Conveyor option in the Transfer In dialog box of the Enter module. The Access module causes an entity to request or access space on a conveyor at a specific location, normally its current station location. The Convey module is used to convey the entity to its destination once it has successfully accessed the required conveyor space. You are essentially requesting that Arena Access and Convey an entity when you select the Access Conveyor option in the Transfer Out dialog box of the Leave modules. The Start and Stop modules cause the conveyor to start and stop its movement, respectively. These two modules can be used to develop your own failure logic or generally control when the conveyor is idle or running.

To develop our new model, we will start with Model 8-4 and replace the Enter and Leave modules for Cell 1 with our new modules, as shown in Figure 9-1.

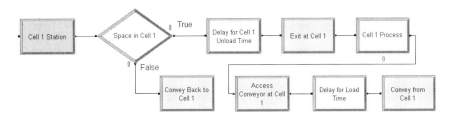

Figure 9-1. New Model Logic for Cell 1

In the Enter module we are replacing, we defined the station, with the name `Cell 1`, and also exited the conveyor. The definition of the station is critical because Arena must know where to send an entity that is being conveyed to `Cell 1`. Therefore, we started our replacement set of modules with a Station module from the Advanced Transfer panel with the specific purpose of defining the entry point to the station `Cell 1`. We also could have used our Enter module for this purpose, but we wanted to show you the Station module. The Station module simply defines the logical entry for entities that are transferred to that station. In our case, the only value provided in the dialog box is the station name, shown in Display 9-1.

When an entity, or part, arrives at station `Cell 1`, it must check the status of the waiting queue before it can determine its fate. We cause this to happen by sending the entity to the Decide module where we check to see if the number of entities in queue `Cell 1 Process.Queue` is equal to 0, shown in Display 9-2. This is the same queue name used in Model 8-4, and it was defined in the Process module. If the queue is currently occupied by another entity, the arriving entity will take the False branch of our module.

In this case, we do not want to exit the conveyor; we instead want to leave the part on the loop conveyor and convey it around the loop and back to the same station. We do this by sending the entity to the Convey module, in Display 9-3, where we convey the entity around the loop and back to Cell 1.

Name	`Cell 1 Station`
Station Name	`Cell 1`

Display 9-1. The Station Module

Name	`Space in Cell 1`
Type	`2-way by Condition`
If	`Expression`
Value	`NQ(Cell 1 Process.Queue) == 0`

Display 9-2. The Decide Module

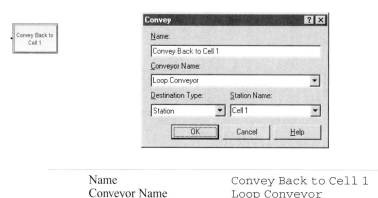

Name	Convey Back to Cell 1
Conveyor Name	Loop Conveyor
Station Name	Cell 1

Display 9-3. The Convey Module

At this point, you might be thinking, "But the entity is already at station Cell 1!" Arena assumes that your intention is to convey the entity and it will send the entity on its way. Of course, it also assumes that it is physically possible to convey the entity to the specified destination, which is the case. Thus, the Convey module will convey the entity on the designated conveyor to the specified destination. If the entity is not already on the conveyor, Arena will respond with a terminating error message.

If the queue at Cell 1 is unoccupied, the entity will satisfy the condition and be sent to the connected Delay module, in Display 9-4, where the entity is delayed in the `Delay for Cell 1 Unload Time` Delay module for the unload time. It is then sent to the Exit module, which removes the entity from the conveyor and releases the conveyor cells it occupied, shown in Display 9-5.

The Entity is sent from the Exit module to the Process module that represents the actual machining operation, shown in Figure 9-1. Once the part finishes its operation, it exits the Process module and is sent to the following Access module to get space on the conveyor, as shown in Display 9-6.

The Access module allocates cells on the conveyor so the entity can then be conveyed to its next destination; it does not actually convey the entity. Thus, if you do not immediately cause the entity to be conveyed, it will cause the entire conveyor to stop until the entity is conveyed.

Name	Delay for Cell 1 Unload Time
Delay Time	Unload Time
Units	Minutes

Display 9-4. The Unload-Delay Module

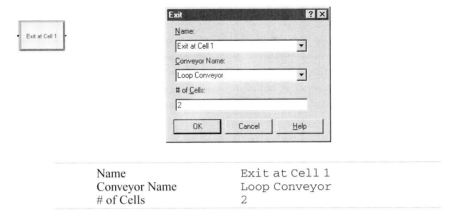

Name	Exit at Cell 1
Conveyor Name	Loop Conveyor
# of Cells	2

Display 9-5. The Exit Module

In the event that the required cells were not available, the entity would reside in the queue (`Access Conveyor at Cell 1.Queue`) until space was accessible. This provides a way for the entity to show up in the animation if it has to wait for available conveyor space.

Having accessed conveyor space, the entity delays for the `Load Time`, then is sent to the Convey module where it is conveyed according to its specified sequence. This completes the replacement modules for Cell 1 in Model 8-4.

You also need to perform the same set of operations for Cell 2. You can simply repeat these steps or make a copy of these modules and edit them individually. We chose to replace only the Enter module, retaining the Leave module, for Cell 2, as shown in Figure 9-2. We'll assume that by now you're comfortable making the required changes.

Name	Access Conveyor at Cell 1
Conveyor Name	Loop Conveyor
# of Cells	2

Display 9-6. The Access Module

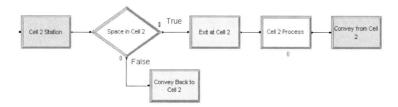

Figure 9-2. New Model Logic for Cell 2

When you delete the Enter and Leave modules, you may find that part of your animation has been deleted too. This is because you received a "free" animated queue with the Leave module. When you are editing or changing a model like this, it is frequently easier to animate the new features using the constructs from the Animate toolbar, so when we developed our model, we deleted the added animation features that were provided by the new modules and animated these new features ourselves. In our model, we deleted the queue and the storage for Cell 1. We simply added the new access queue. When we ran our new model, we could see parts being blocked from entry to Cell 1 starting at about time 275.

9.1.2 Model 9-2: Parts Stay on Conveyor During Processing

Now that you have conquered these new conveyor modules, let's examine another quick problem. Start with the accumulating conveyor model, Model 8-5, presented in Section 8.4.2. Assume that we're trying a new layout that requires that the operations at Cell 2 be performed with the part remaining on the conveyor. Specifically, parts conveyed to Cell 2 do not exit the conveyor, but the part stops at Cell 2, the actual operation is performed while the part sits on the conveyor, and the part is then conveyed to its next destination. Since the conveyor is accumulating, other parts on the conveyor will continue to move unless they are blocked by the part being operated on at Cell 2.

We could implement this new twist by replacing the current Enter and Leave modules for Cell 2 with a set of modules similar to what we did for Model 9-1. However, since the entities arriving at Cell 2 do not exit or access the conveyor, there is a far easier solution. We'll modify the Transfer In option on the Enter module by selecting the None option. In this case, the entity will not exit the conveyor. This will cause the entity to reside on the conveyor while the operation defined by the server takes place. However, if this is the only change we make, an error will occur when the entity attempts to leave Cell 2. In the Transfer Out dialog box of the Leave module, we had previously selected the Access Conveyor option that caused the entity to try to access space on the conveyor, and since the entity will remain on the conveyor, Arena would become confused and terminate with a runtime error. This is easily fixed by selecting the None option in the Transfer Out dialog box in the Logic section of the Leave module. These are the only model changes required.

You can test your new model logic by watching the animation. We suggest you fast-forward the simulation to about time 700 before you begin to watch the animation. At this point in the simulation, the effects of these changes become quite apparent.

9.2   More on Transporters

In Chapter 8, we presented the concepts for modeling Arena transporters and conveyors using the functionality available in the high-level modules found in the Basic Process panel. In Section 9.1, we further expanded that functionality for conveyors by using modules from the Advanced Transfer panel to modify and refine the models presented in Chapter 8. The same types of capabilities also exist for Transporters. Although we're not going to develop a complete model using these modules, we'll provide brief coverage of their functions and show the sequence of modules required to model several different situations. This should be sufficient to allow you to use these constructs successfully in your own models. Remember that online help is always available.

Let's start with the basic capabilities covered in Section 8.3. The fundamental transporter constructs are available in the Transfer In and Transfer Out section of the Enter and Leave modules. Consider the process of requesting a transporter and the subsequent transfer of the transporter and entity to its next station or location. The Request and Transport modules found in the Advanced Transfer panel also provide this capability.

The Request module provides the first part of the Transfer Out section of the Leave module and is almost identical to it. You gain the ability to override the default Velocity, but you lose the ability to specify a Load Time. However, a Load Time can be included by then directing the entity to a Delay module. The Request module actually performs two activities: allocation of a transporter to the entity and moving the empty transporter to the location of that entity, if the transporter is not already there. The Transport module performs the next part of the Transfer Out activity by initiating the transfer of the transporter and entity to its next location. Now let's consider a modeling situation where it would be desirable to separate these two functions. Assume that when the transporter arrives at the entity location there is a loading operation that requires the assistance of an operator. If we want to model the operator explicitly, we'll need these new modules. The module sequence (Request – Process – Transport) required to model this situation is shown in Figure 9-3.

You can also separate the two activities of the Request module by using the Allocate and Move modules. The Allocate module allocates a transporter to the entity, but leaves the transporter at its current location. The Move module allows the entity that has been allocated a transporter to Move the empty transporter anywhere in the model. When using the Request module, the transporter is automatically moved to the entity location. Consider the situation where the empty transporter must first pick up from a staging area a fixture required to transport the entity. In this case, we need to allocate the transporter, send it to the staging area, pick up the fixture, and finally send it to the entity's location. The module sequence (Allocate – Move – Process – Move) required to model these activities is shown in Figure 9-4.

Figure 9-3. Operator-Assisted Transporter Load

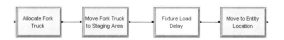

Figure 9-4. Fixture Required for Entity Transfer

Of course, you could include all kinds of embellishments for this activity. Suppose that only a portion of the entities need this fixture. We could easily include a Decide module between them to check for this condition, as shown in Figure 9-5.

Our two examples using the Allocate and Move modules both resulted in the transporter being moved to the location of the entity. Now let's assume that whenever your transporter is freed, you want it to be roving around the system looking for work. You have a predefined path that the transporter should follow. This path would be very similar to a sequence of stations, except it would form a closed loop. Each time the unallocated transporter reached the next station on its path, it would check to see if there is a current request from somewhere. If there is no request, the transporter continues on its mindless journey. If there is a request, the transporter proceeds immediately to the location of that request.

To model this, you would create a single entity (let's call it the loop entity) that would attempt to allocate the transporter with a very low priority (high number for the priority). Once allocated, the empty transporter is moved to the next station on its path. Upon arrival at this station, the transporter is freed so it can respond to any current request. The loop entity is directed to the initial Allocate module where it once again tries to allocate the transporter. This assumes that any current request for the transporter has a higher priority than the loop entity. Remember that the default priority is 1, with the lowest value having the highest priority.

Once a transporter arrives at its destination, it must be freed. The Free module provides this function. You only need to enter the transporter name in the dialog box, and in some cases, the unit number. Two additional modules, Halt and Activate, allow you to control the number of active or available transporters. The Halt module causes a single transporter unit to become inactive or unavailable to be allocated. The Activate module causes an inactive transporter to become active or available.

9.3   Entity Reneging

9.3.1   Entity Balking and Reneging

Entity *balking* occurs when an arriving entity or customer does not join the queue because there is no room, but goes away or goes someplace else. Entity *reneging* occurs when an entity or customer joins a queue on arrival but later decides to jump out and

Figure 9-5. Checking for Fixture Requirement

leave, probably regretting not having balked in the first place. Let's consider the more complex case where it's possible to have both customer balking and customer reneging. First, let's define the various ways that balking and reneging can occur.

In most service systems, there is a finite system capacity based on the amount of waiting space. When all the space is occupied, a customer will not be able to enter the system and will be balked. This is the simplest form of entity balking. Unfortunately, most systems where balking can occur are much more complicated. Consider a simple service line where, theoretically, the queue or waiting-line capacity is infinite. In most cases, there is some finite capacity based on space, but a service line could possibly exit a building and wind around the block (say the waiting line for World Series or Super Bowl tickets). In these cases, there is no concrete capacity limit, but the customers' decisions to enter the line or balk from the system are based on their own evaluation of the situation. Thus, balking point or capacity is often entity-dependent. One customer may approach a long service line, decide not to wait, and balk from the system—yet the next customer may enter the line.

Entity reneging is an even more complicated issue in terms of both modeling and representation in software. Each entity or customer entering a line has a different tolerance threshold with respect to how long to wait before leaving the line. The decision often depends on how much each customer wants the service being provided. Some customers won't renege regardless of the wait time. Others may wait for a period of time and then leave the line because they realize that they won't get served in time to meet their needs. Often the customers' decisions to remain or leave the line are based on both the amount of time they have already spent in the line as well as their current place in the line. A customer may enter a line and decide to wait for ten minutes; if he is not serviced by that time, he plans to leave the line. However, after ten minutes have elapsed, the customer may choose to stay if he is the next in line for service.

Line switching, or *jockeying*, is an even more complicated form of reneging that often occurs in supermarket checkout lines, fast-food restaurants, and banks that do not employ a single waiting line. A customer selects the line to enter and later re-evaluates that decision based on current line lengths. After all, we invariably enter the slowest-moving line, and if we switch lines, the line we left speeds up and the line we enter slows down. We won't cover the logic for jockeying.

9.3.2   Model 9-3: A Service Model with Balking and Reneging

Let's look at balking and reneging in the context of a very simple model. Customers arrive with EXPO(5) interarrival times at a service system with a single server—service time is EXPO(4.25). All times are in minutes. Although the waiting line has an infinite capacity, each arriving customer views the current length of the line and compares it to his *tolerance* for waiting. If the number in the line is greater than his tolerance, he'll balk away from the system. We'll represent the customer-balking tolerance by generating a sample from a triangular distribution, TRIA(3, 6, 15). Since our generated sample is from a continuous distribution, it will not be an integer. We could use one of the Arena math functions to convert it to an integer, but we're only interested if the number in the waiting line is greater than the generated tolerance.

There are two ways to model the balking activity. Let's assume that we create our arrivals, generate our tolerance value from the triangular distribution, and assign this value to an entity attribute. We could send our arrival to a Decide module and compare our sample value to the current number in the waiting line, using the Arena variable NQ. If the tolerance is less than or equal to the current number in queue, we balk the arrival from the system. Otherwise, we enter the waiting line. An alternative method is to assign our tolerance to a variable and also use this same variable for the server queue capacity. We then send our arrival directly to the server. If the current number in the queue is greater than or equal to the tolerance, which is equal to the queue capacity, the arrival will be balked automatically. Using the second method, it's possible for the queue capacity to be assigned a value less than the current number in queue. This method works because Arena checks only the queue-capacity value when a new entity tries to enter the queue. Thus, the current entities remain safely in the queue, regardless of the new queue-capacity value. We will use this second method when we develop our model.

To represent reneging, assume that arriving customers who decide not to balk are willing to wait only a limited period of time before they renege from the queue. We will generate this renege tolerance time from an ERLA(15, 2) distribution (Erlang), which has a mean of 30, and assign it to an entity attribute in an Assign module. Modeling the mechanics of the reneging activity can be a challenge. If we allow the arrival to enter the queue and the renege time is reached, we need to be able to find the entity and remove it from the queue. At this point in our problem description, you might want to consider alternative methods to handle this. For example, we could define a variable that keeps track of when the server will next be available. We generate the entity-processing time first and assign it to an attribute in an Assign module. We then send our entity to the Decide module where we check for balking. If the entity is not balked, we then check (in the same Decide module) to see if the entity will begin service before its renege time. If not, we renege the entity. Otherwise, we send the entity to an Assign module where we update our variable that tells us when the server will become available, then send the entity to the queue. This model logic may seem complicated, but can be summarized as follows:

```
Define Available Time = Time in the future when server will be available

Create arrival
     Assign Service Time
     Assign the Time in the future the activity would renege, which is equal
           to Tolerance Time +  TNOW
     Assign Balk Limit

Decide
     If Balk Limit > Number in queue
           Balk entity
     If Renege Time < Available Time
           Renege entity
     Else
           Assign Available Time = MX(Available Time , TNOW) + Service Time
     Send entity to queue
```

Note the use of the "maximum" math function, MX.

There is one problem with this logic that can be fixed easily—our number in queue is not accurate because it won't contain any entities that have not yet reneged. We can fix this by sending our reneged entities to a Delay module where they are delayed by the renege time; we also specify a Storage (e.g., Renege_Customers). Now we change our first Decide statement as follows:

```
If Balk Limit > Number in queue + NSTO(Renege_Customers)
```

We have to be careful about our statistics, but this approach will capture the reneging process accurately and avoid our having to alter the queue.

Let's add one last caveat before we develop our model. Assume that the actual decision of whether to renege is based not only on the renege time, but also on the position of the customer in the queue. For example, customers may have reached their renege tolerance limit, but if they're now at the front of the waiting line, they may just wait for service (i.e., renege on reneging). Let's call this position in the queue where the customer will elect to stay, even if the customer renege time has elapsed, the customer *stay zone*. Thus, if the customer stay zone is 3 and the renege time for the customer has expired, the customer will stay in line anyway if they are one of the next three customers to be serviced.

We'll generate this position number from a Poisson distribution, POIS(0.75). We've used the Poisson distribution because it provides a reasonable approximation of this process and also returns an integer value. For those of you with no access to Poisson tables (you mean you actually sold your statistics book?), it is approximately equivalent to the following discrete empirical distribution: DISC(0.472, 0, 0.827, 1, 0.959, 2, 0.993, 3, 0.999, 4, 1.0, 5). See Appendix D for more detail on this distribution.

This new decision process means that the above logic is no longer valid. We must now place the arriving customer in the waiting line and evaluate the reneging after the renege time has elapsed. However, if we actually go ahead and place the customer in the queue, there's no mechanism to detect that the renege time has elapsed. To overcome this problem, we'll make a duplicate of each entity and delay it by the renege time. The original entity, which represents the actual customer, will be sent to the service queue. After the renege-time delay, we'll have the duplicate entity check the queue position of the original entity. If the customer is no longer in the service queue (that is, the customer was served), we'll just dispose of the duplicate entity. If the customer is still in the queue, we'll check to see if that customer will renege. If the current queue position is within the customer stay zone, we'll just dispose of the duplicate entity. Otherwise, we'll have the duplicate entity remove the original entity from the service queue and dispose of both itself and the original entity. This model logic is outlined as follows:

```
Create Arrivals
      Assign Renege Time = ERLA(15,2)
      Assign arrival time: Enter System = TNOW
      Assign Stay Zone Number = POIS(0.75)
      Assign Balking Tolerance = TRIA(3,6,15)

Create Duplicate entity
      Original entity to server queue
              If Balk from queue
                      Count balk
                      Dispose
              Delay for Service Time = EXPO(4.25)
              Tally system time
              Dispose

      Duplicate entity
              Delay by Renege Time
              Search queue for position of original entity
              If No original entity
                      Dispose
              If Queue position <= Stay Zone Number
                      Dispose
              Remove original entity from queue and Dispose
                      Count Renege customer
                      Dispose
```

In order to implement this logic, we obviously need a few new features, such as the ability to search a queue and remove an entity from a queue. As you would suspect, Arena modules that perform these functions can be found in the Advanced Process and Blocks panels. Our completed Arena model (Model 9-3) is shown in Figure 9-6.

We start our model with a Create module that creates arriving customers, which are then sent to the following Assign module where the values we'll need later are assigned, as in Display 9-7.

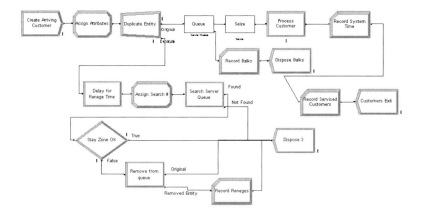

Figure 9-6. The Service Model Logic

Name	Assign Attributes
Type	Attribute
Attribute Name	Enter System
New Value	TNOW
Type	Attribute
Attribute Name	Renege Time
New Value	ERLA(15, 2)
Type	Variable
Variable Name	Server Queue Capacity
New Value	TRIA(3, 6, 15)
Type	Attribute
Attribute Name	Stay Zone
New Value	POIS(0.75)
Type	Variable
Variable Name	Total Customers
New Value	Total Customers + 1
Type	Attribute
Attribute Name	Customer #
New Value	Total Customers

Display 9-7. The Assign Module

We send the new arrivals to a Separate module. This module allows us to make duplicates (clones) of the entering entity. The original entity leaves the module by the exit point located at the right of the module. The duplicated entities leave the module by the exit points below the module. In this case, we only need to enter the name for our module, accepting the remaining data as the defaults.

The duplicates are exact replicas of the original entity (in terms of attributes and their values) and can be created in any quantity. If you make more than one duplicate of an entity, you can regard them as a batch of entities that will all be sent out of the same (bottom) exit. Note that if you enter a value n for # of duplicates, $n+1$ entities actually leave the module—n duplicates from the bottom connection point and 1 original from the top.

The original entity (customer) is sent to a Queue – Seize module sequence (from the Blocks panel) where it tries to enter queue Server Queue to wait for the server. In order to implement this method, we need to be able to set the capacity of the queue that precedes the Seize to the variable Server Queue Capacity. We accomplish this by attaching the Blocks panel to our model and selecting and placing a Queue module. (Note that when you click on the Queue module, there is no associated spreadsheet view; this will be the case for any modules from the Blocks or Elements panels.) Before you open the Queue dialog box, you might notice that there is a single exit point at the right of the module. Now we double-click on the module and enter the label, name, and capacity, as shown in Display 9-8. After you close the dialog box, you should see a second exit point near the lower right-hand corner of the module. This is the exit that the balked entities will take.

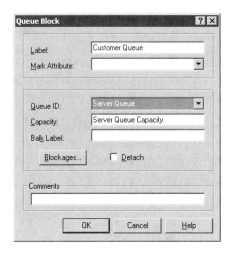

Label	Customer Queue
Queue ID	Server Queue
Capacity	Server Queue Capacity

Display 9-8. The Queue Module from the Blocks Panel

The capacity of this queue was entered as the variable `Server Queue Capacity`, which was set by our arriving customer as his tolerance for not balking, as explained earlier. If the current number in queue is less than the current value of `Server Queue Capacity`, the customer is allowed to enter the queue. If not, the customer entity is balked to the Record module where he increments the balk count and is then sent to a Dispose module, where he exits the system (see Figure 9-6). A customer who is allowed to enter the queue waits for the resource `Server`.

We need to follow this Queue module with a Seize module. When you add your Seize module, make sure it comes from the Blocks panel and not from the Advanced Process panel. The module in the Advanced Process panel automatically comes with a queue and Arena would become quite confused if you attempted to precede a seize construct with two queues. We then double-click on the module and make the entries shown in Display 9-9.

When the customer seizes the server, it is sent to the following Process module. This Process module uses a `Delay Release` Action, which provides the service delay and releases the server resource for the next customer. A serviced customer is sent to a Record module where the system time is recorded, to a second Record module where the number is counted, and then to the following Dispose module.

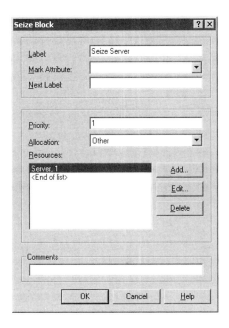

Label	Seize Server
Resource ID	Server
Number of Units	1

Display 9-9. The Seize Module from the Blocks Panel

The duplicate entity is sent to a Delay module where it is delayed by the renege time that was assigned to the attribute Renege Time. After the delay, the entity enters an Assign module where the value of the attribute Customer # is assigned to a new variable named Search #. The attribute Customer # contains a unique customer number assigned when the customer entered the system. The entity is then sent to the following Search module, from the Advanced Process panel. A Search module allows us to search a queue to find the *rank*, or queue position, of an entity that satisfies a defined search condition. A queue rank of 1 means that the entity is at the front of the Queue (the next entity to be serviced). In our model, we want to find the original customer who created the duplicate entity performing the search. That customer will have the same value for its Customer # attribute as the variable Search # that we just assigned.

The Search module (Display 9-10) searches over a defined range according to a defined condition. Typically, the search will be over the entire queue contents, from 1 to NQ. (Note that the search can be performed backward by specifying the range as NQ to 1.)

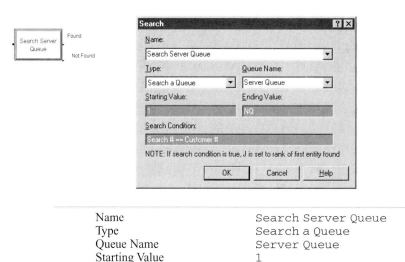

Name	Search Server Queue
Type	Search a Queue
Queue Name	Server Queue
Starting Value	1
Ending Value	NQ
Search Condition	Search # == Customer #

Display 9-10. The Search Module

However, you may search over any range that your model logic requires. If you state a range that exceeds the current number in the queue, Arena will terminate with a runtime error. Arena will assign to the variable J the rank of the first entity during the search that satisfies the condition and send the entity out the normal exit point (labeled Found). If the condition contains the math functions MX or MN (maximum or minimum), it will search the entire range. If attributes are used in the search condition, they will be interpreted as the attribute value of the entity in the search queue. If the queue is empty, or no entity satisfies the condition, the entity will be sent out the lower exit point (labeled Not Found). Ordinarily, you're interested in finding the entity rank so you can remove that entity from the queue (Remove module) or make a copy of the entity (Separate module). The Search module can also be used to search over entities that have been formed as a temporary group using a Batch module. In addition, you can search over any Arena expression.

For our model, we want to search over the entire queue range, from 1 to NQ, for the original entity that has the same Customer # value as the duplicated entity initiating the search. If the original entity, or customer, is no longer in the queue, the entity will exit via the Not Found exit point and be sent to the following Dispose module. If the original entity is found, its rank in the queue will be saved in the variable J. The entity is then sent to the following Decide module. The check at this module is to see if the value of J is less than or equal to the value of the attribute Stay Zone. If this condition is True, it implies that the position of the customer in the queue is good enough that he chooses to remain in the line. In this case, we dispose of the duplicate entity (see Figure 9-6). If the condition is False, we want to renege the original customer. Therefore, we send the duplicate entity to the following Remove module.

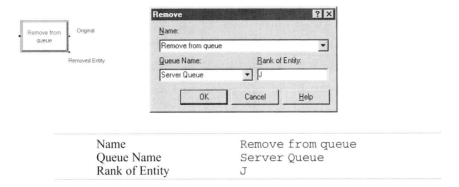

Name	Remove from queue
Queue Name	Server Queue
Rank of Entity	J

Display 9-11. The Remove Module

The Remove module allows us to remove an entity from a queue and send it to another place in our model. It requires that you identify the entity to be removed by entering the queue identifier and the rank of that entity. If you attempt to remove an entity from an undefined queue or to remove an entity with a rank that is greater than the number of entities in the specified queue, Arena will terminate the run with an error. In our model, we want to remove the customer with rank J from queue `Server Queue`, as shown in Display 9-11.

If you look at the Remove module, you'll see two exit points on the right side. The entity that entered the Remove module will depart from the upper exit point; in our model, it is sent to the same Dispose module we used for the first branch of our Decide module. The customer entity removed from the server queue will depart by the lower exit point and is sent to a Record module to count the number of customers that renege. The entity is then sent to the Dispose module.

We set our Replication length to 2000 minutes. The result of the summary report for this model shows that we had 8 balking and 41 reneging customers with 332 serviced customers.

9.4 Holding and Batching Entities

In this section, we'll take up the common situation where entities need to be held up along their way for a variety of reasons. We'll also discuss how to combine or group entities and how to separate them later.

9.4.1 Modeling Options

As you begin to model more complex systems, you might occasionally want to retain or hold entities at a place in the model until some system condition allows these entities to progress. You might be thinking that we have already covered this concept, in that an entity waiting in a queue for an available resource, transporter, or conveyor space allows us to hold that entity until the resource becomes available. Here we're thinking in more general terms; the condition doesn't have to be based on just the availability of a resource, transporter, or conveyor space. The conditions that allow the entity to proceed can be based on any system conditions; for example, time, queue size, etc. There are two different methods for releasing held entities.

The first method holds the entities in a queue until they receive permission or a signal to proceed from another entity in the system. For example, consider a busy intersection with a policeman directing traffic. Think of the cars arriving at the intersection as entities being held until they are allowed to proceed. Now think of the policeman as an entity eventually giving the waiting cars a signal to proceed. There may be ten cars waiting, but the policeman may give only the first six permission to proceed.

The second method allows the held entities themselves to evaluate the system conditions and determine when they should proceed. For example, think of a car wanting to turn across traffic into a driveway from a busy street with oncoming traffic; unfortunately, there's neither a traffic light nor a policeman. If the car is the entity, it waits until conditions are such that there is no oncoming traffic within a reasonable distance and the driveway is clear for entry. In this case, the entity continuously evaluates the conditions until it is safe to make the turn. If there's a second car that wants to make the same turn directly behind the first, it waits until it is at the front of the line and then performs its own evaluation. We'll illustrate both methods in Model 9-4 below.

There are also situations where you need to form *batches* of items or entities before they can proceed. Take the simplest case of forming batches of similar or identical items. For example, you're modeling the packing operation at the end of a can line that produces beverages. You want to combine or group beverages into six-packs for the packing operation. You might also have a secondary operation that combines 4 six-packs into a case. In this illustration, the items or entities to be grouped are identical, and you would most likely form a *permanent* group (that is, one that you'd never want to take apart again later). Thus, six entities enter the grouping process and one entity, the six-pack, exits the process. However, if you're modeling an operation that groups entities that are later to be separated to continue individually on their way, you would want to form a *temporary* group. In the first case, you lose the unique attribute information attached to each entity. In the second case, you want each entity departing the operation to retain the same attribute information it held when it joined the group. So when you're modeling a grouping operation, you need to decide whether you want to form a temporary or permanent group. We'll discuss both options in Model 9-4 below.

9.4.2   Model 9-4: A Batching Process Example

Randomly arriving items are into batches before being processed. You might think of the process as an oven that cures the arriving items in batches. The maximum size of the batch that can be sent to the process depends on the design capacity. Let's assume that each item must be placed on a special fixture for the process, and these fixtures are very expensive. The number of fixtures determines the process capacity. Let's further assume that these fixtures are purchased in pairs. Thus, the process can have a capacity of 2, 4, 6, 8, etc. In addition, we'll assume that the process requires a minimum batch size of 2 before it can be started.

Arriving items are sent to a batching area where they wait for the process to become available. When the process becomes available, we must determine the batch size to process, or cause the process to wait for the arrival of enough additional items to make a viable batch. Here's the decision logic required:

```
Process becomes available
       If Number of waiting items ≥2 and ≤Max Batch
               Form batch of all items
               Set Number of waiting items to 0
               Process batch
       Else if Number of waiting items > Max Batch
               Form batch of size Max Batch
               Decrement Number of waiting items by Max Batch
               Process batch
       Else if Number of waiting items < 2
               Wait for additional items
```

As long as there are items available, the items are processed. However, if there are insufficient items (< 2) for the next batch, the process is temporarily stopped and requires an additional startup-time delay before the next batch can be processed. Because of this additional startup delay, we may want to wait for more than two items before we restart the process.

We want to develop a simulation model that will aid us in designing the parameters of this process. There is one design parameter (the process capacity or Max Batch) and one logic parameter (restart batch size) of interest. In addition, we would like to limit the number of waiting entities to approximately 25.

We'll design our simulation model and then use OptQuest for Arena to search for the best solution based on minimizing the number of restarts.

The completed model that we'll now develop is shown in Figure 9-7. You might note that there is one new module: Batch from the Basic Process panel. You might also note that we don't have a resource defined for the process; it's not required as the model will limit the number of batches in the oven to 1. Let's start with the Variable module and *Run > Setup > Replication Parameters* dialog box and then proceed to the model logic. We'll discuss the Statistic data module later.

We use the Variable module to define the two parameters that we'll use in OptQuest: the maximum batch size or process capacity, Max Batch; and the restart batch size, Restart. These variables are initially set to 10 and 4, respectively. The *Run > Setup > Replication Parameters* dialog box specifies a replication length of 10,000 (all times are in minutes).

Now let's look at the item arrival process—the Create – Hold modules at the upper left of Figure 9-7. The Create module is used to generate arrivals with exponential interarrival times. We've used a value of 1.1 as the interarrival mean. No other entries are required for this module. The arriving items are then sent to the Hold module from the Advanced Process panel. The Hold module holds entities until a matching *signal* is received from elsewhere in the model. The signal can be based on an expression or an attribute value. Different entities can be waiting for different signals, or they can all be waiting for the same signal. When a matching signal is received, the Hold module will release up to a maximum number of entities based on the Limit (which defaults to infinity), unless the signal contains additional release limits. This will be explained when we cover the Signal module.

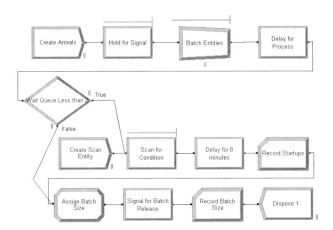

Figure 9-7. The Batch Processing Model

In our example, all entities will wait for the same signal, which we have arbitrarily specified as Signal 1. We have defaulted the Limit to infinity since a limit will be set in the Signal module. We have also requested an Individual Queue, `Hold for Signal.Queue`, so we can obtain statistics and plot the number in queue for our animation.

Now let's consider the conditions at the start of the simulation. There is nothing being processed; therefore, nothing should happen until the arrival of the fourth item, based on the initial value of `Restart`. As arriving items do not cause a signal to be sent, some other mechanism must be put into the model to cause the start of the first batching operation and process.

This mechanism can be found in the Create – Hold – Delay – Record, etc., sequence of modules found toward the center of Figure 9-7. The `Create Scan Entity` Create module has Max Arrivals set to 1, which results in only a single entity being released at time 0. This entity is sent directly to the Hold module that follows.

The `Scan for Condition` Hold module, with the type specified as Scan for Condition, allows us to hold an entity until the user-defined condition is true; at that time, the entity is allowed to depart the module. The waiting entities are held in a user-defined queue (the default) or in an internal queue. If an entity enters a Hold module that has no waiting entities, the condition is checked and the entity is allowed to proceed if the condition evaluates to true. If there are other entities waiting, the arriving entity joins the queue with its position based on the selected queue-ranking rule, defaulted to FIFO. If there are entities waiting, the scan condition is checked as the last operation before any discrete-event time advance is triggered anywhere in the model. If the condition is true, the first entity in the scan queue will be sent to the next module. Arena will allow that entity to continue until it's time for the next time advance. At this time, the condition is checked again. Therefore, it's possible for all waiting entities to be released from the Hold module at the same time, although each entity is completely processed before the next entity is allowed to proceed.

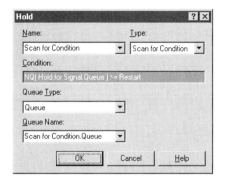

Name	Scan for Condition
Type	Scan for Condition
Condition	NQ(Hold for Signal.Queue) >= Restart

Display 9-12. The Hold Module (Scan for Condition)

For our model, we've entered a condition that requires at least four items to be in the wait queue preceding our Hold module before the condition is satisfied (Display 9-12); note that you can use the Arena Expression Builder if you right-click in this field. At that time, the entity will be sent to the following `Delay for 8 Minutes` Delay module. As you begin to understand our complete model, you should realize that we've designed it so there will never be more than one entity in the scan queue.

Now, at the start of a simulation run, the first arriving item will be created and sent to the queue `Hold for Signal.Queue`. At the same time, the second Create module will cause a single entity to arrive and be placed in the queue `Scan for Condition.Queue`. Nothing happens until the first queue has four items. At that time, the entity is released from the second Hold module and is sent to the Delay module where it incurs an 8-minute delay, accounting for the process restart time. Be aware that during this delay additional items may have arrived. The entity is then sent to a Record module where the number of startups, counter `Startups`, is incremented by 1. The new process batch size, `Batch Size`, is calculated in the following Assign module, in Display 9-13. Recall that, at least for the first entity, we know there are at least four items in the wait queue. Thus, our process batch size is either the number in the wait queue or the maximum batch size if the number waiting is greater than the process capacity.

Name	Assign Batch Size
Type	Variable
Variable Name	Batch Size
New Value	MN(NQ(Hold for Signal.Queue), Max Batch)

Display 9-13. Assigning the Next Process Batch Size

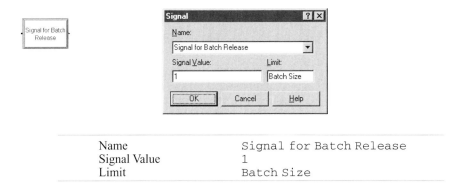

Name	Signal for Batch Release
Signal Value	1
Limit	Batch Size

Display 9-14. The Signal Module

Having calculated the next batch size and assigned it to a global variable, `Batch Size`, we send the entity to a Signal module, seen in Display 9-14. This module broadcasts a signal with a value of 1 across the whole model, which causes the entities in the wait queue, up to a maximum `Batch Size`, to be released. This entity then enters a Record module where the next batch size is tallied, and then the entity is disposed.

You can have multiple Signal and multiple Hold modules in your model. In this case, a Signal module will send a signal value to each Hold module with the Type specified as `Hold for Signal` and release the maximum specified number of entities from all Hold modules where the Signal matches.

Now the process has undergone a startup delay, and the first batch of items has been released to the `Batch Entities` Batch module following the `Hold for Signal` Hold module. The Batch module allows us to accumulate entities that will then be formed into a permanent or temporary batch represented by a single entity. In our example, we have decided to form a permanent batch. Thus, the unique attribute values of the batched entities are lost because they are disposed. The attribute values of the resulting representative entity can be specified as the Last, First, Product, or Sum of the arriving individual batched entities. If a temporary batch is formed, the entities forming the batch are removed from the queue and are held internally to be reinstated later using a Separate module. Entities may also be batched based on a match criterion value. If you form a permanent batch, you can still use the Separate module to recreate the equivalent number of entities later. However, you've lost the individual attribute values of these entities.

Normally, entities arriving at a Batch module will be required to wait until the required batch size has been assembled. However, in our model, we defined the batch size to be formed to be exactly equal to the number of items we just released to the module; see Display 9-15. Thus, our items will never have to wait at this module.

The entity that now represents the batch of items is directed to the `Delay for Process` Delay module where the process delay occurs. Note that this process delay depends on the batch size being processed: 3 minutes plus 0.6 minute for each item in the batch.

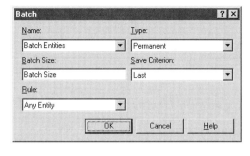

Name	Batch Entities
Batch Size	Batch Size

Display 9-15. Forming the Process Batch

The processed batch is sent to the following `Wait Queue Less Than 2` Decide module where we check whether there are fewer than two items in the wait queue. If so, we must shut down the process and wait for more items to arrive. We do this by sending the entity to the previously discussed `Scan for Condition` Hold module to wait for enough items to restart the process. If there are at least two waiting items, the entity is sent to the `Assign Batch Size` Assign module where we set the next batch size to start the next batch processing.

Since the preceding few paragraphs were fairly complex, let's review the logic that controls the batching of items to the process. Keep in mind that the arriving entities are placed in a Hold queue where they are held until a signal to proceed is received. The first process is initiated by the second Create module that creates only one control entity. This entity is held in the Scan queue until the first four items have arrived. The value 4 is the initial value of the Variable `Restart`. This control entity causes the first batch to be released for processing, and it is then disposed. After that, the last batch to complete processing becomes a control entity that determines the next batch size and when to allow the batch to proceed for processing.

Before we're ready to run our model in OptQuest, we need to define an output statistic in the Statistic data module that will allow us to limit the average number of entities waiting in the queue `Hold for Signal.Queue`. We've given it the name `Max Batch Value`. The expression is `DMAX(Hold for Signal.Queue.NumberInQueue)`. The function `DMAX` will return the maximum value of an Arena time-persistent statistic. The statistic of interest is automatically generated as part of the output report. The name of the statistic is the queue name followed by `.NumberInQueue`. The interested reader can find this type of information in the help topic "Statistics (Automatically Generated by SIMAN)."

We set up OptQuest to vary the variable `Max Batch` from 4 to 14 in discrete increments of 2. The `Restart` variable was varied from 2 to 10 in discrete increments of 1.

The objective was to minimize the number of startups with a requirement that our `Max Batch Value` statistic not exceed 25. The best solution found was:

> Max Batch = 10
> Restart = 8
> Max Batch Value = 24.9
> Startups = 30.6.

9.5 Overlapping Resources

In Chapters 4-8, we concentrated on building models using the modules available from the Basic Process panel, the Advanced Process panel, and the Advanced Transfer panel. Even though we used these modules in several different models, we still have not exhausted all the capabilities.

As we developed models in the earlier chapters, we were not only interested in introducing you to new Arena constructs, but we also tried to cover different modeling techniques that might be useful. We have consistently presented new material in the form of examples that require the use of new modeling capabilities. Sometimes the fabrication of a good example to illustrate the need for new modeling capabilities is a daunting task. We have to admit that the example that we are about to introduce is a bit of a stretch, not only in terms of the model description, but also the manner in which we develop the model. However, if you bear with us through this model development, we think that you will add several handy additions to your toolbox.

9.5.1 System Description

The system we'll be modeling is a tightly coupled, three-workstation production system. We have used the words "tightly coupled" because of the unique part-arrival process and because there is limited space for part buffering between the workstations.

We'll assume an unlimited supply of raw materials that can be delivered to the system on demand. When a part enters the first workstation, a request is automatically forwarded to an adjoining warehouse for the delivery of a replenishment part. Because the warehouse is performing other duties, the replenishment part is not always delivered immediately. Rather than model this activity in detail, we'll assume an exponential delivery delay, with mean of 25 (all times are minutes), before the request is acted upon. At that point, we'll assume the part is ready for delivery, with the delivery time following a UNIF(10, 15) distribution. To start the simulation, we'll assume that two parts are ready for delivery to the first workstation.

Replenishment parts that arrive at the first workstation are held in a buffer until the workstation becomes available. A part entering the first workstation immediately requests a setup operator. The setup time is assumed to be EXPO(9). Upon completion of the setup, the part is processed by the workstation, lasting TRIA(10, 15, 20). The completed part is then moved to the buffer between Workstations 1 and 2. This buffer space is limited to two parts; if the buffer is full, Workstation 1 is blocked until space becomes available. We'll assume that all transfer times between workstations are negligible, or occur in 0 time.

There are two almost-identical machines at Workstation 2. They differ only in the time it takes to process a part: The processing times are TRIA(35, 40, 45) for Machine 2A and TRIA(40, 45, 50) for Machine 2B. A waiting part will be processed by the first available machine. If both machines are available, Machine 2A will be chosen. There is no setup required at this workstation. A completed part is then transferred to Workstation 3. However, there is no buffer between Workstations 2 and 3. Thus, Workstation 3 must be available before the transfer can occur. (This really affects the system performance!) If Machines 2A and 2B both have completed parts (which are blocking these machines) waiting for Workstation 3, the part from Machine 2A is transferred first.

When a part enters Workstation 3, it requires the setup operator (the same operator used for setup at Workstation 1). The setup time is assumed to be EXPO(9). The process time at Workstation 3 is TRIA(9, 12, 16). The completed part exits the system at this point.

We also have failures at each workstation. Workstations 1 and 3 have a mean Up Time of 600 minutes with a mean Down Time for repair of 45 minutes. Machines 2A and 2B, at Workstation 2, have mean Up Times of 500 minutes and mean Down Times of 25 minutes. All failure and repair times follow an exponential distribution. One subtle but very important point is that the Up Time is based only on the time that the machines are processing parts, not the elapsed time.

Now to complicate the issue even further, let's assume that we're interested in the percent of time that the machines at each workstation are in different states. This should give us a great amount of insight into how to improve the system. For example, if the machine at Workstation 1 is blocked a lot of the time, we might want to look at increasing the capacity at Workstation 2.

The possible states for the different machines are as follows:

> Workstation 1: Processing, Starved, Blocked, Failed, Waiting for setup operator, and Setup
> Machines 2A and 2B: Processing, Starved, Blocked, and Failed
> Workstation 3: Processing, Starved, Failed, Waiting for setup operator, and Setup.

We would also like to keep track of the percent of time the setup operator spends at Workstations 1 and 3. These states would be: WS 1 Setup, WS 2 Setup, and Idle.

These are typical measures used to determine the effectiveness of tightly coupled systems. They provide a great deal of information on what are the true system bottlenecks. Well, as long as we've gone this far, why not go all the way! Let's also assume that we'd like to know the percent of time that the parts spend in all possible states. Arranging for this is a much more difficult problem. First, let's define the *system* or *cycle time* for a part as starting when the delivery is initiated and ending when the part completes processing at Workstation 3. The possible part states are: Travel to WS 1, Wait for WS 1, Wait for setup at WS 1, Setup at WS 1, Process at WS 1, Blocked at WS 1, Wait for WS 2, Process at WS 2, Blocked at WS 2, Wait for setup at WS 3, Setup at WS 3, and Process at WS 3. As we develop our model, we'll take care to incorporate the resource states. But,

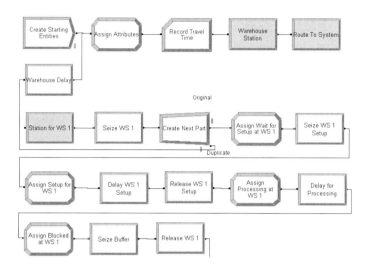

Figure 9-8. Part Arrival and Workstation 1

we'll only consider the part states after we have completed the model development. You'll just have to trust that we might know what we're doing.

9.5.2   Model 9-5: A Tightly Coupled Production System

In developing our model, we'll use a variety of different modules. Let's start with the modules for the arrival process and Workstation 1, which are shown in Figure 9-8.

Since you have seen almost all the modules that we'll be using in this model, we won't provide all the displays. However, we will provide sufficient information for you to re-create the model on your own. Of course, you can always open `Model 09-05.doe` and follow along.

Let's start with the initial arrivals that provide the first two parts for the system. All the remaining arrivals will be based on a request issued when a part enters the machine at the first workstation. The single Create module found at the upper left of Figure 9-8 has three entries; a Name, a Batch Size of 2, and Max Arrivals of 2. This causes two entities, or parts, to be created at time 0. The Create module then becomes inactive—no more entities are created by it for the rest of the simulation. These two parts are sent to an Assign module where we make two attribute assignments. We assign the value of TNOW to the attribute `Enter System` and assign a value generated from UNIF(10, 15) to the attribute `Route Time`. The value assigned to `Enter System` is the time the part entered the system, and the `Route Time` is the delivery time from the warehouse to the first workstation. Since we are interested in keeping detailed part-status information, we send these parts to a Record module where we tally the `Delivery Time` based on the expression `Route Time`, which we assigned in the previous Assign module. The part is then sent to the following Station module where we define the Station `Warehouse`. Next we use the Route module to route from the station `Warehouse` to the station `WS 1 Station` using the `Route Time` we previously assigned.

StateSet

States		
	State Name	**AutoState or Failure**
1	Processing	
2	Starved	IDLE
3	Blocked	
4	Failed	WS 1_3 Failure
5	Waiting for setup	
6	Setup	

Display 9-16. The WS 1 StateSet States

Upon completion of this transfer, the part arrives at the Station module, `Station for WS 1`. The following Seize module attempts to seize 1 unit of resource WS 1. We now need to take care of three additional requirements. First, we need to specify the resource states and make sure that the statistics are kept correctly. Second, we need to make a request for a replenishment part. Finally, we need to have a setup occur before the part is processed.

Let's start by defining our resources using the Resource data module. We enter six resources: `WS 1`, `WS 2A`, `WS 2B`, `WS 3`, `Setup Operator`, and `Buffer`. The first five have capacity of 1 and the `Buffer` resource has a capacity of 2.

Back in Section 4.2.4, we showed you how to use Frequencies to generate frequency statistics on the number in a queue. In order to get frequency data on our resources, we first need to define our StateSets. We do this using the StateSet data module from the Advanced Process panel. We entered five StateSets: `WS 1 StateSet`, `WS 2A StateSet`, `WS 2B StateSet`, `WS 3 StateSet`, and `Setup Operator StateSet`. The states for the `WS 1 StateSet` are shown in Display 9-16.

We then entered our two failures (`WS 1_3 Failure` and `WS 2 Failure`) using the Failure data module. Now we need to go back to the Resource data module and add our StateSets and Failures. When we added our failures, we selected `Ignore` as the Failure Rule.

Now let's go back to our model logic in Figure 9-8. Having dealt with the entry to WS 1, we need to seize the WS 1 resource, shown in Display 9-17.

Next we need to take care of the part-replenishment and the setup activity. The part that has just Seized the workstation resource enters the following Separate module, which creates a duplicate entity. The duplicate entity is sent to the `Warehouse Delay` Delay module, which accounts for the time the part-replenishment request waits until the next part delivery is initiated, lasting `EXPO( 25 )`. We then send this entity to the same set of modules that we used to cause the arrival of the first two parts. The original entity exits the Separate module to an Assign module where we assign the state of the resource WS 1 to `Waiting for Setup Operator`, in Display 9-18.

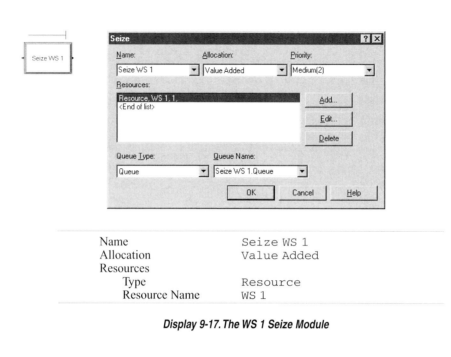

Name	Seize WS 1
Allocation	Value Added
Resources	
Type	Resource
Resource Name	WS 1

Display 9-17. The WS 1 Seize Module

Name	Assign Wait for Setup at WS 1
Assignments	
Type	Other
Other	STATE(WS 1)
New Value	Waiting for Setup

Display 9-18. Resource State Assignment

It is then directed to the `Seize WS 1 Setup` Seize module where it requests the setup operator, in Display 9-19. Note that we specify the state of the setup resource as `WS 1 Setup`. This will allow us to obtain frequency statistics on the `Setup Operator Resource`.

Name	Seize WS 1 Setup
Resources	
Type	Resource
Resource Name	Setup Operator
Resource State	WS 1 Setup

Display 9-19. Seizing the Setup Operator

By now, you may be scratching your head and asking, "What are they doing?" Well, we warned you that this problem was a little contorted, and once we finish the development of Workstation 1, we'll provide a high-level review of the entire sequence of events. So let's continue.

In the following `Assign Setup for WS 1` Assign module, we set the `WS 1` resource to the `Setup` state and then we delay in the following `Delay WS 1 Setup` Delay block for the setup activity. Upon completion of the setup, we release the `Setup Operator` resource, assign the `WS 1` resource state to `Processing`, and undergo the processing delay. After processing, we assign the `WS 1` resource state to `Blocked`. At this point, we're not sure that there is room in the buffer at Workstation 2. We now enter the `Seize Buffer` Seize module to request buffer space, from the resource `Buffer`. Once we have the buffer resource, we release the WS 1 resource and depart for Workstation 2.

Now (whee!—we feel like we're out of breath after running through this logic), let's review the sequence of activities for a part at Workstation 1. We start by creating two parts at time 0; let's follow only one of them. That part is time stamped with its arrival time, and a delivery time is generated and assigned. The delivery time is recorded and the part is routed to Workstation 1. Upon entering Workstation 1, it joins the queue to wait for the resource `WS 1`. Having seized the resource, it exits the Seize module, where it duplicates the replenishment part, which is sent back to where the first part started. The part then assigns the server resource state to `Waiting for setup` and queues for the setup operator. Having seized the setup operator, setting its state to `Setup`, it assigns the `WS 1` resource state to `Setup`, delays for the setup, releases the setup operator, assigns the `WS 1` resource state back to `Processing`, and delays for processing. After processing, the part queues to seize one unit of the resource `Buffer` (capacity of 2), setting the server state to `Blocked` during the queueing time. After seizing the buffer resource, it releases the `WS 1` resource and exits the Workstation 1 area with control of one unit of the resource `Buffer`.

The modules for Workstation 2, which has two machines, are shown in Figure 9-9.

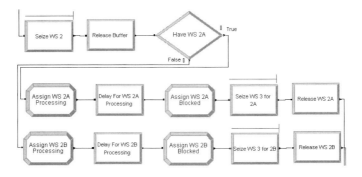

Figure 9-9. Workstation 2 Modules

Recall that parts arriving at Workstation 2 have control of one unit of the buffer resource. Thus, as soon as the part seizes one of the two machines, the buffer must be released because there is now a free space in front of the workstation. This action is accomplished in the following `Release Buffer` Release module.

Upon seizing an available machine and releasing the buffer resource, the part enters the `Have WS 2A` Decide module, which is used to determine which machine resource has been seized, based on the `WS 2 Resource` attribute value assigned in the first Seize module when the resource was allocated to the part. If the part was allocated resource `WS 2A`, the first resource in the set, this attribute would have been assigned a value of 1. We use this logic to determine which machine we have been allocated and send the part to logic specifically for `WS 2A` or `WS 2B`. The middle five modules in Figure 9-9 are for `WS 2A` and the bottom five are for `WS 2B`. Since there is no setup at Workstation 2, the first Assign module assigns the state of the machine resource to `Processing` and directs the part to the following Delay module, which represents the part processing. The last Assign module assigns the machine resource state to `Blocked`.

The part then attempts to seize control of resource `WS 3` in the following Seize module. If the resource is unavailable, the part will wait in queue `Seize WS 3.Queue`, which we defined in the Queue module as being a shared queue between the two Seize modules. Now, if there are two waiting parts, one for each machine, we want the part from `WS 2A` to be first in line. We accomplish this by selecting the ranking rule to be Low Attribute Value based on the value of attribute `WS 2 Resource`. This causes all entities in the queue to be ranked according to their value for the attribute `WS 2 Resource`. This assures that `WS 2A` will receive preference.

Once a part has been allocated the `WS 3` resource, it will be transferred to that machine in 0 time. The modules we used to model Workstation 3 are shown in Figure 9-10. You might notice that these modules look very similar to those used to model Workstation 1, and they are essentially the same. An entering part (it already has been allocated the resource `WS 3`) first assigns the workstation resource state to `Waiting for setup` and then attempts to seize the setup operator. After being allocated the setup operator resource, the resource `WS 3` state is set to `Setup`, the part is delayed for setup, the setup operator is released, the resource WS 3 state is set to Processing, and the part is delayed for the process time. Upon completion, the WS 3 resource is released, the part cycle time is recorded, and the entity is disposed.

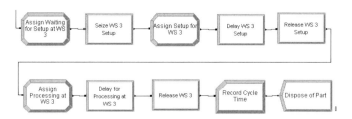

Figure 9-10. Workstation 3 Modules

Statistic - Advanced Process							
	Name	Type	Frequency Type	Resource Name	Report Label	Output File	Categories
1	WS 1 Resource	Frequency	State	WS 1	WS 1 Resource		0 rows
2	WS 2A Resource	Frequency	State	WS 2A	WS 2A Resource		0 rows
3	WS 2B Resource	Frequency	State	WS 2B	WS 2B Resource		0 rows
4	WS 3 Resource	Frequency	State	WS 3	WS 3 Resource		0 rows
5	Set Operator Resource	Frequency	State	Setup Operator	Set Operator Resource		0 rows

Display 9-20. The Statistic Module for Frequency Statistics

Now that we've completed our model logic, we need to request the frequency statistics for our workstations and setup operator. We want frequency statistics based on the resource states. The required inputs for the Statistic module to obtain these frequencies are shown in Display 9-20.

Recall that when we were describing our problem, we also wanted to output the percent of time that the parts spend in all possible states. Obtaining this information for resource states is fairly easy. We simply define the resource states and Arena collects and reports this information using the Frequencies feature. Unfortunately, this use of Frequencies is valid only for resources.

Collecting the same type of information for part or entity status is difficult because the part states span numerous activities over several resources. This creates a modeling problem: What is the best way to obtain this information? Before we show you our approach, we'll discuss several alternatives, including their shortcomings.

Our first thought was to define a variable that we could change based on the current part state. Then could request frequency statistics on that variable, much as we did for the part-storage queue in the rework area of Model 4-2 in Section 4.2.4. Of course, this would require us to edit our model to add assignments to this variable whenever the part status changed. Although this sounds like a good idea, it falls apart when you realize that there can be multiple parts in the system at the same time. It would work just fine if we limited the number of parts in our system to one. Since this was not in the problem description, we decided to consider alternate ways to collect this information.

Our second idea was to assume that all the required information was already being collected (a valid assumption). Given this, we need only assemble this information at the end of the run and calculate our desired statistics. Arena automatically calls a wrap-up routine at the end of each run. We could write user code (see Section 9.2) that would perform this function. One drawback is that it would not give us this information if we decided to look at our summary report before the run ended. This looked like it could be a lot of work, so we explored other options.

For our third approach, we decided to consider adding additional output statistics to the Statistic data module. This would not allow you to collect additional statistics; however, this does allow you to calculate additional output values on statistics already being collected by the model. Since the model is currently collecting all the information we need, although not in the correct form, we will use this option to output information on part status. First we entered a Replication Length of 50,000 time units in the *Run > Setup > Replication Parameter* dialog box.

Set Operator Resource	Number Obs	Average Time	Standard Percent	Restricted Percent
IDLE	2,166	9.1514	39.64	39.64
WS 1 Setup	1,668	9.0429	30.17	30.17
WS 3 Setup	1,666	9.0603	30.19	30.19

WS 1 Resource	Number Obs	Average Time	Standard Percent	Restricted Percent
Blocked	76	19.0329	2.89	2.89
Failed	54	47.5478	5.14	5.14
Processing	1,668	15.0505	50.21	50.21
Setup	1,668	9.0429	30.17	30.17
Starved	2	6.5662	0.03	0.03
Waiting for setup	662	8.7386	11.57	11.57

WS 2A Resource	Number Obs	Average Time	Standard Percent	Restricted Percent
Blocked	616	16.1455	19.89	19.89
BUSY	1	28.8279	0.06	0.06
Failed	49	30.3818	2.98	2.98
Processing	714	49.3767	70.51	70.51
Starved	205	16.0090	6.56	6.56

WS 2B Resource	Number Obs	Average Time	Standard Percent	Restricted Percent
Blocked	422	24.3547	20.56	20.56
Failed	51	22.7702	2.32	2.32
Processing	522	67.4234	70.39	70.39
Starved	177	19.0171	6.73	6.73

WS 3 Resource	Number Obs	Average Time	Standard Percent	Restricted Percent
Failed	59	40.4641	4.77	4.77
Processing	1,666	12.2982	40.98	40.98
Setup	1,666	9.0603	30.19	30.19
Starved	636	11.2011	14.25	14.25
Waiting for setup	507	9.6755	9.81	9.81

Figure 9-11. The Tightly Coupled System Frequencies Report

If we ran our model at this point, we'd get the results shown in Figure 9-11. Let's temporarily focus our attention on the frequency statistics for WS 1 Resource. It tells us that the workstation was processing parts only 50.21% of the time, with a large amount of non-productive time spent in the Waiting for setup and Setup states.

9.5.3   Model 9-6: Adding Part-Status Statistics

Before we continue with our model development, let's describe what we need to do in order to obtain the desired information on part status. What we want is the percent of time that parts spend in each of the previously defined part states: Travel to WS 1, Wait for WS 1, Wait for Setup at WS 1, Setup at WS 1, Process at WS 1, Blocked at WS 1, Wait for WS 2, Process at WS 2, Blocked at WS 2, Wait for Setup at WS 3, Setup at WS 3, and Process at WS 3.

All the information we need to calculate these values is already contained in our summary output. Let's consider our first part state, Travel to WS 1. The average delivery time per part was 12.517, tallied for a total of 1672 parts. The cycle time was 225.07 for 1665 parts. You might note that there were still seven parts (1672-1665) in the system when the simulation terminated. We'll come back to this later. So if we want the percent of time an average part spent traveling to Workstation 1, we could calculate that value with the following expression:

```
((12.517 * 1672) / (225.07 * 1665)) * 100.0
```

or 5.5848%. We can use this approach to calculate all of our values. Basically, we compute the total amount of part time spent in each activity, divide it by the total amount of part time spent in all activities, and multiply by 100 to obtain the values in percentages. Since the last two steps of this calculation are always the same, we will first define an expression, `Tot`, to represent this value (Expression data module). Note that by using an expression, it will be computed only when required. That expression is as follows:

```
TAVG(Cycle Time)  *  TNUM(Cycle Time)/100
```

TAVG and TNUM are Arena variables that return the current average of a Tally and the total number of Tally observations, respectively. The variable argument is the Tally ID. In this case, we have elected to use the Tally name as defined in our Record module. In cases where Arena defines the Tally name (for example, for time-in-queue tallies), we recommend that you check a module's drop-down list or Help for the exact name.

The information required to calculate three of our part states is contained in Tallies: `Travel to WS 1`, `Wait for WS 1`, and `Wait for WS 2`. The expressions required to calculate these values are as follows:

```
TAVG(Delivery Time) *  TNUM(Delivery Time)/Tot
TAVG(Seize WS 1.Queue.WaitingTime) *
    TNUM(Seize WS 1.Queue.WaitingTime)/Tot
TAVG(Seize WS 2.Queue.WaitingTime) *
    TNUM(Seize WS 2.Queue.WaitingTime)/Tot
```

The information for the remaining part states is contained in the frequency statistics.

As you would expect, Arena also provides variables that will return information about frequencies. The Arena variable FRQTIM returns the total amount of time that a specified resource was in a specified category, or state. The complete expression for this variable is

```
FRQTIM(Frequency ID, Category)
```

The Frequency ID argument is the frequency name. The Category argument is the category name. Thus, our expression for `Setup at WS 1` becomes

```
FRQTIM(WS 1 Resource, Setup)/Tot
```

where `WS 1 Resource` is the name of our previously defined frequency statistic (defined in the Statistic data module) and `Setup` is the category name for our setup state for the `WS 1` resource (defined in the StateSet data module).

If we define all our expressions in this manner, the nine remaining expression are:

```
Wait for Setup at WS 1      FRQTIM(WS 1 Resource, Waiting for Setup)/
                               Tot
Setup at WS 1               FRQTIM(WS 1 Resource, Setup)/Tot
Process at WS 1             FRQTIM(WS 1 Resource, Processing)/Tot
Blocked at WS 1            FRQTIM(WS 1 Resource, Blocked)/Tot
Process at WS 2             FRQTIM(WS 2A Resource, Processing)/Tot +
                           FRQTIM(WS 2B Resource, Processing)/Tot
Blocked at WS 2            FRQTIM(WS 2A Resource, Blocked)/Tot +
                           FRQTIM(WS 2B Resource, Blocked)/Tot
Wait for Setup at WS 3      FRQTIM(WS 3 Resource, Waiting for Setup)/
                               Tot
Setup at WS 3              FRQTIM(WS 3 Resource, Setup)/Tot
Process at WS 3            FRQTIM(WS 3 Resource, Processing)/Tot
```

You should note that there could be two parts being processed or blocked at Workstation 2. Thus, we have included terms for both the 2A and 2B resources at Workstation 2.

Now that we have developed a method, and the expressions, to calculate the average percent of time our parts spend in each state, we will use the Statistic data module to add this information to our summary report. The new statistics that were added to the Statistic data module are shown in Display 9-21.

Before we proceed, let's address the discrepancy between the number of observations for our Delivery Time and Cycle Time statistics on our summary report. Because of the methods we use to collect our statistics, and the fact that we start our simulation with the system empty and idle, this discrepancy will always exist. There are several ways to deal with it. One method would be to terminate the arrival of parts to the system and let all parts complete processing before we terminate the simulation run. Although this would yield statistics on an identical set of parts, in effect we're adding a shutdown period to our simulation. This would mean that we would have potential transient conditions at both the start and end of the simulation, which would affect the results of our resource statistics.

We could increase our run length until the relative difference between these observations would become very small, thereby reducing the effect on our results. Although this could easily be done for this small textbook problem, it could result in unacceptably long run times for a larger problem. If we wanted to be assured that all part-state statistics

6	Travel to WS 1	Output	TAVG(Delivery Time)*TNUM(Delivery Time)/TOT	Travel to WS 1
7	Wait for WS 1	Output	TAVG(Seize WS 1.Queue.WaitingTime)*TNUM(Seize WS 1.Queue.WaitingTime)/Tot	Wait for WS 1
8	Wait for WS 2	Output	TAVG(Seize WS 2.Queue.WaitingTime)*TNUM(Seize WS 2.Queue.WaitingTime)/Tot	Wait for WS 2
9	Wait for Setup at WS 1	Output	FRQTIM(WS 1 Resource,Waiting for Setup)/Tot	Wait for Setup at WS 1
10	Setup at WS 1	Output	FRQTIM(WS 1 Resource,Setup)/Tot	Setup at WS 1
11	Process at WS 1	Output	FRQTIM(WS 1 Resource,Processing)/Tot	Process at WS 1
12	Blocked at WS 1	Output	FRQTIM(WS 1 Resource,Blocked)/Tot	Blocked at WS 1
13	Process at WS 2	Output	FRQTIM(WS 2A Resource,Processing)/Tot +FRQTIM(WS 2B Resource,Processing)/Tot	Process at WS 2
14	Blocked at WS 2	Output	FRQTIM(WS 2A Resource,Blocked)/Tot + FRQTIM(WS 2B Resource,Blocked)/Tot	Blocked at WS 2
15	Wait for Setup at WS 3	Output	FRQTIM(WS 3 Resource,Waiting for Setup)/Tot	Wait for Setup at WS 3
16	Setup at WS 3	Output	FRQTIM(WS 3 Resource,Setup)/Tot	Setup at WS 3
17	Process at WS 3	Output	FRQTIM(WS 3 Resource,Processing)/Tot	Process at WS 3

Display 9-21. The Summary Statistics for Part States

Output	Value
Blocked at WS 1	0.3881
Blocked at WS 2	5.4040
Process at WS 1	6.6783
Process at WS 2	18.7632
Process at WS 3	5.4595
Setup at WS 1	4.0051
Setup at WS 3	4.0136
Travel to WS 1	5.5551
Wait for Setup at WS 1	1.5395
Wait for Setup at WS 3	1.3023
Wait for WS 1	36.5635
Wait for WS 2	10.3425

Figure 9-12. The Appended Summary Report

were based on the same set of parts, we could collect the times each part spends in each activity and store these times in entity attributes. When the part exits the system, we could then Tally all these times. Although this is possible, it would require that we make substantial changes to our model, and we would expect that the increased accuracy would not justify these changes.

If you step back for a moment, you should realize that the problem of having summary statistics based on different entity activities is not unique to this problem. It exists for almost every steady-state simulation that you might construct. So we recommend that in this case you take the same approach that we use for the analysis of steady-state simulations. Thus, we would add a Warm-up Period to eliminate the start-up conditions. Because our system is tightly coupled, it will not accumulate large queues, and the number of parts in the system will tend to remain about the same. Because the warehouse delay is modeled as exponential, it is possible that there could be no parts in the system, although this is highly unlikely. At the other extreme, there can be a maximum of only eight parts in the system. Thus, by adding a warm up to our model, we will start collecting statistics when the system is already in operation. This will reduce the difference between the number of observations for our Tallies. For now, let's just assume a Warm-up Period of 500 minutes and edit the *Run > Setup > Replication Parameters* dialog box to include that entry. If this were a much larger system that could accumulate large queues, you might want to reconsider this decision.

If you now run the model, the information shown in Figure 9-12 will be added to the summary report.

9.6   A Few Miscellaneous Modeling Issues

Our intention in writing this tome was not to attempt to cover all the functions available in the Arena simulation system. There are still a few modules in the Basic Process, Advanced Process, and Advanced Transfer panels that we have not discussed. And there are many more in the Blocks and Elements panels that we have neglected entirely. However, in your spare time, we encourage you to attach the panels that you seldom use and place some modules. Using the Help features will give you a good idea of what we have omitted. In most cases, we suspect that you'll never need these additional features. Before we close this chapter, we'd like to point out and briefly discuss a few features we have not covered. We don't feel that these topics require an in-depth understanding— only an awareness.

9.6.1 Guided Transporters

There is an entire set of features designed for use with guided transporters. These features are useful not only for modeling automated guided vehicle (AGV) systems, but also for warehousing systems and material-handling systems that use the concept of a moving cart, tote, jig, fixture, etc. Interestingly, they're also great for representing many amusement-park rides. Because this topic would easily fill an additional chapter or two, we've chosen not to present it in this book. However, you can find a complete discussion of these features in Chapter 9 of Pegden, Shannon, and Sadowski (1995).

9.6.2 Parallel Queues

There are also two very specialized modules from the Blocks panel, QPick and PickQ, that are seldom used; but if you need them, they can make your modeling task much easier. The QPick module can be used to represent a process where you want to pick the next entity for processing or movement from two or more different queues based on some decision rule. Basically, the QPick module would sit between a set of detached Queue modules and a module that allocates some type of scarce resource (for example, Allocate, Request, Access, or Seize modules). Let's say that you have three different streams of entities converging at a point where they attempt to seize the same resource. Furthermore, assume that you want to keep the entities from each stream separate from each other. The modules required for this part of your model are shown in Figure 9-13. Each entity stream would end by sending the entity to its Detached queue (more on a Detached queue shortly). The link between the QPick and Queue modules is by module Labels. When the resource becomes available, it will basically ask the QPick module to determine from which queue to select the next entity to be allocated the resource. Also note that you must use the Seize module from the Blocks panel for this to work, not the Seize module from the Advanced Process panel. When using modules from the Blocks panel, Arena does not automatically define queues, resources, etc. Thus, you may have to place the corresponding modules from the Basic Process or Elements panels to define these objects.

The PickQ module can be used to represent a process where you have a single arrival stream, and you will use some decision rule to pick between two or more queues in which to place the entity. Let's assume that you have a stream of arriving entities that are to be loaded onto one of two available conveyors. The modules required for this part of your model are shown in Figure 9-14 (the Access modules are from the Advanced Transfer panel). Note that you can't specify internal queues for the Access modules, and the decision as to which conveyor the entity is directed to is based on characteristics of the queue, not the conveyors. The PickQ module can be used to direct entities to Access, Seize, Allocate, Request, or Queue modules, or any module that is preceded by a queue.

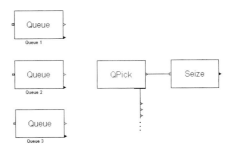

Figure 9-13. Using a QPick Module

Now let's address the issue of a Detached queue. If you use the Queue module from the Blocks panel, you're given the option of defining the queue as being Detached. This means that the queue is not directly linked to a downstream module. Such a queue can be indirectly linked, as was the case with the QPick module, or there might be no obvious link. For example, you may want to use a set of entities whose attribute values hold information that you might want to access and change over the course of a simulation. If this is all you want to do, it might be easier to use a Variable defined as a matrix, or perhaps an external database. The advantage of using a queue is that you can also change the ranking of the entities in the queue. You can access or change any entity attribute values using Arena variables. You can also use the Search, Copy, Insert, Pickup, and Remove modules from the Blocks panel (or the Search, Remove, Pickup and Dropoff modules from the Advanced Process panel) to interact with the entities in the queue. Note that if queue ranking is important, Arena only ranks an entity that enters the queue. Thus if you change an entity attribute that is used to rank the queue, you must remove the entity from the queue and then place it back into the queue.

9.6.3 Decision Logic

There are situations where you may require complex decision logic based on current system conditions. Arena provides several modules in the Blocks panel that might prove useful. The first is a set for the development of if-then-else logic. The If, ElseIf, Else, and EndIf modules can be used to develop such logic. We will not explain these modules in

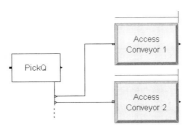

Figure 9-14. Using a PickQ Module

detail here, but we encourage you to use Help if you implement logic with these modules. Although these modules will allow you to develop powerful logic, you need to be very careful to ensure that your logic is working correctly. These modules are designed to work only when all of the modules between an If and its matching EndIf (including any ElseIf or Else modules inside) are graphically connected in Arena. This allows you to use many of Arena's modules inside If/EndIf logic, such as Assign, Seize, and Delay, but precludes use of those that don't permit graphical connections, such as Route, Convey, and Transport. There is also a set of modules to implement do-while logic: the While and EndWhile modules. The same warnings given previously also apply for these modules.

If you really need to implement this type of logic, there are several options. The easiest is to use the Label and Next Label options to connect your modules rather than direct graphical connections. Although this works, it doesn't show the flow of logic. An alternative is to write your logic as an external SIMAN *.mod* file and use the Include module from the Blocks panel to include this logic in your module. Unfortunately, this option is not available for use with the academic version software. The safest and most frequently used option is to use a combination of Decide and Branch modules to implement your logic. This always works, although the logic may not be very elegant.

9.7   Exercises

9-1   Packages arrive with interarrival times distributed as EXPO(0.46) minutes to an unloading facility. There are five different types of packages, each equally likely to arrive, and each with its own unload station. The unloading stations are located around a loop conveyor that has room for only two packages in the queue at each unload station. Arriving packages enter an infinite-capacity queue to wait for space on the loop conveyor, and 2 feet of conveyor space is required per package. Upon entering the loop conveyor, the package is conveyed to its unload station. If there is room in the queue at its unload station, the package is automatically diverted to the queue, which takes no time. The package then waits for a dedicated unloader (dedicated to this unload station, not to this package) to be unloaded, which takes time distributed as EXPO(2) minutes. If a package arrives at its station and finds the queue full, it is automatically conveyed around the loop and tries the next time. Each station is located 10 feet from the next, and the conveyor speed is 12 feet per minute. Animate your model.

(*a*)   Run the simulation for 500 minutes, collecting statistics on the time in system (all package types together, not separated out by package type), and the number of packages in queue at the package-arrival station.

(*b*)   Which will have more impact on average time in system: increasing the conveyor speed to 15 feet per minute or increasing each unloading-queue capacity to 4?

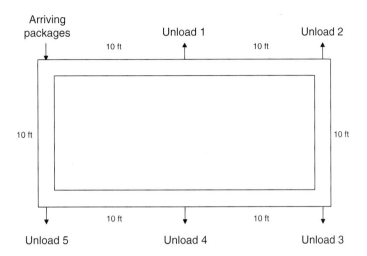

9-2   A merging conveyor system has a main conveyor consisting of three segments, and two spur conveyors, as depicted in the figure below. Separate streams of packages arrive at the input end of each of the three conveyors, with interarrival times distributed as EXPO(0.75) minutes. Incoming packages queue to wait for space on the conveyor. Packages arriving at the input end of the main conveyor are conveyed directly to the system exit. Packages arriving at the two spur lines are conveyed to the main conveyor, where they wait for space. Once space is available, they exit the spur line and are conveyed to the system exit. All conveyor segments are 20 feet (the main conveyor has three such segments) and all three conveyors are accumulating and move at 15 feet per minute. Each package requires 2 feet of space on a conveyor. When packages reach the exit-system point, there is an unload time of 0.2 minute, during which time the package retains it space on the main conveyor.

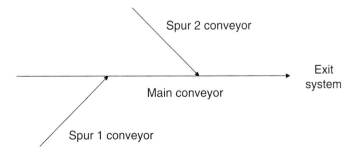

Develop a model and animation of this system and run it for 300 minutes, observing statistics on conveyor status (such as number of packages conveying and accumulated on each segment) as well as times in system for packages.

9-3   A small automated power-and-free assembly system consists of six workstations. (A power-and-free system could represent things like tow chains and hook lines.) Parts are placed on pallets that move through the system and stop at each workstation for an operation. There are 12 pallets in the system. A blank part is placed on an empty pallet as part of the operation at Workstation 1. The unit (part and pallet) is then moved progressively through the system until all the operations are completed. The final assembled part is removed from the pallet at Workstation 6 as part of the operation there. The power-and-free system moves at 4 feet per minute. Each pallet requires 2 feet of space. The distance between adjacent workstations is 10 feet, except that Workstation 6 (the final assembly operation) and Workstation 1 (the beginning assembly operation) are 20 feet apart. The figure below indicates how the workstations are arranged. The operation times at each workstation are WEIB(3.31, 4.2) minutes.

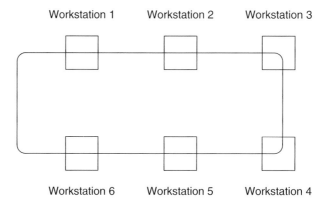

Develop a simulation and animation of this system and run it for 1,500 minutes, observing statistics on the production per hour. (HINT: Model the power-and-free system as an accumulating conveyor. At the start of your simulation, load the empty pallets at Workstation 6. These pallets become a permanent part of the system.) Look at the effect on hourly production of the number of pallets. Would more (or fewer) than 12 pallets be preferable? Is there something like an optimal number of pallets? Be sure to back up your statements with a valid statistical analysis.

9-4   A small production system has parts arriving with interarrival times distributed as TRIA(6, 13, 19) minutes. All parts enter at the dock area, are transported to Workstation 1, then to Workstation 2, then to Workstation 3, and finally back to the dock, as indicated in the figure below. All part transportation is provided by two carts that each move at 60 feet per minute. The distances from the dock to each of Workstations 1 and 3 are 50 feet, and the distances between each pair of workstations are also 50 feet. Parts are unloaded from the cart at Workstations 1 and 3 for processing, but parts get processed on the cart at Workstation 2. Processing times are $8 + \text{WEIB}(4.4, 4.48)$ minutes, $\text{UNIF}(8, 11)$ minutes, and $\text{TRIA}(8, 11, 18)$ minutes for Workstations 1, 2, and 3, respectively. Assume that all load and unload times are negligible.

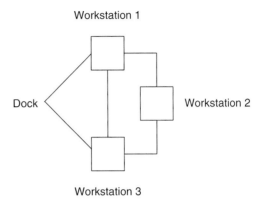

Develop a simulation and animation of this system and run it for 10,000 minutes, observing statistics on cart and workstation utilizations, as well as part cycle times.

9-5   A special-order shop receives orders arriving with interarrival times distributed as $\text{EXPO}(30)$—all times are in minutes. The number of parts in each order is a $\text{UNIF}(3, 9)$ random variable (truncated to the next smallest integer). Upon receiving the order, the parts are immediately pulled from inventory and sent to the prep area (this transfer takes zero time) where they undergo an individual prep operation, the time for which is distributed as $\text{TRIA}(2, 3, 4)$. After the prep operation, the parts are transferred (this transfer takes 4 minutes) to a staging area to wait for the final order authorization. The final order authorization takes an amount of time distributed as $\text{UNIF}(180, 240)$, after which the parts are released to be processed, which requires an amount of time distributed as $\text{TRIA}(3, 4, 6)$. After processing, the parts are assembled into a batch and sent to the packer (zero transfer time) to be packed, which takes an amount of time distributed as $\text{TRIA}(8, 10, 14)$ for the batch. The packed parts exit the system. During the order-authorization process, 4% of the orders are canceled. These parts are removed from the staging queue and sent back to inventory (zero transfer time). Develop a simulation model for this system and run it for 20,000 minutes with a warm-up of 500 minutes. Observe statistics on the number of canceled orders, the number of canceled parts, the time in system for shipped orders, and resource utilizations. Also, use Frequencies to determine the number of racks required to hold the parts in the staging area (each rack can hold 25 parts). (HINTS: Use Hold/Signal for order authorization and Search/ Remove to cancel orders.)

9-6   A food-processing system starts by processing a 25-pound batch of raw of product, which requires $1.05 + \text{WEIB}(0.982, 5.03)$ minutes. Assume an infinite supply of raw product. As soon as a batch has completed processing, it is removed from the processor and placed in a separator where it is divided into one-pound units. The separation process requires 0.05 minute per pound. Note that this is not a "batch process," but rather in each 0.05 time unit a one-pound unit is completed. These one-pound units are sent to one of the three wrapping machines. Each wrapping machine has its own queue, with the unit being sent to the shortest queue. The wrappers are identical machines, except for the processing times. The wrapping times are constants of 0.20, 0.22, and 0.25 minute for wrappers 1, 2, and 3, respectively. The wrapped products are grouped, by wrapper, into batches of six for final packaging. Once a group of six items is available, that group is sent to a packer that packs the product, which takes $0.18 + \text{WEIB}(0.24, 4.79)$ minutes. These packages exit the system. Assume all transfer times are negligible.

  (**a**)   Develop a simulation and animation of this system and run it for 1,000 minutes, observing statistics on resource utilizations, queues, and the total number of packages shipped.

  (**b**)   Modify your model to include wrapper failures with the following characteristics for all wrappers: EXPO(20) for uptimes and EXPO(1) for repair times.

  (**c**)   Add a quality check to your model from part (b). Every half hour, a scan is made, starting with the first wrapper queue, for products that are more than 4 minutes old; that is, that exited the processor more than 4 minutes ago. Any such items are removed from the queue and disposed. It takes 3 seconds to find and remove each item. Keep track of the number of units removed.

9-7   A small automated system in a bakery produces loafs of bread. The dough-making machine ejects a portion of dough every UNIF(0.5, 1.0) minute. This portion of dough enters a hopper to wait for space in the oven. The portions will be ejected from the hopper in groups of four and placed on the oven-load area to wait for space on the oven conveyor. There is room for only one group of portions at the oven-load area. The oven conveyor has 10 buckets, each 1 foot long, and moves at 0.35 feet per minute.

  (**a**)   Develop your model using the Hold (Scan for Condition), Signal, and Hold (Wait for Signal) modules for your logic to control the group of portions. Run it for 3,000 minutes and observe statistics on the number of portions in the hopper, oven utilization, and the loaf output per hour.

  (**b**)   Modify your model from part (a) by replacing the Hold (Scan for Condition) module with logic developed based on the Decide module.

9-8   Customers arrive, with interarrival times distributed as EXPO(5)—all times are in minutes—at a small service center that has two servers, each with a separate queue. The service times are EXPO(9.8) and EXPO(9.5) for Servers 1 and 2, respectively. Arriving customers join the shortest queue. Customer line switching occurs whenever the difference between the queue lengths is 3 or more. At that time, the last customer in the longer queue moves to the end of the shorter queue. No additional movement, or line switching, in that direction occurs for at least the next 30 seconds. Develop a model and animation of this system and run it for 10,000 minutes. Observe statistics on the number of line switches, resource utilization, and queue lengths.

9-9 A small cross-docking system has three incoming docks and four outgoing docks. Trucks arrive at each of the three incoming docks with interarrival times distributed as UNIF(35, 55)—all times are in minutes. Each arriving truck contains UNIF(15, 30) pallets (truncated to the next smaller integer), which we can assume are unloaded in zero time. Each pallet has an equal probability of going to any of the four outgoing docks. Transportation across the dock is provided by three fork trucks that each travel at 60 feet per minute. Assume that the distance between any incoming dock and any outgoing dock is 50 feet. Also assume that the distance between adjacent incoming docks (and adjacent outgoing docks) is 15 feet.

(*a*)   Develop a model in which the fork trucks remain where they drop off their last load if there are no new requests pending.

(*b*)   Modify your model so that free fork trucks are all sent to the middle (Dock 2) incoming dock to wait for their next load.

(*c*)   Modify your model so that each fork truck is assigned a different home incoming dock and moves to that dock when there are no additional requests pending.

Compare the results of the above three systems, using the pallet system time as the primary performance measure. Be sure to back up your comparison with a proper statistical analysis.

9-10 Develop a model and animation of a Ferris-wheel ride at a small, tacky county fair. Agitated customers (mostly small, over-sugared kids who don't know any better) arrive at the ride with interarrival times distributed as EXPO(3) minutes and enter the main queue. When the previous ride has finished and the first customer is ready to get off, the next five customers (or fewer if there are not five in the main queue) are let into the ride area. As a customer gets off the Ferris wheel, the new customer gets on. Note that there can be more current riders than new customers, or more new customers than current riders. It requires UNIF(0.05, 0.15) minute to unload a current rider and UNIF(0.1, 0.2) to load a new rider. The Ferris wheel has only five single seats spaced about 10 feet apart. The wheel rotates at a velocity of 20 feet per minute. Each customer gets five revolutions on the wheel, and no new riders are allowed to board until the ride is complete. Riders who get off the wheel run to the exit, which takes 2 minutes. We're are not sure if they want to get away or to be first in line at the next tacky ride. (HINTS: Use a conveyor to represent the Ferris wheel itself. Use Wait/Signal to implement the loading and unloading rules.) By the way, the Ferris wheel was developed by the American engineer G.W.G. Ferris, who died in 1896; its German name is *Riesenrad*, which translates roughly as "giant wheel."

CHAPTER 10

Arena Integration and Customization

In this chapter, we introduce the topics of integrating Arena models with other applications and building customized Arena modules. We will illustrate these concepts with a very simple model of a call center (much simpler than the one introduced in Chapter 5).

Our first topic, in Section 10.1, presents a model in which we read scheduled arrival times from an external file and then write performance data to a file. This will illustrate several different ways to incorporate data from outside sources (such as a text file) into an Arena model. In Section 10.2, we introduce two Microsoft® Windows® operating system technologies that Arena exploits for integrating directly with other programs—ActiveX®Automation and Visual Basic® for Applications (VBA)—and we describe how VBA has been incorporated into Arena. We present this material with the assumption that either you are familiar with VBA programming or you will access other sources to become so; our treatment of this topic focuses on what Arena provides to put VBA to use. Then in Section 10.3, we show you how these technologies can be used to create a custom user interface. Section 10.4 continues on the topic of VBA, enhancing the call-center model to record individual call data and to chart the call durations in Microsoft® Excel. We close this chapter in Section 10.5 with an overview of how you can design your own modules to augment Arena's standard modeling constructs. When you finish this chapter, you should have an idea of the types of features you can employ to integrate Arena with other desktop applications and a few ways to create custom Arena interfaces.

10.1 Model 10-1: Reading and Writing Data Files

We'll start with a very simple (trivial, in fact) model of a call center. Then we'll enhance that model in several interesting ways. Our call center has a single arrival stream of randomly generated calls and a single agent processing calls, after which the calls are disposed. The call-center manager has estimated that the calls come in an average of 1.1 minutes apart according to an exponential distribution and that the call durations tend to be around 0.75 minute with a range of 0.3 to 1.1 minutes. To model the system, we use a single Create module, a single Process module, and a single Dispose module as shown in Figure 10-1.

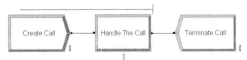

Figure 10-1. Simple Call-Center Model

The data for the three modules are shown in Display 10-1 and the Replication Parameters should be set in *Run > Setup > Replication Parameters* as illustrated in Display 10-2. After creating this model, we can run and view the results.

Although the call-center model is quite simple, we have a few opportunities to make it more realistic. The call-center manager has just informed us that he has some historical call-arrival times for a typical day. Let's use that record of the actual calls received over a period of time to generate entities into our model instead of creating entities based on sampling from a probability distribution. We might take this approach as part of the validation of our model. If the results of a simulation run that's driven by the recorded arrivals from the actual system closely correspond to the actual system performance over that period of time, then we have greater faith in the correctness of our model logic. Or we might use this logic because we want to perform runs based on a specific pattern of arrivals, such as a fixed but irregular schedule of trucks arriving to loading docks at a distribution center.

We'll start in Section 10.1.1 with the easy case of reading our arrival data from a text file. Then in Section 10.1.2, we'll look at reading similar data from other sources.

(Create module)	
Name	`Create Call`
Entity Type	`Arrival Entity`
Type	`Random (Expo)`
Value	`1.1`
Units	`Minutes`
(Process module)	
Name	`Handle The Call`
Action	`Seize Delay Release`
Type	`Resource`
Resource Name	`Agent`
Delay Type	`Triangular`
Units	`Minutes`
Minimum	`0.3`
Value	`0.75`
Maximum	`1.1`
(Dispose module)	
Name	`Terminate Call`

Display 10-1. The Create, Process, and Dispose Modules

| Replication Length | `8` |
| Time Units | `Hours` |

Display 10-2. Replication Parameters

```
1.038
2.374
4.749
9.899
10.52
17.09. . .
```

Figure 10-2. Call-Time Data for Revised Call-Center Model

10.1.1 Model 10-2: Reading Entity Arrivals from a Text File

To make this modification to the call-center model, we'll need a file containing the arrival times over the period to be studied, and we'll need to replace the entity-creation model logic to use the data in this file. To simplify the required logic, we'll structure the data file to be an ASCII text file containing values that correspond to simulation times, assuming our run starts at time 0 minute. The first few values of this file (`Model 10-02 Input.txt`) are shown in Figure 10-2. Since we have the luxury of being textbook authors rather than real-life problem solvers, we'll skip over the details of how this ASCII text file was created. In practice, you'll probably find that the information you can obtain isn't quite so conveniently stored, but with creative application of spreadsheet or database software, you usually can export the raw data points into simulation-ready values in a text file.

We need to decide how to use the historical data stored in the text file. We are faced with two issues: the mechanics of transferring data from the file into the model, and then how to use the values to generate entities at the appropriate times. We'll first look at the logic required to create entities at the appropriate times, covering the details of reading the data when we reach that part of the model logic.

Thus far, our approach for creating entities has been to use a Create module, which makes a new entity every so often throughout the run based on the interarrival time. We saw in our hand simulation (remember the pain of Chapter 2?) that at each entity creation, the "current arriving" entity is sent into the model and the "next arriving" entity is placed on the calendar to arrive at the appropriate future time. The data in the Time Between Arrivals section of the dialog box determine how far into the future the next arriving entity is to be scheduled; most often, this involves sampling from a probability distribution.

To establish the call arrivals from our data file, however, we can't formulate a simple expression for the time between arrivals. Instead, we'll use a control entity that mimics the "current arriving" and "next arriving" entity logic directly in our model, as shown in Figure 10-3. The first four modules shown there replace the Create module from our original call-center model.

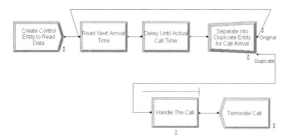

Figure 10-3. Logic for Generating Entities from a File

The Create module will generate just a single entity, as shown in Display 10-3. For this Create module, Arena will create a single entity at the beginning of each replication, then will "turn off" the entity creation stream, since it reached the number of entity creations specified in the Max Arrivals field (namely, 1).

The created entity enters the ReadWrite module, per Display 10-4, where it reads the next value from the data file and assigns this value to the Call Start Time entity attribute. The ReadWrite module, which can be found on the Advanced Process panel, reads one or more values from a source external to Arena and assigns these values to model variables. The Arena File Name, Arrivals File, is used as a *model* identifier for the file. It shouldn't be confused with the actual name of the file on your hard disk (or wherever the file is stored), which we'll define later in the File data module.

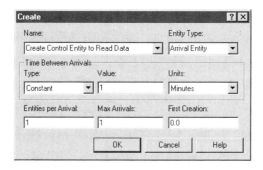

Name	Create Control Entity to Read Data
Entity Type	Arrival Entity
Type	Constant
Value	1
Units	Minutes
Max Arrivals	1

Display 10-3. The Modified Create Module

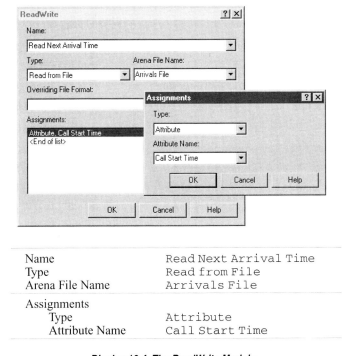

Name	Read Next Arrival Time
Type	Read from File
Arena File Name	Arrivals File
Assignments	
Type	Attribute
Attribute Name	Call Start Time

Display 10-4. The ReadWrite Module

After the entity reads the value from the data file, it moves to the Delay module (Display 10-5) to wait until time `Call Start Time`, so that the actual entity representing the call will arrive at the model logic at the appropriate time. Because the values in the data file represent the absolute number of minutes from the beginning of the run for each call rather than the interarrival times, the Delay Time is specified as `Call Start Time - TNOW` so that the entity will delay until the time stored in the `Call Start Time` attribute.

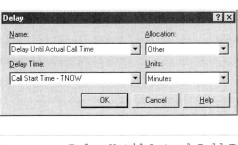

Name	Delay Until Actual Call Time
Delay Time	Call Start Time - TNOW
Units	Minutes

Display 10-5. The Delay Module

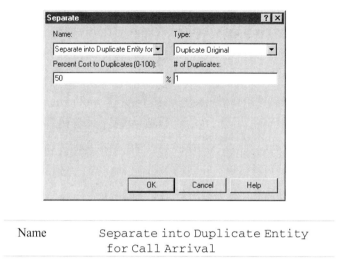

Name	Separate into Duplicate Entity for Call Arrival

Display 10-6. The Separate Module

When the control entity completes the delay, it's time to create the actual call entity and dispatch it to the system logic. The Separate module from the Basic Process panel works nicely for this need, shown in Display 10-6. It sends the control entity (the module exit point labeled Original) back to the ReadWrite module to obtain the next call time from the data file and creates one duplicate of itself that's sent into the model logic via the exit point labeled Duplicate (as previously shown in Figure 10-3). This duplicate entity represents the arrival of a new call, which will move through the rest of the model logic to process the call.

Our final change to the model is to specify the required information about the data file. To do so, we edit the File data module found in the Advanced Process panel, as in Display 10-7. When we typed `Arrivals File` as the Arena File Name in the ReadWrite module, Arena automatically created a corresponding entry in the File data module. To complete the required information, we enter `Model 10-02 Input.txt` as the Operating System File Name. This tells Arena to access `Model 10-02 Input.txt`, stored wherever the model file is located, whenever a ReadWrite module references the Arena `Arrivals File`. (You could give a full path for the file, such as `C:\My Documents\Cool Stuff\Model 10-02 Input.txt`, but if you would decide to send the model and data file to someone else, they'd have to have the same folder structure. Depending on your taste in folder names, this might or might not be a good idea.) We left the other options at their default values, including the file type as `Free Format`, which indicates that the `Model 10-02 Input.txt` file contains text values. We also retained the default end-of-file action as `Dispose` so that the control entity will be disposed of after it reads the last value in the file, effectively terminating the arrival stream of entities to the model.

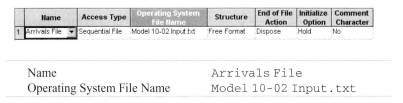

	Name	Access Type	Operating System File Name	Structure	End of File Action	Initialize Option	Comment Character
1	Arrivals File	Sequential File	Model 10-02 Input.txt	Free Format	Dispose	Hold	No

Name	Arrivals File
Operating System File Name	Model 10-02 Input.txt

Display 10-7. The File Data Module

Using the data values in Figure 10-2 to step through the logic for the first two calls, our control entity will first read a value of 1.038 into its `Call Start Time` attribute after it's created at simulation time 0 (the start of the replication). It will delay for 1.038 – 0 time units, leaving the Delay module at time 1.038. It creates a duplicate at that time, sending it to the Queue module to begin the actual processing of a call. The control entity returns to the ReadWrite module, reading a value of 2.374 from the data file and over-writing its `Call Start Time` attribute with this value. The control entity proceeds to the Delay module, where it delays for 2.374 – 1.038 = 1.336 time units, which causes Arena to place the control entity on the event calendar with an event time that's 1.336 time units into the future (or an actual event time of 2.374). When it's the control entity's turn to be processed again, at simulation time 2.374, Arena will remove it from the event calendar and send it to the Separate module, where it will spawn a call entity with `Call Start Time` attribute value 2.374, representing the second call to enter the system. This will continue until all of the values in `Model 10-02 Input.txt` have been read by the control entity.

The simulation run will terminate under one of two conditions. If the run's ending time specified in the Run Setup dialog box occurs before all of the calls listed in the data file have been created, Arena will end the run at that time; under no conditions can a model run for longer than is defined. On the other hand, our model may terminate earlier than the specified run length if all of the calls listed in the data file have been created and have completed processing, leaving the event calendar empty. (Remember that the control entity is disposed after it reads the last data value.) If Arena encounters a condition where there are no additional entities on the event calendar and no additional other time-based controls to process, it will terminate the run after the last entity leaves the model.

10.1.2 *Model 10-3 and Model 10-4: Reading and Writing Access and Excel Files*
What if instead of a text file the call-arrival data were in a Microsoft® Access database or Microsoft® Excel spreadsheet file? We'll explore how both of those can be handled easily.

Let's assume first that you were provided an Access database containing the call-arrival information, with a table named ArrivalTimes to store the data. The first few values of this file (`Model 10-03 Input.mdb`) are shown in Figure 10-4.

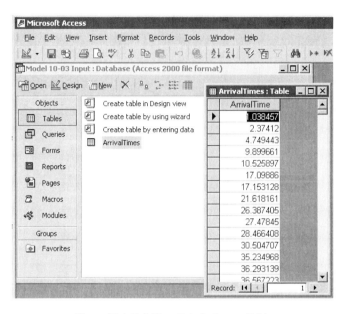

Figure 10-4. Call-Time Data in Access Table

Starting with `Model 10-02.doe` that we created in the last section, we only have to make a few small changes to read from the database. The model logic that we crafted to read and use those data to control entity creation is still useful. We need only to change some details about where we get the data.

This time we will start by describing the new data file. To do so, we'll edit the existing File data module found in the Advanced Process panel, as in Display 10-8. We can leave the Arena File Name set as `Arrivals File`. Recall that this is the name used inside model logic (in this case, the ReadWrite module) to specify this file—there is no reason to change this. But we do need to change the Operating System File Name to `Model 10-03 Input.mdb`.[1] We'll leave most of the other options at their default values, including the end-of-file action as `Dispose` so that the control entity will be disposed of after it reads the last value in the file as before.

You will notice that the Access Type field has a drop-down list of choices; this time, we'll choose `Microsoft Access (*.mdb)`. When you make this selection, you may see a difference in the display of fields or columns because database files require different information from text files. Specifically, you should see a new column labeled Recordsets above a button that initially says 0 rows. Clicking that button loads the Recordsets Editor as in Display 10-9.

---

[1] When using files with the extension *.mdb* (the default for Access files), be careful not to name the file the same as your model. Arena automatically stores its output in a file named `ModelName.mdb` and will be unhappy to see any file structure there except its own.

	Name	Access Type	Operating System File Name	End of File Action	Initialize Option	Recordsets
1	Arrivals File	Microsoft Access (*.mdb)	Model 10-03 Input.mdb	Dispose	Hold	0 rows
	Double-click here to add a new row.					

Name	Arrivals File
Access Type	Microsoft Access (*.mdb)
Operating System File Name	Model 10-03 Input.mdb

Display 10-8. The File Data Module

As you read from a database, you are reading a set of records from a table or range. A single database may have many different sets of records. Arena identifies each set by defining a Recordset Name. Each Recordset Name must be linked to a table in the database. You could select any valid symbol name for the Recordset Name, but for simplicity, we will type in the same as the table name, `ArrivalTimes`. Click the *Add/Update* button to complete the definition of the Recordset. Under Table Name, if you click on the arrow, you'll see a drop-down list of all tables currently defined in the database. In this case, select the only one defined—`ArrivalTimes`, as shown in Display 10-9. If you would like to view some sample data from a Recordset, first highlight the Recordset in the left column, then click on the *View* button. When you are done, click *Close* to close the viewer. Finally click *OK* to exit the Recordset Editor.

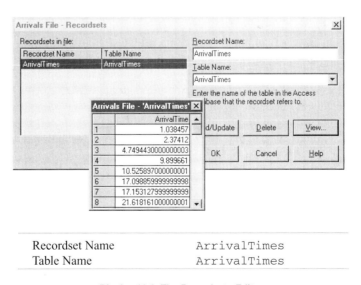

Recordset Name	ArrivalTimes
Table Name	ArrivalTimes

Display 10-9. The Recordsets Editor

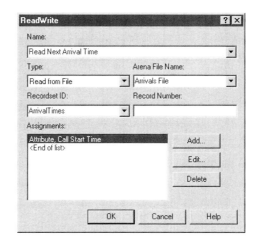

Recordset ID	ArrivalTimes

Display 10-10. The ReadWrite Module

You may wonder why we changed the file properties before we changed the ReadWrite module. If you open the ReadWrite module now (as in Display 10-10), you will see that it looks a bit different. Essentially, it tailors itself to collect the information required for the file type being read. In this case, it recognizes that we are reading from a database and that we need to identify the Recordset; we'll select `ArrivalTimes` from the drop-down list. If you want to control the record read in each iteration, enter an optional expression in the Record Number field. However, we'll leave it blank so it reads the records sequentially by default, starting with the first record. Note that we do not have to change the Arena File Name or the Assignments. Click *OK* and we are done.

That's it! Save your newly created model, or look for the `Model 10-03.doe` file included on your CD. We're now ready to run the model and have it read its data from the Access file instead of the text file.

If you stayed with us through the first part of this section, we have good news for you. You already know most of what's needed to read Excel worksheet files. You may notice that the rows and columns of an Excel spreadsheet closely resemble the rows and columns of a database table. Excel is not a relational database management system, but with a few caveats, we can treat an Excel worksheet much like we treat an Access database file.

In Excel, a rectangular named range—a set of cells (rows and columns)—is the equivalent of the rows and columns of an Access table. If we want to treat it like a database (which we do here), the named range must refer to a rectangular set of at least two cells. You can create an Excel named range by highlighting a set of cells and then select *Insert > Name > Define*, and type a name. For example, if you highlight cells A3 through C9 and follow the procedure above, naming your workbook selection `MyRange`, you

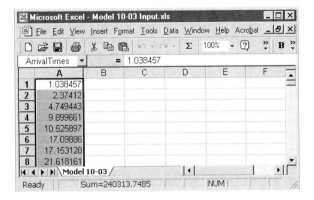

Figure 10-5. Call-Time Data in Excel Worksheet

would have three columns and seven rows that look much like an Access table. If you look at the drop-down list in the name box (typically, in the upper-left region of the menu bar), you'll find all of the names of cell ranges that you have defined. Again, this is equivalent to all of the tables in an Access database.

Let's resume with `Model 10-03.doe` that we just created above and modify it to read from Excel instead of Access. The sample data file provided, `Model 10-03 Input.xls`, is illustrated in Figure 10-5. You can see that the range `ArrivalTimes` is highlighted both in the name box and in the worksheet itself. Again, we'll start by describing the new data file in the existing File data module, as in Display 10-11. We can leave nearly everything set as it was for Access except the File Name and Access Type. Change the Operating System File Name to `Model 10-03 Input.xls`. On the Access Type drop-down list, choose `Microsoft Excel (*.xls)`.

Let's again click the *Recordsets* button to load the Recordsets Editor. We still have a Recordset named `ArrivalTimes` defined, but since we chose a new file, the link has been deleted. Select the previously defined Recordset on the left and let's restore the link. The drop-down list under Named Range will show you all of the named ranges defined in that file. As luck would have it (okay, we planned ahead), the Excel spreadsheet has a named range just like the table we used previously. Select `ArrivalTimes` and click *Add/Update* to record the change. If you would like to view some sample data from

	Name	Access Type	Operating System File Name	End of File Action	Initialize Option	Recordsets
1	Arrivals File	Microsoft Excel (*.xls)	Model 10-03 Input.xls	Dispose	Hold	1 rows
	Double-click here to add a new row.					

Name	`Arrivals File`
Access Type	`Microsoft Excel (*.xls)`
Operating System File Name	`Model 10-03 Input.xls`

Display 10-11. The File Data Module

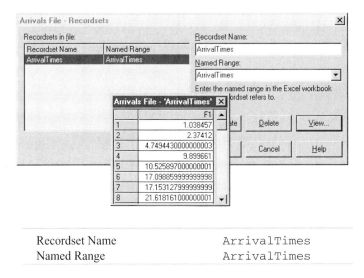

| Recordset Name | ArrivalTimes |
| Named Range | ArrivalTimes |

Display 10-12. The Recordsets Editor

a Recordset, first highlight the Recordset in the left column, then click the *View* button to see something similar to Display 10-12. When you are done, click the *Close* button to close the viewer. Finally, click *OK* to exit the Recordset Editor.

Since we have already revised the Read/Write module to read from ArrivalTimes Recordset, no additional changes are required. We are now ready to run the model, this time reading the data from the Excel file.

Now that you know how to read Access and Excel files, we are ready to write to them as well. There is basically one change necessary to write instead of read data. In the Type operand of the ReadWrite module, choose `Write to File` instead of `Read from File`. Everything else works just as described in the read information above. Of course, there is almost always a catch. Here, it's that when you are writing to Access or Excel files, the skeleton file must exist. ActiveX® Data Objects (ADO), which is a Microsoft technology that provides high-performance data access to a variety of data stores, gets its file structure and formatting information from the file. Specifically, the table or named range must already exist. An additional constraint for Excel files is that if the range is not formatted as numeric with some initial data, then all of the data will be written as strings. The simple solution is to put some formatted sample data in the range before writing. For more detailed information, look in the Arena help topic "Excel: Read/Write Limitations Using ADO."

We'll make one final enhancement to our call-center model to allow us to record the processing times in the same Excel file from which we read the Arrival times, but in a new worksheet. We have a new data file called `Model 10-04 Call Data.xls`, which contains the same information as `Model 10-03 Input.xls`, but also contains a new range called `ProcessTimes` that was defined in a new worksheet called `OutputData`. The `ProcessTimes` range has been formatted as numeric and has a

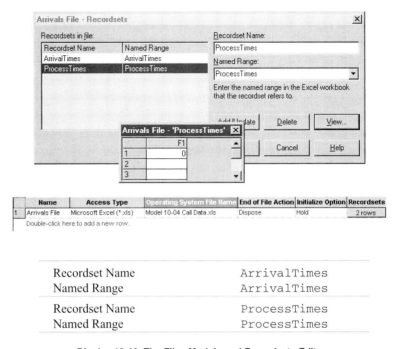

	Name	Access Type	Operating System File Name	End of File Action	Initialize Option	Recordsets
1	Arrivals File	Microsoft Excel (*.xls)	Model 10-04 Call Data.xls	Dispose	Hold	2 rows
	Double-click here to add a new row.					

Recordset Name	ArrivalTimes
Named Range	ArrivalTimes
Recordset Name	ProcessTimes
Named Range	ProcessTimes

Display 10-13. The Files Module and Recordsets Editor

zero in the first cell. Just to illustrate the possibility, we have also added the skeleton of a plot to our new worksheet. Although it is empty now, when the data are written to the worksheet, they will be picked up automatically and displayed in the plot.

First, we need to change the File data module to specify the new file name and use the Recordsets Editor to specify that we have a second range in that file. With these changes, the module should look like Display 10-13.

We need to adjust our logic so that after call processing is complete, we can write the elapsed processing time to the file. Add a ReadWrite module immediately before the call termination as shown in Figure 10-6.

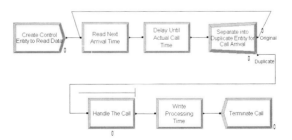

Figure 10-6. Logic for Writing Process Time Data

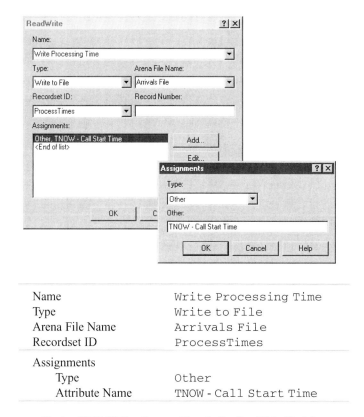

Name	Write Processing Time
Type	Write to File
Arena File Name	Arrivals File
Recordset ID	ProcessTimes

Assignments	
Type	Other
Attribute Name	TNOW - Call Start Time

Display 10-14. Writing Process Time in the ReadWrite Module

We want to write to the same file, `Model 10-04 Call Data.xls`, that we read from in the previous ReadWrite module. But this time, we'll specify `ProcessTimes` for the Recordset ID. The value we want to write is a calculated expression, so we need to specify Type of `Other`. In the Other field, we will enter our expression for processing time, which subtracts Call Start Time from the current time (`TNOW`) as shown in Display 10-14.

We are now ready to run. This model will read the same data as before, but now, as each call is completed, its processing time will be written to the OutputData worksheet. When the run is complete, the worksheet will look something like Figure 10-7.

10.1.3 Advanced Reading and Writing

Although life would be easy if data always came in a single form, that's seldom the case. Instead, we often receive data that are not tailored to our use. In a simple case, it could be a text file with excess data in it. For example, what if our data file looked like Figure 10-8?

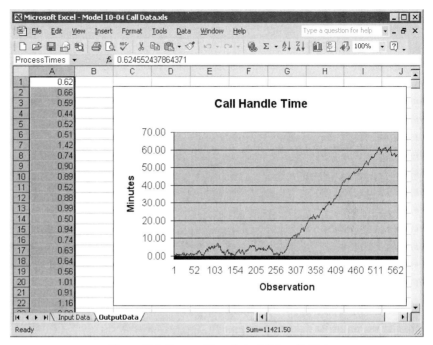

Figure 10-7. Process-Time Output Data with Plot

These values certainly look a bit odd as the first two lines look like they each contain a word and two numbers. The third line looks like it contains a word and a single number. But in fact, it is the same file that was presented in Figure 10-2, except that it contains a description in columns 1-8. To read this type of file, you would use a formatted read. A simple way to do this is to modify the File module for the model we created in Section 10.1.1. In the File data module, select an Access Type of Sequential File. For the Structure, we'll type in a FORTRAN or C-style format describing that we'll skip past the first eight columns and then read our data; see Display 10-15.

```
Part 1  1.038
Part 27 2.374
Part 5194.749
Part 67 9.899
Part 72 10.52
Part 16217.09 . . .
```

Figure 10-8. More Call-Time Data for Revised Call-Center Model

	Name	Access Type	Operating System File Name	Structure	End of File Action	Initialize Option	Comment Character
1	Arrivals File	Sequential File	Model 10-02 Input.txt	"(8x,F8.3)"	Dispose	Hold	No

Display 10-15. File Module for Formatted Data

By the way, what could you do if your data were not in the first columns of the table? If you have the luxury of changing the file, the obvious choice would be to define the range or table so that it encompasses exactly the data you need to read. For example, if you had a spreadsheet with a named range defined as columns A through C, but your data were only in column C, the best approach would be to define a new range of only column C. Unfortunately, in many cases, it's not possible to change the file; in those cases, the best choice may be simply to read and discard the extraneous data. You can do this by defining a throwaway variable and reading the extraneous data into that. To continue our example, you could define a variable called `JunkValue` and populate the Read module as shown in Display 10-16 to bypass the first two columns of the range.

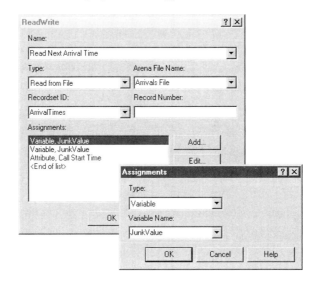

Name	Read Next Arrival Time
Type	Read from File
Arena File Name	Arrivals File

Assignments	
Type	Variable
Attribute Name	JunkValue

Assignments	
Type	Variable
Attribute Name	JunkValue

Assignments	
Type	Attribute
Attribute Name	Call Start Time

Display 10-16. The ReadWrite Module to Skip Extraneous Data

```
Provider=Microsoft.JET.OLEDB.4.0;
Data Source=C:\Documents\Book1.xls;
Extended Properties=""Excel 8.0; HDR=Yes;""
```

Display 10-17. Excel with Headings Using ADO

```
Driver={SQL Server};
Server=RSI-Joe;
Database=BizBikes;
Uid=BizWareUser;
pwd=MyPassword
```

Display 10-18. SQL Commands Using ADO

Don't despair if your data are in a more complicated Excel or Access file than we've described, if the data require more sophisticated data access techniques, or if the data are from another package. Microsoft has a solution to help—ADO, and the good news is that you have used it already. The Excel and Access interfaces used in the example above were implemented using ADO, and Arena did most of the work to make it easy to do common tasks. However, if you happen to encounter something a little more complex, the effort to learn a little ADO can pay off big dividends. We won't delve into complex ADO, but we'll show you a couple of examples just to whet your appetite.

If you specify an Access Type of `Active Data Objects (ADO)` in the File data module, a new field (Connection String) will appear. The connection string is used to customize the ADO interface by presenting information such as the provider, file name, and any other tidbits required to complete the connection. Let's say that you want to ignore the built-in support for Excel and customize your own, perhaps because you want to support column headings. You could enter the connection string specified in Display 10-17. Or if you want to use the driver for SQL Server to connect to a database `BizBikes` in Microsoft SQL Server `RSI-Joe`, you might use a connection string like the one in Display 10-18. In both cases, these strings would be entered as one continuous string in the Connection String field of the File data module.

It's important for you to know that file input/output (I/O) may affect execution speed and can be slow compared to other processing tasks. In many cases, it may not matter if access is slow, because the amount of time spent reading or writing data is often trivial compared to the time required to execute the rest of the model. But I/O speed may be important with very high volumes of data. In such cases, you might want to specify the data to be in the fastest or smallest form possible. Although ADO generally competes well on ease of use and flexibility, it often has problems with execution speed. The winner of the speed and file-size competition is often binary files. Binary files store the data in machine-readable form that the average person cannot easily decipher. For that reason, it is a poor choice if people will interact directly with the data. But for passing data between programs on the same platform, it can generate files that are very small and fast to read and write. For large or frequently accessed files, this could be an important consideration. Most computer languages support reading and writing binary files. You can even use Arena for a one-time conversion of files into binary forms that are then later reused frequently during runs for experimentation.

To experiment with this option, in the File data module, select an Access Type of `Sequential` and a Structure of `Unformatted`.

10.2 VBA in Arena

In this section, we introduce you to the concepts surrounding VBA, which is embedded in Arena. You can use this technology to write custom code that augments your Arena model logic. Sections 10.3 and 10.4 provide a few illustrations of how Arena and VBA work together.

10.2.1 Overview of ActiveX Automation and VBA

Arena exploits two Microsoft Windows technologies that are designed to enhance the integration of desktop applications. The first, ActiveX® Automation, allows applications to control each other and themselves via a programming interface. It's a "hidden" framework provided by Windows, accessible through a programming language (such as Visual Basic®) that has been designed to use the ActiveX capabilities. In fact, if you've created a macro in Excel, you've utilized this technology, whether you realized it or not. Excel macros are stored as VBA code that uses an ActiveX interface to cause the Excel application to do things such as format cells, change formulas, or create charts.

The types of actions that an application supports are defined by what's called an *object model*. The designers of the application (e.g., the developers of Excel or Arena) built this object model to provide an interface so that programming languages can cause the application to do what a user normally would do interactively with a mouse and keyboard. The object model includes the following:

- a list of application *objects* that can be controlled (e.g., Excel worksheet, chart, cell);
- the *properties* of these objects that can be examined or modified (e.g., the name of a worksheet, the title of a chart, the value of a cell); and
- the *methods* (or actions) that can be performed on the objects or that they can perform (e.g., delete a worksheet, create a chart, merge cells).

When you install an application that contains an object model, its setup process registers the object model with the Windows operating system (i.e., adds it to the list of object models that are available on your computer). Then, if you use a programming language and want to utilize the application's functionality, you can establish a reference to its object model and program its objects directly. We'll see how this works when we write VBA code in Arena to send data to Microsoft Excel. Many desktop applications can be automated (i.e., controlled by another application), including Microsoft Office, AutoCAD®, and Visio®. To create the program that controls the application, you can use programming languages like C++, Visual Basic, or Java.

The second technology exploited by Arena for application integration embeds a programming language (VBA) directly in Arena to allow you to write code that automates other applications without having to purchase or install a separate programming language. VBA is the same language that works with Microsoft Office, AutoCAD, and Arena. It's also the same engine behind Visual Basic. When you install

Arena, you also receive this full Visual Basic programming environment, accessed via *Tools > Macro > Show Visual Basic Editor*.

These two technologies work together to allow Arena to integrate with other programs that support ActiveX Automation. You can write Visual Basic code directly in Arena (via the Visual Basic Editor) that automates other programs, such as Excel, AutoCAD, or Visio. In our enhancement to the call-center model, we'll create a new Excel worksheet, populate it with data during the simulation run, and automatically chart the data, all without "touching" Excel directly. You also can write VBA code to automate Arena, such as to add animation variables, get the values of simulation output statistics, or change the values of module operands.

When you write Visual Basic code using VBA in Arena, your code is stored with the Arena model (*.doe*) file, just as VBA macros in Excel are stored in the spreadsheet (*.xls*) files. You have at your disposal the full arsenal of VBA features, including:

- Visual Basic programming constructs, such as Sub, Function, Class, If, Elseif, Endif, While, Wend, Do, On Error, and Select Case;
- UserForms (often referred to as dialog boxes) with an assortment of fancy controls like toggle buttons, scroll bars, input boxes, command buttons, and the ever-popular spin buttons;
- code-debugging tools like watches, breakpoints, and various step options; and
- comprehensive help.

To open the *Visual Basic Editor* in Arena, select *Tools > Macro > Show Visual Basic Editor*, or click the corresponding button on the Integration toolbar ( ). This opens a separate window that hosts all of the VBA code, forms, debugging interface, and access to VBA help, as illustrated in Figure 10-9.

This window is a "child window" of the Arena window, so that if you quit from Arena, the Visual Basic Editor window will automatically follow its parent by closing as well. (If only all children were so well behaved.) Furthermore, changes in the Arena window, such as opening a new model, are reflected in the information provided in the Visual Basic Editor window. This relationship reflects the architecture of VBA integrated into host applications such as Arena—the VBA code is owned by the parent document and cannot be accessed except through the host application.

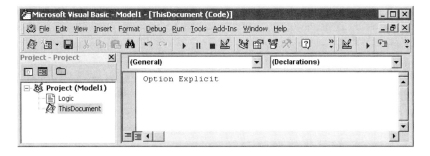

Figure 10-9. Visual Basic Editor Window

10.2.2 Built-in Arena VBA Events

The Project panel at the left side of the Visual Basic Editor window shows a list of open models, each containing a list of *Arena Objects* that starts with a single entry called *ThisDocument*. The ThisDocument object gives the VBA project access to various events within the Arena model. To add code for an event procedure, you select the *ModelLogic* object in the Visual Basic Editor, and choose the desired event (e.g., RunBegin) in the procedure list on the right.

Arena's built-in VBA events fall into three broad categories. (The actual function names all are preceded by ModelLogic_, such as ModelLogic_DocumentOpen.)

- Pre-run events (e.g., DocumentOpen, DocumentSave)
- Arena-initiated run events (e.g., RunBegin, RunBeginSimulation, RunEndReplication)
- Model/user-initiated run events (e.g., UserFunction, VBA_Block_Fire, OnKeystroke)

The events displayed in the procedure list provide the complete set of locations where your VBA code can be activated. So one of the first decisions you face is which event(s) you'll use so that your code can be called at the proper time to do whatever you desire. The pre-run events such as DocumentOpen and DocumentSave provide opportunities for VBA code to be executed under certain user-initiated actions (e.g., opening and saving a model). However, most of the support for VBA events in Arena centers around the simulation run. Whenever you start a run (e.g., by selecting *Run > Go*), a sequence of Arena actions and VBA events occurs (Figure 10-10).

Arena automatically calls each of these events (listed in bold in Figure 10-10) as the run begins, is performed, and terminates. If you have not written VBA code for an event, then no special actions will take place; Arena will behave as if the event didn't exist.

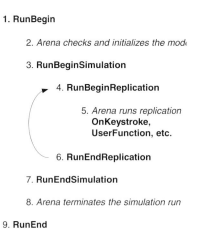

1. RunBegin

    2. *Arena checks and initializes the mode*

3. RunBeginSimulation

    4. RunBeginReplication

        5. *Arena runs replication*
        OnKeystroke,
        UserFunction, etc.

    6. RunEndReplication

7. RunEndSimulation

    8. *Arena terminates the simulation run*

9. RunEnd

Figure 10-10. Simulation-Run VBA Events

(That's what's been happening all along.) However, if you open the Visual Basic Editor and type VBA code in one or more of these events, then Arena will execute that code during the run, as we'll see in the next two sections of this chapter.

Figure 10-10 also points out an important aspect of the timing of these calls, regarding the type of data that are available to your VBA code. While you're editing the Arena model, the simulation variables, counters, statistics, etc., are defined as part of the model's structure via the information you've provided in modules, but they haven't yet been created for use in the simulation run (e.g., there's no average value of a number-in-queue yet, no state for a resource). So when the model's in this *edit state*, you can only work with the values of module operands (the fields in module dialog boxes and spreadsheet cells). In Model 10-5, we will utilize this approach to modify values in the Max Arrivals field of two Create modules.

When you start a run, Arena checks and initializes the model, translating the information you've supplied in the modules into the format required for performing the simulation and placing the model in a *run state*. During the run, the values of variables, resource states, statistics, etc., can be examined and changed through Arena's run controller and through VBA code. This use of VBA is illustrated in Model 10-6, which sends the values of run statistics to Microsoft Excel. At the end of the run, Arena destroys this information and returns the model to an edit state.

Returning to Arena's built-in events, we'll briefly describe each of the steps shown in Figure 10-10. Note that the items in Figure 10-10 that are listed in bold correspond to VBA events available in Arena; the phrases shown in italics simply describe notable steps in Arena's run processing.

1. `ModelLogic_RunBegin` is called.

   At this point, VBA code can make changes to the structural data of the model—that is, the values in module operands—and have the changes be included in the simulation run. However, RunBegin cannot change runtime simulation values (e.g., variables, entity attributes) directly, because the simulation has not yet been initialized. You might use this event to query for inputs that will overwrite what's stored in modules (e.g., a probability at a Decide module). This event's use is illustrated later in Model 10-5.

2. Arena checks the model and initializes the simulation to a run state.

   This process takes place behind the scenes, with Arena verifying that your model is ready to run. After the model is checked and initialized in this step, Arena ignores what's contained in the modules, working instead with the simulation runtime data that change dynamically once the run has started. At this point, all variables have been assigned their initial values (e.g., the numbers you typed in the Variable spreadsheet), resources are at their initial capacities, but no entities have yet entered the model.

3. `ModelLogic_RunBeginSimulation` is called.

   Here, Arena gives you the chance to insert VBA code that will be executed only once at the beginning of the simulation run. (Of course, if Arena detected errors in your model in Step 2, this event won't be called; you first have to fix your mistakes to allow proper initialization of the model.) When the VBA code in RunBeginSimulation is being executed, Arena holds off on starting the simulation run. So at this point, you can safely load lots of data from outside sources like Excel, Access, or Oracle®; display a UserForm to ask whoever's running your model for some sage advice, such as how many shifts they want to run in today's simulation; or set up the headers in an Excel file that will be storing some detailed simulation results (as we'll see in Model 10-6). Often this code will end up assigning values to variables in the Arena model, though it also could create entities, alter resource capacities, or dozens of other things that you'll find if you delve into the depths of the Arena object model. After your RunBeginSimulation VBA code is finished doing its work, Arena moves to the next step.

4. `ModelLogic_RunBeginReplication` is called.

For each replication that you've defined for this run, Arena will call `RunBeginReplication` at the beginning of the replication (i.e., before any entities have entered the model). The types of things you might do here are similar to those described for `RunBeginSimulation`—except that whatever you define in `RunBeginReplication` will be repeated at the beginning of each replication. We'll see in Model 10-6 how these two events can be used together.

5. Arena runs the simulation.

Next, the model run is performed. In this step, entities are created and disposed, resources are seized and released, etc.; basically all that stuff you've learned in your suffering through the previous chapters of this book is put to work. During the run, Arena provides a number of opportunities for you to activate VBA code, such as the following.

- `ModelLogic_UserFunction`—called whenever the UF variable is referenced in Arena logic. This event might be used to perform complex calculations for a process delay or decision criterion.
- `ModelLogic_VBA_Block_Fire`—called when an entity passes through a VBA module (from the Blocks panel) in model logic. This event is used in Model 10-6 to write detailed information to Excel as each entity departs the model.
- `ModelLogic_OnKeystroke`—called whenever the user strikes a key during the simulation run. For example, your VBA code might display a UserForm with summary model status whenever the "1" key is pressed.
- `ModelLogic_OnClearStatistics`—called whenever statistics are cleared, such as when simulation time reaches the value entered for the warm-up period in the Run Setup dialog box. You might write VBA code here to set certain model variables or to send an entity into the model to activate specialized logic that augments the standard Arena statistics initialization.

6. `ModelLogic_RunEndReplication` is called if the simulation reaches the end of a replication.

This event typically stores VBA code that either writes some summary information to an external file, increments some global variables, or both. Note that it's only called when the run reaches the end of a replication; if the run is stopped for some other cause (such as the user clicking the *End* button on the Run toolbar), then this event is not executed.

7. `ModelLogic_RunEndSimulation` is called.

Regardless of how Arena ended the run, the VBA code in this event will be executed. When `RunEndSimulation` is called, the runtime simulation data are still available, so final statistics values, resource states, etc., can be accessed by the VBA code here. `RunEndSimulation` typically contains logic to write custom statistics to files, spreadsheets, or databases, as well as "clean-up" code to close files or display end-of-run messages.

8. Arena terminates the simulation run.

This is the counterpart to Step 2. Arena clears all of the runtime simulation data and returns the model to an edit state.

9. `ModelLogic_RunEnd` is called.

Finally, the `RunEnd` VBA event is called, providing a balance to the `RunBegin` event in Step 1. The VBA code in this event cannot access any information from the run that was just performed (because nasty old Step 8 took it all away), but all other VBA functions are available. If you wanted, you might display a UserForm in `RunEnd` that asks whether the user wants to run the simulation again, so that we'd be sent automatically back to Step 1 to start the process over again!

10.2.3 Arena's Object Model

After you've decided where to locate your code, you'll probably need to access information from Arena. The object model (also referred to as a *type library*) provides the list of objects and their properties and methods that are available to your code. All of the information that your code needs from Arena and all of the actions your code causes Arena to perform will be carried out by utilizing elements from the Arena object model.

An *object* is an item that you can control from your code. Its characteristics, referred to as *properties*, can be examined in VBA, and many of the properties also can be changed by your code. Examples of object properties include line colors for rectangles, polylines, etc.; positions (*x* and *y*); and identifiers for animation objects such as stations and queues. Most objects also have *methods* that you can invoke to cause an action to be performed; for example, activating a window or creating a new polygon. Also, most object types are grouped into *collections*, which are simply one or more objects of the same type (usually). There's a collection of Modules in the Arena object library, for instance, that holds all of the module instances (which are objects of type Module) in a submodel window.

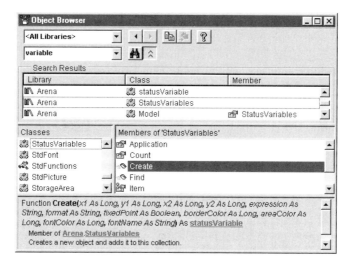

Figure 10-11. The VBA Object Browser

If you'd like to explore the object library, VBA provides a handy browser, opened by selecting *View > Object Browser* in the Visual Basic Editor window or pressing the *F2* key. This allows you to navigate through object libraries, exploring objects and their properties and methods. Figure 10-11 shows the browser's display of the StatusVariables collection object and its Create method in the Arena object library.

Arena's object model contains the following categories of objects.

- *Model-window objects*—all of the items that can be placed in an Arena model window or viewed via its spreadsheet. Modules, connections, lines, polygons, text, the animation objects, and interaction-related objects such as named views fit into this category.

- *SIMAN object*—the special object type that provides access to information about the running simulation, such as variable values, resource capacities, queue lengths, and the current simulation time.

- *Structural objects*—the objects that are used to access general Arena functions, such as the Application, Module Definition, and Panel objects.

When you write code that mimics what normally would be performed interactively (i.e., via the mouse and keyboard), you'll be working mostly with objects that fall into the model-window category. These objects also constitute the majority of the objects that you'll see if you browse the object library, because Arena has a large number of objects that you can work with in a model.

For example, the code in Figure 10-12 adds ten animation variables to the model containing this VBA code. The first object variable referenced in the code is named oModel. It's declared as type Arena.Model, which defines it as coming from the Arena object-model-type library and being of type Model. (We use the lowercase 'o' as a variable-name prefix as a way of indicating that it's an object variable as opposed to

```
Dim oModel As Arena.Model
Dim i As Integer
Dim nX As Long

' Add the status variables to this Arena model
Set oModel = ThisDocument.Model

nX = 0              ' Start at x position 0
For i = 1 To 10
    ' Add a status variable to the model window
    oModel.StatusVariables.Create nX, 0, _
      nX + 400, 150, "WIP(" & i & ")", "**.*", False, _
      RGB(0, 0, 255), RGB(0, 255, 255), RGB(0, 0, 0), "Arial"
    ' Move over 500 world units for next position
    nX = nX + 500
Next i
```

Figure 10-12. Sample Code to Create Ten Status Variables

an integer or other data type.) We set the oModel variable to point to the model window that contains the VBA code by using a special object provided by Arena's object model, called ThisDocument, and getting its Model property. Then, within our loop, we add a new animation variable by using the Create method of the StatusVariables collection, which is followed by a number of parameters that establish the variable identifiers as WIP(1) through WIP(10)—formed by concatenating the string "WIP(" with the loop index variable, i, and closing with ")" to complete the name—and their positions, sizes, and text characteristics.

The second major area of content in the object library falls under the main object type, SIMAN. This object, which is contained in a Model object, provides access to all of the information about a running simulation model. Using the properties of the SIMAN object, your VBA code can find out almost anything about the simulation. If you click on the SIMAN object in the browser, you'll see that there's an extensive list of functions at your disposal.

Figure 10-13 shows a snippet of code that displays a message querying the user for a variable value (using the InputBox function from Visual Basic, which is displayed below the code), then assigning a new value to the Mean Cycle Time model variable. The answer supplied to the InputBox is stored in a local string variable, sNewValue. Then the oSIMAN variable is set to point to the SIMAN object contained in ThisDocument.Model. (SIMAN is the simulation engine that runs Arena models.) Any information we want to get from the running simulation or change in the run will be accessed by using this oSIMAN variable. The final two lines of code assign the value to the variable. We first use the SymbolNumber function to convert the name of the variable to the internal index used by the SIMAN engine. Finally, the VariableArrayValue property is changed to whatever was typed in the InputBox.

Because of the natures of these two categories of objects, they typically are used in different sections of code. The model-window objects most often are incorporated in code that's either hosted outside of Arena or in utility VBA functions to do things like lay out tables of variables (as in Figure 10-12) or automatically build an Arena model (as in the Visio interface that's distributed with Arena).

```
Dim oSIMAN As Arena.SIMAN
Dim nVarIndex As Long
Dim sNewValue As String

' Prompt for a new value
sNewValue = InputBox("Enter the new average cycle time:")

' Assign their answer to the Mean Cycle Time variable
Set oSIMAN = ThisDocument.Model.SIMAN
nVarIndex = oSIMAN.SymbolNumber("Mean Cycle Time")
oSIMAN.VariableArrayValue(nVarIndex) = sNewValue
```

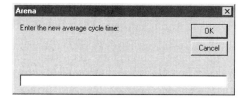

Figure 10-13. Code to Assign New Variable Value During Run

The SIMAN object, on the other hand, is active only during the simulation run. It's typically used in the built-in events described in Section 10.2.2. If you write any code that uses the SIMAN object, it must be executed while the model is in a run state (that is, between the `RunBegin` and `RunEnd` events). The properties and methods that are made available through the SIMAN object provide access to all of the variables and statistics related to your running model (e.g., number in queue, average tally value, current simulation time) as well as a few modeling actions to create entities and send them into the model. The SIMAN object does not provide any modeling capabilities or simulation data beyond what's available in an Arena model. This automation interface is simply an alternate way to get at or change the information in your model. We'll see how this can be useful in Section 10.4 when we send data from the Arena model to Excel.

10.2.4 Arena's Macro Recorder

As you can see from Sections 10.2.1–10.2.3, VBA provides significant power and flexibility, but also can be somewhat difficult to learn. Many products that support VBA also provide a macro recording capability to give you a jump start. A *Macro* is a series of VBA commands or instructions stored in a Visual Basic module that can be run whenever you need to perform a task. The *Macro Recorder* is a tool that makes it easy to generate macros by recording the commands associated with actual tasks. In this section, we'll explore how to use macros to help us build VBA solutions.

Let's say that we want to write some code to place some modules. While we could certainly explore the Arena object model described in Section 10.2.3, we'll instead generate some sample code that we can customize. We'll start by turning on macro recording by selecting *Tools > Macro > Record Macro* or by selecting the *Start/Pause/ Resume* button (●❙❙) on the Record Macro toolbar. This brings up the Record Macro dialog box where we give our macro a meaningful name and a description. Complete the Macro name and Description in this dialog box as illustrated in Display 10-19. (Note that VBA macro names cannot contain spaces.)

Display 10-19. Starting to Record a Macro

As soon as you select *OK*, the macro begins to record (more on that in a moment), and the Record Macro toolbar changes to (●ıı ■) to indicate that a macro recording session is in progress. The *Start/Pause/Resume* button is pressed in to indicate that you are currently recording your actions. Pressing the button again pauses the recording, but does not end the session; simply press the button again to resume. When you are recording or pausing a session, the *End* button is available (an indication that a macro is being recorded) and may be pressed at any time to stop the recording.

Once you activate the macro recording, be sure to perform only those steps that you want to record. The actions defined in the Arena object model may be recorded, but some actions, like the module dialog box entries, are not. If you want to change module data, you must do so using the spreadsheet interface. Because we want to explore how to place a module and set its operands, that is exactly what we'll do next. Select a Create module from the Basic Process panel and place it in the model window. Then use the spreadsheet interface to set the Name, Entity Type, and Value as shown in Figure 10-14.

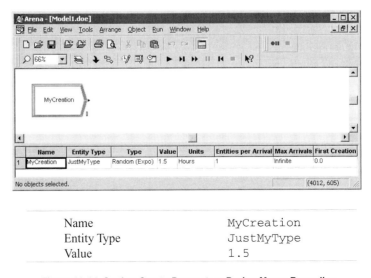

Name	MyCreation
Entity Type	JustMyType
Value	1.5

Figure 10-14. Setting Create Parameters During Macro Recording

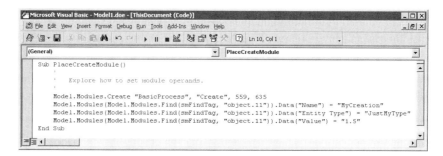

Figure 10-15. Code Generated During Macro Recording

Press the *End* button on the Macro Record toolbar to stop the recording and automatically store the generated code in the VBA module named `PlaceCreateModule`. Load the Visual Basic Editor (*Tools > Macro > Show Visual Basic Editor*) and then double-click on ThisDocument to see the code we just generated, as shown in Figure 10-15. It is a subroutine with the name we gave our macro, followed by the description we entered. The next line does the actual work of placing the module at the specified *x* and *y* coordinates. The next three lines set the operands we changed in the spreadsheet interface. And finally, the very last line closes the subroutine and ends the macro.

Let's close the VB Editor and return to our model to try our new code. Select *Tools > Macro > Run Macro* to open the Run Arena Macro dialog box (pictured in Display 10-20), which provides a list of all of the macros we have defined (only one so far) and gives us some options to run. Select the *Run* button. Although it may appear that nothing happened, in fact, it placed a new Create module at the exact *x/y* coordinates of the old one, just as we told it to do. If you drag aside the module on the top, you'll see our original module hiding directly beneath. A closer look shows that the parameters of the new module are still at their default values. Can that be right? We told it to place a Create module and it did. So far, so good. But let's look again at the code in Figure 10-15.

Most VBA commands work on the currently selected object (`Model. ActiveView.Selection`). In our case, the module we created was still active, and its tag is `object.11`. In order to set the operand of the module we just created, each of the next three lines executes an `smFindTag` on "`object.11`." So the macro seems to have recorded correctly exactly what we did. But let's walk through what happens when we run the macro. The first line places a new Create module exactly where we placed the original—we've already seen that work. The second line finds `object.11`, the *original* object, and sets its operand. The macro works as designed, but not as intended. If we'd go back to the model, delete both modules, and run our macro again, what would you expect to happen? We'd get an error because we'd be looking for an object that didn't exist.

The macro we just created is obviously not too useful to execute, so you might wonder where its value lies. While there are some applications where you can record a macro and then use it repeatedly without changes, in most cases, this will not be true. Typically, the value is in looking at the code that was generated. Although the macro above was trivial, it provides a lot of good information. We can see how to place a

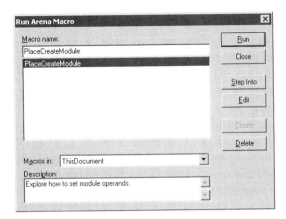

Display 10-20. Running the PlaceCreateModule Macro

module and what parameters are required. We know how to find an object using its tag. We know how to set the operands in a module, and we know the names of the three specific operands we wanted to change. In short, that simple macro gave us a big jump start to writing some real code.

The primary value of the macro recorder is to make it easier to learn and prototype VBA model extensions. Once you see how it's done, you can use those same techniques to solve other problems. There is no limit to the number of macros you can create, and you can mix and match techniques from several different ones as needed. In short, the macro-recording capability, used in combination with the other techniques you learn in this chapter, can be a powerful addition to your "bag of tricks."

10.3 Model 10-5: Presenting Arrival Choices to the User

In Section 10.1, we exploited Arena features to integrate data into a model. Next, we'll turn our attention to how we can interact with someone who's running an Arena model.

In Model 10-5, we'll set up a mechanism for when the model runs that allows a choice to be made to use either call arrivals generated from a random process (as in the original Model 10-1) or from a file (as in Model 10-2). At the beginning of a run, we'll display a *UserForm* (VBA's term for a dialog box) providing the options, as shown in Figure 10-16.

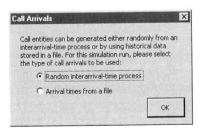

Figure 10-16. Visual Basic UserForm

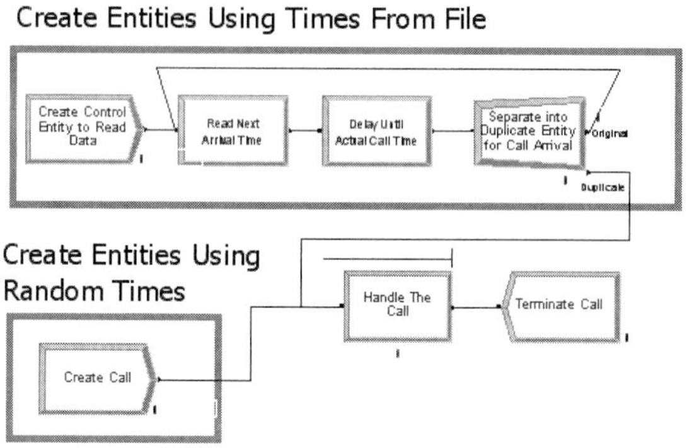

Create Entities Using Times From File

Create Entities Using
Random Times

Figure 10-17. Call-Creation Logic for Model 10-5

The user selects the desired option by clicking on the appropriate option button, then clicks *OK* to allow the simulation run to commence. We'll write VBA code to make the required changes to the model so that only one of the call arrival types actually sends entities into the model. And then at the end of this section, we have a surprise for you that will add a bit of pizzazz to the model with a quick bit of VBA code. Don't peek ahead, though...you'll have to make your way through the "fun with forms" material first!

10.3.1 Modifying the Creation Logic

First, we need to set up the model so that it can generate entities from either the random arrival process of Model 10-1 or the input file of Model 10-2. To do so, we'll place both sets of logic in the same model, connecting to the Process module that's the beginning of the call logic, as in Figure 10-17.

If we're using the random-process-arrivals logic, then we'll set the Max Arrivals field in its Create module to `Infinite` (as was the case in Model 10-1) and the Max Arrivals in the file logic's Create module to `0`, indicating that no entities should enter the model there. In the other case, for arrivals from the file, we'll set the Max Arrivals of the random-process-arrivals Create module to `0` and the value in the Create module for the file logic to `1`. (Remember that its logic creates only a single entity.) Display 10-21 shows the Create module for turning off entities at the random-process-arrival Create module.

To verify that the model works properly, you can edit the Create modules to set their Max Arrivals values, then run the simulation and watch the counters next to the two Create modules. After we add the VBA code at the beginning of the run, we won't need to worry about the Max Arrivals fields; the proper values will be filled in by VBA, based on the user's selection for the type of arrivals.

Before continuing, we're going to make two additional model changes that will be exploited by the VBA code to find the proper Create modules. Though you probably didn't realize it, each object in an Arena model window has an associated *tag*. This tag isn't shown anywhere when the objects are displayed; it is only accessible through the Properties dialog box, which you open by right-clicking on the object and selecting the Properties option, as shown in Figure 10-18 for the Create Arrivals module.

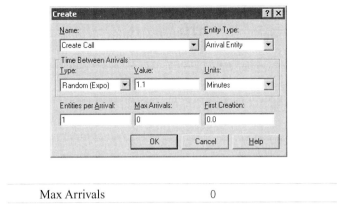

Max Arrivals	0

Display 10-21. Create Module with Max Arrivals of 0

Arena assigns these tags as you add objects, using a format "Object.*nnn*," where *nnn* is an integer that increases as each new object is placed in the model (e.g., Object.283). We'll change the tag for the Create module that uses the random process to `Create from random process` as shown in Figure 10-18, and we'll change the tag of the file-arrivals Create module to `Create from file`. Right-click on each Create module and select the Properties option from the menu, then type the appropriate tags. Later, when our VBA code needs to find these particular modules, we'll use a Find method to search for a matching tag value, which we'll know is unique since we're the ones who assigned them to these modules.

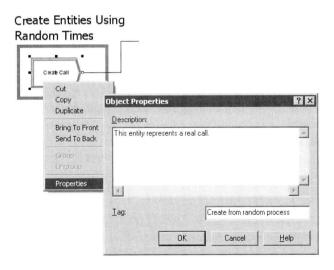

Figure 10-18. Changing the Module Tag via the Properties Dialog Box

10.3.2 Designing the VBA UserForm

Now that the model is designed to trigger either type of call creation, we'll redirect our attention to VBA to give the modeler a choice each time a simulation run begins. We'll first draw the form that will be presented when the run is started, then we'll write some VBA code to act on the selection made in the form.

To add a UserForm (which we'll refer to as a *form*), open the Visual Basic Editor and select the *Insert > UserForm* menu option. This adds a new object to our VBA project and opens a window in which you can draw the form, lay out controls (the things with which a user can interact), add pictures and labels, and so on. We'll skip over the details of how forms are designed in VBA; the VBA help topic under the Microsoft Forms Design Reference topic can aid your understanding of the basic concepts, toolbars, and features of Microsoft Forms.

For the VBA code to access the form, it needs to have a name. We'll call it `frmArrivalTypeSelection` by typing in the Name field of the Properties box, as in Figure 10-19.

To design the form's contents, drag controls from the Control Toolbox onto the form. Drag a label onto the form and type its Caption field (in the Properties box) to match the descriptive text in Figure 10-20 (`Call entities can...`). Next, drag an option button onto the form below the label. Change its Name to `optFromRandomProcess`, its Caption to `Random interarrival-time process`, and its Value to `True` so that it is the default selection. Then add another option button named `optFromFile`, with `Arrival times from a file` as its caption and a default value of `False`. The final control is the command button at the bottom of the form. Name it `cmdOK` with a caption of `OK`, and set its Default property to `True` so that the Enter key will act like the command button had been clicked.

This UserForm is now part of the Arena model; when you save the model file, the form will be saved with it. However, it's not of much help to us until VBA knows what to do with the form. We'll tackle that effort next.

Figure 10-19. UserForm in VBA Project

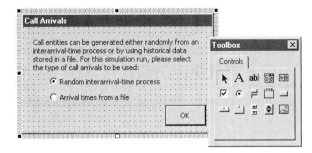

Figure 10-20. UserForm Layout

10.3.3 Displaying the Form and Setting Model Data

We want the VBA code for this model to turn on and off the two Create module arrivals, based on which type of calls the user actually wants to run through the simulation. For these changes to work, we need to set the Max Arrivals fields in the two Create modules before the run begins. By doing so, the values will be compiled into the model just as though the modules had been edited by typing values in the module dialog boxes. If you recall the steps outlined in Section 10.2, the `RunBegin` event gives us the perfect opportunity to make these changes. It's called at the beginning of every simulation run before the model is checked, so the changes to the module data made by our VBA code will be included in the run.

Of course, we'll need to display the form to the user, who will select which type of arrival is desired. We'll do that by inserting a single line of code in the `RunBegin` event. You can get to the `RunBegin` event by double-clicking on `ThisDocument` in the Project box, then selecting `ModelLogic` in the left drop-down list and `RunBegin` in the list on the right, as shown in Figure 10-21.

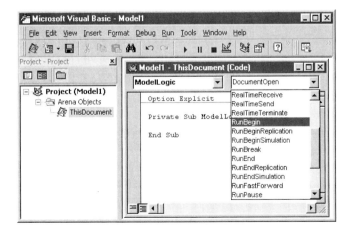

Figure 10-21. Adding the ModelLogic_RunBegin Event to the VBA Project

```
Option Explicit
Private Sub ModelLogic_RunBegin()
    ' Display the UserForm to ask for the type of arrivals
    frmArrivalTypeSelection.Show

    Exit Sub
End Sub
```

Figure 10-22. ModelLogic_RunBegin Code to Display Form

The code to display the form, shown in Figure 10-22, can be read by starting at the right and reading toward the beginning—we're going to "Show" the object (in this case, a UserForm) named `frmArrivalTypeSelection`.

This line of code will display the form and pass program control to it whenever the simulation run begins. After the Show event is executed, VBA is suspended until the user takes some action that initiates a VBA event for the form. Also, Arena is paused until the VBA code returns from the `RunBegin` event, allowing time for the user to decide what to do and for VBA to complete its `RunBegin` work. So if you were to run the model at this point, Arena's VBA would immediately display the form and wait for someone to tell it to go away so that the run could start. We haven't written any code to do that yet, so it may appear that you're stuck. Luckily, the VBA forms have a close box (the small button at the top right of the dialog box) that you can click to make the form go away and let Arena start its run.

The next time we want VBA code to intervene, though, is after the user has made the arrival-type selection and clicked the *OK* button. While the form is displayed, the user can change his mind as often as he wants by clicking the option buttons. But when he has given his final answer and clicked *OK*, we want our VBA code to be able to set the appropriate values in the Create modules, have the form disappear, and have the run begin.

VBA UserForms have predefined events similar to those in Arena. We'll insert our code in the event that's tied to the user's clicking on the *cmdOK* button in our UserForm. Open the form design window by double-clicking on the `frmArrivalTypeSelection` entry in the Project box, then double-click on *OK*. The Visual Basic Editor automatically opens the code window that's associated with the UserForm and inserts the beginning and ending statements for the `cmdOK_Click( )` event, as shown in Figure 10-23.

```
Private Sub cmdOK_Click()
End Sub
```

Figure 10-23. UserForm cmdOK_Click Event

In this event, we'll write all of the code to take action on the user selection. Our code will need to do the following:

- Locate each of the two Create modules
- Set the Max Arrivals field value in each Create module
- Add the code for the fun surprise (and you thought we forgot!)
- Close the UserForm

It's helpful to understand that in the Arena object model, usually code is designed to mimic the actions you would take interactively with a mouse and keyboard. So if you think about the steps you'd be taking to modify the Create modules, your first action would be to find and open a Create module and edit its fields, and then you would repeat these steps for the second Create module.

To describe the code for the `cmdOK_Click` event, we're first going to present smaller sections of code in Figures 10-24, 10-25, and 10-28 so that you can refer to the lines of VBA code more easily as you read the narration. Figure 10-29 repeats all the code in the complete `cmdOK_Click` procedure, so that you can see how it all fits together.

We'll start with code to locate each of the two Create modules so that we can assign the Max Arrivals fields to the appropriate values. We'll use the tag values that we assigned in Section 10.3.1 to find the Create modules from our VBA code, as in Figure 10-24. (We included the `Dim` statements for the variables used in this section of code, although it's usual—but not required—to place all variable declarations at the top of the function, as we do in the complete code listing of Figure 10-25.) The line that sets the `nCreateRandomProcessIndex` variable tells VBA to `Find` in the collection of `Modules` contained in the `oArrivalsModel` the item whose tag value is `Create from random process`. (The `smFindTag` constant dictates to `Find` based on matching tags instead of module names.) Similarly, the `nCreateFileIndex` variable finds the Create module for creating from a file. After each index is evaluated, it is used

```
Dim nCreateRandomProcessIndex As Long
Dim oCreateRandomProcessModule As Arena.Module
Dim nCreateFileIndex As Long
Dim oCreateFileModule As Arena.Module
Dim oModel As Arena.Model
Set oModel = ThisDocument.Model

' Find the two Create modules
nCreateRandomProcessIndex = _
    oModel.Modules.Find(smFindTag, _
        "Create from random process")
If nCreateRandomProcessIndex = 0 Then
    MsgBox "No module with tag 'Create from random process'"
    frmArrivalTypeSelection.Hide
    Exit Sub
End If
Set oCreateRandomProcessModule = _
    oModel.Modules(nCreateRandomProcessIndex)

nCreateFileIndex = _
    oModel.Modules.Find(smFindTag, "Create from file")
If nCreateFileIndex = 0 Then
    MsgBox "No module with tag 'Create from file'"
    frmArrivalTypeSelection.Hide
    Exit Sub
End If
Set oCreateFileModule = oModel.Modules(nCreateFileIndex)
```

Figure 10-24. Locating the Create Modules via Tags from VBA

```
' Set the Max Arrivals fields
   If optFromRandomProcess.value = True Then
      oCreateRandomProcessModule.Data("Max Batches") = "Infinite"
      oCreateFileModule.Data("Max Batches") = "0"
   Else
      oCreateRandomProcessModule.Data("Max Batches") = "0"
      oCreateFileModule.Data("Max Batches") = "1"
   End If
```

Figure 10-25. Updating the Create Modules with the User Selection

to point the corresponding variable (oCreateRandomProcessModule or oCreateFileModule) to the Create module. The If statements that test whether the returned index values equal 0 are there to provide a message if no module has the needed tag value (e.g., if the Create module had been deleted).

At this point, we have two variables pointing to the Create modules, so we simply need to check the option buttons in the UserForm to find out what the user selected and make the assignments to the Max Arrivals fields. To retrieve the setting of an option button, the VBA code tests the Value property, seen in Figure 10-25. If the button named optFromRandomProcess was selected when *OK* was clicked, its Value is True. In this case, we want to allow the Create module for random processes to generate entities (i.e., set its MaxArrivals value to Infinite) and turn off the other Create module for creating entities from a file. Otherwise, we want to set the random-process-arrivals Create module to a maximum of 0 and the from-file Create module to a maximum of 1 (so that our control entity enters the model to read arrival times from the file). To modify the value of a field (also referred to as an *operand*) in a module, the Data method is used in the VBA code.

If you're paying close attention (as we're sure you must be), you may have noticed that "Max Arrivals" doesn't appear anywhere in this code. Instead, the field in the Create module that defines the maximum number of arrivals is called "Max Batches." In Arena modules, the text that appears in the module's dialog box is called the *prompt*, and it may or may not match the underlying *name* of the operand in the module's definition. The Data method of a Module object in VBA requires this operand name. To find it, you look up the module in a text file that's installed with Arena (that is, wherever the Arena program file is located on your hard disk). In this case, the Create module came from the Basic Process panel, so we open the file named BasicProcess.txt and find the Create module, which is conveniently located at the top of the file. Looking down the list of prompts in Figure 10-26, we see that the operand whose prompt is "Max Arrivals" has an operand name of Max Batches (don't ask us why), so we entered Max Batches as the argument for the Data method.

As a refresher, let's review how we got here. Someone started a simulation run, perhaps by clicking the *Go* button on the Run toolbar. Arena called the VBA event, ModelLogic_RunBegin, where our code called the Show method for the frmArrivalTypeSelection form. Whoever started the run then had the opportunity to select the type of arrivals to be used for this run by clicking the option buttons. (VBA's forms take care of permitting only one option button to be selected, by the way.)

```
Module: Create

       Operands Contained in Dialog 'Create':

       Operand Name              Prompt
       Name                      Name
       Entity Type               Entity Type
       Interarrival Type         Type
       Schedule                  Schedule Name
       Value                     Value
       Expression                Expression
       Units                     Units
       Batch Size                Entities per Arrival
       Max Batches               Max Arrivals
       First Creation            First Creation
```

Figure 10-26. BasicProcess.txt *Listing of Operands*

While the form was displayed, the Arena run was suspended, patiently waiting for our VBA code to finish its work.

Eventually (we hope), someone clicked the *OK* command button on the form, which caused VBA to call the cmdOK_Click event in the frmArrivalTypeSelection code window. This code opened the Create and Direct Arrivals submodel, found the two Create modules, and entered the appropriate values into the Max Arrivals fields. And remember that all of this took place in RunBegin, which is called prior to the model being checked and compiled, so our changes to the module data *will* apply for the run. We needed to place our code there, rather than in RunBeginSimulation or RunBeginReplication, because those two events are called after the modules have been compiled and the run initialized, where we can access the runtime data only through the SIMAN part of the Model object. (Now might be a good time to page back to Figure 10-10 for a refresher on the order of Arena VBA events.)

And now, the moment you've been waiting for...the surprise! To close out our work at the beginning of the simulation run, we'll add some drama to our project (or waken the sleepy modeler) by playing some stirring music via a sound file placed in the model window. For this addition, we need to insert the sound file into the Arena model, give it a unique tag so that our code can find it, then write the VBA code to cause the sound to play.

To place the file in the model, return to the Arena model window. Then open your file browser (e.g., Explorer or Microsoft® Outlook®), locate the sound file, and drag it into the Arena model, as illustrated in Figure 10-27.

The sound object in the Arena model is an embedded file containing a copy of the file we originally had on our disk. This file also could have been placed in the Arena model using the Windows clipboard or via the *Edit > Insert New Object* menu option in Arena. The object itself is just like a sound file stored on your disk; double-click on it to play its inspiring melody. To provide the sound object with a unique tag, right-click on its icon in the Arena model window, select Properties from the menu, and enter Mission Possible for the Tag value.

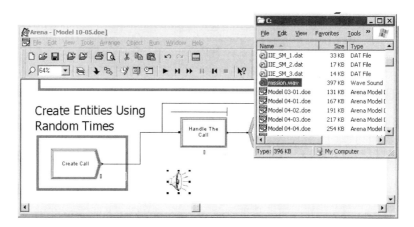

Figure 10-27. Dragging and Dropping a File Into Arena

Finally, return to the Visual Basic Editor window to insert the code that will activate the sound file at the beginning of the run, shown in Figure 10-28. We once again use the `Find` method, this time from within the `Embeddeds` collection of the `Model` object, to search for the desired object. If we find the object with tag `Mission Possible`, then we activate it (in the case of a sound file, playing the sound) using the `Do` method of the collection item at the appropriate index.

After you've run the simulation a few times, you may find that you have a compelling need to disable the sound file. If so, feel free to turn off your computer speakers. Or you can comment out the offensive lines of code by placing the statement, `If 0`, before the first line (where we retrieve the value of `nSoundFileIndex`) and the statement, `End If`, after the final line (the `Do` method). Later, if you want to restore the excitement to your model, you can simply change the `0` to a `1` in the `If` statement, a handy trick for code that you're not sure you always want to utilize.

Now that we've completed all of our work, including the surprise, our last step is to close the UserForm and exit the Click method of the *OK* button, so that Arena can begin the simulation run. These lines of code are at the bottom of Figure 10-29, which lists the complete code for the `cmdOK_Click` event.

After the form is hidden and the `cmdOK_Click` procedure is exited, VBA code control returns to the `ModelLogic_RunBegin` subroutine after the call to the `Show` event for the form. Back in Figure 10-22, we saw that `RunBegin` simply exits the

```
Dim nSoundFileIndex As Long

' Play the sound file
nSoundFileIndex = _
    oModel.Embeddeds.Find(smFindTag, "Mission Possible")
If nSoundFileIndex > 0 Then _
    oModel.Embeddeds.Item(nSoundFileIndex).Do
```

Figure 10-28. Code to Play Sound File

subroutine. At this point, Arena checks the model, initializes the simulation run, and begins the first replication, all while entertaining you with the inspiring music of our embedded sound file.

```
Private Sub cmdOK_Click()
    ' Set the Max Arrivals field in each of the Create modules
    ' based on the selection in the UserForm
    Dim nCreateRandomProcessIndex As Long
    Dim oCreateRandomProcessModule As Arena.Module
    Dim nCreateFileIndex As Long
    Dim oCreateFileModule As Arena.Module
    Dim oModel As Arena.Model
    Dim nSoundFileIndex As Long
    Set oModel = ThisDocument.Model

    ' Find the two Create modules
    nCreateRandomProcessIndex = _
        oModel.Modules.Find(smFindTag, _
            "Create from random process")
    If nCreateRandomProcessIndex = 0 Then
        MsgBox "No module with tag 'Create from random process'"
        frmArrivalTypeSelection.Hide
        Exit Sub
    End If
    Set oCreateRandomProcessModule = _
        oModel.Modules(nCreateRandomProcessIndex)

    nCreateFileIndex = _
        oModel.Modules.Find(smFindTag, "Create from file")
    If nCreateFileIndex = 0 Then
        MsgBox "No module with tag 'Create from file'"
        frmArrivalTypeSelection.Hide
        Exit Sub
    End If
    Set oCreateFileModule = oModel.Modules(nCreateFileIndex)

    ' Set the Max Arrivals fields
    If optFromRandomProcess.value = True Then
        oCreateRandomProcessModule.Data("Max Batches") = "Infinite"
        oCreateFileModule.Data("Max Batches") = "0"
    Else
        oCreateRandomProcessModule.Data("Max Batches") = "0"
        oCreateFileModule.Data("Max Batches") = "1"
    End If

    ' Play the sound file
    nSoundFileIndex = _
        oModel.Embeddeds.Find(smFindTag, "Mission Possible")
    If nSoundFileIndex > 0 Then _
        oModel.Embeddeds.Item(nSoundFileIndex).Do

    ' Hide the UserForm to allow the run to begin
    frmArrivalTypeSelection.Hide
    Exit Sub

End Sub
```

Figure 10-29. Complete cmdOK_Click Event VBA Code

10.4 Model 10-6: Recording and Charting Model Results in Microsoft Excel

We've already seen in Section 10.1.2 how we can write data to an existing worksheet and even take advantage of existing plots. In this section, we'll learn how to use VBA in Arena to take more complete control, including creating the worksheet, formatting the data, and creating the chart. We'll see how to use VBA to record information about each departing call in an Excel spreadsheet. Our objective will be to create an Excel file that lists three pieces of data for each completed call: the call start time, end time, and duration. We also want to chart the call durations to look for any interesting trends, such as groups of very short or very long calls. Figure 10-30 shows a sample of the model's results.

To build the logic necessary to create the spreadsheet and chart the results, we'll need to perform three tasks. First, we need to write the VBA code necessary to create the Excel file. We'll open the file at the beginning of the simulation run and will write new headers on the data worksheet so we can examine results for each eight-hour day. (Worksheets are the tabbed pages within an Excel file.) Second, we need to modify the model logic and write the necessary VBA code so that each time a sales call is completed its entity "fires" a VBA event to write the pertinent values to the worksheet. Finally, we'll write the VBA code to create a chart of the call durations at the end of each replication and to save the file at the end of the simulation run.

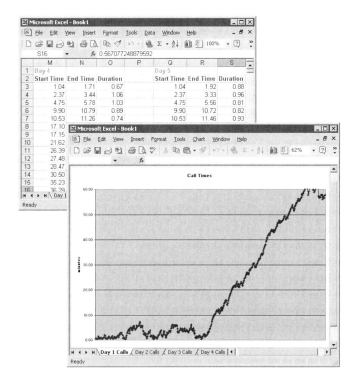

Figure 10-30. Microsoft Excel Results

The following sections describing these steps only highlight the concepts and list the corresponding code. If you're interested in developing a deeper understanding of these materials, explore Arena's help related to VBA and the Arena object model, and examine the SMARTs library models related to these topics. We also recommend any of the numerous books related to developing Excel solutions with VBA for reading about automating Excel. And there are many other resources at your disposal for learning Visual Basic, including a variety of books, CD tutorials, training courses, and Web sites.

10.4.1 Setting Up Excel at the Beginning of the Run

In order to store the call data during the run, we'll first create the Excel file, just as if we had run Excel and started a new file. At the start of each new day (i.e., replication), we also want to write headers for the data columns. To do this, we'll use two of Arena's built-in VBA events: `RunBeginSimulation` and `RunBeginReplication`. In `RunBeginSimulation`, we'll place the start-up code to run Excel and create the new workbook (i.e., file). And as you might expect, `RunBeginReplication` is where we'll write the headers, since we need a new set for each day (replication) of the run.

Figures 10-31, 10-33, and 10-34 list the VBA code for Model 10-6. To view this code in the Arena model, open the `Model 10-06.doe` file, which is Model 10-2 plus the VBA code described in this section. Select *Tools > Macro > Show Visual Basic Editor*, and double-click on the ThisDocument item in the Visual Basic project toolbar. Note that when the corresponding functions for the built-in events are created in the code window, the `ModelLogic_` prefix is added. We'll leave the prefix off for simplicity in our discussions, as Arena's VBA event list does.

The global declarations section, shown in Figure 10-31, consisting of the lines that are outside any procedure (that is, before the line defining the procedure `ModelLogic_RunBeginSimulation`), defines variables that are global to all procedures via a series of `Dim` statements (Visual Basic's variable declaration syntax, short for "dimension"). Also, we included the `Option Explicit` statement to tell VBA that all variables must be declared (i.e., with `Dim` statements); we recommend using this option to prevent hard-to-find coding errors due to mistyped variable names.

We declare a global variable `oSIMAN` that will be set in `RunBeginSimulation` to point to the model's SIMAN data object and will be used in the remaining procedures to obtain values from the running simulation. The variable type `Arena.SIMAN` establishes that the `oSIMAN` variable is a SIMAN object variable from the Arena object

```
Option Explicit

' Global variables
Dim oSIMAN As Arena.SIMAN, nArrivalTimeAttrIndex As Long
Dim nNextRow As Long, nColumnA As Long, nColumnB As Long, nColumnC As Long

' Global Excel variables
Dim oExcelApp As Excel.Application, oWorkbook As Excel.Workbook, _
    oWorksheet As Excel.Worksheet
```

Figure 10-31. VBA Global-Variable Declarations for Model 10-6

library. The other variables in the first two `Dim` statements are used to keep track of other values that are needed in more than one of the procedures. We'll describe them as we examine the code that uses them.

The remaining global variables—whose data types begin with `Excel.`—are declared to be object variables from the Excel object library. Because Excel is an external application (as opposed to Arena, which is the application hosting our VBA code), a *reference* to the Excel library must be established by clicking on the Microsoft Excel Object Library entry in the References dialog box of the Visual Basic Editor window, which is opened via *Tools > References*, as shown in Figure 10-32. This reference will allow you to use ActiveX Automation calls to control Excel. Note that it also requires that Excel be installed on the computer that's running this model. If you don't have Excel, you'll be able to open and edit this model, but you won't be able to perform simulation runs.

Examining the `RunBeginSimulation` code in Figure 10-33, we first set the `oSIMAN` variable equal to some interesting series of characters containing dots. When you're reading code that exploits objects, it's often helpful to read the object part from right to left, using the dots as separators between items. For example, the statement `Set oSIMAN = ThisDocument.Model.SIMAN` can be thought of as something like, "Set the `oSIMAN` variable to point to the `SIMAN` property of the `Model` object contained in `ThisDocument`." Without delving too deeply into the details of the Arena object model or Visual Basic, let's just say that `ThisDocument` refers to the Arena model containing this `RunBeginSimulation` code; `Model` is its main object, providing access to the items contained in the model; and `SIMAN` is an object that's part of the Model and you can use it to query or modify the model's runtime data (such as resource states, variable values, and entity attributes).

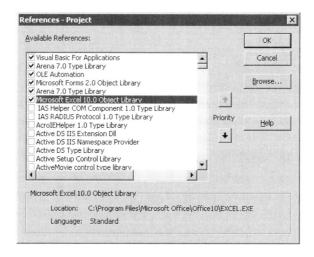

Figure 10-32. Tools > References Dialog Box from VBA

```
Private Sub ModelLogic_RunBeginSimulation()
    ' Set the global SIMAN variable
    Set oSIMAN = ThisDocument.Model.SIMAN

    ' Set global variable to store Arrival Time attribute index
    nArrivalTimeAttrIndex= oSIMAN.SymbolNumber("Call Start Time")

    ' Start Excel and create a new spreadsheet
    Set oExcelApp = CreateObject("Excel.Application")
    oExcelApp.Visible = True
    oExcelApp.SheetsInNewWorkbook = 1
    Set oWorkbook = oExcelApp.Workbooks.Add

    Set oWorksheet = oWorkbook.ActiveSheet
    With oWorksheet
        .Name = "Call Data"
        .Rows(1).Select
        oExcelApp.Selection.Font.Bold = True
        oExcelApp.Selection.Font.Color = RGB(255, 0, 0)
        .Rows(2).Select
        oExcelApp.Selection.Font.Bold = True
        oExcelApp.Selection.Font.Color = RGB(0, 0, 255)
    End With
End Sub
```

Figure 10-33. RunBeginSimulation VBA Code for Model 10-6

After we've established the oSIMAN variable to point to the SIMAN run data, we use it to store the index value of the Call Start Time attribute in one of our VBA global variables—nArrivalTimeAttrIndex—so that it can be used throughout the run to retrieve the individual attribute values from entities as they depart the model. The SymbolNumber function from Arena's object model is used to convert the name of some model element, such as an attribute, to its Arena runtime index (an integer between 1 and the number of those elements contained in the model). We use this for our attribute named Call Start Time and store the attribute index in our VBA global variable so that we require Arena to perform this calculation only once. The code that is executed during the run (in the VBA_Block_1_Fire event, which we'll cover shortly) will pass this index to the EntityAttribute property function to retrieve the actual value of the attribute.

Next, we start Excel using the CreateObject ActiveX Automation call and make it visible by setting the application's Visible property to True. Then we create a new Excel file (also referred to as a *workbook*) by automating Excel with the oExcelApp.Workbooks.Add method; you can think of this as "Add a new item to the Workbooks collection of the oExcelApp application." The oWorkbook variable will store a pointer to the newly created Excel workbook; we'll save the workbook to a file at the end of the simulation run. The remaining lines of code in RunBeginSimulation set our oWorksheet variable to point to the worksheet in the newly created Excel workbook and establish some of its characteristics. If you will be working with Excel, we suggest that you explore its macro-recording capabilities. You often can create a macro in Excel and paste its code into Arena, with minor modifications; this is much quicker than trying to dig through the Excel object model on your own for just the right code magic to perform some task.

```
Private Sub ModelLogic_RunBeginReplication()
    Dim nReplicationNum As Long, i As Integer

    ' Set variables for the columns to which data is to be written
    nReplicationNum = oSIMAN.RunCurrentReplication
    nColumnA = (4 * (nReplicationNum - 1)) + 1
    nColumnB = nColumnA + 1
    nColumnC = nColumnA + 2

    ' Write header row for this day's call data and
    '  set nNextRow to 3 to start writing data in third row
    With oWorksheet
        .Activate
        .Cells(1, nColumnA).value = "Day " & nReplicationNum
        .Cells(2, nColumnA).value = "Start Time"
        .Cells(2, nColumnB).value = "End Time"
        .Cells(2, nColumnC).value = "Duration"
        For i = 0 To 2
            .Columns(nColumnA + i).Select
            oExcelApp.Selection.Columns.AutoFit
            oExcelApp.Selection.NumberFormat = "0.00"
        Next i
    End With
    nNextRow = 3
End Sub
```

Figure 10-34. RunBeginReplication VBA Code

While the `RunBeginSimulation` event is called only once at the beginning of the run, `RunBeginReplication` is called by Arena at the beginning of each replication. Its code for Model 10-6 is listed in Figure 10-34. We use this event to write the column headers for our new day's data. To figure out the columns into which we'll write data, we use the current replication number (retrieved from the SIMAN part of Arena's object model via the `oSIMAN` global variable we initialized in `RunBeginSimulation`) and do the proper calculation to get to the three columns we want. (Refer to Figure 10-30 to see how the columns are laid out, with three data columns and then a blank one for each replication, which represents a day's worth of calls.)

The lines of code between the `With oWorksheet` and `End With` statements utilize the Excel object model to assign values to various cells in the worksheet and to format them to match the design shown in Figure 10-30. We'll leave it to you to explore the details of Excel automation via Excel's help. Our final statement sets the global `nNextRow` variable to start with a value of 3; this value will be used to determine the row to which each call's results are to be written during the run.

10.4.2 Storing Individual Call Data Using the VBA Module

Our next task requires both changing the model logic and writing some VBA code. Let's first make the model modification. Just before a call is finished (that is, prior to leaving the model via the Dispose module), it should trigger VBA code to write its statistics—call start time, call completion time, and call duration—to the next row in the Excel worksheet. The triggering of the VBA code in this case isn't as predictable as the first two events we examined, which are defined to be called at the beginning of the run and at the beginning of each replication. Instead, the dynamics of the simulation model will dictate

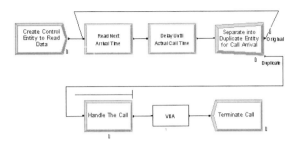

Figure 10-35. VBA Module

when the VBA code should be executed. Arena provides a VBA module on the Blocks panel for just this purpose. When you place this module, Arena will fire the VBA code that you've written for the VBA module instead of including some predefined logic in the module itself. Figure 10-35 shows the modified call-center model logic using an instance of the VBA module, inserted just before disposing the sales call entity.

When you place a VBA module, Arena assigns it a unique value that's used to associate a particular VBA module with its code in the Visual Basic project; these numbers are integers starting at a value of 1 and increasing by one with each newly placed VBA module. To edit the code for a VBA module, you return to the Visual Basic Editor and select the appropriate item from the object list in the code window for the `ThisDocument` project entry, as shown in Figure 10-36.

The event associated with our VBA module is named `VBA_Block_1_Fire()`, where the `1` matches the Arena-provided number of the VBA module. Its code is shown in Figure 10-37.

When a call entity completes its processing and enters the VBA module, the code in the `VBA_Block_1_Fire` event is invoked. The first two lines of the procedure retrieve information from the running simulation (via the `oSIMAN` variable that we set in `RunBeginSimulation`). First, we store the value of the active entity's attribute with index value `nArrivalTimeAttrIndex` in a local variable `dCreateTime`. The `nArrivalTimeAttrIndex` global variable was declared outside of all of the

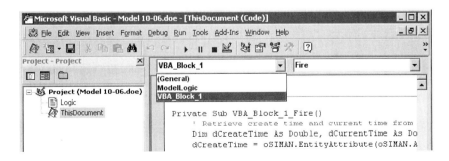

Figure 10-36. Visual Basic Editor Selection of VBA Module 1 Event

```
Private Sub VBA_Block_1_Fire()
    ' Retrieve create time and current time from SIMAN object data
    Dim dCreateTime As Double, dCurrentTime As Double
    dCreateTime = oSIMAN.EntityAttribute(oSIMAN.ActiveEntity, nArrivalTimeAttrIndex)
    dCurrentTime = oSIMAN.RunCurrentTime

    ' Write the values to the spreadsheet
    With oWorksheet
        .Cells(nNextRow, nColumnA).value = dCreateTime
        .Cells(nNextRow, nColumnB).value = dCurrentTime
        .Cells(nNextRow, nColumnC).value = dCurrentTime - dCreateTime
    End With

    ' Increment the row variable
    nNextRow = nNextRow + 1
End Sub
```

Figure 10-37. VBA_Block_1_Fire Code

procedures (so is available to all of them), and its value was assigned in RunBeginSimulation to be the index of the attribute named Call Start Time (using the SymbolNumber function). Next, we store the current simulation time in a local variable dCurrentTime. These values are used to store information about this entity in the spreadsheet, using the nNextRow variable to determine the row, which we then increment. After this code is executed for a particular entity, the entity returns to the model and enters the Dispose module, where it is destroyed.

10.4.3 Charting the Results and Cleaning Up at the End of the Run

Our final task is to create charts of the call durations for each replication and to save the spreadsheet file at the end of the simulation run. While we could do all of the charting at the end of the run since the data will exist on the data worksheet, we'll instead place the code in RunEndReplication to build the charts as the run proceeds.

After the final entity has been processed at the end of each replication, Arena calls the VBA RunEndReplication procedure. In our model, we'll chart the data contained in the Duration column of the worksheet for the replication (day) just completed, showing a line graph of the call lengths over that replication. We'll skip discussion of the charting code. If you're interested in exploring Excel's charting features, we recommend browsing the online help and using the macro recorder to try different charting options.

Finally, at the end of the simulation run, the RunEndSimulation procedure is called; ours will simply save the Excel workbook to file Model 10-06.xls in the same folder as the model (by using ThisDocument.Model.Path), leaving Excel running. Figure 10-38 shows the code for these two routines.

When you run this model, you'll see a copy of Excel appear on your desktop at the beginning of the run, followed by a series of numbers being added to the worksheet as call entities depart the model. Finally, the charts will be created at the end of each replication.

As you might expect, Arena's integration with other applications isn't limited to writing data to Excel and creating charts. If you've installed an application that can be automated, the VBA interface in Arena will allow you to do whatever is possible through the external application's object model, such as creating a chart in an Excel spreadsheet or cataloging run data in an Oracle database.

```
Private Sub ModelLogic_RunEndReplication()
    ' Chart today's sales call data on a separate chart sheet
    oWorkbook.Sheets("Call Data").Select
    oWorksheet.Range(oWorksheet.Cells(3, nColumnC), _
        oWorksheet.Cells(nNextRow, nColumnC)).Select
    oExcelApp.Charts.Add

    ' Format the chart
    With oExcelApp.ActiveChart
        .ChartType = xlLineMarkers
        .SetSourceData Source:=oWorksheet.Range(oWorksheet.Cells(3, _
            nColumnC), oWorksheet.Cells(nNextRow, nColumnC)), PlotBy:=xlColumns
        .SeriesCollection(1).XValues = ""
        .Location Where:=xlLocationAsNewSheet, _
            Name:="Day " & oSIMAN.RunCurrentReplication & "Calls"
        .HasTitle = True                   ' Title and Y axis
        .HasAxis(xlValue) = True
        .HasAxis(xlCategory) = False    ' No X axis or Legend
        .HasLegend = False
        .ChartTitle.Characters.Text = "Call Times"
        .Axes(xlValue).MaximumScale = 60
        .Axes(xlValue).HasTitle = True
        .Axes(xlValue).AxisTitle.Characters.Text = "minutes"
    End With
End Sub

Private Sub ModelLogic_RunEndSimulation()
    ' Save the spreadsheet
    oExcelApp.DisplayAlerts = False            ' Don't prompt to overwrite
    oWorkbook.SaveAs ThisDocument.Model.Path & "Model 10-06.xls"
End Sub
```

Figure 10-38. VBA Code for End of Replication and End of Run

10.5 Creating Modules Using the Arena Professional Edition: Template 10-1

Now we'll take a brief look at how you can customize Arena by building new modules using Arena's template building capabilities. If you are using the student version of Arena (which is what's on the CD accompanying this book), you won't be able to try this on your own since the template-building features aren't included. However, template building *is* part of the Research or Educational Lab packages, either of which may be available through your academic institution. Either way, though, you'll be able to *see* in an Arena model the use of the module that we describe in this section.

To present this quick tour of template creation, we'll walk through the steps to build a very simple module, showing you some of the windows and dialog boxes along the way. While we'll end up with a usable (and potentially useful) module, we'll only touch on a small portion of Arena's template-building features. This should be sufficient to fulfill our objective of raising your awareness of the Arena's capabilities. For a more thorough treatment of building your own modules, refer to the *Arena Template Developer's Guide* via *Help > Product Manuals*.

10.5.1 The Create from File Module

For our template, we'll revisit the modification to the call-center model, `Model 10-02.doe`, in which we developed logic to read arrival times from a file for the call creation times. To accomplish the task, we used four modules—Create, ReadWrite, Delay, and Separate—plus the File data module, which worked together to generate entities into the model at the appropriate times. If we were doing a fair amount of modeling that might utilize this trace-driven approach for entity creation, it might be handy to "package" these modules (and the appropriate data to make things work right) into a single module. That way, we wouldn't have to remember the trick of how we got the entities into the model at the right time, and our models themselves would be a little less complicated to view.

To build this module, which we'll call `Create from File`, we'll copy the model logic that we already built in Model 10-2 into what's called a *template file*. Then we'll define an *operand*—a field that shows up in the dialog box when you double-click on the module—that allows the file name to be changed whenever an instance of our `Create from File` module is used. (We wouldn't want to be so presumptuous as to think that all of our modules will read from a file named `Model 10-02 Input.txt` like the original one did.) We'll also draw a picture to be displayed in the template toolbar, and we'll arrange the objects that are to show up when the module is placed in a model window. These four simple steps—defining the logic, operands, panel icon, and user view—are the basics of building templates in Arena.

Before we take a more careful look at each of these steps, let's see the end result—one of these `Create from File` modules placed in a model window—as shown in Figure 10-39. Our guess is that you'll think, "Well, it looks like any other module to me." And, as a matter of fact, it *is* just like any other module. When you use Arena's template-design tools to create your own modules, the end result is a template panel object (*.tpo*) file just like those you've been using—`BasicProcess.tpo`, `AdvancedProcess.tpo`, etc.

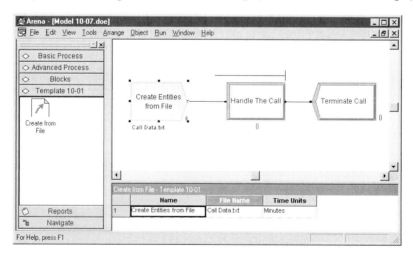

Figure 10-39. Create from File Module Used in a Model

An important part of the philosophy behind Arena is to allow easy creation of personalized modeling constructs that work with the standard templates.

10.5.2 The Template Source File: Template 10-01.tpl

To examine the definition of the `Create from File` module, we'll look at the `Template 10-01.tpl` file containing its logic, operands, etc. If you're running the Research or Educational Lab version of Arena, you can open this file via the standard *File > Open* menu option; just change the entry in the Files of type field to `Template Files (*.tpl)` and select `Template 10-01.tpl`. This opens a template window, listing all of the modules contained in the template. (For our template, we have just the one lonely `Create from File` module.) From this window, the buttons on the Template Development toolbar (in Figure 10-40) open the various other windows that define modules, preview the module dialog box, and generate the template panel object (*.tpo*) file for use in a model.

10.5.3 The Panel Icon and User View

Before we dig into the innards of the `Create from File` module, we'll briefly discuss the first two things that someone using the module sees. To utilize the module in an Arena model, the template panel containing it needs to be attached to the Template toolbar, as was shown in Figure10-39. On the toolbar, Arena displays a picture drawn by the template's creator for each module, referred to as the module's *panel icon*. In the case of our `Create from File` module, the picture depicts a dog-eared page with an arrow (representing, to the best of the limited artistic abilities of this module's designer, the concept of creating entities from a file).

When a modeler places an instance of the module in a model window, the graphic objects that are added to the window are collectively referred to as the module's *user view*. Each module has at least a handle, which typically is the module name surrounded by a box. Most modules also have one or more entry or exit points to connect them with other modules. Referring to Figure 10-39, the `Create from File` module shows the Name operand (some description of the module that the modeler types in its dialog box) inside a box with a pointy side, and it has a single exit point (connecting it to the Process module).

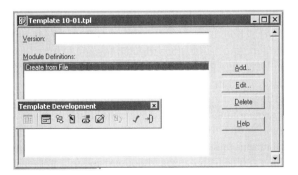

Figure 10-40. Template Window and Toolbar

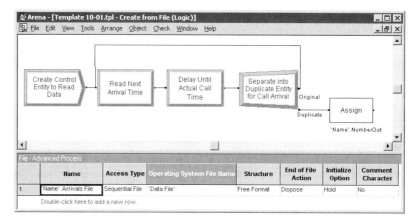

Figure 10-41. Logic Window in Module Definition

10.5.4 The Module Logic and Operands

The heart of what a module provides when it's placed in a model window is the actual logic to be performed during the simulation run. When you're building a module in your own template, you define this logic just as you build a model, by placing and connecting modules from other templates. The logic underlying our `Create from File` module contains the modules we used in Model 10-2 plus an Assign module to count the number of entities that left the module. Figure 10-41 shows this logic, which is placed in the logic window of the module's definition. We copied the four modules and their connections from Model 10-2 and pasted them into the logic window. Then we added the Assign module from the Blocks panel, connecting it to the exit point labeled Duplicate from the Separate module.

Each time our `Create from File` module is placed in a model, this underlying logic will be included so that entities are generated and sent into the model based on the data in the text file. The astute reader might wonder, "What file? Where in the model?" This is where the module's operands come into play. If we just used the original logic from Model 10-2, then we could only use the `Create from File` module to generate entities from a file called `Model 10-02.txt`. Instead, we'll add an operand to our module called `File Name` that permits a new file name to be entered each time the module is placed. This file name will be added to the model as part of the File data, using

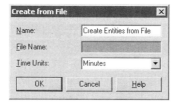

Figure 10-42. Module Dialog Box When Placed in Model

a special type of operand designed for this purpose. We'll also present an operand called the Name, which will be displayed in the user view as part of the flowchart, and one called Time Units to define the type of information in the data file.

The Create from File dialog with which the user will interact is shown in Figure 10-42. In the template, we design that dialog and, in the dialog design window, define its operands, as shown in Figure 10-43. The dialog design window has four sections. The center area is the dialog design window, which displays the dialog form layouts for the module and is equivalent to the background screen in a word processor where you might type. The other three sections are:

- the *Toolbox* (left side) to add user-interface controls to dialog forms,
- the *Operand Explorer* (upper right) to navigate through all of the dialog objects in the module definition, and
- the *Design Properties* (lower right) to edit the properties of one or more selected objects.

Every dialog design window must have at least one dialog form, which is provided by default. This represents the dialog that is first displayed if the user double-clicks on the module in an Arena model window. You may resize a dialog form here by simply selecting it and dragging its sizing handles. Notice that the dialog design window has an upper and lower part separated by a sliding splitter bar. The upper part is for things the user sees (e.g., dialogs, operands). The lower part is for hidden operands—a place to store data that the user does not directly enter. We will see an example of that in a moment. An empty dialog form is neither interesting nor useful, so let's see how we can populate it to make it helpful.

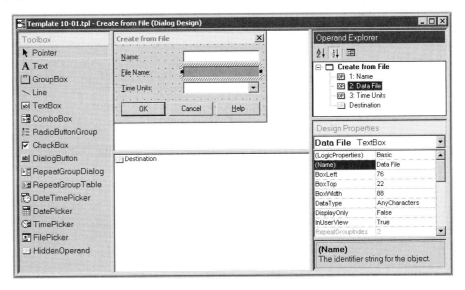

Figure 10-43. Operand Window in Module Definition

The Toolbox provides an interface for graphically adding user interface controls (e.g., text boxes, combo boxes, or dialog buttons) and static graphics (e.g., text, lines, or group boxes) to a dialog form. Items can be added to the dialog form by simply clicking on the desired item in the Toolbox, then dragging it to the desired location on the dialog form. After placement, it can be graphically relocated and resized as needed. If you have too much information to place in a single dialog form, then one of the controls is a dialog button that creates an associated sub-dialog form.

While a dialog form containing controls and static graphics is more interesting, it doesn't really become useful until we define the properties of the controls. For that we move to Design Properties, which also has two parts. The upper part is for navigating through the properties and viewing/editing values for each property. The lower part of this window provides helpful information about the highlighted property. The properties displayed will change based on the control and options selected, but here you will find all the data necessary to define the behavior of each control.

There are two ways to navigate between controls. The first is simply to click on a control to select it. This may be the most convenient procedure when all your controls are on a single, relatively simple dialog form. When you have either complex or nested dialog forms, you may find it more convenient to navigate using the Operand Explorer. Just click on an operand and its dialog form will be brought up (if necessary), the control will be selected in the dialog form, and the properties of that control will be displayed in the Design Properties. Note that there is one significant difference between the Operand Explorer and the dialog form. In the dialog form, operands are typically identified by their Text property, which is how the user identifies a field. In the Operand Explorer, operands are identified by their Name property, which is how controls are identified within the module definition.

Now back to our template. Referring to Figure 10-43, you'll see the dialog design window for our dialog form with the `Data File` operand selected. In Figure 10-44, we take a closer look at the Design Properties for that operand. We named the operand `Data File` (which will be referenced in the File data module); accepted the default Basic type; allowed for any characters as input; left the `Default Value` empty; set the `In Userview` option as `True` to display the file name under the module handle in the model window; and checked the `Required` option as `True`, so that a modeler must provide a non-blank value when editing the module. We also specified a `Text` property of `&File Name`, which is displayed in the dialog form. The "&" before the "F" signifies that this operand can be reached in the Arena dialog by typing "Alt-F."

In the dialog form, we also added a TextBox named `Name`, which appears in the model window and will define part of the Arena File Name in the ReadWrite and File modules. We added a ComboBox named `Time Units` to provide choices of Seconds, Minutes, Hours, and Days in a drop-down list; its value is passed to the Delay module in the logic window. And finally, we added a HiddenOperand, `Destination`, into the lower part of the main window—it is not added to the dialog form because it will not be directly visible to the user. This is an Exit Point type, which is referenced in the Next Label field of the Assign module in the logic window to establish the flow of entities through the model that contains an instance of the `Create from File` module. When

Figure 10-44. Design Properties for File Name Operand

Arena generates the SIMAN model (e.g., when you Check the model), it will start with the Create module in the logic window (since Create is a special module that starts the flow of entities into a model); follow the connections through the ReadWrite, Delay, Separate, and Assign modules; establish the loop back for the primary entity to leave the Separate and return to the ReadWrite module; and, via the `Destination` operand of the Assign module (because it's of the special Exit Point type), connect the Assign module to whatever the `Create from File` connects with, such as the Process module in Figure 10-39.

You might think of the Exit Point operand type as an "elevator up" in hierarchy, connecting from the underlying logic of the `Create from File` module to the main logic of the model in which it's included. And, not surprisingly, there's an Entry Point operand type that's an "elevator down," establishing a connection from logic in the higher level of hierarchy into modules that are contained in another module's underlying logic window.

The values that a user enters in the module dialog box are passed down to the modules in the logic window via *operand referencing*. If some field of a module in the logic window is enclosed in back quotes (`` ` ``), then its actual value comes from the module's dialog box—in particular, the operand whose name matches the name inside the back quotes. In our case, we'll define the Arena File Name in the ReadWrite module's dialog box and the File module spreadsheet to be `` `Name` ``.ArrivalsFile (as shown in Figure 10-45). If someone typed `June26Data`, for example, as the module Name in the Create from File module, then the Arena file name used for the simulation run (through the ReadWrite and File modules) would be `June26Data.ArrivalsFile`. This operand referencing also is the mechanism used to dictate that connections from the Assign module should leave via the Destination operand; we typed `` `Destination` `` in the Next Label field of the Assign module in the logic window to establish this exit point from the module.

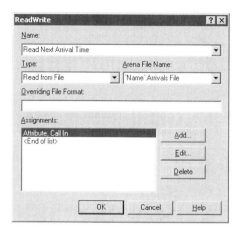

Figure 10-45. Operand Referencing By Logic Window Module

This completes our definition of the `Create from File` module, allowing someone who uses it in his or her model to specify the name of the file containing the arrival times and to connect it to another module. The underlying logic also establishes some fixed logical elements (that is, aspects that can't be changed by a modeler). In particular, each entity that's sent into the model by this `Create from File` module will have an attribute named `Call Start Time` that has a value of its arrival time, because the ReadWrite module still makes that assignment. When you design a module, you need to decide which aspects you want to permit a user to change (as in the file name in our example) and which you want to protect from modification (for example, the attribute name and assignment).

Most modules that you're accustomed to using contain numerous operands and often include complex logic. These modules, like those in the Arena template's panels, were created using the Arena template-building capabilities. Their panel icons, user views, operands, and underlying logic employ standard template-design capabilities of the software. Exploring the features of these modules can help you grasp the potential of Arena templates. While we've touched on only the basic architecture and product capabilities here, we'll close with some ideas about how custom templates might be employed by individuals and throughout organizations.

10.5.5 Uses of Templates

Templates may be developed to address a wide range of needs. Some might be conceived for use by a large targeted market, such as the Arena Contact Center Template or the Arena Packaging Template. Others might be more like a utility, such as the simple example presented in this chapter.

The most ambitious Arena templates are those that are developed for a commercial market, typically targeted at a particular industry, such as high-speed production or customer-relationship contact centers. There are two main advantages of industry-focused templates. First, the template can use terminology that is appropriate for the industry, minimizing the abstraction needed for a modeler to translate a system into the software. More importantly, through Arena's hierarchy, a template can be built that is fully customized to represent accurately the elements of systems in the industry. The designer of the template has the capabilities at hand to mimic exactly the behavior of equipment, people, parts, components, etc., providing whatever spectrum of options is appropriate for the variations of these system elements. Furthermore, through the ActiveX Automation technology supported by Arena, wizards and other utilities can be created that work in cooperation with a particular template for generating specialized graphs and reports, loading data directly from external databases, etc.

Many of the templates that are developed using Arena aid modelers in representing a particular system, facility, or process. While they may not be "commercial-grade," these templates have many of the same goals as the industry templates and provide many of the same benefits. Here, though, the focus might be narrower than a commercial template. For example, a template might be built for use in analyzing truck-loading schemes or for representing dispatching rules of incoming calls to technicians. These application-focused templates benefit from Arena's hierarchical structure in the same ways as industry-focused templates: the interface presented to a modeler can be customized to be very familiar (both in terms of graphical animation and the terminology presented to the user); and the elements of the target application environment can be represented accurately. In some cases, a modeler might build these templates just for his or her own individual use, if the same type of problem is likely to be modeled repeatedly. In other cases, templates might be created for use among modelers in a common group, and application templates can be shared among different modeling groups throughout an enterprise.

For an individual modeler, Arena templates afford the opportunity to reuse modeling techniques that are learned in the process of building models. In the evolution of programming tools, reusable code was captured in procedures/subroutines/functions; later, object-oriented tools allowed the full characteristics of "objects" represented in the software to be defined for reuse. A module can be thought of as analogous to an object (in object-oriented software)—the module allows you to capture the complete characteristics of a process that you want to model in a self-contained package that you may reuse and that may be customized in each use. The `Create from File` module sketched out in this section is an example of this type of template, where once a modeling technique was established and tested, its implementation details were "hidden" inside the definition of the module. Later uses of the same technique require little knowledge of its approach or implementation details and have less risk of error since the embedded logic has already been tested.

10.6 Real-time Integration

So far our discussions about integration have primarily dealt with getting information from a file (or putting it into a file) at the beginning (or end) of a run or perhaps incrementally while a simulation is running. But there are many interesting simulation

Figure 10-46. Process Controller

uses that can result from a simulation sharing data with other software during a run. In particular, there are many potential uses in the process control world—the world of low-level computers (see Figure 10-46) and devices responsible for driving and monitoring virtually everything that moves in an automated facility. We will discuss a few applications:

- Unit testing process controller logic
- Comprehensive control system testing
- Operator interface design
- Operator training/certification
- Testing high-level software

The term *real-time*, as we will be using it here, simply means concurrent. More specifically, two or more software programs are running and interacting concurrently (or in real-time) and, for our purposes, at least one of those programs is an executing simulation model. The programs may be running on the same processor or on different processors that are somehow connected (say a serial cable, a network, or carrier pigeons). The programs may be very similar (for example, two simulation programs) or can be quite different (for example, Arena communicating with a hardware controller).

The interaction generally consists of useful information being sent in one direction, typically with at least a *response* (acknowledgment of receipt) being returned to the sender. In many cases, the interaction has useful information being exchanged in both directions. The interaction can be specifically programmed custom to the applications such as you might do using Arena's VBA or C++ user code interface, or it can be accomplished using common standards for data interchange like *OLE for Process Control* (OPC). Using high-level approaches like OPC allows the interaction to be almost invisible to the users.

The timing of the interaction is important depending on the reason for the interaction. In some cases, a response time (typically the time between when a message is sent and when its acknowledgment is received) of a few seconds or even minutes is acceptable; in other cases, responses are expected in a few milliseconds. Obviously the response-time requirement can affect the selection of the software, hardware, and communication mechanisms used.

A final aspect of real-time simulation is that of the simulation time clock. In most cases, each program interfaced has its own time clock, or at least some concept of time. In the case of two simulation programs running concurrently, it is important that their time clocks be (at least periodically) synchronized. In some cases, one program might have additional constraints. For example, most process controllers have a strong concept of time but cannot run in accelerated time. When interfacing to a process controller, it may be necessary to synchronize the simulation time clock with a global system clock time.

There are many possible uses for simulation and soon we will survey a few of them. Some of those uses involve *emulation*, commonly defined as the process of modeling the performance of a computer device or program. Since emulation is clearly a subset of simulation and it is often a bit fuzzy when something changes from simulation to emulation, we will use the more generic term simulation in our discussions.

Now let's look a little more into each of the applications we introduced previously.

- **Unit-testing process controller logic:** This verifies that the logic and I/O in controller code behaves in the way it was intended. Traditionally this is attempted by connecting a controller to a physical test stand (shown in Figure 10-47) containing dials, switches, lights, and other devices so that engineers can set up various configurations and try the results. Sometimes a set of independent logic is written on the controller with the sole purpose of exercising the main controller code. Either method is difficult, time consuming, and often unreliable. A simulation can represent a set of devices (e.g., switches, valves, motors) that are the essence of the plant floor. When a controller is interfaced to the simulation, it behaves exactly as though it is controlling the real system and the simulation will react similarly to the real system. In addition, the simulation can introduce both deterministic and stochastic failures and unusual situations to challenge the controller and ensure that it can handle those situations.
- **Comprehensive control-system testing:** This validates that a set of controllers works together to address correctly all the situations that are likely to occur in the live system. While some of the traditional techniques used in unit testing are also used here, they are even less effective when applied to an entire system. Control

Figure 10-47. Example of a Physical Test Stand

programs inevitably fail to account for some unforeseen sequence of events or interaction of components, but evaluating system performance under a variety of scenarios and conditions is one strength of simulation. One or more simulations can be used in a way similar to that described above to provide assurances that as a whole the system works and that devices interact appropriately.

- **Operator interface design:** More and more processes are being converted from traditional dials and switches to control through a graphical user interface (GUI). A bank of physical controls can be replaced by a single touch-screen monitor with a sophisticated menu system. Designing these systems to be used effectively by an operator can be a daunting task—often done with paper mock-ups of the proposed system. At the late design/early implementation phase, simulation can be used to drive the proposed GUI and interact much like the real system would. Operators can then exercise the GUI to try both common and uncommon tasks and look for bugs and improvements well before system delivery.

- **Operator training/certification:** Initial operator training is often accomplished on the actual production system. This would be somewhat akin to training a 747 pilot at the controls of a real 747. We obviously don't do that in aviation due to safety and cost considerations, and we should not do it in manufacturing for similar reasons. Just as in aviation, simulation can be used to represent the real system and allow operators and maintenance people to interact with the simulation as practice for similar interactions with a real system. Also similar to aviation, trainees can accomplish accelerated training, practice non-routine events, and even earn certifications, all without touching or affecting the real system. Some of this training can be completed before the actual system is even built.

- **Testing high-level software:** High-level software (HLS) like manufacturing execution systems, scheduling, and batch management software is difficult to test because of the complexity and variations of the configurations and applications. Most testing is done on the live system. But the HLS system can be simulated and its reaction to routine and non-routine operation can be evaluated before deployment. By also proactively simulating realistic drivers to the system (MRP output, for example), the HLS response under various scenarios can be tested and validated.

Most of the applications affect the commissioning time of new systems. Commissioning time is the period from when a new system is delivered until it is running at full capacity. During this period, all the capital has been spent for the system but it is not yet producing its design capacity due to issues like system debugging and personnel training. The lost productivity during commissioning time is a major expense associated with every new system. According to a 2006 ARC Advisory Group survey[2], it is common to spend 15 percent or more of a project budget on startup costs—and that does not include the significant cost of lost production and delayed time-to-market. Using simulation in the ways we've described can reduce the effort associated with

---

[2] ARC Advisory Group is a research and advisory firm in manufacturing and supply chain solutions.

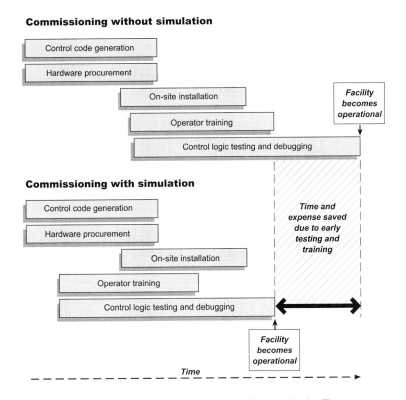

Figure 10-48. Impact of Simulation on Commissioning Time

commissioning and shift the timing of this effort earlier in the process when changes are easier and less expensive to implement (detailed in Figure 10-48).

You might wonder what features Arena has available to implement real-time simulation. To start with, you can use VBA routines to interface with other programs. You can also use the C/C++ user-coded functions in a similar fashion. With either of these approaches, you can use any communication mechanism you choose. A completely different approach is to use ActiveX to automate Arena from the other program. In this case, the other program would be controlling Arena, rather than working as peers.

Arena Real-Time (Arena RT) is a product developed with real-time applications in mind. Arena RT adds several useful features to Arena. First, it allows the simulation time clock to be synchronized with the computer clock so that the model can run in real-time or any multiple of real-time. This is useful not only for the above applications, but also to speed up (or slow down) very slow (or fast) processes for training and evaluation purposes. As an alternative to synchronizing with the computer clock, you can instead use any external time-keeper. This feature is especially useful in High-Level Architecture (HLA) and similar applications where multiple simulation models are being synchronized.

Arena RT also includes constructs to integrate external behavior into simulation logic (search for "TASKID" in Arena's Help). This feature is integrated into a schema for inter-process communication based on sending and receiving messages to communicate action requests and status changes. The communication is implemented in user-customizable code with a Windows Sockets example. There is an evaluation mode for all of the above Arena features. Refer to the Examples folder (inside the Arena 10.0 folder) and look for the model RT Real Time Clock for a simple example of running in real time, and look at the model RT Execution Mode for an example of the messaging.

The next release of Arena RT is expected to have several major enhancements. OPC[3], mentioned above, is a popular standard for communications in the control world. It works by defining OPC Servers that manage data, and OPC Clients that supply and consume data to/from OPC Servers. Arena version 11.5 will have native ability to be an OPC Client, an OPC Server, or both (in case you would like to see Arena talk to itself). This provides an easy mechanism to share data. With a few clicks, all the variables in an Arena model can be shared with other OPC applications, with no code changes. And if you choose to define Tags manually (the data items being shared), you can do so with the Rockwell Tag Browser, which is widely used across all Rockwell products and allows a tag to be defined once, then easily referenced in other applications.

10.7 Summary and Forecast

In this chapter, we have examined a number of topics that are part of a theme of customizing different aspects of Arena modeling and integrating Arena with other applications. We took a whirlwind tour of VBA, saw how Arena and Microsoft Office can work together, and built a custom modeling construct of our own with Arena's template-building tools. We finished with a look at some real-time applications and their benefits. The material in this chapter represents the "tip of the iceberg," with the intent of tickling your imagination for what's possible and arming you with enough fundamental knowledge of the features to go off and explore on your own.

If you choose to continue, you'll find that Chapter 11 takes on the topic of modeling continuous-change processes, such as the flow of liquid or bulk materials. Then we turn our attention to some important topics that underlie and support good simulation studies in Chapters 12 and 13.

10.8 Exercises

10-1 Starting with Model 10-1, modify the logic to store the time between arrivals and the processing time (excluding any waiting time) in attributes. Then use the ReadWrite module from the Advanced Process panel to write this pair of attributes to an Excel or Access file (or a text file if you don't have either of these applications installed).

10-2 Start with Model 10-2 and use the data file you generated in Exercise 10-1 to specify both call-arrival logic and call-handling time. Note that since the data are time

---

[3] Additional information about OPC can be obtained from the OPC Foundation at http://www.opcfoundation.org/

between arrivals rather than absolute arrival time, slightly different logic will be required. How do your results compare to those in Exercise 10-1?

10-3 Build a simple, single-server queueing model with entity interarrival times of EXPO(0.25) minutes. Using the ReadWrite module, prompt and query at the beginning of the simulation run for the server's process time mean (give a default value of 0.2 minutes). Use this value to establish a uniform distribution on [*a*, *b*] where *a* is 10% below the entered mean and *b* is 10% above the entered mean. Run the simulation until 300 entities have departed the model.

10-4 Create the model described in Exercise 10-3, replacing the ReadWrite modules with a VBA form that's displayed at the beginning of the simulation run.

10-5 Using the single-server model from Exercise 10-4, add logic to play a sound or display a message whenever the number of entities in the service queue exceeds some threshold value. Allow the modeler to establish this threshold in the form that's displayed at the beginning of the run.

10-6 Present a VBA form at the end of the simulation run reporting the average and maximum queue lengths for the product queues in Model 10-1. If you have a charting program (e.g., Excel) installed on your computer, also draw a bar graph of the average values.

10-7 Create a model that writes 1000 records to a file. Each record will include the record number and ten random samples, each between 0 and 100,000. In *Run > Setup > Reports*, disable the normal model output by selecting SIMAN Summary Report as the default output and select Disable generation of report database. Enable *Run > Run Control > Batch Run (No Animation)* so that we can execute as fast as possible. Repeat this experiment using a text file, a binary file, an Excel file, an Access file, an XML file, and no file (just generate and discard the random samples). Compare the execution time and the file sizes for each trial. If you have a fast machine, repeat with 10,000 and 100,000 records.

CHAPTER 11

Continuous and Combined Discrete/Continuous Models

So far, whether you realized it or not, all of our modeling has been focused on discrete systems—that is, processes in which changes to the state of the system occur at isolated points in time called *events*. In Arena, these event times are managed by the event calendar, with entities moving through the flowcharted process to define when changes in the system state will occur and what changes should take place (for example, a resource is released). In Chapter 2, our hand simulation accounted for all of these transitions, defining all of the events that could change the system and their exact occurrence times. These discrete-event models can capture many types of systems, from service processes (such as the automotive repair center in Chapter 5) to manufacturing systems (such as the small manufacturing system in Chapter 7).

In other cases, though, the state of the system might change continuously over time. Consider a brewery, where various ingredients (hops, barley, water) are mixed, heated, stored, and transferred in pipes among tanks. In this system, the process of emptying a tank involves a flow of product. To model the volume of beer in the tank, you'd want to be able to represent its rate of change over time and allow this rate to produce a smooth, continuous flow of lager into the target receptacle (preferably one that's destined for your next social gathering). The level of the tank at some future time can be calculated by applying the rate of change to the starting level; simple arithmetic is all that's needed, assuming that no intervening event occurs, such as a power outage, technical malfunction, or large sampling from a passing tour group. You may want to capture these discrete events that interact with this continuous process, such as closing a valve to interrupt the emptying process and set the rate of change of the tank's level to zero.

More complex processes may involve rates of change that depend on other continuous processes, in which case the future level value can't be calculated so simply. In these cases, integration algorithms are used to determine the levels based on the relationship between the continuous processes. The temperature of water in an aquarium is an example of this type of system. It changes continuously over time, depending on factors such as the heat from sunlight, the change in temperature of the room in which it is displayed, and the cycling on and off of a heater. These elements—heat from the sun, room temperature, and warmth generated by the heater—also change continuously. To capture these processes accurately in a simulation model, specialized continuous-modeling approaches are used that allow these relationships to be defined as mathematical equations that are incorporated in the integration algorithm.

In this chapter, we'll turn our attention to these processes in which the state of the system can change continuously over time. In Section 11.1, we discuss the nature of continuous processes and introduce Arena's terminology and constructs for modeling the continuous aspects of these systems via a simple model. We barge forward into the interface between continuous processes and discrete model logic in Section 11.2 and present a model of a coal-loading operation to illustrate how this works in Arena. More complex continuous systems, where the rate of change can depend on other aspects of the process, are discussed in Section 11.3 and are explored via a soaking-pit furnace model.

To give credit where it is due, much of the discussion of continuous and combined discrete/continuous modeling concepts in Section 11.3, including the soaking-pit furnace example, is based on materials from Pegden, Shannon, and Sadowski (1995). Some of this chapter's exercises also came from this source.

After you've digested the material in this chapter, you should understand how simulation models of continuous processes differ from those of discrete models. And once you've successfully completed working through this material, we hope you have an appetite for putting these powerful modeling approaches to work!

11.1 Modeling Simple Discrete/Continuous Systems

When modeling a continuous-change process, the two primary elements of interest are the value that's changing and its rate of change over time. Arena provides two types of variables called *levels* and *rates* to represent these values. For each pair (a level and a rate), Arena applies the defined rate of change to approximate a continuous change in the value of the level. The discrete-event portion of the model (that is, the modules you're used to using to model a process) also can assign new values to levels and rates.

In this section, we introduce a small example, Model 11-1, which illustrates how Arena's continuous and combined discrete/continuous modeling features work.

11.1.1 Model 11-1: A Simple Continuous System

Let's look at a very simple continuous process where some liquid product—carpet cleaning liquid—is being poured into a tank at a fixed rate of 10 volume units per minute. We'll build into our model a single level (called `Tank Volume`) and its corresponding rate (called `Tank Input Rate` and having an initial value of `10`), and we'll add an animation plot[1] of the value of the `Tank Volume`, Figure 11-1. We also need a Continuous module, which establishes some of the settings required for continuous models.

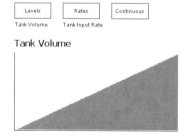

Figure 11-1. Simple Continuous Model

---

[1] For plots like this that display values, we enable the Non-Stepped option to show changes as a continuous line rather than discrete steps.

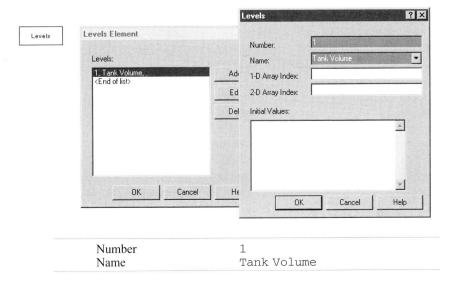

Number	1
Name	Tank Volume

Display 11-1. Levels Module Defining Tank Volume

Display 11-1 shows the entries for defining the level variable, `Tank Volume`, in the Levels module from the Elements panel. We'll match each level in the model with its corresponding rate variable by assigning them unique numbers (`1`, in this case). Arena will match a rate and level by these numbers to know which rate value should be applied to change the value of a level variable during the run. These numbers must be assigned so there's a one-to-one correspondence between levels and rates.

The `Tank Input Rate` is defined in the Rates module from the Elements panel, shown in Display 11-2. It is given number `1` to match it with the Tank Volume level and is assigned an initial value of `10` so that the rate of change of the Tank Volume is 10 volume units per base time unit (which we'll establish as minutes in the *Run > Setup > Replication Parameters* dialog box).

As you can see from the plot in Figure 11-1, the value of the Tank Volume (a continuous level) changes smoothly over time. And, in the absence of any discrete-event logic to change the value of the level or rate, the Tank Volume would continue to rise at this constant rate—10 volume units per time unit—until the end of the simulation run.

When a model contains a continuous component, Arena must augment the discrete time advance that's driven by the event calendar with logic to track the change in continuous variables over time. While Arena can't truly advance time continuously, it can approximate a continuous change in level values by making a series of small steps between the usual discrete events. At each of these continuous-update times, Arena calculates new values for the level variables based on their rates of change. To do so, Arena integrates the rate values to determine the corresponding new values of the levels.

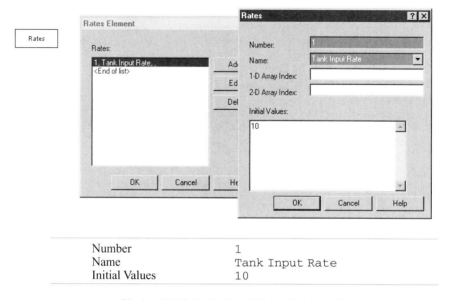

Number	1
Name	Tank Input Rate
Initial Values	10

Display 11-2. Rates Module Defining Tank Input Rate

The Continuous module from the Elements panel establishes settings that are needed to configure Arena's continuous calculations. Display 11-3 shows the Continuous module for Model 11-1. In the case of simple, constant-rate models, only a few of these fields are pertinent. The Number of Dif. Equations defines how many differential equations are to be evaluated for this model. In constant-rate systems, each level/rate pair should be counted among the differential equations so that Arena will calculate new values via the continuous time updates. For Model 11-1, we leave this field at its default value (blank), which will set the number of these equations equal to the number of level/ rate pairs defined by the Levels and Rates modules (in our case, one). We don't have any state equations (which we'll tell you about in Section 11.3), so we default that field. The Minimum Step Size and Maximum Step Size fields dictate to Arena how often it should update the continuous calculations. For this model, we'll leave them at their defaults of 1.0 minute. (Note that the units are from the Base Time Units selection in the *Run > Setup* dialog box.) You can see the effect of the step size on the time required to run the simulation by changing both to something smaller (e.g., 0.001); it takes Arena much longer to perform the same run, because it is calculating the continuous-variable updates more often. We also accept the default for the Save Point Interval, which relates to continuous-statistics calculations; we're not using those yet. In the Method field, we leave the default, Euler, which is the appropriate integration algorithm to be used when the rates remain constant between continuous-time updates, as in our model. And finally, we'll ask Arena to generate a Warning message if any crossing threshold tolerances are violated; we'll see more about this in Model 11-2, as described in Section 11.1.2.

Minimum Step Size	1.0
Maximum Step Size	1.0
Method	Euler
Cross Severity	Warning

Display 11-3. Continuous Module for Model 11-1

11.1.2 Model 11-2: Interfacing Continuous and Discrete Logic

In this carpet-cleaning-liquid model, what if we wanted to cause the amount of liquid in the tank to fill to a threshold and then empty at a constant rate (e.g., roughly representing the life of new puppy owners)? Here, we need to interface discrete, process logic with the continuous model.

Let's use a very simplistic approach to illustrate the idea. We'll keep the initial Tank Input Rate value at 10, so that the volume changes at a rate of 10 volume units per minute. We'll add model logic to create an entity at the beginning of the run, delay for 10 minutes to let the volume reach 100, and then assign the Tank Input Rate to -10. This will change the calculation of the Tank Volume to *decrease* at a rate of 10 volume units per minute, emptying the tank. After making the assignment, our entity delays again for 10 minutes to allow the tank to empty, then assigns the Tank Input Rate back to +10 to start the refilling process. (This version of the model is Model 11-02a.doe.)

Figure 11-2 shows the logic and a plot of the Tank Volume.[2] This works fine, as long as we can calculate the delay times required until the tank would fill or empty—that is, there aren't any other events influencing the volume in the tank (the level) or the fill/empty rate.

---

[2] If you build and run this model, you'll see that the plot finishes at a blinding pace. To see the plot build more slowly, open the *Run > Setup > Speed* dialog box and check the option to Update Simulation Time Every 0.1 time unit.

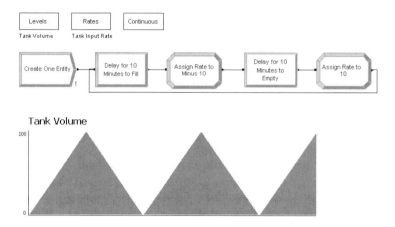

Figure 11-2. Control Loop to Empty and Refill Tank

If we're concerned about other events that might affect the level or rate of flow, we can model the process by "watching" the level of the tank in place of the fixed-time delays. For this approach, we'll use the Detect module from the Blocks panel. The Detect module provides a point at which an entity is created in the discrete-event portion of the model, analogous to a Create module. The timing of the entity creations is dictated by watching for the value of a continuous-level variable to cross a threshold value, in contrast to the Create module's predefined series of time-based arrivals. Whenever the specified threshold is crossed, Arena creates an entity, which proceeds out of the Detect module into the flowchart logic with which you are so intimately familiar.

In our carpet-cleaning-liquid model, we want to do something whenever the tank fills (i.e., the Tank Volume reaches 100 in the positive direction) and whenever it empties (i.e., reaches 0 going down). The model logic in Figure 11-3 generates this sequence of events during the run and, not coincidentally, results in the same plot of the Tank Volume as Figure 11-2. (It's named Model 11-02b.doe.)

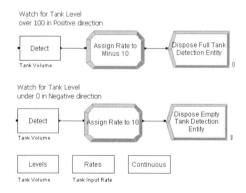

Figure 11-3. Fill and Empty Logic Using Detect Modules

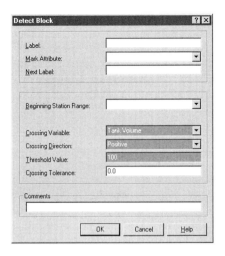

Crossing Variable	Tank Volume
Crossing Direction	Positive
Threshold Value	100
Crossing Tolerance	0.0

Display 11-4. Detect Module Watching for Full Tank

The Detect module at the top of Figure 11-3 is watching for the `Tank Volume` continuous level to pass a value of 100 going in the positive direction, seen in Display 11-4.

During the simulation run, whenever Arena detects this event, an entity is created and dispatched from the Detect module to the Assign module, where it begins emptying the tank by setting the `Tank Input Rate` value to -10, shown in Display 11-5. The entity is then disposed.

The second set of logic starts with a Detect module that's watching the `Tank Volume` to pass 0 in the negative direction. Its entities are created whenever the tank empties. Then, they assign the `Tank Input Rate` a value of `10` to begin filling the tank again and finally are disposed. Throughout the simulation run, Arena will create entities at the Detect modules whenever the tank fills or empties, resulting in a repeating pattern of filling and emptying approximately every 10 minutes.

Returning to the Detect module in Display 11-4, there was an additional field, the Crossing Tolerance, which we defined to be `0.0`. This quantity defines an acceptable error range for Arena's determination of the crossing value. We mentioned earlier that Arena can't truly advance time continuously; instead, it carves time into small steps, whose size is defined in the Continuous module. Because the continuous-value calculations are performed only at these steps in time, there's the potential that Arena could miss the exact time at which a value crossed a threshold defined in a Detect module.

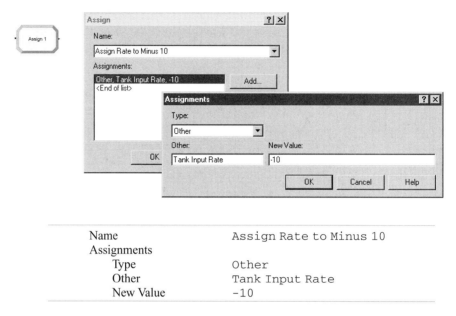

Name	Assign Rate to Minus 10
Assignments	
Type	Other
Other	Tank Input Rate
New Value	-10

Display 11-5. Assign Module Changing Tank Input Rate

Since we are using the Euler integration method, Arena can calculate within numerical round-off accuracy the exact time of the event being detected. So, in theory, there is no reason for error, and you should accept none. Notice the we said *within numerical round-off accuracy*. When you specify a Minimum Step Size of 0.0 or a Crossing Tolerance of 0.0, Arena actually uses a very small number internally. Although behind the scenes, there is a very small overshoot, from our perspective, this captures most events dead on. For this reason, these are generally the best settings to use with the Euler integration method.

In some cases, particularly RKF or other integration methods, you need to enter these parameters carefully. Consider a Detect module that is looking at the value of a level to cross a threshold of 100 in the positive direction with a tolerance of 0.1. Figure 11-4 illustrates a case where the level changed from a value of 99.7, which is less than the threshold of 100, at one continuous-time update (time 3.31) to a value of 100.6, which exceeds the threshold plus the tolerance, at the next time update (time 3.32).

In such a case, where Arena was unable to perform the required continuous calculations at the accuracy you specified, you can define how you want the model to behave via the Cross Severity field in the Continuous module. The default option is Warning, in which case Arena will display a warning message telling you that a Detect module exceeded the crossing tolerance, but will allow you to continue the run. You can choose Fatal (which fortunately describes how the Arena run should be treated, not the modeler!) to tell Arena to generate an error message and end the run if a crossing threshold is passed. Or if you want to ignore these types of errors, choose No, and Arena will just continue running the model as if you had a very large tolerance for error.

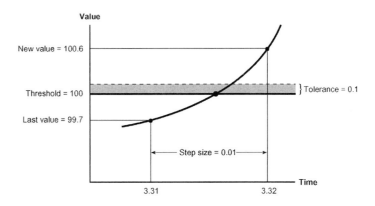

Figure 11-4. Relationship Between Crossing Tolerance and Time Updates

This discussion can help us determine what values to establish for the two primary continuous-system settings—the step size in the Continuous module and the crossing threshold(s) in the Detect module(s). To decide on a value for the step size, you'll be trading off accuracy of your model with run speed. As you decrease the step size, you typically will generate more accurate results, because Arena will be recalculating the continuous-change variables more often. However, the simulation run will take longer as you decrease the step size due to the increased frequency of calculations being performed.

After you've selected a value for the step size, you can calculate the tolerance that you'll need to supply to Arena for your Detect modules. Referring to Figure 11-4, we can identify a useful relationship among the continuous-time step size (the distance along the time axis between calculations), the rate of change of the level value (the slope of the line), and the tolerance value (the distance between values of the level). If we take the maximum absolute rate that will be encountered during the run (assuming that we know this quantity) and multiply it by the maximum time-step size, the result is the maximum change that can occur in the level's value in any single time-step.

For example, if our maximum rate of change is 100 volume units per hour and our step size is 0.01 hours, then the level can't change by more than 1 volume unit between Arena's calculations of the continuous variables. So if we've set our step size at 0.01, then we can enter a value of 1 in the Detect modules that are watching that level and avoid missing any crossings.

11.2 A Coal-Loading Operation

In this section, we'll build a somewhat larger (and hopefully more interesting) model of a combined discrete/continuous system—a coal-loading operation along the banks of a lazy river.

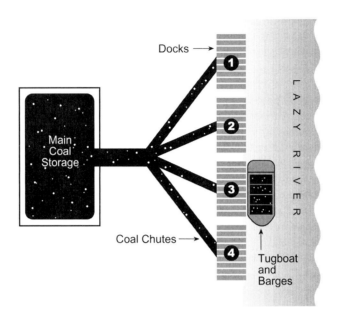

Figure 11-5. Coal-Loading Facility

11.2.1 System Description

At this coal-loading facility, coal is loaded from a main storage yard down chutes into barges at the facility's four docks. The rate at which coal can be discharged from the storage area is fixed at 2400 tons per hour, but it can be split into as many as four parallel chutes for filling barges at the docks, shown in Figure 11-5.

When a tugboat arrives at the loading facility, it waits until one of the four docks is available. Then, a dock crew ties the tugboat and its barges to the dock and sets up for loading, taking TRIA(2, 5, 10) minutes. The total amount of coal required to fill the barges varies based on the number of barges and their sizes. (The tugboat and its barges will be treated as a single unit in this model.) Table 11-1 lists the distributions of tonnage requirements for traffic at this site.

After the filling operation is finished, the dock crew unties the tugboat and its barges and the tug chugs away, freeing the dock. This process takes TRIA(3, 4.5, 7.5) minutes, depending on how much the tug and dock crews chat. The loading facility operates 24 hours a day, but the arrival of the tugboats varies throughout the day as shown in Table 11-2. For the purposes of this analysis, we'll assume that the dock crew is sufficiently staffed to handle all demand.

In evaluating this operation, we're interested in the tugboats that can't begin loading immediately. Whenever all of the facility's docks are occupied, tugs and their barges will be waiting upriver at tie-downs. We'd like to know something about the number of tugs waiting for a dock, as well as how long it takes to process tugs through the facility.

Table 11-1. Distribution of Barge Capacities

Capacity (tons)	Percentage
300	12
400	3
450	13
500	7
550	8
600	12
650	24
700	3
800	11
1000	7

Table 11-2. Tugboat Arrivals

Time Period	Average Number of Arriving Tugboats per Hour
12:00 AM – 2:00 AM	0.50
2:00 AM – 6:00 AM	1.00
6:00 AM – 8:00 AM	2.00
8:00 AM – 12:00 PM	3.50
12:00 PM – 1:00 PM	1.75
1:00 PM – 3:00 PM	2.75
3:00 PM – 4:00 PM	4.00
4:00 PM – 6:00 PM	5.00
6:00 PM – 8:00 PM	4.50
8:00 PM – 9:00 PM	2.50
9:00 PM – 10:00 PM	1.00
10:00 PM – 12:00 AM	0.50

11.2.2 Modeling Approach

If you recall our earlier discussion of simulation concepts (Section 2.3.7), we said that events occur at an instant of simulated time and cause changes in the system state (e.g., attributes, variables, statistics). And earlier in this chapter, we described continuous processes as those that change the state of the system continuously over time. Using these as a basis for analyzing our coal-loading operation, we can categorize its activities as follows:

- **Tugboat Arrival:** *Discrete event* initiating logic to model demand for loading.
- **Preparation for Coal Loading:** *Discrete event* for a tugboat entering a dock.
- **Beginning of Loading:** *Discrete event* beginning the loading of the tugboat's barges and changing the distribution of the coal among the docks.

- **Coal Loading**: *Continuous process* during which coal flows from the storage area into one or more docked barges. Note that the rate of coal being delivered to a particular dock may change due to the occurrence of other discrete events (beginning of loading, barges full).
- **Barges Full:** *Discrete event* occurring when a tugboat's barges have been filled. The timing of this event for each individual entity is dictated by its Beginning of Loading event and the duration of the Coal Loading continuous operation, as well as intervening events caused by the arrival and departure of other tugs.
- **Tugboat Departure:** *Discrete event* terminating the tugboat's processing in the simulation model.

We already know how to model the discrete events (and you should, too, if you've been paying attention to the earlier chapters) using familiar concepts such as entities, resources, queues, variables, and so on. And in fact, it almost feels like we could model the whole thing using what we've already learned. The logic might look something like:

- Create arriving tugboat
- Queue for and seize a dock
- Assign capacity requirement for tugboat (tons)
- Delay for docking and coal-loading preparation
- Delay to fill barges
- Delay for untying and leaving dock
- Release dock
- Dispose tugboat entity

So why is this model here in the continuous chapter when it appears that we can do everything with the concepts and modules we've already covered? Let's take a closer look at the sequence of delays required to model the filling operations to see if anything interesting will crop up to make this problem fit here.

The first step in loading a tugboat's barges was described as requiring a work crew (which we've conveniently assumed away as a constraint for this model) and taking TRIA(2, 5, 10) minutes. It seems like a Delay module should work just fine—just hold the tugboat entity in the model until the proper amount of simulated time (something between 2 and 10 minutes) elapses.

Next, the tugboat's barges are filled with coal. We know the capacity requirements in tons (Table 11-1), and we were told that the coal comes out of the storage area at a rate of 2400 tons per hour. That should be easy, too—just divide the capacity (tons) by the rate (tons/hr) and we'll have the loading time in hours, which we could enter in a Delay module. For example, if a tugboat requires 600 tons, it should take 600/2400 or 0.25 hours (15 minutes) to complete the loading.

Finally, we're told that the tugboat is untied and chugs away, taking TRIA(3, 4.5, 7.5) minutes. This clearly is another Delay module before the tugboat entity releases the Dock resource and departs the system.

But wait, there was another twist to the tale. We said that while the rate of coal coming out of the storage area is fixed at 2400 tons/hour, it's then split evenly among whichever docks are occupied by barges being loaded. This complicates our second delay calculation. When a tugboat has finished its first delay and is ready for loading, we can't be sure how long the loading operation will take. If it's the only tug in the docks, then it will start with a loading rate of the full 2400 tons/hour. But, if other tugboats arrive while this one's being loaded, the rate coming in will be decreased: first to 1200 tons/hour (2400 split evenly into two streams), then to 800 tons/hour (when three are being loaded), and possibly all the way down to 600 tons/hour if all four docks become busy before our first tugboat is filled to capacity.

Each of these occurrences is a discrete event (the beginning of the loading event in the previous list), so we could try to capture the logic necessary to adjust delay times for tugboats in the docks when new tugs arrive. However, this might become a little complicated, requiring some creative use of Arena's advanced discrete-event capabilities. (Give it a try for a nice little challenge.)

Instead, we can capture the flow of coal being split into multiple streams and filling the barges using the continuous modeling features of Arena. We'll still have all of the discrete-event parts of the model, but where we get to the thorny issue of how long a tugboat stays in dock for the actual loading operation, we'll interface our discrete model with continuous calculations performed by Arena.

11.2.3 Model 11-3: Coal Loading with Continuous Approach

From our classification of the system's activities as discrete or continuous in Section 11.2.2, we concluded that the only portion to be modeled as a continuous process is the actual loading of coal into the barges from the storage area. Everything else in the model will be represented as discrete processes, some of which will interface with the continuous portion of the system.

We'll first define the continuous-change levels and rates for our system in the Levels and Rates modules. Because we have four similar processes—filling operations at each of the four docks—in this system, we'll use arrays for the levels and rates, so that our model can index into the array based on the dock number assigned to a tugboat for its filling operation. Display 11-6 shows the entries for the Levels module, where we add a single level named `Barge Level`. We define four level variables, numbered 1 though 4, by establishing a starting Number of `1` and a 1-D Array Index of `4`. They'll be named `Barge Level(1)` through `Barge Level(4)`. During the simulation run, these will be treated independently; we used an array simply for the convenience of indexing when we make assignments in model logic.

Number	1
Name	Barge Level
1-D Array Index	4

Display 11-6. Levels Module Defining Four Barge Levels

Number	1
Name	`Barge Rate`
1-D Array Index	4

Display 11-7. Rates Module Defining Four Barge Rates

The Rates module contains similar entries, as shown in Display 11-7. These will also be numbered as rates 1 through 4 to match the `Barge Level` variables and will be referenced in the model as `Barge Rate(1)` through `Barge Rate(4)`.

We'll also add a Continuous module to establish the settings for the run's continuous calculations, shown in Display 11-8. Since we are still using the Euler integration method, we set the Minimum Step Size to `0.0`, and when we get to Crossing Tolerance, we'll leave that at `0.0` as well. We'll set Maximum Step Size to `100` (any large number will do) and leave the remaining fields at their default values so that Arena will perform continuous-update calculations for all of our level variables and will generate a warning if any detect-crossing-thresholds are violated.

The first few activities in the model will get the tugboats and their barges into a dock and begin filling the barges, seen in Figure 11-6.

The Create module generates entities of entity type `Empty Barge` according to an arrival schedule named `Barge Traffic Schedule`, in Display 11-9. The created entity will represent a tugboat and its barges, which are viewed as a single point of demand for loading. A Schedule module also is added to define the arrival pattern described in Table 11-2.

The barge entity next enters a Process module, Display 11-10, where it attempts to seize one of the docks. We'll model the docks as individual resources (`Dock 1`, `Dock 2`, `Dock 3`, `Dock 4`), formed into a set called `Docks`. In this process, we use the `Seize Delay` action so that the tugboat seizes a dock, but doesn't yet release it. It needs to hold the dock resource until after its barges have been filled and it has cleared the dock. We also store the index of the individual dock assigned to the arriving barge to an attribute named `Dock Number`. After seizing a dock, the tugboat delays in the Process module for the tie-down operation, which takes TRIA(2, 5, 10) minutes.

Minimum Step Size	0.0
Maximum Step Size	100

Display 11-8. Continuous Module

Figure 11-6. Barge Arrival Logic

Name	Create Arriving Barge
Entity Type	Empty Barge
Time Between Arrivals	
Type	Schedule
Schedule Name	Barge Traffic Schedule

Display 11-9. Create Module for Arriving Empty Barges

Name	Seize Dock and Prepare for Filling
Action	Seize Delay
Resources	
Type	Set
Set Name	Docks
Save Attribute	Dock Number
Delay Type	Triangular
Units	Minutes
Minimum	2
Value (Most Likely)	5
Maximum	10

Display 11-10. Process Module for Entering a Dock

After the tugboat entity has seized a dock and delayed for its tie-down time, it enters an Assign module to initiate the flow of coal into the barges. Remember that the actual rate of coal coming into a barge at a dock will depend on how many other docks are currently loading coal. Whenever the number of active docks (i.e., with barges being loaded) changes, the 2400 tons/hour of coal leaving the storage area needs to be adjusted to maintain an even distribution among those docks. We'll use a global variable, Filling Docks, to keep track of the number of docks that are loading. Then, whenever a new barge begins loading or reaches capacity, we'll adjust the loading rates in the active docks to be 2400/Filling Docks to keep our even distribution of coal.

In the Assign module, Display 11-11, the tugboat first increments the number of docks that are currently being filled (Filling Docks) by one. It then sets the rate of flow of coal flowing into its dock by assigning a new value to the rate variable at this dock, Barge Rate(Dock Number). And, so that we can later collect statistics on filling times, we'll assign the current simulation time to a global variable, Beginning Fill Time(Dock Number). Later in the model, when a full barge departs, we'll record the time that the barge spent in dock. Finally, the tugboat determines its size by taking a sample from a discrete probability distribution using the values provided in Table 11-1. This quantity is assigned to a variable, Barge Capacity(Dock Number), which we'll use as the threshold value for a Detect module that's watching for full tugboats at each dock. We also opened the Variable module on the Basic Process panel and defined the Beginning Fill Time and Barge Capacity variables, each with four rows.

Name	Begin Filling Barge
Assignments	
Type	`Variable`
Variable Name	`Filling Docks`
New Value	`Filling Docks + 1`
Assignments	
Type	`Other`
Other	`Barge Rate(Dock Number)`
New Value	`2400 / Filling Docks`
Assignments	
Type	`Other`
Other	`Beginning Fill Time(Dock Number)`
New Value	`TNOW`
Assignments	
Type	`Other`
Other	`Barge Capacity(Dock Number)`

New Value	`DISC( 0.12,300, 0.15,400,`
	`0.28,450, 0.35,500, 0.43,550,`
	`0.55,600, 0.79,650, 0.82,700,`
	`0.93,800, 1.00,1000 )`

Display 11-11. Assign Module to Begin Filling Barge

After beginning the proper flow of coal into the dock that the arriving tugboat entered, we next need to adjust the flow rate into any other docks that are currently loading barges. For example, if there were two occupied docks loading barges at the time of a new arrival, they each would have been receiving flow of 1200 tons/hour. When the new tugboat arrives and begins being filled, the flow to each of the three occupied docks needs to be changed to 800 tons/hour (the total of 2400 tons/hour divided equally among the three docks).

To adjust the rates, we'll use an Assign module that changes the fill rate of barges in docks that are currently filling (i.e., have a `Barge Rate` greater than 0) to be our new rate for all of the filling docks, `2400/Filling Docks`.

The Assign module, `Change Filling Rate On Barge Arrival`, uses an expression to assign a new value to each of the four `Barge Rate` variables, seen in Display 11-12. These expressions multiply two quantities together to assign a new `Barge Rate`. If a particular barge (for example, the barge at Dock 1) is currently being filled, its `Barge Rate` will be positive (i.e., `Barge Rate(1)>0` will evaluate to 1). This will result in its `Barge Rate` being assigned a value of `1 * (2400/Filling Docks)`, adjusting its rate based on the new number of docks that are filling. On the other hand, if a dock does not hold a filling barge, then its `Barge Rate` will be 0, and the expression `Barge Rate(1)>0` will evaluate to 0. This results in an assignment of `0 * (2400/Filling Docks)`, which keeps the `Barge Rate` at 0.

Name	Change Filling Rate on Barge Arrival
Assignments	
Type	Other
Other	Barge Rate(1)
New Value	(Barge Rate(1)>0) * (2400/Filling Docks)
Assignments	
Type	Other
Other	Barge Rate(2)
New Value	(Barge Rate(2)>0) * (2400/Filling Docks)
Assignments	
Type	Other
Other	Barge Rate(3)
New Value	(Barge Rate(3)>0) * (2400/Filling Docks)
Assignments	
Type	Other
Other	Barge Rate(4)
New Value	(Barge Rate(4)>0) * (2400/Filling Docks)

Display 11-12. Adjust Rate to Filling Dock

After making these assignments, we then can dispose of the tugboat entity, since all of the required information has been assigned to variables. The logic that we're about to describe will take care of the remainder of the tugboat's activities.

At this point in the logic, the arriving tugboat has completed its docking (i.e., seized a dock, delayed for the tie-up time) and has adjusted the continuous-level variables to represent proper distribution of the flow of coal loading into the docks. Now, the tugboat simply needs to wait until its barges are full and then perform the departure activities (untying from and releasing the dock).

To determine when the barges are full, we will use a Detect module, watching for the continuous Barge Level values at each dock to exceed their corresponding Barge Capacity values. This will create an entity when a barge is full, essentially replacing the arriving barge entity that we disposed of earlier. Figure 11-7 shows the logic for initiating these full-barge entities into the model.

The Detect module, Display 11-13, looks at all four of the continuous-level variables by defining a station range from 1 to 4. For a Detect module with a station range, Arena watches the values of the Crossing Variables (in this case, all four of the indexed Barge Level variables) throughout the simulation run. As is the case with the Search module, the Arena index variable, J, is used to indicate places in the module where the range values should be used. In this Detect module, we want Arena to watch all four Barge

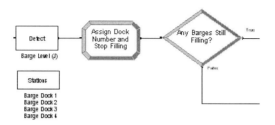

Figure 11-7. Detect Logic for Full Barges

`Level` variables vs. all four `Barge Capacity` values. Whenever one of the barge levels passes its Threshold Value (the value of the `Barge Capacity` variable with a corresponding index) in the Crossing Direction (positive), an entity is created and leaves the Detect module. Arena also assigns the index that was detected (e.g., 2 if `Barge Level(2)` passed `Barge Capacity(2)` in the positive direction) to the special attribute, Entity.Station, of the newly created entity. This attribute is one of those that is provided automatically by Arena for each entity (see Section 7.1.1). It is used in continuous modeling with the Detect module to allow the convenience of watching a number of level variables with a single module, as in our example. Since we are still using the Euler integration method, we will use a Crossing Tolerance of 0.

We also add a Stations module from the Elements panel to define four stations. Though they aren't directly used in the model for station transfers, they are required to allow the Detect module to search across the station range.

When an entity is created by this Detect module, it enters the Assign module shown in Display 11-14. Here we assign the entity's type to `Full Barge`, which will be used later in the model so that we can send the arriving barges and these full barges to the appropriate dispose logic. Then we use the `Entity.Station` attribute, which was initialized by Arena at the Detect module, to assign the `Dock Number` attribute a value of 1 through 4 for indexing into our arrayed variables. The next assignment terminates the flow of coal to this dock, setting the rate of flow, `Barge Rate(Dock Number)`, to 0. Then we reset the `Barge Level` variable for this dock to 0. We also set the `Barge Capacity(Dock Number)` variable to 0. Finally, we decrease the number of `Filling Docks` by one so that the rate adjustment calculations won't include this dock.

Beginning Station Range	1
Ending Station Range	4
Crossing Variable	`Barge Level(J)`
Crossing Direction	`Positive`
Threshold Value	`Barge Capacity(J)`
Crossing Tolerance	0

Display 11-13. Detect Module for Full Barges

Name	Assign Dock Number and Stop Filling
Assignments	
Type	`Attribute`
Attribute Name	`Entity.Type`
New Value	`Full Barge`
Assignments	
Type	`Attribute`
Attribute Name	`Dock Number`
New Value	`Entity.Station`
Assignments	
Type	`Other`
Other	`Barge Rate(Dock Number)`
New Value	`0`
Assignments	
Type	`Other`
Other	`Barge Level(Dock Number)`
New Value	`0`
Assignments	
Type	`Other`
Other	`Barge Capacity(Dock Number)`
New Value	`0`
Assignments	
Type	`Variable`
Variable Name	`Filling Docks`
New Value	`Filling Docks - 1`

Display 11-14. Assignments for Full Barge

At this point in the model, we have an entity that was created by the Detect module because the filling operation finished at one of the docks. In the Assign module, we stopped the flow of coal to that dock. Now, we need to adjust the flow to the other docks, if any are filling barges, to distribute the 2400 tons per hour evenly among them. First, we'll use a Decide module, shown in Display 11-15, to check whether any docks are filling barges.

Name	`Any Barges Still Filling?`
Type	`2-way by Condition`
If	`Expression`
Value	`Filling Docks > 0`

Display 11-15. Decide Module Checking Whether Any Barges Are Filling

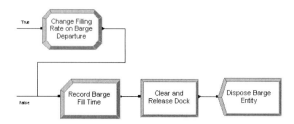

Figure 11-8. Dispose Barge Entities

If there are some docks occupied by barges being filled, then we need to adjust their filling rates. We'll connect the True exit point of the Decide module to a copy of the module named Change Filling Rate on Barge Arrival (name it Change Filling Rate on Barge Departure, keeping all other operands the same). If there are no occupied docks with filling in process, then the entity will leave via the False exit point of the Decide module. We'll connect this to the logic shown in Figure 11-8, which takes care of disposing the full-barge entity. (The connection from the Decide module's False exit point is the bottom-most line in the figure.)

To remove a full barge from the model, we need to collect statistics on how long it took to fill the barge, delay for the time to untie the barges and leave the dock, and dispose the entity.

The three modules that remove a full barge from the system start with the Record module, Record Barge Fill Time, seen in Display 11-16. This is used to tally how long it took to perform just the filling operation, not including the time required to tie up the tugboat or to untie it. Recall that when an arriving, empty barge began its filling operation (Figure 11-6), it assigned the beginning time of the filling operation to a variable named Beginning Fill Time with the appropriate Dock Number index (Display 11-11). To determine how long it took to fill the tugboat's barges, we'll subtract that beginning time from the time of departure of the tugboat, which is the current simulation time (TNOW) when the entity reaches the Record module. We'll name the Tally statistic Barge Fill Time so that we can view its results on the User-Specified area of the Arena reports.

Name	Record Barge Fill Time
Type	Expression
Value	TNOW − Beginning Fill Time(Dock Number)
Tally Name	Barge Fill Time

Display 11-16. Record Module for Barge Fill Time

Name	Clear and Release Dock
Action	Delay Release
Resources	
Type	Set
Set Name	Docks
Selection Rule	Specific Member
Set Index	Dock Number
Delay Type	Triangular
Units	Minutes
Minimum	3
Value (Most Likely)	4.5
Maximum	7.5

Display 11-17. Process Module to Untie Tug and Release Dock

Next, the entity representing the full barge delays for the time required to untie it from the dock and releases the dock. A Process module is used for these operations, as in Display 11-17. After this process is completed, our processing is done, so we destroy the barge at the Dispose module `Dispose Barge Entity`.

To complete the model, we'll establish the run parameters for our analysis. Since this operation runs 24 hours a day, 7 days a week, it seems clear that we'll need to analyze it as a non-terminating system. So we'll need to establish an appropriate warm-up time, run length, and number of replications. After performing some pilot runs, we decided that a Warm-Up Period of 5 days brings us to steady state (see Section 7.2.1). We'll add to that 200 days of simulated time for each replication (for an overall replication length of 205 days), and we'll perform 15 replications for our analysis. We select Hours as our Base Time Units to match the time units we used for the continuous rate variables (an important item to remember!). These *Run > Setup* settings are shown in Display 11-18.

The animation for this model is fairly simple, as shown in Figure 11-9. We added a plot for each of the docks showing the value of the `Barge Capacity` variable and the `Barge Level` variable on the same axis. This gives us a visual display of when barges began and ended filling operations, not including the time to tie and untie the tugboats at the docks. We also plotted the number of barges waiting for docks.

Number of Replications	15
Warm-up Period, Time Units	5, Days
Replication Length, Time Units	205, Days
Hours Per Day	24
Base Time Units	Hours

Display 11-18. Replication Parameters for Model 11-3

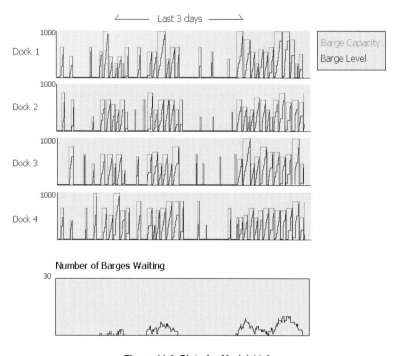

Figure 11-9. Plots for Model 11-3

To perform the simulation run, we checked the option for *Batch Run (No Animation)* under *Run > Run Control* to perform the run as fast as possible. After our 15 replications have completed, Arena's Category Overview report provides summary statistics across the replications.

The Queue section of the report shows that the average waiting time for a dock (in the queue named `Seize Dock and Prepare for Filling.Queue`) was about 0.4 hours (24 minutes), and the maximum observed value across all of the replications was about 7 hours. (Our 15 replications resulted in a half width of 0.01 for the mean, so we should feel good about the precision of the average predicted in this analysis.) The Number Waiting statistic for the queue reports that there was about one tugboat waiting for a dock, on average, with a maximum of 30 waiting at some stressful point during our replications.

In the User Specified section of the Category Overview report, we find the additional Tally statistic that we added to the model using a Record module. The average barge fill time was about 0.62 hours (just under 40 minutes), the shortest recorded fill time was 0.125 hours (7.5 minutes), and the longest was 1.6 hours. For a sanity check, let's think through these numbers to see if they make sense (always a good idea). The shortest fill time we would expect for this system would be a 300-ton barge (our smallest size) receiving the total 2400 tons/hour rate for its complete filling time. This would result in a fill time of 300/2400 = 0.125 hours, which matches our minimum observed value (so we must have had a small barge that was filled all by itself at some point). Looking at the maximum, if a 1000-ton barge was filled at the lowest fill rate of 600 tons/hour (2400 divided among the four docks), its fill time would have been 1000/600 or 1.67 hours. Our observed maximum of 1.6 hours is less than the longest possible (but fairly close), so again, our results seem reasonable.

11.2.4 Model 11-4: Coal Loading with Flow Process

Let's now look at using the Flow Process template, which facilitates modeling bulk material-holding areas, sensor detection, control logic, and continuous and semi-continuous flow between those holding areas. Does this sound familiar? Well, in fact, this template provides an easier, more straightforward way of modeling systems like the coal operations we just modeled. In this section, we'll start with a brief introduction to the Flow Process template and then show how these modules can be used in our coal model.

Flow Process has seven modules that all follow the paradigm of tanks and regulators, but it may be useful to think of these in the broadest possible sense. The Tank module represents a holding area where material is stored and defines the regulators that control flow in and out of that holding area. A regulator is a monitored input/output from the tank over which the maximum flow rate can be adjusted by the Regulate module. The Flow module creates a temporary flow connection into and/or out of a tank, or between two tanks. The entity-based flow logic of this module is ideal for representing batch processing operations. The Sensor module is somewhat similar to the Detect module discussed in Section 11.1.2; it allows you to sense and act on changes in the tank level. And finally the last three modules allow you to treat regulators in a way similar to resources—that is, seizing, releasing, and grouping into a set.

A related feature is a Level object of type Flow, which provides an easy-to-use mechanism for animating pipes, conveyors, and other devices that carry flow. The level provides an animation indicator of both direction and relative rate of flow.

You might think that the constructs just introduced are similar to and somewhat redundant to other Arena constructs. To some extent, you would be right. But the continuous constructs discussed earlier in this chapter were designed to handle complex systems and can introduce unneeded complexity in common batch-processing systems. Not only does Flow Process allow easier modeling of these systems, but it is also more efficient (i.e., it can execute faster), more accurate, and provides some sophisticated capabilities.

In Section 11.2.2 when we first described this coal-loading model, it appeared simple to model with discrete concepts. Then, as we got into more detail on the continuous aspects, the logic got a little more complex. This time we will follow the same basic procedure that we did in the earlier model, except that with the Flow Process approach, it will be a linear process description without the need for most of the data modules. We will model the coal-storage area as a tank (remember, a tank is just a holding area). We will consider each of the barge docks to be a regulator; that is, a regulated output from the coal inventory. A barge can be represented by an entity that steps through the process sequentially (as shown in Figure 11-10). We will use these basic steps:

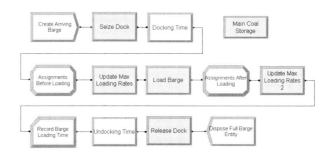

Figure 11-10. Coal Model Logic Using Flow Process

- Create arriving barge
- Seize a dock (Seize Regulator module)
- Delay for docking and coal-loading preparation
- Prepare for filling (Assign and Regulate modules)
- Fill barge (Flow module)
- Cleanup from filling (Assign and Regulate modules)
- Delay for untying and leaving dock
- Release dock (Release Regulator module)
- Dispose barge entity

We start by representing the Main Coal Storage with a Tank module from the Flow Process template, shown in Figure 11-11. For now we will give it 10,000 units of capacity and specify that it is full (later we will deal with replenishment). We will also specify that it has one regulator for each of the four barge docks. Separately, we create a new set containing all four of these regulators by using the Regulator Set data module from Flow Process.

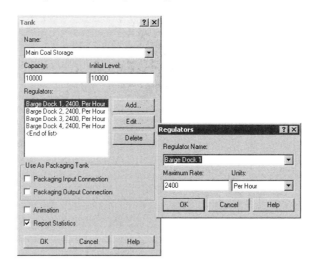

Figure 11-11. Tank Module Representing Coal Storage

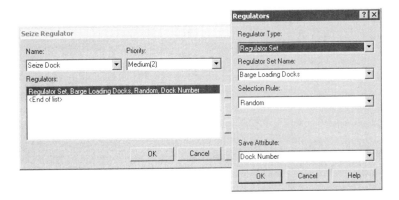

Figure 11-12. Seize Regulator Module Representing a Dock

The Create module is identical to what was used previously. Instead of the Process module that we used to seize the dock and delay for the preparation time, this time we will use two modules: a Seize Regulator from Flow Process (see Figure 11-12) and a Delay from Advanced Process.

The assignments that we must make to prepare for filling are similar to what we did before, but are less complex because a single entity represents each barge, so we can use attributes instead of variable arrays, where appropriate. The Assign that we previously used to change filling rates is replaced by a Regulate module from Flow Process (Figure 11-13).

Most of the "magic" occurs in the Flow module from Flow Process. We use this to actually move the coal onto the barge. Here (shown in Figure 11-14) we specify that we want to Remove material from the Regulator Set named `Barge Loading Docks` using the dock specified in the attribute `Dock Number`. We will stop transferring when we have transferred the quantity of material specified in the attribute named `Barge Capacity`.

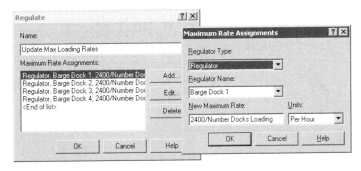

Figure 11-13. Regulate Module to Set Maximum Loading Rates

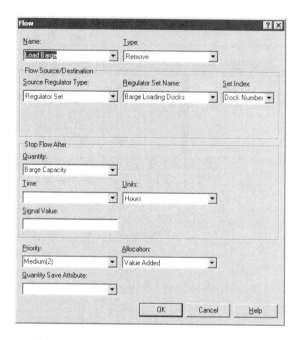

Figure 11-14. Flow Module to Transfer Coal to Barge

This simple application barely scratches the surface of the Flow module capabilities. Flow can transfer into, out of, or between tanks, and it automatically handles the cases of either tank becoming full, empty, or situations in which the flow is otherwise restricted. It also handles priority allocation between regulators contending for the same tank capacity. Options in the dialog allow for terminating the flow based on a signal or elapsed time. And Flow gracefully handles a flow that is abnormally terminated for any reason. But while all this is happening, on the outside, Flow behaves like a simple delay—an entity enters the Flow and waits there until the specified transfer is completed or terminated, then the entity exits to the next module.

After the material transfer in the Flow module is complete, we again use the Assign and Regulate modules to clean up after the transfer. (Note that the Regulate modules are not routinely necessary with Flow, but in this case, we are adjusting the maximum flow rates on all regulators each time any transfer is started or stopped.) Since the Flow Process modules assume that material is being moved from one location to another, we need to add some sort of replenishment logic to ensure that we do not run out of coal. We will take the simplest approach by adding an entry to `Assignments After Loading` to assign `TankLevel(Main Coal Storage)` to `10000`, essentially refilling the pile after each barge is serviced. The rest of the model is similar to our previous one, again using the Delay and Release Regulator modules instead of the Process module we used before.

If you run the completed model `Model 11-04.doe`, you will find that the results are similar to the results of `Model 11-03.doe`. Although `Model 11-04.doe` has no animation, it has many good opportunities for animation. If you look in the Arena folder in the \Examples\FlowProcess subfolder, you will find a model named Coal Loading.doe. This is the same problem and approach discussed in Model 11-04.doe, with some embellishments. The major change to the model (that will affect the results) is that we no longer assume that the coal inventory is infinite. Instead, we have periodic deliveries by train to replenish the inventory. We also have a fixed maximum inventory amount that can limit deliveries, and of course, if the pile ever empties, it will limit shipments. Another change is that we model the barge movement in the immediate vicinity. The rest of the changes are cosmetic, including animation of the trains and barges, animation of coal transfers using flow levels, and some "fun" animation including passenger trains and an occasional water skier.

11.3 Continuous State-Change Systems

In Sections 11.1-11.3, we examined continuous processes in which the rate of change of the continuous variables (e.g., liquid in a tank, coal in a barge) was constant or changed at discrete points in time. Many physical systems fall into this category, particularly where some bulk or liquid product is being manufactured or transferred among containers. Arena is well equipped (and now, so are you) to analyze these processes accurately using the features we've presented thus far.

We'll now turn our attention to slightly more complex continuous processes, where the rate values also change continuously over time. Many types of processes fall into this category, particularly when physical activities involving temperature or chemical reactions take place. Other cases, such as studies of large populations (e.g., for spread of disease), also call for the type of modeling approach we'll discuss here.

To introduce you to the new concepts required to model these processes, we'll use a simple example from the metals industry, where we need to model the temperature of a furnace and of the ingots being heated inside. We'll examine the conceptual approach required to capture this process, and we'll discuss the algorithms used by Arena to simulate these types of systems. When you've mastered this material, you should be armed with enough knowledge to identify the appropriate approach for simulating continuous systems and to employ Arena to do the job successfully.

11.3.1 Model 11-5: A Soaking-Pit Furnace

This system contains a soaking-pit furnace, where steel ingots are heated so they can be rolled in the next stage of the process (Ashour and Bindingnavle, 1973). Ingots arrive to this process approximately every 2.25 hours on average, following an exponential distribution for interarrival times. The furnace itself has space for at most nine ingots, which can be loaded and unloaded independently. If the furnace is full when a new ingot arrives, the ingot is held in a storage area until a space becomes available in the furnace. Each ingot is heated by the furnace to a temperature of 2200 degrees, then is removed and departs the system.

When an ingot is loaded into the furnace, its temperature is uniformly distributed between 300 and 500 degrees. Because it is cooler than the temperature of the furnace, it reduces the furnace temperature. For the purposes of our example, we'll assume that this change takes place immediately and equals the difference between the furnace temperature at the time the ingot is inserted and the ingot temperature, divided by the current number of ingots in the furnace.

The heating process in the furnace causes the temperature to change by twice the difference between 2600 and the current furnace temperature. The rate of change of an ingot's temperature while it's in the furnace is 0.15 times the difference between the furnace temperature and the ingot temperature.

For our study, we'd like to predict how many ingots are waiting to be loaded into the furnace and the range of the furnace temperature as it varies with ingots being loaded and removed.

11.3.2 Modeling Continuously Changing Rates

To model the furnace process, we'll use many of the concepts and constructs that were appropriate for capturing the coal-loading operation in Section 11.2. The elements that change continuously over time—the level variables for this model—are the furnace temperature and the individual temperatures of the ingots in the furnace. We'll define one level for the overall furnace temperature and nine additional level variables representing the temperature of the ingots in the nine furnace positions (corresponding to its capacity of nine ingots). These levels are analogous to the level variables we established in the coal-loading operation to monitor the amount of coal loaded into a tugboat's barges.

Where things become interesting is when we look at the rates for this process—how the temperature changes over time in the furnace and for heating the ingots. Previously, our rates held constant values over time, though we could introduce instantaneous changes in the rates by assigning new values to them (for example, immediately dividing the flow of coal differently among the four docks when a barge began or finished its filling operation). In the furnace system, though, the rate at which the furnace reheats to its target temperature of 2600 depends on its current temperature. Similarly, the rate at which an ingot is heated depends on both the temperature of the furnace and the ingot temperature. What complicates the determination of any of the rates or temperatures at a point in time is that they're all changing continuously over time.

Systems such as these are modeled using *differential equations*, which are mathematical equations that involve the derivative (rate of change) of one or more variables. If we denote the value of some system variable (e.g., a temperature) as x, then its derivative is denoted by dx/dt and defines its rate of change with respect to time.[3] In Arena, a level defines a system variable, x, and a rate defines its derivative, dx/dt. When we can't directly define a quantity's value over time, we use this continuous-modeling approach to calculate it indirectly via a differential equation.

The models that we've seen thus far in this chapter used simple differential equations, where the derivative (or rate) was a numerical value that changed only at discrete points in time (via model assignments). You might recall that when we added Continuous

---

[3] Hopefully, this looks somewhat familiar from your extensive background in mathematics. If not, we'll try to keep things simple and will count on your initiative to find a good math text for a deeper explanation of how this all works.

modules to our model, we saw a field called Number of Dif. Equations (e.g., Section 11.1.1). We were fortunate enough to be able to defer much discussion of that item, because Arena's default is to calculate an equation for each level/rate pair that are defined in the model. Now we'll discover a bit more about what it really means.

For the furnace example, we need to solve a differential equation for each of our ten continuous variables. If we denote as F the temperature of the furnace, then its rate of change follows the differential equation:

$$dF / dt = 2.0 \times (2600 - F).$$

Each of the ingots in the furnace changes temperature according to the equation:

$$dP_j / dt = 0.15 \times (F - P_j)$$

where P_j is the temperature of the ingot currently occupying the jth position in the furnace. Nine such equations will be needed to define the ingot temperatures completely.

11.3.3 Arena's Approach for Solving Differential Equations

Except for certain special classes of problems, differential equations are difficult to solve mathematically. Because of this, special numerical techniques have been developed to approximate the values of the continuous variables over time. (Note that these approaches generally apply only to systems of first-order differential equations—in other words, those involving the continuous variable or its rate but no higher-order derivatives. Higher-order differential equations must be converted to a series of first-order equations.)

Arena uses these techniques to calculate the value of a continuous-change variable at a new point in time based on its previous value, the change in time, and its estimation of the rate of change over that time step. Figure 11-15 illustrates how this works, where the curve is plotting the value of the continuous variable (level) over time, and the dotted line indicates the approximated rate over the time step.

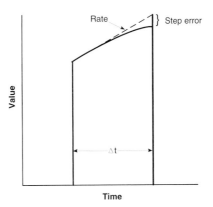

Figure 11-15. Numerical Approximation for Continuous-Change Variables

The actual duration of the time step between calculations, Δt, depends on the integration approach selected. Arena provides two built-in methods for numerically integrating continuous systems. The Euler method is a simple approach that works well for continuous processes where the rate of change remains constant between time steps. We utilized this approach for the previous models in this chapter. It uses a fixed time step defined by the Maximum Step Size field of the Continuous module.

For other continuous processes, in which one or more rates change continuously over time, Arena provides a more advanced set of algorithms under the RKF (for Runge-Kutta-Fehlberg) selection in the Continuous module. These approaches for solving differential equations use a variable step size, which is automatically adjusted each time Arena recalculates the continuous values. When the rate is close to linear, a larger step size (closer to the maximum) can yield good accuracy. However, when the rate is non-linear, smaller step sizes are necessary to provide the same quality of estimates (though at the penalty of more frequent calculations, which can cause slower runs).

In the Continuous module, to provide to Arena all the required information for this RKF approach, you supply two error values—the relative and absolute error quantities that are acceptable for each step in the integration. Using these quantities, Arena will determine whether a time-step size is acceptable by calculating the total error as (absolute error) + (value of the continuous variable) × (relative error).

We've mentioned that when the rate changes continuously, we develop differential equations that Arena calculates at each continuous time step. To add these to an Arena model, it's necessary to use one of Arena's interfaces to user-written routines (in C, C++, VBA, etc.). The equations are coded in a special, built-in routine called `ModelLogic_UserContinuousEquations` in VBA or `cstate` in C/C++. Because we presented the VBA interface in Chapter 10, we'll use VBA to code the equations for the furnace and ingot temperature changes.

11.3.4 Building the Model

To capture the activities of the ingots moving through the furnace, we'll have both discrete and continuous portions in our model, just as in the coal-loading operation model of Section 11.2. The continuous portion of the model, Figure 11-16, includes a Continuous module; individual Level variables for the furnace temperature and for each ingot position in the furnace; Rate variables to model the rate of temperature change for ingots and the furnace; and a new module, CStats, which we'll use to collect a statistic on the furnace temperature.

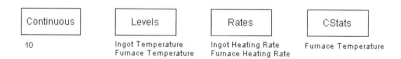

Figure 11-16. Continuous-Related Modules

Number of Dif. Equations	10
Minimum Step Size	0.01
Maximum Step Size	0.1
Save Point Interval	0.1
Method	RKF
Absolute Error	0.00001
Relative Error	0.00001
Severity	Warning
Cross Severity	Warning

Display 11-19. Continuous Module for Soaking-Pit Model

The Continuous module, Display 11-19, defines the parameters for the continuous portion of our model. We have ten differential equations, one for each of the nine ingot temperatures plus one for the furnace temperature. Because our rates change continuously over time, we select the RKF integration Method with a Minimum Step Size of 0.01, a Maximum Step Size of 0.1, and values of 0.00001 for the Absolute and Relative Errors. We'll also use a Save Point Interval of 0.1, which establishes the maximum time that can elapse before continuous statistics (CStats) are recorded.

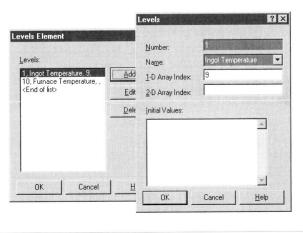

Levels		
Number	1	
Name	Ingot Temperature	
1-D Array Index	9	
Number	10	
Name	Furnace Temperature	

Display 11-20. Levels Module Entry Defining Ingot Temperature Levels

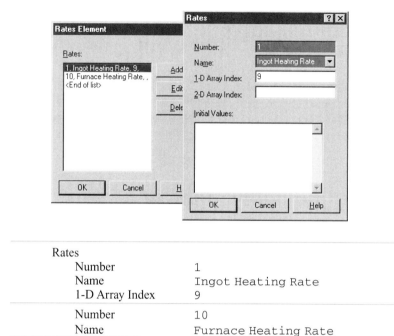

Rates
    Number   1
    Name   Ingot Heating Rate
    1-D Array Index   9

    Number   10
    Name   Furnace Heating Rate

Display 11-21. Rates Module Entry Defining Heating Rates

In the Levels module, we'll call the continuous Level variables representing the temperature of the ingots in the furnace Ingot Temperature and will establish 9 of them (one for each furnace position), shown in Display 11-20. They'll have 9 corresponding Rate variables called Ingot Heating Rate, so that Ingot Heating Rate(1) will be the rate of change of Ingot Temperature(1), etc.

For the overall furnace temperature, we'll define Level number 10 to be called Furnace Temperature and Rate number 10 to be called Furnace Heating Rate. Display 11-21 shows the main dialog box for the Rates module, as well as its entry for the Ingot Heating Rate.

The CStats module from the Elements panel defines time-persistent statistics to be collected on continuous variables. We'll add a single statistic on the furnace temperature, in Display 11-22. The SIMAN Expression is the Level variable on which we want to collect statistics, namely Furnace Temperature. We define the Name to be Temperature of Furnace, which will appear as the label of our statistic on Arena's user-defined report. In the Report Database Classification section, we specify "Continuous" for the Data Type, "User Specified" for the report Category, and "Furnace Temperature" for the Identifier on the report. Arena will list the average, half-width, minimum, and maximum values for the Furnace Temperature on the user-specified report. (The double-quotes around each of these report database classification entries are required by Arena in this module.)

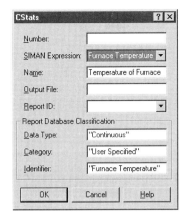

SIMAN Expression	Furnace Temperature
Name	Temperature of Furnace
Data Type	"Continuous"
Category	"User Specified"
Identifier	"Furnace Temperature"

Display 11-22. CStats Module Collecting Furnace Temperature Statistic

Because we're confident that you're quite adept at modeling discrete processes in Arena by now, we'll move through the process flow rather quickly so that we can get to the hot stuff—capturing the temperatures of all of these items accurately. Figure 11-17 shows the discrete process flow for creating the ingot entities. To generate ingots into the system, the Create module creates an entity randomly (exponential interarrival distribution) with a mean of 2.25 hours. The entity seizes one of the nine positions in the furnace via a set of resources, storing the selected position in an attribute named Ingot Number. Then it moves to the Set Ingot and Furnace Temperatures Assign module, where it increments a variable named Number in Furnace by one; assigns the level variable Ingot Temperature(Ingot Number) to a sample from a UNIF(300,500) distribution; and assigns the Furnace Temperature to its new value, Furnace Temperature - (Furnace Temperature - Ingot Temperature(Ingot Number)) / Number In Furnace.

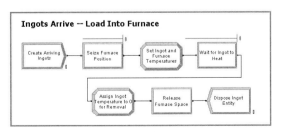

Figure 11-17. Ingot Creation Logic

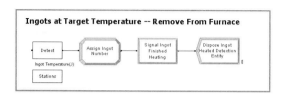

Figure 11-18. Logic to Remove Ingots from Furnace

At this point, the ingots are ready to be heated. We use a Hold module, `Wait for Ingot to Heat`, keeping the entity in a queue to wait for a signal that matches its `Ingot Number` attribute (which has a value between 1 and 9 defining its position in the furnace). This signal will be sent by an entity created at a Detect module that watches for the value of the ingot temperature to cross the target temperature (2200 degrees); we'll cover this logic shortly.

After the ingot has been heated, it leaves the Hold module and enters the `Assign Ingot Temperature to 0 for Removal` module, where it decreases the `Number in Furnace` variable by one and sets the `Ingot Temperature(Ingot Number)` level value to 0 (indicating that the position is now empty). The ingot then releases the resource it previously had seized and is disposed.

Figure 11-18 shows the logic for detecting the ingots heating to 2200 degrees. The Detect module watches the nine crossing (level) variables using the expression `Ingot Temperature(J)` for the station range 1 to 9. (The Stations module defines the nine stations.) Its threshold is `2200` degrees and crossing direction is `Positive`, using a crossing tolerance of 5. When an ingot has reached its target temperature, the `Assign Ingot Number` module assigns the `Ingot Number` attribute equal to the value of the `Entity.Station` attribute (which the Detect module automatically assigns, based on which `Ingot Temperature` level exceeded the threshold). The entity then sends a signal code equal to the `Ingot Number`, indicating that its position in the furnace has been cleared, and the entity finally is disposed.

11.3.5 Defining the Differential Equations Using VBA

We have completed the portion of the model that deals with the ingots—getting them into the system, loading them into the furnace, and removing them when they've reached the target temperature. Now we'll complete the model by defining the differential equations that will dictate the rates of change of the ingot and furnace temperatures.

To enable this feature in VBA, we need to check the Continuous Equations option found by clicking the Advanced Settings button on the Run Control tab from *Run > Setup*, shown in Figure 11-19. This will establish that during the simulation run, Arena should call the VBA code for user continuous equations at each integration update.

In the Visual Basic Editor, the VBA code is typed in the `ModelLogic_ UserContinuousEquations` subroutine, shown in Figure 11-20. First, we set the `oSIMAN` variable to point to the SIMAN data, which will give us access to the rate and level variables. Next, we store the value of the `Furnace Temperature` level variable (which is level number 10) in the VBA variable, `dFurnaceTemp`, using a call to the

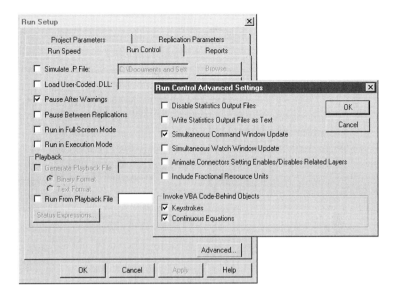

Figure 11-19. Enabling VBA Continuous Equations

$\texttt{LevelValue}$ function in the Arena object library. The For loop cycles through for values of $\texttt{nIngotNum}$ from 1 through 9. It calls the $\texttt{RateValue}$ function to set the values of the nine rate variables (that change the ingot temperatures) equal to 15% of the difference between the current furnace temperature and the individual ingot's temperature. This corresponds to the formula, $dP_j/dt = 0.15 \times (F - P_j)$, presented in Section 11.3.2. Finally, we update the rate of change of the furnace temperature—rate value number 10—by taking twice the difference between 2600 and the current furnace temperature ($\texttt{dFurnaceTemp}$), representing the previously described formula, $dF / dt = 2.0 \times (2600 - F)$.

```
Private Sub ModelLogic_UserContinuousEquations()
    Dim oSIMAN As Arena.SIMAN
    Dim dFurnaceTemp As Double
    Dim nIngotNum As Long

    Set oSIMAN = ThisDocument.Model.SIMAN

    ' Set heating rate for each ingot
    dFurnaceTemp = oSIMAN.LevelValue(10)
    For nIngotNum = 1 To 9
        oSIMAN.RateValue(nIngotNum) = _
            0.15 * (dFurnaceTemp - oSIMAN.LevelValue(nIngotNum))
    Next nIngotNum

    ' Set heating rate for furnace
    oSIMAN.RateValue(10) = 2 * (2600 - dFurnaceTemp)
End Sub
```

Figure 11-20. Differential Equations Coded in VBA

Whenever you have differential equations, you'll use this approach. Formulate the equations in terms of levels and rates, and then create the code to implement the formulas in VBA or C/C++. During the simulation run, the routine will be called many times, updating the values of the level and rate variables as the continuous integration steps are performed. The settings in the Continuous module for the minimum and maximum step size for the RKF algorithm determine how frequently these updates are performed. With smaller values for these step sizes, the continuous variables are calculated more often, which can result in greater accuracy. However, as you might guess, this accuracy comes at the expense of longer run times since the VBA code must be executed at each of the time steps.

With regard to run times, when you run this model, you may notice that it takes a long time to run. Part of this is due to the nature of continuous problems in general. In addition, interpreted VBA code is much slower than compiled C++ code. As mentioned above, the VBA code must be executed at every time advance, and time advances are short to keep good accuracy—the combination puts a heavy load on interpreted code. When run time becomes an issue, you will find it will improve dramatically when these algorithms are coded in C++.

In examining the results for ten replications of 5100 hours each (with 100 hours of warm-up time), we can see on the Queue section of the Category Overview report that the average waiting time for cold ingots that couldn't be loaded into the furnace was 0.3 hours, and the maximum waiting time was 11.3 hours. Also, on average, there were 0.15 cold ingots waiting, though at some point during the run there were as many as nine in the queue (from the Maximum Value column of the Number Waiting table). We find the statistic on the furnace temperature on the User Specified section of the Category Overview report. There, the Continuous table lists our CStat information, which shows that the average furnace temperature was 2514 degrees, the minimum was 301 degrees, and the maximum was 2600 degrees.

11.4 Summary and Forecast

Chapter 11 concludes the modeling topics of this book. In it, we examined Arena's framework for modeling continuous-change systems. We also put together the discrete modeling concepts of our earlier chapters with continuous modeling, illustrating how these concepts are applied together in two examples. Further, we built a model based on Flow Process constructs that blend these constructs.

When you are preparing to analyze a system in which fluids or other bulk materials are being produced or handled, the continuous-modeling and Flow Process approaches should be considered. Some systems may be modeled using both continuous and discrete concepts, due to the nature of the material being modeled. Arena's structure is designed to model any combination of discrete and continuous processes, using Flow Process modules or the Detect module and assignments of continuous rate and level variables illustrated in this chapter.

The remaining two chapters are not directly about modeling, but discuss important topics that are critical for performing good simulation studies. Chapter 12 rounds out our treatment of the probabilistic underpinnings and statistical issues in simulation, which we began in Section 2.6, Section 4.6, Chapter 6, and Section 7.2. Then in Chapter 13, we'll provide you with some sage observations and advice about how to conduct simulation studies, including modeling, design, analysis, and (gasp!) dealing with people.

11.5 Exercises

11-1 Build a discrete-event model that changes the value of the volume in a tank as described for Model 11-2b using a maximum step size of 0.01 minutes. Record time-persistent statistics on the volume in the tank, and compare the average reported volume for an 800-hour run with the results from the continuous model (11-2b). Also compare the computing time required to perform the simulation run for the discrete case vs. the continuous approach. (Note: For this comparison, clear the Update Simulation Time Every option in *Run > Setup > Run Speed* from Model 11-2b, and run both models in batch mode. To show computing time, select SIMAN Summary Report under *Run > Setup > Reports*.)

11-2 The owner of a franchise of gas stations is interested in determining how large the storage tank should be at a new station. Four gas pumps, all dispensing the same grade of fuel, will be installed to service customers. Cars arrive according to an exponential distribution with a mean of 0.8 minute. (We'll assume that this is uniform throughout the station's hours of operation.) Their time at the pump (from start to finish) follows a triangular distribution with parameters 2, 2.8, and 4 minutes. The cars require varying amounts of fuel, distributed according to a triangular distribution with parameters 4, 7.5, and 15 gallons.

Refill trucks arrive according to a uniform distribution with a minimum interarrival time of 6.75 hours and a maximum of 8.25 hours. They carry enough fuel to refill the storage tank and do so at a rate of 300 gallons per minute.

If the storage tank empties before a refill truck arrives, the pumps are closed until the storage tank contains 100 gallons (from its next refill). For purposes of this analysis, assume that cars that are in-process when the tank empties can complete their service and that waiting cars will stay at the station until the pumps reopen. However, any cars that arrive while the pumps are closed will drive by to find another place to fill up.

Determine (to the nearest 100 gallons) the tank capacity that will result in fewer than 0.1% of cars balking due to closed pumps.

11-3 An earnest analyst for Grace Enterprises, the owner of the coal-loading operation described in Model 11-3, has become concerned that assuming that coal will always be available to load into the barges might not be realistic. She would like to refine the estimates of loading times and numbers of waiting barges by incorporating into her model the delivery of coal by train to the storage yard.

Trains are scheduled to arrive every eight hours throughout the day and night, and they're usually on time. Each train carries 12,000 tons of coal, which is unloaded into the storage yard at a rate of 7,500 tons per hour. The storage yard can hold 17,000 tons of coal; for purposes of this analysis, cancel a train's delivery if the yard is full at the train's scheduled arrival time.

Modify Model 11-3 to incorporate the availability of coal in the storage yard so that barges will wait at the dock until coal is available for loading. Compare the average and maximum number of barges waiting and the loading time of barges with the results from Model 11-3.

11-4 O'Hare Candy Company, maker of tasty sweets, is preparing to install a new licorice production facility and needs to determine the rates at which equipment should run. In particular, they are interested in the cutting/wrapping machines, as they are prone to frequent breakdowns.

At this facility, there are three identical parallel lines fed by a single kitchen producing continuous strands of licorice. Each line is fed licorice at a rate of 1374 kg/hour and has its own dedicated cut/wrap machine (modeled as a single process). The wrappers cut the licorice into individual pieces and wrap them at a rate of 1500 kg/hour. These machines experience breakdowns of various forms. Analysis of breakdown data from similar equipment has concluded that the frequency is approximately one breakdown per hour with a high degree of variability that can best be represented by an exponential distribution. The time to repair a breakdown varies quite a bit as well; it can be modeled by a triangular distribution with parameters 3.75, 4.5, and 8.25 minutes.

It is very costly to shut down the kitchen, so surge tables will be located in front of each machine. The current design calls for a surge table capacity of 1000 kg. If the amount of product exceeds this capacity, then the rate at which the kitchen is feeding product to that machine is reduced to 900 kg/hour until less than 700 kg of licorice is on the table.

For the analysis of this system, begin the simulation with all three surge tables full. Analyze the system to evaluate whether the planned surge table capacities are sufficient and whether the system will be able to produce licorice at the required rates.

11-5 Hope Bottling Company operates a bottling plant, handling many types of products. They are interested in analyzing the effective capacity of an orange-juice bottling line as part of their plans for future business expansion.

At this facility, trucks deliver bulk orange juice (2000 gallons per truckload) that is pumped into a surge tank feeding a bottling operation. The trucks arrive at an average rate of 1.75 trucks per hour during the first 8 hours of each day and an average of one truck per hour during the remainder of each day. Upon arrival, they wait in line for a single dock to unload their juice; after unloading, they undock and depart the system. The docking and undocking time for a truck is uniformly distributed between 1 and 2 minutes. During the unloading operation, the juice is pumped from the truck into the surge tank at a rate of 200 gallons per minute. The juice is pumped from the surge tank to the bottling operation at a rate of 48 gallons per minute.

If the level of juice in the surge tank reaches the tank's capacity of 10,000 gallons, the truck-unloading operation is suspended until the level drops by 500 gallons. When the surge tank empties, bottling is stopped until the next truck arrives and begins unloading juice into the surge tank.

The packaging operation (which runs 24 hours per day, 7 days per week) bottles the orange juice into one-gallon containers and then combines these into boxes of 12. The boxes are then grouped into sets of four and placed on a pallet for shipping. Hence, each 48 gallons of juice that are processed through the operation will generate a pallet for shipping, resulting in a maximum production rate of the bottling operation of one pallet every minute. However, the actual production rate may be less than this due to starvation when the surge tank is empty. The bottling operation runs without any operational failures.

Predict the average turnaround time of trucks (i.e., from arrival at the facility to departure) and the number of pallets that will be produced per week. Also look at whether increasing the surge tank capacity to a 20,000-gallon tank would noticeably increase production or decrease the time for trucks to unload.

11-6 For the soaking-pit furnace problem (Model 11-5), use the model to evaluate the performance improvement resulting from preheating arriving ingots so that their temperature is uniformly distributed between 600 and 700 degrees. Assume that the ingots waiting in the cold bank (that is, ones that could not immediately be loaded into the furnace on arrival) have a temperature of 600 degrees.

11-7 Simulate the population dynamics involving the growth and decay of an infectious, but easily curable disease. The disease occurs within a single population, and recovery from the disease results in immunity. The population consists of the following three groups: (1) those who are well, but susceptible; (2) those who are sick; and (3) those who are cured and therefore immune. Although the system's state actually changes discretely, we will assume that we can approximate the system with continuous-change variables describing the size of each group.

We will use the state variables named Well, Sick, and Cured to denote each group's current size. Initially, the Well population size is 1000, the Sick population is 10, and the Cured is 0. The following system of differential equations governs the infection rate, where d/dt indicates the rate of change of the population size.

$$d/dt \,(\text{Well}) = -0.0005 \times \text{Well} \times \text{Sick}$$
$$d/dt \,(\text{Sick}) = 0.0005 \times \text{Well} \times \text{Sick} - 0.07 \times \text{Sick}$$
$$d/dt \,(\text{Cured}) = 0.07 \times \text{Sick}$$

Assuming that the above formulas are based on days, how long will it take until the size of the Well group decreases to 2 percent of its original size? Include a plot of the three populations.

CHAPTER 12

Further Statistical Issues

One of the points we've tried to make consistently in this book is that a good simulation study involves more than building a good model (though good models are certainly important too). In a simulation with stochastic (random) input, you're going to get random output as well. Thus, it's critical to understand how simulations generate this randomness in the input and what you can do about the resulting randomness in the output. We've already blended some of these statistical issues in with our tour through model building and analysis, specifically in Section 2.6, Section 4.6, Chapter 6, and Section 7.2. Part of the point of those sections is that Arena can help you deal with these issues, but you must be aware that they exist.

This chapter discusses additional statistical issues related to both the input and output sides of a simulation. Random-number generators, the source of all randomness in simulations, are discussed in Section 12.1. Then in Section 12.2, we'll talk about how to generate observations on whatever input distributions you decided to use as part of your modeling. Section 12.3 discusses specifying and generating from a particular yet important type of random input, a nonstationary Poisson process (which, by the way, was introduced in Model 5-2). Ways to reduce output variance (other than just simulating some more) are described in Section 12.4. The idea of sequential sampling—that is, deciding on the fly how much simulation-generated data you need—is the subject of Section 12.5. The chapter concludes in Section 12.6 with brief mention of the possibility of using experimental design in simulation. By the time you reach the end of this chapter, you should have a thorough understanding of statistical issues in simulation and know how Arena can help you deal with them.

Obviously, this chapter is a heavy user of foundational material in probability and statistics. We've provided a refresher on these subjects in Appendix C of the book, which you might want to look at before going on. In addition, Appendix D contains a listing of all the probability distributions supported by Arena.

12.1 Random-Number Generation

Hidden deep down in the engine room of any stochastic simulation is a *random-number generator* (RNG) quietly churning away. The sole purpose of such a machine is to produce a flow of numbers that are observations (also known as *draws*, *samples*, or *realizations*) from a continuous uniform distribution between 0 and 1 (see Appendix D) and are independent of each other. In simulation, these are called *random numbers*. This is certainly not the only probability distribution from which you'll want to draw observations to drive your simulations (see Section 4.6), but as we'll discuss in Sections 12.2 and 12.3, generating observations from all other distributions and random processes starts with random numbers.

Any method for generating random numbers on a computer is just some kind of recursive algorithm that can repeat the same sequence of "random" numbers again and again. For this reason, these are often called *pseudorandom*-number generators. Some people have worried philosophically that such methods are fundamentally flawed since part of what it means to be random is to be unpredictable. This might make for an interesting after-dinner debate, but at a practical level, the issue is really not very important. Modern and carefully constructed RNGs generally succeed at producing a flow of numbers that appear to be truly random, passing various statistical tests for both uniformity and independence, as well as satisfying theoretically derived criteria for being "good." Also, it's quite helpful in simulation to be able to regenerate a specific sequence of random numbers; this is an obvious aid in debugging (not to mention grading homework), but is also useful statistically, as we'll describe in Section 12.4.

Unfortunately, there seems to be a common perception that any seemingly nonsensical method will, just because it looks weird, generate "random" numbers. Indeed, some extremely poor methods have been provided and used, possibly resulting in invalid simulation results. Designing and implementing RNGs is actually quite subtle, and there has been a lot of research on these topics. In part because computers have become so fast, there continues to be work on developing new and better RNGs that can satisfy the voracious appetite that modern simulations can have for random numbers.

So exactly how do RNGs work? Historically, the most common form (and the type still built into a lot of simulation software, but not Arena . . . more on this issue below) is called a *linear congruential generator* (LCG). An LCG generates a sequence $Z_1, Z_2, Z_3, \ldots$ of integers via the recursion

$$Z_i = (aZ_{i-1} + c) \bmod m$$

where m, a, and c are constants for the generator that must be chosen carefully, based on both theoretical and empirical grounds, to produce a good flow of random numbers. The "mod m" operation means to divide by m and then return the *remainder* of this division to the left-hand-side as the next Z_i (for instance, 422 mod 63 is 44). As with any recursion, an LCG must be initialized, so there is also a *seed* Z_0 specified for the generator. This sequence of Z_i's will be composed of integers, which is certainly not what we want for a continuous distribution between 0 and 1. However, since the Z_i's are remainders of division of other integers by m, they'll all be between 0 and $m - 1$, so the final step is to define $U_i = Z_i/m$, which will be between 0 and 1. The sequence $U_1, U_2, U_3, \ldots$ are the (pseudo-)random numbers returned for use in the simulation.

As a tiny example (nobody should ever actually *use* this generator), take $m = 63$, $a = 22$, $c = 4$, and $Z_0 = 19$. The recursion generating the Z_i's is thus $Z_i = (22 Z_{i-1} + 4) \bmod 63$. Table 12-1 traces this generator through the first 70 generated random numbers, and you can check some of the arithmetic. (We used a spreadsheet to generate this table.) At first blush, scanning down through the U_i column gives the impression that these look like pretty good random numbers—they're certainly all between 0 and 1 (as they're guaranteed to be by construction), they appear to be spread fairly uniformly over the interval [0, 1], and they are evidently pretty well mixed up (independent). The sample

Table 12-1. Tracing an LCG's Arithmetic

i	$22Z_{i-1}+4$	Z_i	U_i	i	$22Z_{i-1}+4$	Z_i	U_i	i	$22Z_{i-1}+4$	Z_i	U_i
0		19		24	1060	52	0.8254	48	400	22	0.3492
1	422	44	0.6984	25	1148	14	0.2222	49	488	47	0.7460
2	972	27	0.4286	26	312	60	0.9524	50	1038	30	0.4762
3	598	31	0.4921	27	1324	1	0.0159	51	664	34	0.5397
4	686	56	0.8889	28	26	26	0.4127	52	752	59	0.9365
5	1236	39	0.6190	29	576	9	0.1429	53	1302	42	0.6667
6	862	43	0.6825	30	202	13	0.2063	54	928	46	0.7302
7	950	5	0.0794	31	290	38	0.6032	55	1016	8	0.1270
8	114	51	0.8095	32	840	21	0.3333	56	180	54	0.8571
9	1126	55	0.8730	33	466	25	0.3968	57	1192	58	0.9206
10	1214	17	0.2698	34	554	50	0.7937	58	1280	20	0.3175
11	378	0	0.0000	35	1104	33	0.5238	59	444	3	0.0476
12	4	4	0.0635	36	730	37	0.5873	60	70	7	0.1111
13	92	29	0.4603	37	818	62	0.9841	61	158	32	0.5079
14	642	12	0.1905	38	1368	45	0.7143	62	708	15	0.2381
15	268	16	0.2540	39	994	49	0.7778	63	334	19	0.3016
16	356	41	0.6508	40	1082	11	0.1746	64	422	44	0.6984
17	906	24	0.3810	41	246	57	0.9048	65	972	27	0.4286
18	532	28	0.4444	42	1258	61	0.9683	66	598	31	0.4921
19	620	53	0.8413	43	1346	23	0.3651	67	686	56	0.8889
20	1170	36	0.5714	44	510	6	0.0952	68	1236	39	0.6190
21	796	40	0.6349	45	136	10	0.1587	69	862	43	0.6825
22	884	2	0.0317	46	224	35	0.5556	70	950	5	0.0794
23	48	48	0.7619	47	774	18	0.2857				

mean of the U_i's is 0.4984 and the sample standard deviation is 0.2867, which are close to what we'd expect from a true uniform [0, 1] distribution (1/2 and $(1/12)^{1/2} = 0.2887$, respectively).

But there are a couple of things to notice here. First, as you read down through the Z_i's, you'll see that $Z_{63} = 19$, which is the same as the seed Z_0. Then, note that $Z_{64} = 44 = Z_1$, $Z_{65} = 27 = Z_2$, and so on. The Z_i's are repeating themselves in the same order, and this whole cycle will itself repeat forever. Since the U_i's are just the Z_i's divided by 63, the random numbers will also repeat themselves. This *cycling* of an LCG will happen as soon as it hits a previously generated Z_i, since each number in the sequence depends only on its predecessor, via the fixed recursive formula. And it is inevitable that the LCG will cycle since there are, after all, only m possibilities for the remainder of division by m; in other words, the cycle length will be at most m. In our little example, the cycle length actually achieved its maximum, $m = 63$, but had the parameters a and c been chosen differently, the cycle length could have been shorter. (Try changing a to 6 but leave everything else the same.) We weren't just lucky (or persistent) in our choice since there's fully developed theory on how to make parameter-value choices for LCGs to achieve full, or at least long, cycle lengths. Real LCGs, unlike our little example above, typically take m to be at least $2^{31} - 1 = 2{,}147{,}483{,}647$ (about 2.1 billion) and choose the other parameters to achieve full or nearly full cycle length. Though we can remember when 2.1 billion was still a lot, such a cycle length is not as impressive as it once was, given today's computing power. Indeed, using just a garden-variety PC, we can exhaust all of the 2.1 billion possible random numbers from such LCGs in a matter of minutes or at most a few hours, depending on what we do with the random numbers after we generate them. While choosing m even bigger in LCGs is possible, people have instead developed altogether different kinds of generators with truly enormous cycle lengths, and Arena uses one of these (more on this below).

The other thing to realize about the U_i's in Table 12-1 is that they might not be quite as "random" as you'd like, as indicated by the two graphs in Figure 12-1. The left graph simply plots the random numbers in order of their generation, and you'll notice a certain regularity. This is perhaps not so upsetting since we know that the generator will cycle and repeat exactly the same pattern. The pattern for a real generator (with a higher value of m), will not be so apparent since there are far more random numbers possible and since, in many applications, you'll generally be using only a small part of a complete cycle.

But the right graph in Figure 12-1 might be more unsettling. This graph plots the pairs (U_i, U_{i+1}) over the complete cycle, which is of natural interest if you're using the random numbers in pairs in the simulation (such as generating an interarrival time followed up immediately by a random part-type assignment). As you can see, it has an eerie pattern to it that will, in fact, be present for any LCG (as shown in the colorfully named paper, "Random Numbers Fall Mainly in the Planes" by Marsaglia, 1968). A truly random generator should instead have dots haphazardly scattered uniformly over the unit square, rather than being compulsively arranged and leaving comparatively large gaps where no pairs are possible. And this *lattice* structure gets even worse if you make (or imagine) such a plot in higher dimensions (triples, quadruples, etc., of successive random numbers). These kinds of considerations should drive home the point that "designing" good RNGs is not a simple matter, and you should thus be careful when encountering some mysterious undocumented RNG.

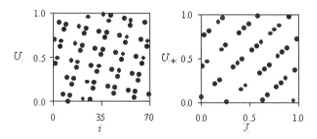

Figure 12-1. Plots for an LCG

The RNG originally in Arena, starting in the early 1980s with SIMAN, was an LCG with $m = 2^{31} - 1$, $a = 7^5 = 16,807$, and $c = 0$; the cycle length of it is $m - 1 = 2.1$ billion. In its day, this generator was acceptable since it had been well tested and in fact delivered a respectable flow of random numbers. However, because computers have gotten much faster, it became apparent that this cycle length was no longer adequate and that this old generator actually ran the risk of "lapping" itself around the entire cycle within a few hours or maybe even minutes of simulating (which could obviously do serious damage to the validity of your results).

So a new RNG has been installed in Arena (and is in the version of Arena on the CD that comes with this book), based on research by L'Ecuyer (1996, 1999) and L'Ecuyer, Simard, Chen, and Kelton (2002). While it uses some of the same ideas as LCGs (particularly the modulo-division operation), it differs in that (1) it involves two separate component generators that are then combined, and (2) the recursion to get the next values looks back beyond just the single preceding value. This kind of generator is called a *combined multiple recursive generator* (CMRG). First it starts up the two separate recursions, which you can think of as operating in parallel at the same time:

$$A_n = (1403580 \, A_{n-2} - 810728 \, A_{n-3}) \bmod 4294967087$$

$$B_n = (527612 \, B_{n-1} - 1370589 \, B_{n-3}) \bmod 4294944443$$

Then it combines these two values at the nth step as follows:

$$Z_n = (A_n - B_n) \bmod 4294967087$$

Finally, this RNG delivers the nth random number U_n as either

$$Z_n / 4294967088, \text{ if } Z_n > 0,$$

or

$$4294967087 / 4294967088, \text{ if } Z_n = 0.$$

To get the two recursions going, the generator must define a six-vector of seeds, composed of the first three A_n's and the first three B_n's.

The rather scary constants involved in this generator have been carefully chosen, based on the papers cited above, to ensure two very desirable properties. First, the statistical properties of the random numbers produced are extremely strong; we get good distribution of the generated points, like the right-hand plot in Figure 12-1, but up through a 45-dimensional cube rather than just the two dimensions of Figure 12-1. Second, while this generator will cycle (like LCGs), the length of this cycle is a stunning 3.1×10^{57}, rather than the 2.1×10^9 for the old generator. And the running speed on a per-random-number basis of the new generator is only slightly slower than the old one.

Let's put into perspective the difference between the old and new cycle lengths. If the old generator can be exhausted in ten minutes (which it can on a 2 GHz PC if we just generate the random numbers and throw them away), the new generator would keep this machine busy for about 2.78×10^{40} *millennia*. Now we know you're thinking "sure, but in 1982 they thought 2.1 billion was a lot and it would last forever"—but even under Moore's "law" (which observes that computers double in speed about every year and a half), it will be something like 216 years before a typical computer will be able to exhaust this new generator in a year of nonstop computing. So we're good for a while, but we admit, not forever.

It turns out to be quite useful to be able to separate the cycle of a generator into adjacent non-overlapping *streams* of random numbers, which we can effectively think of as being separate "faucets" delivering different streams of random numbers. To define a stream, we need to specify the seed vector (six numbers in the Arena generator) and take care that successive streams' seed vectors are "spaced apart" around the cycle so that the streams are long. The Arena generator has facility for splitting the cycle of 3.1×10^{57} random numbers into 1.8×10^{19} separate streams, each of length 1.7×10^{38}. Each stream is further subdivided into 2.3×10^{15} substreams of length 7.6×10^{22} apiece. Our weary 2 GHz PC would take about 669 million years to exhaust one of the substreams; under Moore's law, in 49 years, it will take a month to exhaust a substream—but 187 billion millennia to exhaust a stream.

You can specify which stream is used whenever you ask for an observation from a distribution in Arena by appending the stream number to the distribution's parameters; for instance, to generate an exponential observation with mean 6.7 from stream 4, use EXPO(6.7, 4). If you don't specify a stream number, Arena defaults it to 10. (Since Arena does some of its own random-number generation, for instance in generating nonstationary arrival patterns and in a "chance"-type Decide module, it uses stream 10 for that, so you should avoid using stream 10 if you're specifying your own streams.) The idea of using separate streams of random numbers for individual purposes in a simulation (for instance, stream 1 for interarrival times between parts, stream 2 for part types, stream 3 for processing times, etc.) comes in quite handy for variance reduction, discussed in Section 12.4. Arena does not actually store all the seed vectors, but rather uses a fast on-the-fly method to compute them when you reference the corresponding stream; for this reason, you should probably use the streams in order 1, 2, 3, etc., (but skip 10) so that Arena does not have to compute a bunch of seed vectors for intervening streams that you're not going to use.

Finally, if you're making multiple replications of your model (see Chapter 6 and Section 7.2.2), Arena will automatically move to the beginning of the next substream within all the streams you're using (even if you're just using the default stream 10) for the next replication. As we'll see in Section 12.4, this is important for synchronizing random-number use across variations of a model in order to improve the precision of your comparisons between alternatives.

While we feel that Arena's current RNG is extremely good, you can choose to use the old one (though we don't recommend it) if you really need to for some legacy reason. To do so, you need only place a Seeds module from the Elements panel into your model; you will need to edit this module to avoid a run-time error, but it doesn't matter what seed you specify for what stream (as long as you identify a stream you're not using) since its mere presence tells Arena to use the old generator. Online help gives you further information about what this Seeds module does for the old generator. If you want to use the new generator (which we strongly recommend), just make sure you don't have a Seeds module present in your model.

12.2 Generating Random Variates

In Section 4.6, we discussed how you can select appropriate probability distributions to represent random input for your model. Now that you know how to generate random numbers—that is, samples from a uniform distribution between 0 and 1—you need to transform them somehow into draws from the input probability distributions you want for your model. In simulation, people often refer to such draws as *variates* from the distribution.

The precise method for generating variates from a distribution will, of course, depend on the form of the distribution and the numerical values you estimated or specified its parameters to be, but there are some general ideas that apply across most distributions. Because implementation is a bit different for discrete and continuous random variables, we'll consider them separately.

12.2.1 Discrete

Let's start by considering discrete random variables (see Section C.2.2 in Appendix C). To take a simple concrete example, suppose you want to generate a discrete random variate X having possible values -2, 0, and 3 with probability mass function (PMF) given by

$$p(x) = P(X = x) = \begin{cases} 0.1 & \text{for } x = -2 \\ 0.5 & \text{for } x = 0 \\ 0.4 & \text{for } x = 3 \end{cases}$$

Since the probabilities in a PMF have to add up to 1, we can divide the unit interval $[0, 1]$ into subintervals with widths equal to the individual values given by the PMF, in this case, $[0, 0.1)$, $[0.1, 0.6)$, and $[0.6, 1]$. If we generate a random number U, it will be distributed uniformly over the whole interval $[0, 1]$, so it will fall in the first subinterval with probability $0.1 - 0 = 0.1$, in the second with probability $0.6 - 0.1 = 0.5$, and in the third subinterval with probability $1 - 0.6 = 0.4$. Thus, we would set X to its first value, -2, if U falls in the first subinterval, which will happen with probability $0.1 = p(-2)$, as desired. Similarly, we set X to 0 if U falls in the second subinterval (probability 0.5), and set X to 3 if U falls in the third subinterval (probability 0.4). This process, which is pretty obviously correct for generating X with the desired distribution, is depicted in Figure 12-2.

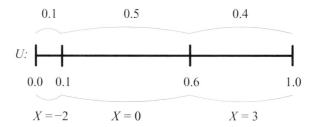

Figure 12-2. Generating a Discrete Random Variate

Another way of looking at this algorithm is that it *inverts*, in a sense, the cumulative distribution function (CDF) of X, $F(x) = P(X \leq x)$, as illustrated in Figure 12-3. First, generate a random number U, then plot it on the vertical axis, then read across (left or right) until you "hit" one of the jumps in the CDF, and finally read down and return X as whatever x_i ($= -2$, 0, or 3 in our example) you hit. In the example shown, U falls between 0.6 and 1, resulting in a returned variate $X = 3$. This is clearly the same algorithm as described above and shown in Figure 12-2, but sets things up for generating continuous random variates below.

This method, looking at it in either of the above ways, clearly generalizes to any discrete distribution with a finite number of possible x_i's. In fact, it can also be used even if the number of x_i's is infinite. In either case, the real work boils down to some kind of search to find the subinterval of [0, 1] in which a random number U falls, then returning X as the appropriate x_i. If the number of x_i's is large, this search can become slow, in which case there are altogether different approaches to variate generation. We won't cover these ideas here.

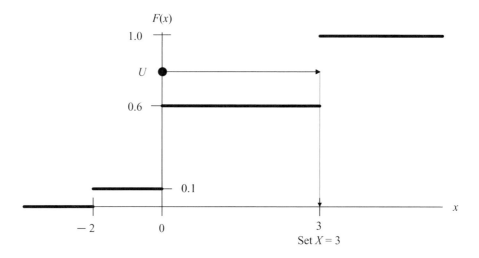

Figure 12-3. Generating a Discrete Random Variate via Inversion of the CDF

Arena has built in the Poisson distribution, as well as any user-defined finite-range discrete distribution (see Appendix D). In both cases, the above algorithm is used to generate the variates.

12.2.2 Continuous

Now let's consider generation of variates from a continuous probability distribution (Section C.2.3 in Appendix C). In this case, we can't think in terms of the probability of getting (exactly) a particular value returned since this probability will always be zero. Instead, we need to think in terms of the returned X being *between* two values. As a specific example, take the exponential distribution with mean $\beta = 5$, which has probability density function (PDF)

$$f(x) = \begin{cases} (1/5)e^{-x/5} & \text{for } x > 0 \\ 0 & \text{for } x \leq 0 \end{cases}$$

and CDF

$$F(x) = \begin{cases} 1 - e^{-x/5} & \text{for } x > 0 \\ 0 & \text{for } x \leq 0 \end{cases}$$

To generate a variate X from this distribution, start (as usual) by generating a random number U. Then set U equal to the CDF (evaluated at the unknown X) and solve for X in terms of the (now known) value of U:

$$\begin{aligned} U &= 1 - e^{-X/5} && \text{(ignore the probability-zero event that } U = 0) \\ e^{-X/5} &= 1 - U \\ -X/5 &= \ln(1 - U) && \text{(ln is the natural logarithm; i.e., base } e) \\ X &= -5 \ln(1 - U) \end{aligned}$$

(Obviously, replacing the 5 with any value of $\beta > 0$ gives you the general form for generating an exponential variate.) This solution for X in terms of U is called the *inverse CDF* of U, written $X = F^{-1}(U)$ {$= -5 \ln(1 - U)$ in our example}, since this transformation "undoes" to U what F does to X.

The inverse CDF algorithm for this example is shown in Figure 12-4. (Note the similarity to the discrete case in Figure 12-3.) First generate a random number U, plot it on the vertical axis, then read across and down to get X, which is clearly the solution to the equation $U = F(X)$.

To see why this algorithm is correct, we need to demonstrate that the returned variate X will fall to the left of any fixed value x_0 with probability equal to $F(x_0)$, for this is precisely what it means for the CDF of X to be F. From Figure 12-4, you see that, since F is an increasing function, we'll happen to get $X \leq x_0$ if and only if we happen to draw a U that is $\leq F(x_0)$; Figure 12-4 depicts this as happening for this value of U. Thus, the events "$X \leq x_0$" and "$U \leq F(x_0)$" are equivalent, so must have the same probability. But since U is uniformly distributed on [0, 1], it will be $\leq F(x_0)$ with probability $F(x_0)$ itself, since $F(x_0)$ is between 0 and 1. Thus, $P(\text{returned variate } X \text{ is } \leq x_0) = F(x_0)$, as desired.

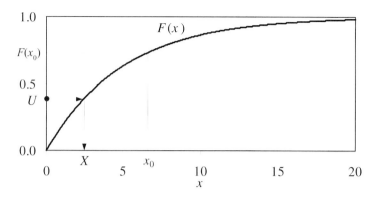

Figure 12-4. Generating a Continuous Random Variate via Inversion of the CDF

Though you may think we're all wet, there's actually an intuitive appeal to this algorithm. What we want is for the returned X's to follow the density function $f(x)$, so we want a lot of X's where $f(x)$ is high and not very many where $f(x)$ is low (see Section C.2.3 in Appendix C). Now the CDF $F(x)$ is the (indefinite) integral of the PDF $f(x)$; in other words, $f(x)$ is the derivative (slope function) of $F(x)$. Thus, where $f(x)$ is high, the slope of $F(x)$ will be steep; where $f(x)$ is low, $F(x)$ will be increasing only slowly (i.e., with shallow slope). In our exponential example, $f(x)$ starts out at its highest point at $x = 0$ (see the Exponential entry in Appendix D) and then declines; thus, $F(x)$ increases steeply just to the right of 0, and then its slope flattens out as we move to the right, which is where $f(x)$ becomes smaller. Now, put yourself in Figure 12-4 and stand just to the left of the vertical axis. Pick up a garden hose that will spray U's uniformly toward the right, and turn it on. As your U's hit the $F(x)$ curve, they will drip through it down to the horizontal axis, landing to define your returned X's. Your uniformly sprayed U's will be more likely to hit the $F(x)$ curve where it rises steeply (in this exponential case, early in its ascent) since it's a bigger target from where you're standing. Just a few of your U's (the really high ones) will hit $F(x)$ out on its upper-right portion. Thus, if you look at where your X drips landed, they will be more dense on the left part (where $f(x)$ is tall and $F(x)$ is rising steeply), and will be sparser on the right part (where $f(x)$ is low and $F(x)$ is rising shallowly). This is what we want for this distribution.

The inverse CDF idea works, in principle, for any continuous distribution. But, depending on the particular distribution, implementing it may not be easy. Some distributions, unlike the exponential example above, have no closed-form formula for the CDF $F(x)$, so that a simple formula for generating the X's is not possible (a notable example of this is the normal distribution). In many such cases, though, numerical methods can be used to approximate the solution of $U = F(X)$ for X to a very high degree of accuracy (with error less than computer roundoff error). There are also completely different approaches to generating variates that are sometimes used; we won't go into these here.

Most of the continuous distributions supported by Arena (see Appendix D) use the inverse CDF method for variate generation, in some cases by highly accurate numerical approximation. A few of the distributions, though, use other methods if they are particularly attractive in that case; refer to online help under the topic "Distributions" to see exactly what Arena does to generate samples from each continuous distribution.

12.3 Nonstationary Poisson Processes

A lot of systems have some kind of externally originating events affecting them, like customers arriving, calls coming in, cars approaching an intersection, or accidents occurring in a plant. Often, it's appropriate to model this kind of *event process* as being random with some continuous probability distribution for interevent times, implying some discrete distribution for the number of events occurring in a fixed interval of time. If the process governing the event occurrences is stationary over the time frame of the simulation, you can just decide on the right interevent-time distribution and generate the events during the simulation as we've done in many models in the book (for example, with the Time Between Arrivals area in the Create module from the Basic Process panel).

However, many systems experience time-varying, or *nonstationary* patterns of events—the lunch stampede at a fast-food restaurant, morning and evening rush hours in traffic systems, mid-afternoon peaks of calls coming in to a call center, or a rash of accidents when the moon is full. While you might be tempted to ignore these patterns and make the events occur at some kind of "average" rate throughout your simulation, doing so could lead to seriously inaccurate results if there is much variation in the actual pattern. For instance, if we average the freeway load over 24 hours, there's little doubt that a small number of lanes would appear to be adequate in the model; in fact, though, rush hours would be impossible messes. So modeling nonstationary external events can be a critical part of valid modeling in general.

Actually, we've already discussed a situation exactly like this, Model 5-2. The external events are the incoming call arrivals, and we specified the observed rates (in arrivals per hour) for each 30-minute period in the `Arrival Schedule` line of the Schedule module.

The usual way to represent time-varying event patterns like this is by what's called a *nonstationary Poisson process* (NSPP). To use such a process, you need to specify a *rate function*, $\lambda(t)$, that changes with time (t), having the rough interpretation that $\lambda(t)$ is high for times t when lots of events are happening, and low when things are quiet. More precisely, the definition of an NSPP is that events occur one at a time, are independent of each other, and the number (count) of events occurring during and interval of time $[t_1, t_2]$ is a Poisson random variable (see Appendix D) with expected value given by

$$\Lambda(t_1, t_2) = \int_{t_1}^{t_2} \lambda(t)dt$$

which is large over time intervals where $\lambda(t)$ is high and small when $\lambda(t)$ is low.

If you want to use an NSPP in a simulation, there are two issues: how to form an estimate of $\lambda(t)$, and then how to generate the arrivals in accordance with your estimated rate function. The estimated rate function we used in Model 5-2 was piecewise constant, with (possible) changes of level occurring every 30 minutes. This approach is probably the most practical one since it's quite general and easy to specify (though you must be careful about keeping the time units straight). There are, however, more sophisticated ways to estimate the rate function, including amplitudes and periodicities, with firm grounding in statistical theory; see, for instance, Leemis (1991); Johnson, Lee, and Wilson (1994); and Kuhl and Wilson (2000, 2001).

As long as you stick with a piecewise-constant estimated rate function, Arena has a built-in generation method, the use of which was illustrated in Model 5-2. The algorithm used is a speeded-up variation of a method attributed to Cinlar (1975, pp. 94–101). If you're interested in the details, see Arena's Help topic "Non-Stationary Exponential Distribution."

12.4 Variance Reduction

As we've indicated in several places (including Sections 2.6, 4.6, 7.2, and Chapter 6), simulations using random variates from probability distributions as part of the input will in turn produce random output. In other words, there is some *variance* associated with the output from a stochastic simulation. The more variance there is, the less precise are your results; one manifestation of high variance is wide confidence intervals. So output variance is the enemy, and it would be nice to get rid of it, or at least reduce it. One (bad) way to eliminate variance in the output is to purge all the randomness from your inputs, perhaps by replacing the input random variables with their expected values. However, as we demonstrated in Section 4.6.1, this might make your output nice and stable but it also will usually make it seriously wrong.

So, barring major violence to your model's validity, the best you can really hope for is to *reduce* the variance in your output performance measures. One obvious way to do this is by just simulating more. For terminating models, this implies more replications (since extending the length of a replication would make the model invalid in the terminating case); in Section 6.3, we gave a couple of formulas from which you can approximate the number of replications you'll need to bring a confidence-interval half width down to a value small enough for you to live with. For steady-state models, you could also make more replications if you're taking the truncated-replications approach to analysis, as discussed in Section 7.2.2; or you could just make your (single) replication longer if you're taking the batch-means approach (Section 7.2.3). In Section 12.5, we'll discuss all of this in detail, including how you can get Arena to "decide" on the fly how much simulating to do.

But what we aspire to in this section is a free lunch. Getting more precise results by more simulation work is not tricky, but there are some situations where you can achieve the same thing without doing any[1] more work. What usually enables this is the fact that, unlike in most physical experiments, you're in control of the randomness in a simulation experiment since you can control the random-number generator, as discussed in Section 12.1. This allows you to induce certain kinds of correlations that you can exploit to your advantage to reduce the variance, and thus imprecision, of your output. These kinds of schemes are called *variance-reduction techniques* (or sometimes variance-reduction

---

[1] Well, hardly any.

strategies). In most cases, you need to have a thorough understanding of your model's logic and how it's represented in Arena in order to apply such methods.

Variance-reduction techniques can be quite different from each other and have been classified into several broad categories. We'll discuss only the most popular one of them in detail, in Section 12.4.1, and will briefly describe some others in Section 12.4.2.

12.4.1 Common Random Numbers

Most simulation studies involve more than just one *alternative* (or *scenario*) of a model. Different alternatives could be determined by anything from just an input-parameter change to a wholesale revision of the system layout and operation. In these situations, you're usually not interested so much in the particular values of the output performance measures from the individual alternatives, but rather in their *differences* across the alternatives. These differences are measures of the effect of changing from one alternative to another.

For example, take Model 7-2, the model of a small manufacturing system with the average total WIP (work in process) output we developed in Section 7.2.1. Let's consider the model exactly as it is as a "base case." We'll keep the Replication Length at 5 Days and keep the Number of Replications at 10; we're implicitly viewing this as either a terminating simulation, or a steady-state simulation using the truncated-replications approach but with no Warm-up Period needed. Suppose there's an opportunity to take on 3.5% more business, with the same mix of part types, sequences, and processing times, which translates into a decrease in the mean interarrival time between parts from 13 minutes to 12.56 minutes (the mean arrival *rate* 1/12.56 is 3.5% more than the mean arrival *rate* 1/13). Call the base-case model alternatives A and the increased-arrival model alternative B. We made the following modifications to Model 7-2 to produce what we'll call Model 12-1:

- Made appropriate documentation changes in *Run > Setup > Project Parameters*.
- In the Statistic module, eliminated the Output File entry `Total WIP History.dat` for the existing statistic with Name `Total WIP`, since we don't need to save the within-run path of this quantity now.
- In the Statistic module, added a row for a new statistic with Name `Avg Total WIP`, Type `Output`, Expression `DAVG(Total WIP)`, Report Label `Avg Total WIP`, and Output File `Avg Total WIP.dat` to save to that file the time-average WIP for each replication to use in the Output Analyzer below.

Figure 12-5 shows the entries in the modified Statistic module. Of course, we made sure that *Run > Run Control > Batch Run (No Animation)* was selected since we're interested only in numerical results and not the animation.

Statistic - Advanced Process					
	Name	Type	Expression	Report Label	Output File
1	Total WIP	Time-Persistent	EntitiesWIP(Part 1) + EntitiesWIP(Part 2) + EntitiesWIP(Part 3)	Total WIP	
2	Avg Total WIP	Output	DAVG(Total WIP)	Avg Total WIP	Avg Total WIP.dat

Figure 12-5. The Completed Statistic Module for Model 12-1

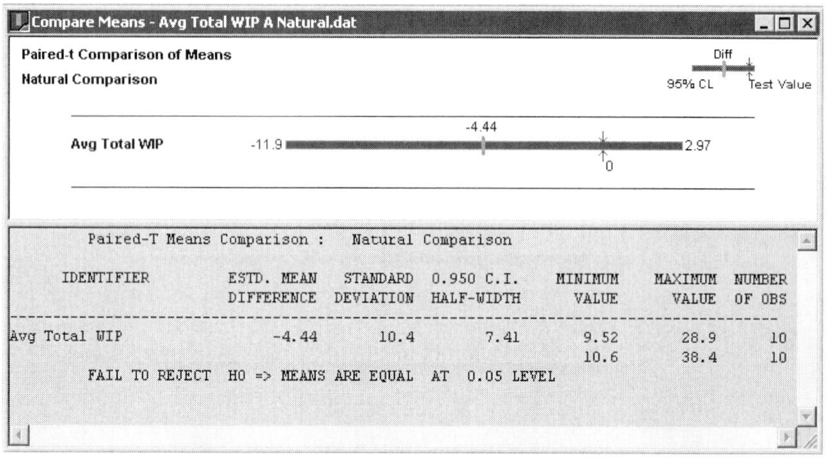

Figure 12-6. Compare Means Results for the Natural Comparison

The natural thing to do here is run alternative A, make the change to get from alternative A to B (change Value in the Create module from 13 to 12.56), run alternative B (of course, you'd remember to change the Output File name out in the operating system after run A so as not to overwrite it in run B), and use *Analyze > Compare Means* in the Output Analyzer, accepting the Paired-t Test default there since we did not do anything to cause us to use separate random numbers in run B; See Section 6.4 for a discussion of the mechanics and further explanation. This results in Figure 12-6 from the Compare Means function in the Output Analyzer. Remembering that the difference is in the direction A – B, it's intuitive that the point estimator of the mean is negative, since alternative B is the higher-arrival model so we'd expect WIP to be bigger. However, the confidence interval on the expected difference contains zero, so we can't conclude that there's a statistically significant difference. Of course, making more than the ten replications could change this, but we're trying to be smarter here.

And here's one way to be smarter. To estimate the difference between the average total WIPs produced by alternatives A and B, it makes intuitive sense to simulate both alternatives under conditions that are as similar as possible, except for the model change we made. In simulation, what the "conditions" boil down to are the generated variates from the input distributions, like interarrival and service times. If we re-use the same random numbers in run B as we did in run A, we might hope to get the same generated variates. In that way, when we look at the difference in the results, we'd know that it's due to the difference in the models rather than due to the random numbers having bounced differently in the two alternatives.

If you think about it, we actually *did* use the same random numbers above in run B as we did in run A, but since we drew on just one fixed sequence (stream) of random numbers for everything, the usage of these random numbers is extremely likely to get mixed up after a little while between runs A and B, due to the difference in the paths followed by alternatives A and B in terms of generating variates. This doesn't make our statistical comparison in Figure 12-6 wrong (so long as we stick with the Paired-t Test rather than the Two-Sample-t Test), but it doesn't accomplish the goal of generating the same variates for the same purposes in runs A and B either. We need to be even smarter (as well as have a good random-number generator with lots of really long streams and substreams, which Arena has, as described in Section 12.1).

Here's how to be even smarter. In this model, there are 14 sources of randomness—interarrival times, part indices, and 12 processing-time distributions as defined in Table 7-1. For our comparison, we'd like to run both alternatives A and B with the "same" external loads. In this case, it means blank parts arriving at the same times, assigned the same part indices, and experiencing the same processing times at each of the stops along their ways. When we ran this comparison above, producing Figure 12-6, we didn't do anything to try to get any of this to happen. True, the same random-number stream (the default, stream 10) was used for everything throughout both alternatives, and this stream started the first of the ten replications from the same seed for both alternatives. But due to the change made in the model between the alternatives, this fixed sequence of random numbers will be used in a different order if, at any point during the runs, there is a difference in the order of execution in which the 14 places draw the variates they need. This causes the "external loads" to differ at this point, and likely from then on, which is not the effect we want.

Instead, we need to *synchronize* the use of the random numbers across the model alternatives, or at least do so as far as possible given the model logic and the changes between the alternatives. One approach (though not the only one) to this end is to *dedicate* a stream of random numbers to each of the 14 places in the model where variates are generated. This is like piping in separate "faucets" of random numbers to each random-variate generator and other situations (like chance-type Decide modules and generation of nonstationary Poisson processes, neither of which occurs in this model) where a random number is needed. In this way, you can usually get reasonable synchronization, though in complex models it might not ensure that everything is matched up across the alternatives (which does not make your model in any way incorrect or invalid, but it is not quite as close a match-up across the alternatives as you might like in order to have the "same" external loads presented to both alternative models). However, there will still probably be at least some variance-reduction benefit even though the matchups might not be quite perfect. This is the synchronization approach we'll take in our example below.

A different way to attempt random-number synchronization, which might work better in some models, is to assign to each entity, immediately upon its arrival, attribute values for all possible processing times, branching decisions, etc., that it might need on its entire path through the model. When the entity needs one of these values during its life in the model, such as a processing time somewhere, you just read it out of the appropriate attribute of the entity rather than generate it on the spot. This might come closer for some models to the ideal of having the "same" external conditions, but it can require a lot more computer memory if you need to hold lots of attributes for lots of entities at the same time. If Arena has to use your disk drive to extend memory temporarily (called *virtual memory*), there can be a significant increase in execution time as well since disk access is much slower than memory access. In this situation, you might be just as well off using this extra computer time doing more simulation (either more replications or a longer single run). We won't carry out this synchronization approach in our example below, but will leave it for you as Exercises 12-1 and 12-2.

Sometimes achieving full, complete, and certain synchronization in complex models is just impractical, in which case you might consider matching up what you can, and generating the rest independently. When you use the same random numbers across simulated alternatives, synchronized in some way, you're using a variance-reduction technique called *common random numbers* (CRN), also sometimes called *matched pairs* or *correlated sampling*.

To implement the first of the above two synchronization approaches to CRN (piping in separate faucets of random numbers to the separate sources of randomness in a model) in Model 12-1, we modified it into what we'll call Model 12-2. To each of the 14 variate-generation expressions, we added an extra parameter (see Section 12.1), assigning a unique stream number, as given in Table 12-2. We actually defined a Variable for each stream number, initialized them all in the Variable module, and used the variable names in the variate-generation expressions; this made it easier to carry out further experiments later. We skipped over stream 10 since that's Arena's default stream and is used for the built-in nonstationary-Poisson-process generator and any Chance-type Decide modules; fortunately, we have neither in our model so it's easy to maintain proper synchronization. (See Exercise 12-7 for a way to trick chance-type Decide modules into using whatever random-number stream you want.) In this model, the processing times are assigned for Cell 1 via an Expression, and for the other cells via assignments in the Sequences; Table 12-2 also indicates where the required change was made.

Table 12-2. Stream Assignment and Usage for Model 12-2

Stream Number	Usage	Variable Name	Location of Change
1	Interarrival Times	S Interarrivals	Create module
2	Part Indices	S Part Index	Assign module
3	Part Type 1 at Cell 1	S P1 C1	Expression module
4	Part Type 1 at Cell 2	S P1 C2	Sequence module
5	Part Type 1 at Cell 3	S P1 C3	Sequence module
6	Part Type 1 at Cell 4	S P1 C4	Sequence module
7	Part Type 2 at Cell 1	S P1 C1	Expression module
8	Part Type 2 at Cell 2 (first visit)	S P1 C2	Sequence module
9	Part Type 2 at Cell 4	S P1 C4	Sequence module
11	Part Type 2 at Cell 2 (second visit)	S P1 C2 Again	Sequence module
12	Part Type 2 at Cell 3	S P1 C3	Sequence module
13	Part Type 3 at Cell 2	S P1 C2	Sequence module
14	Part Type 3 at Cell 1	S P1 C1	Expression module
15	Part Type 3 at Cell 3	S P1 C3	Sequence module

These stream assignments dedicate a separate random-number stream to each source of randomness in the model, and they will be in effect for the running of both alternatives A and B. Recall that we're doing ten replications of each alternative, and that the Arena random-number generator will automatically advance to the next substream within each stream for each new replication. In this way, we're assured that the matchup of random numbers across alternatives will stay synchronized beyond just the first replication, even if the alternatives disagree about how many random numbers from a stream are used for a given replication. Since it will take at least 669 million years for a typical machine of today to exhaust a substream, we can be pretty sure that we won't "lap" ourselves in terms of substream usage from one replication to the next.

What we set up above is a fairly carefully synchronized random-number allocation for CRN. What we did in Section 6.4 and to produce Figure 12-6 above, using the same stream (10) for everything might be described as *using* the same random numbers across the alternatives, but in a disorganized, haphazard, and mostly unsynchronized way, diluting or even erasing the effect. It's possible to simulate the alternatives using completely different, and thus independent random numbers across the alternatives, but you have to work to make this happen by bumping up the stream numbers in Table 12-2 for, say, alternative B (but not A) so that streams 1–15 are not used. Thus, in a way, CRN is the "default" for the way most people would naturally run comparisons, but it's extremely unlikely that proper synchronization of random-number usage will just happen on its own unless your model is awfully simple.

Proceeding as in Section 6.4 and what we did to produce Figure 12-6 above, we made ten replications of alternatives A and B using Model 12-2 for both, with the stream assignments discussed above, resulting in synchronized CRN. We then invoked the *Analyze > Compare Means* menu option in the Output Analyzer except calling for comparisons of A against the CRN run of B, to get the results in Figure 12-7 for a 95% confidence interval on the difference between expected average total WIPs for alternative A minus alternative B (Figure 12-7 actually includes both this synchronized-CRN comparison as well as the unsynchronized "natural"). Looking back at Figure 12-6 (repeated as the top comparison in Figure 12-7), the qualitative conclusion from synchronized CRN is the same as before—increasing the arrival rate increases average total WIP, though the "natural" comparison could not conclude that with statistical significance (at the 0.05 level) while the synchronized CRN comparison can. In terms of the precision of the estimate of the magnitude of this increase, though, the effect of synchronizing CRN is to reduce the confidence-interval half width fairly dramatically, from 7.41 to about 0.711, *without doing any more simulation work than before* (ten replications of both alternative models). Thus, for this model with these comparisons, the benefit from properly synchronized CRN is quite apparent. The magnitude of the effect of CRN is highly model- and situation-dependent, and it is pretty potent in this example. Sometimes, however, it can be less dramatic, but the (small) effort to synchronize stream usage is still probably worthwhile in most cases.

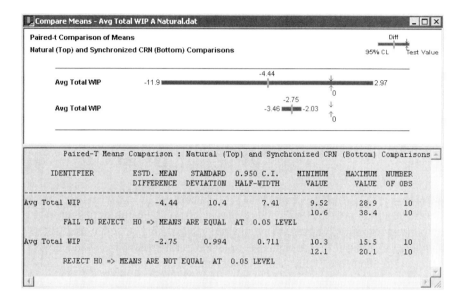

Figure 12-7. Compare Means Results for the Synchronized CRN Comparison

You may have noticed in Figures 12-6 and 12-7, or from the *Analyze > Compare Means* dialog box in Display 6-1, that we've used one of two possibilities for building the confidence intervals on the expected differences, and for testing the null hypothesis that there is no difference between the two expectations. This option is called the *Paired-t* approach (which is the default). This approach takes replication-by-replication differences between the results from the two alternatives, thus "collapsing" the two samples to a single sample on which the analysis is done. The advantage of the Paired-t approach is that it does not require the assumption of independent sampling across the alternatives for statistical validity, thus allowing the use of CRN. The disadvantage is that you wind up with a "sample" that's only half as big as the number of runs you made (in our example going from 20 to 10), resulting in "loss" of degrees of freedom (DF), which has the effect of increasing the confidence-interval half width. The other available option, called the *Two-Sample-t* approach, on the other hand, retains the "full" sample size (20 in our case), but requires that all observations from the two alternatives be independent of each other; this outlaws the use of CRN as well as the "natural" approach since it uses the same random numbers, though it does not synchronize their use. If you're using CRN for your comparison, then you have no choice in the matter—you must use the Paired-t approach. Even though you're suffering the loss of DF, the reduction in variance you're getting from CRN often more than offsets this loss, resulting in a tighter interval. If you're doing completely independent sampling across your alternatives, though, you could use either approach. We made a completely independent set of ten replications of alternative B by changing the stream assignments in Table 12-2 by just adding 100 to each stream number so that they're all different from the streams (1–15, except for 10) used in run A. The resulting half widths of confidence intervals on the expected differences from Compare Means were 4.53 with the Paired-t approach and 4.92 with the Two-Sample-t approach; both are valid, and there's not much difference between them (they're close to what we got from the "natural" comparison), but they aren't even close to being as precise as the 0.711 we got from synchronized CRN.

While the intuitive appeal of synchronized CRN, to "compare like with like," is clear, there's also a mathematical justification for the idea. Let X and Y denote the output random variables for, respectively, alternatives A and B. In our example, X and Y would be the averages over the ten replications of the average total WIP in each replication, in alternatives A and B. What we want to estimate is $E(X) - E(Y) = E(X - Y)$, and our (unbiased) point estimator of it is just $X - Y$. If we make the runs independently, the variance of our independent-samples estimator is

$$Var(X - Y) = Var(X) + Var(Y)$$

since, as random variables, X and Y are independent. If, however, we use synchronized CRN, what we're doing is inducing correlation, hopefully positive, between X and Y. Since correlation has the same sign as covariance (see Section C.2.4 in Appendix C), the variance of our CRN estimator is

$$Var(X - Y) = Var(X) + Var(Y) - 2\, Cov(X, Y),$$

which will be less than the variance of the independent-samples estimator since we're subtracting a (hopefully) positive covariance. So, what's needed to make CRN "work" is that the outputs be positively correlated, and the stronger the better. While you can find examples where the correlation is negative, causing CRN to "backfire," such models are generally contrived pathologies just to make the point. In most cases, CRN will work, sometimes dramatically. It is true, though, that you can't tell how well it will work until you actually do it. And, as you've seen above, CRN in its haphazard, unsynchronized version, is almost automatic—you have to work to *avoid* it. However, you probably won't get much benefit unless you do something to synchronize the random-number usage across the alternatives by assigning random-number streams carefully and with an understanding of how your model works.

12.4.2 Other Methods

In addition to CRN, there are several other variance-reduction techniques, which we'll just briefly mention here; see Chapter 3 of Bratley, Fox, and Schrage (1987) for more detail on these and other methods. Unlike CRN, these techniques apply when you have just a single model variant of interest.

The method of *antithetic variates* attempts to induce negative correlation between the results of one replication and another, and uses this correlation to reduce variance. In a terminating simulation, make the first replication "as usual," but in the second replication, replace the random numbers $U_1, U_2, \ldots$ you used in the first replication by $1 - U_1, 1 - U_2$, etc. This still results in valid variate generation since, if U is distributed uniformly on [0, 1], then $1 - U$ is as well. The idea is that, since a "big" U results in a "small" $1 - U$ (and vice versa), the results from replications one and two will be negatively correlated. Your first observation for statistical analysis, then, is not the result from the first replication, but rather the *average* of the results from the first two replications, which are treated as a pair. You could then go on and make replication three with "fresh" (independent) U's, then re-use these in replication four but in their antithetic form $1 - U$; your second observation for statistical analysis is then the average of the results from replications three and four. Within an antithetic pair, the (hopefully) negative correlation will cause the average of the two output statistics to snap in toward the true expectation more closely than if they were independent. Like CRN, this method requires careful synchronization of random-number usage in your model, probably involving streams and seeds. In Arena, you'd need to carry out some low-level manipulation to implement antithetic variates since there isn't any direct support.

Control variates uses an ancillary "controlling" random variate to adjust your results up or down, as warranted by the control variate. For example, in Model 12-1, if we noted that, in a particular replication, the average of the generated interarrival times of parts happened to be smaller than their expected value (which we'd know since we specified the interarrival-time distribution as part of the model input), then it's likely that we're seeing higher-than-expected congestion measures and thus high average WIP in this replication. Thus, we'd adjust these output measures downward by some amount to "control" for the fact that we know that our parts were arriving faster than "usual." From one replication to another, then, this adjustment will tend to dampen the variation of the results around their (unknown) expectations, reducing variance. In a given model, there are many potential control variates, and there are different ways to select from among them as well as to specify the direction and magnitude of the adjustment to the simulation output. For further information on control variates, see, for example, Bauer and Wilson (1992) or Nelson (1990).

12.5 Sequential Sampling

When you do a simulation, you should always try to quantify the imprecision in your results. If this imprecision is not great enough to matter to you, you're done. But if the imprecision is large enough to be upsetting, you need to do something to reduce it. In Section 12.4, we discussed variance-reduction techniques, which might help. And in Section 6.3, we gave a couple of formulas for approximating the number n of replications you'd need (in a terminating model) to get a confidence-interval half width down to a value small enough for you to live with.

But a rather obvious idea is just to keep simulating, one "step" at a time, until you're happy. In the case of terminating models, a "step" is the next replication; in the case of steady-state models, a "step" is either the next truncated replication (if you're taking that strategy, as in Section 7.2.2); or if you're doing batch means in a single replication (as in Section 7.2.3), extend by some amount the single replication you have going. Then, after this next "step," check again to see, for instance, if the half width of the new confidence interval is small enough. If it is, you can stop; if not, keep going and make the next step. If you can afford to do so, such *sequential sampling* is usually fairly simple, and typically will get you the precision you need. What's even better is that these ideas, while being entirely intuitive, are backed up by solid statistical theory; one consequence is that the actual coverage probability of your confidence interval will approach what it's supposed to be as your smallness demands on the half width get tighter.

In this section, we'll show you some examples of sequential sampling, indicating how you can set things up so that Arena will take care of the checking and stopping for you. We'll consider terminating simulations in Section 12.5.1 and the steady-state case in Section 12.5.2.

12.5.1 Terminating Models

First let's consider a terminating simulation. We'll modify Model 12-2 (with the random-number streams set up) since you never know when you might want to do some variance reduction and synchronize the random numbers. We'll view this as either a terminating simulation or as a steady-state simulation using the truncated-replications approach but with no Warm-up Period needed. As the output performance measure of primary interest (the one whose confidence-interval half width we want to make sure is "small enough"), let's take the average total WIP. We first made a fixed number of replications (25, rather than the original ten to give us a little better chance at getting a decent variance estimate), not really knowing how wide our confidence intervals would be; we got a 95% confidence interval of 13.03 ± 1.28. In Section 6.3, we also discussed two formulas for approximating how many replications would be needed to reduce the half width of a 95% confidence interval to a fixed value; in this example, to reduce it from 1.28 to, say, 0.5, our formulas say to we'd need to make either about 124 or about 164 total replications instead of 25, depending on which formula we used (and keep in mind that these are perhaps-crude approximations subject to uncertainty in the variance estimate from these initial 25 replications).

Let's instead invoke the sequential-sampling idea to get this half width down to 0.5. We have to make a minor change to our model (into what we'll call Model 12-3) to ask it to keep replicating until the across-replications 95% confidence-interval half width for the average total WIP output performance measure falls below 0.5, then stop the replications. Recall from Section 6.3 that if you call for multiple replications, Arena will automatically compute 95% confidence intervals across the replications on quantities you've specified for your reports; we'll use internal Arena variables describing these confidence intervals to get sequential sampling going. The pertinent Arena variables are:

- ORUNHALF(Output ID), the half width of the automatic across-replications confidence interval from however many replications have been completed, where Output ID is the identification of the output measure of interest (in our case, `Avg Total WIP`);
- MREP, the total number of replications we're asking for (initially the Number of Replications field in *Run > Setup > Replication Parameters*);
- NREP, the replication number we're on at the moment (= 1, 2, 3, . . .).

Here's the general strategy. Initially, specify MREP to be some absurdly enormous value in the Number of Replications field in the *Run > Setup > Replication Parameters* dialog box; this gets the replications going and keeps them going until we cut them off ourselves when the half width becomes small enough. Add a chunk of logic to the model to cause a single control entity to arrive at the beginning of each replication, whose job it is to check whether we're done yet—in other words, if ORUNHALF(Avg Total WIP) is smaller than the tolerance we want, 0.5. If not, we need to keep simulating, so this entity just disposes of itself; we'll go ahead and do the replication that's just starting and check again at the beginning of the next replication. On the other hand, if the current half width is small enough, we're ready to quit. However, because we've already begun the current replication (by having the "checking" entity show up) we have to finish it, but before doing so, we'll tell our control entity to reset MREP (the total number of replications we want) to NREP (the number we will have completed at the end of the one that's just starting), which will terminate the replications after this one.

Note that this strategy "overshoots" (by one) the number of replications really required; it turns out that you can't reliably do the termination check at the *end* of a replication, so this is necessary. As a result, you'll usually get a half width that's not only under the tolerance you specify, but probably even a bit smaller due to this extra replication. It's technically possible (though unlikely) that your final half width will be slightly larger than the tolerance you specified if the "extra" replication at the end happens to produce a wild outlier that itself causes the standard-deviation estimate to increase a lot. This whole situation does not seem particularly onerous, though, since in sequential sampling, you're pretty much admitting that you don't really know how many replications to make and that you might have to make a lot of them; thus, overshooting by one is no big deal. Furthermore, the tolerance you specify is probably fairly arbitrary in your mind, so being a little over or under won't matter—(only a truly obnoxious customer would demand that the person behind the deli counter *exactly* hit the half pound of garlic hummus they ordered). The chunk of control logic added (which we put inside a new submodel called Terminating Sequential-Sampling Control Logic) is shown in Figure 12-8. A *submodel* in Arena is just a way of covering up some logic, and what's inside it participates fully in the simulation just as if it were not in a submodel; to open a submodel just double-click on it and to close it, right-click in a blank area within the submodel and select Close Submodel in the popup, or use the Navigate panel in the Project Bar. For more on submodels, go to Arena Help, Index tab, and type in the keyword Submodels.

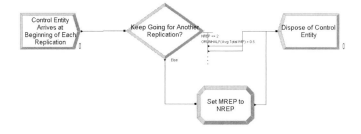

Figure 12-8. Control Logic for Sequential Sampling in Terminating Simulations

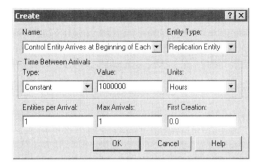

Figure 12-9. Create Module for Control Logic for Sequential Sampling
in Terminating Simulations

The completed dialog box for the Create module in this "control" submodel is shown in Figure 12-9. By setting Max Arrivals to 1, we ensure that there will be only one of these entities showing up, at the beginning of each replication; we also (redundantly) set the Time Between Arrivals to be a constant at a million hours.

After the entity is created, it goes to the Decide module, shown in Figure 12-10. Here it first checks to see if the number of replications, NREP, is less than or equal to 2 (NREP is the number of the replication now beginning). If so, since at least two complete replications are required to form a confidence interval, the entity disposes of itself (via the graphical connection visible in Figure 12-8), meaning that MREP remains at its absurdly high value and we go ahead and do this replication (and begin the next one too). The next check in the Decide module is to see if the current half width ORUNHALF(Total Cost) is still too big; if so, we need to keep going for more replications, so we dispose of the entity as well (again via a graphical connection).

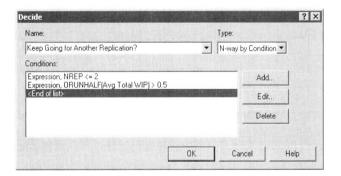

Figure 12-10. Decide Module for Control Logic for Sequential Sampling
in Terminating Simulations

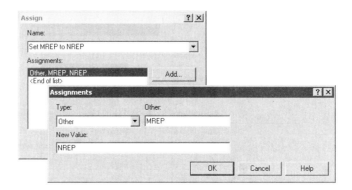

*Figure 12-11. Assign Module for Control Logic for Sequential Sampling
in Terminating Simulations*

If, however, NREP is at least 3 *and* the half width we have on record (the one completed at the end of the previous replication) is at most 0.5, we'll fall through both of the If statements and out the Else exit point, whose graphical connection sends the entity to the Assign module (Figure 12-11), where MREP is set to NREP, the current replication number, causing the simulation to stop at the end of the replication now getting started— (as mentioned earlier, this final replication is technically not needed but will be executed anyway). Note in the Assign module that we selected the Assignment Type to be Other since MREP is a built-in Arena variable with a reserved word as its name. Selecting the Assignment Type to be Variable would result in an attempt to create a new user-defined variable named MREP, which would conflict with the existing Arena variable of the same name (and cause an error).

We made a final change in the Statistic module, to remove the Output File Avg Total WIP.dat since we no longer need to save these values.

We ran this model, and it decided to stop after 232 replications; the User Specified part of the Category Overview Report is in Figure 12-12, showing only the relevant portion. Note that, as advertised, the half width of the confidence interval for expected average total WIP is down to 0.5 (barely—from the .*out* file the half width is actually 0.49699).

Why is the required number of replications, 232 (or, if you prefer, 231due to the extra replication at the end), different from the 124 or 164 that the formulas from Section 6.3 suggested? As we said there, those formulas are only approximations, with errors owing to the fact that one uses the normal distribution rather than *t*-distribution for the critical value, but mostly due to the fact that they're both based on a variance estimate from only

Output	Average	Half Width	Minimum Average	Maximum Average
Avg Total WIP	13.6970	0.50	8.7036	30.0500

Figure 12-12. Report for the Terminating Sequential-Sampling Run

an initial, somewhat arbitrary number of replications (we had used 25). It just turned out this time that the variance estimate from the initial 25 replications was apparently pretty small, so in the end we needed quite a few more replications than what the formulas predicted (it could just as easily have gone the other way).

Sequential sampling, perhaps more aptly termed sequential *stopping*, can be set up for other purposes as well. For instance, in the above run, we demanded that the half width for the confidence interval on only one of the (many) outputs be brought down to under the specified tolerance 0.5; we got just enough replications to satisfy that sole criterion. However, the setup of the above model is general enough to allow easy modification to demand that the half widths on several (or all) of the output measures be "controlled" to be less than separate tolerances for each of them; we'll ask you to look into this in Exercises 12-3 and 12-5.

Another modification would be to ask not that the half width be made smaller than a tolerance, but rather that the half width divided by the point estimate (the average across the replications) be brought down to be less than another kind of tolerance. Note that the half width divided by the point estimate is a dimensionless quantity, and so the tolerance in this case would also be dimensionless, giving it a universal emotional interpretation. For instance, if you specify this tolerance to be 0.10, what you're asking is that the half width of the confidence interval be no more than 10% of the mean; in this case, you could restate the confidence interval as something like "point estimate plus or minus 10%." This is sometimes called a confidence interval with a *relative precision* of 10%. Such a goal might be useful in a situation where you don't have much of an idea what the magnitude of your results will be, making it problematic to specify a sensible (absolute) value under which you'd like the half width itself to fall. Exercises 12-4 and 12-5 ask you to set up this kind of thing. One caution here—if the expected value being estimated is zero (or in some sense close to it) or the point estimate of this expected value is zero or close to it, you can get into difficulty with dividing by a number that's near zero, creating instability; in such a case, it would clearly be better not to use a relative-precision stopping criterion (see Exercise 12-6).

12.5.2 Steady-State Models

Sequential sampling for steady-state models is at least as easy to set up as for terminating models, though naturally the amount of computation time can become frightening if you need to make really long replications and also demand very tight precision. Probably it's prudent to get some notion of how much precision is practical before setting up a sequential-sampling run and just turning it loose.

If you're taking the truncated-replications approach to steady-state analysis, as described in Section 7.2.2, you can do things just as we described above in Section 12.5.1, except now you'd have a Warm-up Period specified in *Run > Setup > Replication Parameters* to carry out the truncation of initial data to ameliorate startup bias. A caution here is that you need to make quite sure that you're warming up long enough to get rid of initialization bias, so err on the side of longer-than-really-necessary warm-ups. The reason for this bit of friendly advice is that if you want a tight confidence interval with this strategy, you'll certainly get it. However, if there's bias in your results due to insufficient warm up, your sequentially-determined confidence intervals will be tightening down around a biased point, meaning that the nice tight interval you get is likely to miss the mark in terms of covering the steady-state expected value. And depressingly, the tighter you make your interval, the worse this problem gets since the bias stays present but the interval gets smaller and thus more likely to miss the steady-state expectation of interest. Thus, the harder you work, the worse off you are in terms of confidence-interval coverage probability.

So, unless you're quite confident that you've pretty much eliminated start-up bias, it might be safer to set up a single long run that you then keep extending until the half width of the resulting confidence interval satisfies your smallness criterion. In this case, the batch-means confidence intervals that Arena attempts to form from a single long run (Section 7.2.3) work quite nicely for this purpose. The key is the *Run > Setup > Replication Parameters* Terminating Condition field, where you specify the half-width smallness criterion (and remove all other replication-stopping devices from your model, such as the Replication Length field in *Run > Setup > Replication Parameters*). The pertinent internal Arena variables for this are THALF(Tally ID), which returns the current half width of the 95% confidence interval on a Tally statistic with Tally ID in its argument, and DHALF(Dstat ID) for DSTAT (time-persistent) output statistics. The batching/rebatching scheme described in Section 7.2.3 takes over and your run will stop as soon as the Terminating Condition is satisfied. If a particular run length along the way is not long enough to form a valid batch-means confidence interval, causing this scheme to conclude "(Insufficient)" or "(Correlated)" as described in Section 7.2.3, the numerical value of the half-width variable is set by Arena to a huge number; this will cause your replication to be extended since the half-width appears too large, which is the behavior you want.

To illustrate this, let's create Model 12-4 by modifying Model 7-4 from Section 7.2.3, which was originally set up for a single long run of 50 days of which the first two days are discarded as a Warm-up Period. Our result from Section 7.2.3 was a 95% batch-means confidence interval of 13.6394 ± 1.38366 on steady-state expected average total WIP. How about extending the run long enough for the half width to drop to 1, for example? Making this change involves only a couple of modifications in *Run > Setup > Replication Parameters*, as shown in Figure 12-13. We've cleared the Replication Length field (its default of "Infinite" appears after closing the dialog box) and filled in the Terminating Condition field with the stopping rule we want, DHALF(Total WIP) < 1.0. We also removed the Output File Total WIP History One Long Run.dat from the Statistic data module since we no longer need to save these values.

Part of the report is in Figure 12-14 and indicates that the desired precision for the Total WIP output measure was indeed achieved, though again only barely, as we'd expect from a sequential stopping rule. To achieve this, the model decided to extend its run length to almost 99 days (142148.4917 minutes, to be more precise), in comparison with the original 50-day run.

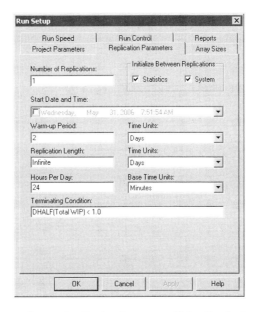

Figure 12-13. The Run > Setup > Replication Parameters Dialog Box for Sequential Sampling in Steady-State Simulation Using the Automatic Batch Means Confidence Intervals

As with sequential sampling for terminating simulations, you can modify the Terminating Condition to include smallness criteria on each of several confidence intervals instead of just one. You can also specify relative-precision stopping rules based on the ratio of the half width to the point estimate; to this end, the Arena variables TAVG(Tally ID) and DAVG(Dstat ID) give the current average of the indicated Tally or Dstat statistic, respectively (see Exercise 12-6).

12.6 Designing and Executing Simulation Experiments

As we've tried to emphasize throughout, a simulation model is a lot more than something to run just once. Rather, you should think of it as a great and convenient test bed for trying out a lot of things and for investigating the effects of various inputs and configurations on various outputs. Thus, simulation is a natural for application of some of the classical experimental-design techniques usually presented for physical rather than simulation-based experiments.

Time Persistent	Average	Half Width	Minimum Value	Maximum Value
Total WIP	12.6314	0.979115017	0.00	34.0000

Figure 12-14. Results for Sequential Sampling on Steady-State Total WIP Using the Automatic Batch Means Confidence Intervals

For instance, consider a model with five different input parameters or configurations. You'd like to know what the *effect* is on the results of changing a parameter or configuration from one *level* to another. Viewing the inputs and configurations as experimental *factors*, you could specify two possible levels for each and carry out a 2^5 factorial design, from which you could measure the main effects of and possible interactions between the input factors on the outputs of interest, which are the *responses* for this simulation-based experiment. Unlike most physical experiments, you could then easily go on and replicate the whole factorial experiment to place confidence intervals around the expected main effects and interactions. Other possibilities include using common random numbers across the design points (as a blocking variable in experimental-design terminology), the use of screening designs to sort out which of many factors are really important, and sophisticated nonlinear and response-surface designs. For more on these and related issues, see Law and Kelton (2000).

12.7 Exercises

12-1 Modify Model 12-2 for a different way to allocate random numbers to support synchronization for CRN, as follows. When a new part arrives, (pre-)generate and store in attributes of this entity its processing-time requirements for all of the cells in its sequence. When a part entity gets to a cell, take its processing time from the appropriate attribute of the entity rather than generating it on the spot. Depending on how you set this up, it might be useful to recall that the attribute Entity.JobStep of an entity is its station number (1, 2, 3, ...) as it makes its way through the sequence. You might also find useful the Attributes element from the Elements panel to define a vector-valued attribute; to assign values to elements of this vector in an Assign module, you'll need to choose the Type to be Other. Use the same random-number-stream allocation as in Model 12-1.

12-2 In Exercise 12-1, is it necessary to have dedicated random-number streams in order to achieve proper synchronization? Discuss this issue, experimenting with some examples if appropriate.

12-3 Modify Model 12-3 from Section 12.5.1 to demand, in addition to the 95% confidence-interval half width on the expected average total WIP being no more than 0.5, that the half width of the 95% confidence intervals on expected average total time in system (for all part types combined) be no more than 4 minutes. One way to do this is to modify the Decide module in Figure 12-10 by adding an additional Condition to keep replicating (i.e., dispose of the checking entity) if any half width is still too big.

12-4 Modify Model 12-3 from Section 12.5.1 to demand instead that the ratio of the 95% confidence-interval half width to the point estimate on the expected average total WIP be less than 0.05, as described at the end of Section 12.5.1; i.e., form a 5% relative-precision confidence interval. The Arena variable ORUNAVG(Output ID) is the average across all completed replications of the output measure with this Output ID.

12-5 Combine Exercises 12-3 and 12-4, as follows. Set up a sequential-sampling run so that you get 5% relative-precision confidence intervals on both the expected average total WIP as well as on the expected total time in system.

12-6 Modify Model 12-4 from Section 12.5.2 to terminate the replication when the ratio of the half width to the midpoint (point estimate) of the automatic batch-means run-time confidence interval on steady-state expected total WIP falls below 0.05; i.e., when the relative precision is 5%. The Arena variable DAVG(Dstat ID) returns the current average of the indicated Dstat (i.e., time-persistent) statistic. Note that the condition you want to check is of the form $H/A < 0.05$, where H is the half width and A is the point estimate, which can be an uncomfortable calculation if $A = 0$ (which it will be at the start of your run). Instead, check the condition in its equivalent form, $H < 0.05\ A$.

12-7 As noted in Section 12.4.1, Chance-type Decide modules use random-number stream 10 for their hypercoin flips, and there's no way to change that in the module. Describe (in words) how you could work around this to use whatever stream you want to accomplish the same thing. Also, build a useless little Arena model (but one that runs) that does the same thing as a Decide module flipping a four-sided hypercoin with probabilities 0.2, 0.4, 0.1, and 0.3 for outbound routes A, B, C, and D, respectively, except using random-number stream 7. Verify that your model produces decisions with probabilities that are about right.

CHAPTER 13

Conducting Simulation Studies

In Section 2.8, we briefly outlined the key ingredients of a simulation study. Now that you've gained some insight into the process of developing and analyzing a simulation model, it's time that we stepped back and discussed the overall activities involved in a typical simulation study. As we proceed with this discussion, we'll assume that you are the analyst—the individual who will perform the simulation study. You may be on a corporate support staff, support staff at the operational level, or be an external consultant. The client is whoever requested the study, which will be focused on a system of some type. The system may produce manufactured goods, fast food, or services. It could also be a system for handling paperwork, a call center, distribution center or system, or any other system that results in a product or service. We will assume that you're able to translate the ideas presented in this chapter (as well as the rest of the book) to your own circumstances.

There are numerous publications on the activities to be discussed in the chapter. Probably the best source is the *Proceedings of the Winter Simulation Conference*, a conference held annually in December. A selection of these include Balci (1990, 1995), Farrington and Swain (1993), Goldsman (1992), Kelton (1996), Kleindorfer and Ganeshan (1993), Musselman (1993), Sadowski (1989, 1993), Sargent (1996), and Seila (1990). Another good source is an article by Banks and Gibson (1996) in *IIE Solutions* magazine.

We'll start by discussing what constitutes a successful simulation in Section 13.1, followed in Section 13.2 by some advice on formulating the problem. Section 13.3 addresses the issue of using the correct solution methodology for the problem. Assuming that simulation is the preferred solution methodology, we continue with the system and simulation specification in Section 13.4. The model formulation and construction activities are discussed in Section 13.5, and the ever-present verification and validation approaches are presented in Section 13.6. Section 13.7 discusses experimentation and analysis, and Section 13.8 provides an overview of the reporting and documentation requirements for a simulation project. We end this chapter with a brief discussion of the Arena run-time capability (Section 13.9), which is a good way to disseminate your work.

13.1 A Successful Simulation Study

Before we start talking about what's involved in a simulation study, we need to address the issue of what defines a successful simulation project. It might seem obvious that if you solve the problem or meet the objectives, you've achieved success; however, that is not always the case. In most instances, the final pronunciation of success or failure will be made by the higher-level management that is paying the bill—and, like it or not, they have a tendency to view the problem or objectives in a different context. Let's illustrate this with a real-life example.

An automotive supplier had developed a highly successful process for producing a specialized set of parts for the automotive OEM (original equipment manufacturer) and after market. The process was implemented using a high-volume, semi-automated manufacturing cell concept. Although the process produced a high-quality, low-cost part, the initial system was not achieving a high utilization for the key, and most expensive, piece of equipment in the process. Because of the variability in the types and quantities of parts being produced and the processing times of those parts, simulation was chosen as an analysis tool. The objective was to re-design the existing system or to design a new production system that would make better use of the key equipment—in other words, a system that would achieve the same quality, but at a lower production cost per part. A secondary objective was to devise a way to allow incremental increases in production volume by adding equipment to the system gradually, rather than just building a new cell.

The simulation study was undertaken, and after an extensive analysis of the existing system, it was determined that it was not possible to achieve the desired results with the current cell concept. The simulation was then used to develop, design, and analyze several new approaches. This resulted in a totally different production system design that ultimately was recommended to management as the production system of the future. This new system design met all the objectives, and the simulation study was pronounced a success by the project team. Management accepted the team's recommendation and authorized the construction of a system using the new design concept. The new system was built, but did not function as projected, causing management to pronounce the simulation study a failure since the new system, which was built based on the results of the simulation, did not meet expectations. In fact, the resulting production output was about 30% lower than projected.

So what happened? Fortunately for the reputation of simulation, the simulation team was directed to conduct a post-analysis of the system. The simulation model of the new system was dusted off and compared to the actual system, which now existed. The simulation was modified to represent the system as it was actually constructed, and much to the surprise of the simulation team, it predicted that the new system would produce exactly the volume that it was currently producing. So they looked at the modifications that were made to the simulation model and found that they could classify these modifications into two groups.

The first group of model modifications was composed of required data changes to the model based on measurements of actual operation times occurring in the new system. As is often the case, several new types of equipment were included in the new system design, and the vendors who provided this new equipment were rather optimistic in their estimates of operation times. This accounted for about a third of the resulting production loss.

The second group of model modifications was composed of changes in the actual system design that was implemented. As it turns out, there were two critical errors made in the new system design. One was caused because the placement of the new system did not allow for sufficient floor space as called for by the simulation design. Thus, the system design was modified to allow it to fit into the available space. The second error was due to changes requested by upper-level management in an attempt to decrease the overall cost of the new system. These changes accounted for the remaining two thirds of the production loss.

So what are the lessons to be learned? First, the simulation team should have recognized that the operation times for the new equipment were only estimates, and they should have conducted a sensitivity analysis of the model outputs to these input data. This would have predicted a potential problem if the actual times were greater than the vendor-supplied estimates. If the times were critical, a penalty clause could have been included in the contract for the new equipment stipulating that it perform as estimated. Second, the simulation team should have tracked the implementation of the new system. This would have allowed them to evaluate the proposed modifications before implementation. It might not have changed the results, but at least it would have predicted the outcome.

Let's return to the initial issue of what defines a successful simulation project. Success seldom means that a good simulation model was developed. It more often means that the simulation study met the objectives set forth by the decision makers. This implies that it is very important to understand which metrics they will use.

If you're asked to undertake a simulation study to redesign a current system, inquire further to gain a better understanding of what is expected. If that inquiry results in a statement that indicates that management is interested in finding out if it is possible to make the system perform better, find out what "better" means. If better means that you are expected to reduce WIP (work in process) by 30%, reduce cycle times by 20%, increase resource utilizations by 15%, and meet all future customer due dates without any capital investment, you at least know that you're facing an impossible mission. We recommend that you elect not to accept the assignment. If, however, "better" means that the primary measures are WIP, cycle times, resource utilizations, and customer due dates balanced against capital investment, you at least have a fighting chance of designing a better system.

Unfortunately, you are seldom *asked* if you want to do the study; in most cases, you are just *told* to do the study. Even under these circumstances, you should identify the decision maker and attempt to define the metrics by which the project will be measured.

If you're about to undertake the first simulation study for a company or facility, it's critical that it result in a success. If the first simulation study is labeled a failure, it is unlikely that a second study will ever be started. If you have a choice of simulation projects, select one that is simple and is almost guaranteed to result in a success. Once you have achieved several successes, you might be able to afford a failure (but not a disaster).

There is also a peculiar dilemma frequently faced by the experienced simulation analyst. If you use simulation as a standard tool through the design of a new system and the new system works as advertised, what have you saved? If you had not used simulation, would the results have been any different? You really don't know. So the tool worked, or at least you think it did, but you have no way of quantifying any savings resulting from the application of the tool.

The one approach that is often suggested, but seldom used, is not to use simulation for the design of the next new system. The assumption here is that the resulting system will not perform as advertised. After it's clear that the system has problems, use simulation to show how the system should have been designed. This would allow you to quantify the savings had you used simulation. This is clearly an extreme approach—and not recommended by us—that's based on numerous assumptions. Of course, there is always the possibility that the system will perform just fine. (It can happen!) This would leave the value of simulation in even greater doubt in the eyes of management.

We close this section by again suggesting that you be aware that *your* definition of success may not be the same as that used by management. Although you may not be able to control management's evaluation, at least you should try to understand what measurements they will use in making their evaluation. Now let's proceed with our discussion of the key activities of a good simulation study.

13.2 Problem Formulation

The first step in any problem-solving task is to define and formulate the problem. In the real world, you are rarely handed some sheets of paper that completely define the problem to be solved. Most often you (the analyst) will be approached by someone (the client) who thinks that you might be able to help them. The initial problem is rarely defined clearly. In fact, it may not even be obvious that simulation is the proper tool. At this point, you need to be able to open a dialog with the requester, or their designate, and begin asking a series of questions that will ultimately allow you to define the problem completely. Often, the final problem you end up addressing is quite different from the one that was initially presented.

Many simulation studies deal with systems that are not meeting the client's expectations. The client wants to know: How do I fix it? Other simulation studies are not focused on a known problem, but are trying to avoid a potential future problem. This is most often the case when you're using simulation to help design a new system. Yet another class of simulation studies is composed of those focused on a system that has been completely designed, but not yet constructed or implemented. In this case, you're being asked if the system will perform as predicted.

So the problems are often put forth in the form of a series of questions: Can you fix it? Will it work? Can you help me make sure it will work? These are really the best kinds of problems because you have the potential to have an impact on the system (and your career). The worst kind of problem is when you're told that the company wants to start using simulation, so they've requested that a simulation be developed for an arbitrary system.

Let's assume that you have at least a vague notion of what the problem is, and hopefully, a better idea of what the system is. It might be a good idea to start with the system. Does the system currently exist, is it a new design, or has it not yet been designed? Knowing this, you can try to *bound* the system for the purpose of the study. Is the system a single or a small number of operations, a large department, the entire facility, or the entire company? Although it would be nice if you could draw walls around the system so that there are no interactions with the other side of the wall (like the Great Wall of China), this is not likely to be the case. However, you might at least try to place initial bounds on the size of the system. Try not to cast these boundaries in stone, as you may have to expand or contract them as you learn more about the problem and system.

Having established some initial boundaries, next try to define the performance metrics. There are really two kinds of metrics with which you should be concerned. The first, and most obvious, are the performance metrics that will be used to measure the quality of the system under study. The second, and maybe the most important, are the performance metrics that will be used to measure the success of the study. Let's concentrate on the first kind of metrics. Although the client might imply that there is only one metric, there are almost always several that need to be considered. For example, the application may be a fast-food restaurant where the client is interested in being assured that any customer who enters the door receives his food within a given period of time. This is most likely not just a performance metric, but a performance objective. Other metrics of interest could be the staffing required, the job assignments, the seating capacity, and the freshness of the food.

Having established how the performance of the system is to be measured, find out if there are current baseline values for these metrics. These values should be available if the system currently exists. If not, there may be similar existing systems that could be used to provide estimates. In the worst case, at least design values should be available. Knowing what the current baseline metrics are (or at least using an estimate), what are the expectations of the client? This type of information should provide some insight as to the magnitude of the problem you have been handed.

By this time, you should have a fairly good understanding of the system (and its size), the performance metrics, and the expectations of the client. The next step is to select a solution methodology—don't assume that you will always use simulation.

13.3 Solution Methodology

We're not going to attempt to describe every problem-solution technique known to humanity and recommend where they should be used. However, we do recommend that you at least consider alternative solution techniques. You should also give some consideration to the cost of using a particular solution technique compared to the potential benefits of the eventual solution. Unfortunately, identifying the best solution methodology is not always an easy task. If you determine that a specific methodology might give you the best answer, but you have never used that technique, this might not be the best time to experiment. Therefore, you might want to rephrase the question: Given the solution methodologies that you feel comfortable using, which will most likely give the most cost-effective solution?

Sometimes the choice is obvious, at least to us. For example, if you're being asked to perform a rough-capacity analysis of a proposed system, and you're given only mean values for all the system parameters and you are only interested in average utilizations, it might be faster (and just as accurate) to use a calculator to determine the answer. At the other extreme, you might be asked to find the set of optimal routes for a fleet of school buses or garbage trucks. Although you could use simulation, there are other tools specifically designed to solve this problem. We would suggest you consider purchasing a product to implement such a tool.

You might first find out how much time you have to analyze the problem. If you need an answer immediately, then simulation is not normally an option. We hedged a little bit on this because there are rare circumstances where simulation can be used to analyze problems quickly. You might have a very simple problem, or a very small system, which allows you to develop a simulation using the Arena high-level constructs in just a few hours.

There are also instances where companies have devoted effort to develop generic simulation models that can be altered quickly. These types of models are typically developed when there are many similar systems within an organization. A generic simulation is then developed that can be used to model any of these systems by simply changing the data. The task of changing the data can be made very easy if the values are contained in an external file or program. This external source might be a text file or a spreadsheet. Instances where this approach has been used include assembly lines, warehousing, fast food, distribution centers, call centers, and manufacturing cells. An even more elegant approach is based on the development of a company-specific template. This method was briefly discussed in Section 10.5.5.

Let's assume that simulation is the correct technique for the problem. Now you need to define the system and the details of the resulting simulation.

13.4 System and Simulation Specification

So far we may have given the impression that a simulation study consists of a series of well-defined steps that need to be followed in a specific order, like a recipe from a cookbook. Although most experienced simulation analysts agree that there are usually some well-defined activities or steps, they are often performed repeatedly in an interactive manner. You may find that halfway through the development of a simulation, conditions suddenly change, new information becomes available, or you gain early insight that spawns new ideas. This might cause you to revisit the problem-formulation phase or to alter the design of your model drastically. This is not all that uncommon, so be prepared to back up and revisit the problem whenever conditions change.

The process of developing a *specification* can take many forms, depending on the size of the study, the relationship between the analyst and client, and the ability of both parties to agree on the details at this early stage. If one individual is playing both roles—client and analyst—this step might be combined with the model formulation and construction phases. Although it is still necessary to define and understand the system completely, the development of a formal simulation specification is probably not required. However, if you find yourself at the other extreme, where the analyst is an external consultant, a formal specification can be very useful for both parties. For the purpose of the following discussion, let's assume we're somewhere between the two extremes. Let's further assume that the client and analyst are not the same person and that a written document is to be developed.

Before we proceed, let's summarize where we are in the process and where we want to go. We've already formulated the problem and defined the objectives of the study. We've decided to use simulation as the means to solve our problem, and we're now ready to define the details of the simulation study that will follow.

A good place to start is with the system itself. If the system exists, go visit the site and walk through it. The best advice is to look, touch, and ask questions. You want to understand what's happening and why it's happening, so don't be afraid to ask questions, even ones that seem insignificant. Often you'll find that activities are performed in a specific way just because that's the way they've always been done, while at other times you'll find that there are very good reasons for a routine. As you learn more about the system, start thinking in terms of how you might model these activities. And even more important, think about whether it's necessary to include certain activities in the simulation model. At this stage of the process, you're a systems analyst, not a simulation modeler. Thus, don't be afraid to make recommendations that might improve the process. Providing early input that will result in only minor improvements can increase your credibility for the tasks to follow.

If the system is a new design, find out if there are similar existing systems that you might tour. At the minimum, you'll obtain a better understanding of the overall process that you're about to simulate. If the system exists only on paper, take a tour of the blue-prints. If there's nothing on paper, develop a process flow diagram or a rough sketch of the potential system. You should do this with the client so that there is total agreement on the specifics of the system.

With an understanding of the system to be modeled, it's now time to gather all interested parties in a conference room and develop your specification. There is no magic formula for such a specification, but generally it should contain the following elements:

- Simulation objectives
- System description and modeling approach
- Animation exactness
- Model input and output
- Project deliverables

The time required to obtain all the necessary information can vary from an hour to a few days. The discussions that yield the details of this specification should include all the interested parties. In most cases, the discussions are focused on the system rather than the simulation because at this point, it is the one common ground. The types of questions that should be asked, and answered, are as follows:

- What is to be included in the simulation model?
- At what level of detail should it be included?
- What are the primary resources of the system?
    - What tasks or operations can they perform?
- Are process plans or process flow diagrams available?
    - Are they up to date?
    - Are they always followed?
    - Under what conditions are they not followed?
- Are there physical, technological, or legal constraints on how the system operates?
    - Can they be changed?
- Are there defined system procedures?
    - Are they followed?
    - Can new procedures be considered?

- How are decisions made?
    - Are there any exceptions?
- Are there data available?
- Who will collect or assemble the data?
    - When will they be available?
    - What form will they be in?
    - How accurate are the data?
    - Will they change, and if so, how will they change?
- Who will provide data estimates if data are not available?
    - How accurate must they be?
    - Will they require that a sensitivity analysis be conducted?
- What type of animation is required?
    - Are there different animations required at different phases of the project?
    - How will the animations be used?
- Who will verify and validate the model, and how will it be done?
    - Are comparative data available?
    - How accurate are the data?
- What kind of output is required?
    - What are the primary performance measures?
    - Can they be ranked or weighted?
- How general should the model be?
    - Will it be revised for other decisions?
- Who will perform the analysis?
    - What type of analysis is required?
    - How confident do you have to be in your results?
- How many scenarios will be considered?
    - What are they?
- What are the major milestones of the study?
    - When do they need to be completed?
- What are the deliverables?

This is not intended to be a complete list, but it should give you a general idea of the detail required.

Normally, these types of discussions are enlightening to both parties. In this type of forum, the analyst is asking questions and recording information. The assumption is that the client has all, or at least most, of the information required. Experience reveals a somewhat different situation. In most cases, the client is a team of three to six individuals representing different levels of interest in the system. They generally have different expertise and knowledge of the system. It is not uncommon to find that these individuals disagree strongly on some of the details of the system that may generally be confined to process descriptions and decision logic, but other areas are not excluded. This should not be surprising, as you're trying to get a complete understanding of the system, as well as consensus on how the system works. If you're faced with disagreement on certain details, we suggest that you stand back and let the client team arrive at a consensus.

If you're developing a specification for the first time, don't expect to get all your answers during the initial meeting. You might want to consider the 70/20/10 rule. Through experience, we have found that about 70% of the time the client team will have the complete answer or the information required. About 20% of the time the team will not know the answer, but they know how to get it (e.g., they may have to ask Dennis, who works on the night shift). The remaining 10% of the time they have no idea what the answer is or where they might find it (or there are several competing answers). This might surprise you, but this is a rather common phenomenon. Don't let this bother you; just make it clear that you need the information and ask when it will be available.

During this meeting, you'll also be exploring the availability, or lack thereof, of data to drive your simulation. Again, don't expect to get all the data you need at this initial meeting. However, it's important to know what data ultimately will be available, and when. You should also make note of what data are from historical sources and what data will be estimates. Finally, it might be advisable to identify the form in which the data will be delivered, as well as who is responsible for collecting, assembling, or observing the data. For a discussion of data issues, see Section 4.6.2.

Before you leave this meeting, you should identify one person from the client team who will serve as the primary contact for the study. When you return to your office, you'll need to organize the information into a document that resembles a specification. We recommend that you do this as soon as possible, while the details are still fresh in your mind. Even then you'll undoubtedly find yourself scratching your head over at least one item in your notes. Something that was very clear during the meeting may now look muddy. No problem! Get on the phone, fax, or e-mail and clarify it with your primary contact.

Once you've developed this document, you should send it to the client for review. It may take several iterations before a final document emerges that's agreed upon and acceptable to both parties. Upon completion, we recommend that both parties sign this final document. If, during the simulation study, you find that conditions change, data are not available, and so forth, you may find it necessary to revise this document.

We would love to be able to provide you with a detailed set of instructions on how to perform this entire task. Unfortunately, each simulation study is different, and circumstances really dictate the amount of detail that's required. We can, however, give you an idea of what a specification looks like. We were fortunate to receive permission from *The Washington Post* to include a functional specification for a simulation study (see Appendix A) that was developed as part of a consulting project conducted in the mid-1990s by the Systems Modeling consulting group. The original specification was developed by Scott Miller, a Systems Modeling consultant, for Gary Lucke and Olivier Girod of *The Washington Post*. Other than a few minor changes made for confidentiality and formatting reasons, it is the original document. We strongly suggest that you take time to read it before you undertake your first specification. We do not suggest that you use this form for all of your specifications, but it should provide an excellent starting point.

Throughout this discussion, we made the assumption that it's possible to define the complete study before you start building the actual model. There are circumstances where this is not possible. These types of projects are open ended in that the complete project is not yet defined or the direction of the project depends on the results obtained in the early phases. Even though it may not be possible to specify the entire project completely, the development of a specification still is often desirable. Of course, it means that you may have to amend or expand the specification frequently. As long as both parties are agreeable, there's no reason to avoid these types of projects. You just have to be willing to accept the fact that direction of the project can change dramatically over its duration. These are often called *spiral projects* in that they tend to spiral up into huge projects or spiral down into no project at all.

For now, let's assume that you have specified the simulation, and it's finally time to start the model.

13.5 Model Formulation and Construction

The nice part of having a complete specification of the simulation model and study is that it allows you to design the simulation model that can easily meet all of the objectives. Before you open a new model window and begin placing modules, we recommend that you spend some time formulating the model design. Some of the things that you want to take into consideration are the data structure or constraints, the type of analysis to be performed, the type of animation required, and your current comprehension of Arena. The more complex the system, the more important the formulation.

For example, consider a simulation of a large warehouse with 500,000 SKUs. You will obviously need to develop a data structure that will contain the necessary information on the changing state of the contents yet be easily accessed by the simulation model. If you're interested only in the number of pickers and stockers required for a given level of activity, it may not even be necessary to include information at the SKU level. You might create a model based on randomly created requests. If you can develop accurate expressions for the frequency and locations of these activities, this type of model might answer your questions. However, if you're interested in the details of the system, you would obviously need to create a much more elaborate model. You also need to understand the animation requirements. If no animation is required, you might model the picker movement and activities as a series of delays. If you need a detailed animation, you'll have to structure your model, most likely using guided transporter constructs, to allow the actual movement and positioning of the pickers on the animation.

You should also consider the potential impact of different scenarios that must be evaluated. Should each scenario be created as a different model, or can a sufficiently general model that requires only data changes be created? Consider our warehouse. If you want to compare a manual picking operation to an automated picking operation, you may require different models (although you may use the same data structure for both models). However, if you simply want to compare different types of layouts, you might develop a general model.

Once you've formulated a modeling approach, you need to consider what constructs you're going to use to build your model. So far we've advocated a top-down approach. This suggests that your first choice should be modules from the Basic Process panel, followed by modules from the Advanced Process and Advanced Transfer panels, and modules from the Blocks and Elements panel only when required. This is a recommended approach for the novice Arena user, but as you gain experience and confidence in your modeling skills, you'll most likely take a different approach. Most experienced modelers prefer to start with the Advanced Process and Advanced Transfer panels and select from the other panels when required. This allows you to create exactly the type of model required and gives you the maximum amount of flexibility.

Finally, you're ready to open a new model window and begin the model construction, which is probably the most fun part of the entire study. Before you begin, we'd like to offer one more piece of advice. As you become more experienced with the Arena system, you'll start to develop habits—some good, some bad. You'll tend to use those constructs with which you're most familiar for *every* model you create. We recall a consultant who created a very detailed model of a complicated assembly system that used a series of overhead power-and-free conveyors to move the assembly through the system (this was a number of years ago). About two days before the model was to be delivered to the client, it developed a bug. The consultant claimed it was caused by extraterrestrial beings, but we suspect that it was a simple modeling error. Two long, sleepless nights later (accompanied by freely offered, unsolicited advice from the other consultants), the error was uncovered. A quick fix was added by using a series of Signal and Hold modules (see Section 9.4) to synchronize the merging of subassembly conveyors, which solved the problem. The model was delivered to the client and the study was ultimately a success.

What's interesting about this extraterrestrial experience is that for the next several years, every model (and we mean *every* model) that the consultant created had at least one pair of Signal and Hold modules. This habit did not result in inaccurate models, but there were often simpler and more direct ways to model the system features. The consultant had stopped being creative in his model building and always used those constructs that saved his career that fateful night. Essentially, he had developed a bad habit, which, by the way, was extremely hard to break.

Another example that comes to mind occurred around the time that the Arena system was first introduced. Prior to that time, the consultants at Systems Modeling developed all their simulation models using the SIMAN language. This required the creation of two separate files—the experiment and model files (see Section 7.1.6). Although there were programs available to aid in developing these files, almost all the consultants developed their models in a text editor. Once their models were running, they'd create the separate animation of the model using the Cinema software. With the release of the Arena system, you could develop your entire model, including the animation, in one file using the point-and-click method with which you're now familiar. The consultants felt very comfortable with their text editors and were very reluctant to change their method of modeling. In fact, they went so far as to try to convince clients that the old way was the best way.

Finally, something akin to an edict was passed down stating, "You *will* use the new Arena system for all future models!" There was a lot of grumbling and complaining, but in a very short period of time, they were all working with the new software. In fact, before long you started to hear such comments as, "How did we ever do this before Arena?" Of course, years later, one individual still claimed that it was easier to develop models in a text editor. We have omitted names to protect the innocent, or the not-so-innocent.

So we recommend that you be open-minded about the methods you use to build models and the constructs that you choose to use. If there are other modelers in your group, ask them how they would approach a new model. You might also consider attending conferences or user-group meetings to find out what your peers are doing.

With that last caution noted, you can start constructing your model. If the model is small, you might place all the required modules, fill in the required data and hope that it works right the first time. (It rarely does.) If the system to be modeled requires a large, complicated model, you might try to partition the model building into phases. Select a portion of the system and build a model of that portion, including at least a rough animation. Once you are convinced it's working correctly, add the next phase. Continue in this manner until the entire model has been created. This approach makes model verification, which is our next topic, much easier.

13.6 Verification and Validation

Once you have a working model, and sometimes even while you are building it, it is time to verify and then validate the model. *Verification* is the task of ensuring that the model behaves as you intended; more colloquially, it's known as debugging the model. *Validation* is the task of ensuring that the model behaves the same as the real system. We briefly introduced these topics in Sections 2.7 and 7.1.6. Let's expand our discussion by first addressing the issue of verification.

You now have a completed simulation model, or at least a completed component, and you'd like to be sure that the model is performing as designed. This may seem to be a simple task, but as your models and the systems they represent become more complicated, it can become very difficult. In larger models, you can have many different simultaneous activities occurring that can cause interactions that were never intended. You'll need to design or develop tests that will allow you to ferret out the offending interactions or just the plain and simple modeling mistakes. If you haven't already developed an animation, we suggest that you complete that task before you start the verification phase. It does not need to be the final animation, but it should have enough detail to allow you to view the activities that are occurring within the system. You might start by checking the obvious.

Consider the system we modeled in Chapter 7, which produced three different parts. Alter your model so that it creates only a single instance of a Part Type 1, and watch that solitary part flow through the system. Repeat this for the other two part types. Change all your model times to constant values, and release a limited number of parts into the system. The results should be predictable. Test the other extreme by decreasing the part interarrival rate and observe what happens as the system becomes overloaded. Change the part mix, processing times, failure rates, etc. What you're trying to do is create a wide variety of different situations where your model logic might just fail. Of course, if you find problems, correct them. You'll have to decide if you need to repeat some of your earlier tests.

Once you have completed the obvious tests, you should try to stand back and visualize what types of scenarios you might consider in your analysis. Replicate these projected scenarios as best you can and see if your model still performs adequately. If you have periods when you're not using your computer, default the replication time and allow the simulation to run for extended periods of time. This type of experiment might be best performed overnight. Carefully review the results from these runs, looking for huge queues, resources not utilized, etc. Basically, you want to ask the following question: Do these results make sense? If you have extended periods of time when you're at your desk, but not using your computer, allow the model to run with the animation active. As time permits, glance at the monitor and see if everything appears to be all right. If you are confident that your model is working correctly, you're ready for the acid test.

Reconvene the client group and show them your simulation. In most cases, this means the animation. If there are problems, this is probably where they will be detected. Once this group grants its blessing, you might try one more experiment before you pronounce your model verified. Ask the group if they'd like to change the model in any way. Individuals who are familiar with the system, but not simulation, tend to think differently. They just might suggest modifications that you never considered.

Before you show your model to the client group, you should give some consideration to the level of detail required in the animation. In most cases, a rough animation showing the basic activities is sufficient. You may have to describe what your animation is showing, but once the individuals get beyond the pretty pictures, they are often able to visualize their system. At the same time, be sensitive to the feedback provided by the group when they first view the animation. You might be surprised at the types of reactions you'll get. We have seen individuals unable to get beyond the fact that your machines are blue and their machines are green. If you detect these types of comments, take the time to make some changes. Ultimately, you want the client group to accept your model as an accurate representation of their system. If you're lucky, they'll become your strongest supporters as you proceed with the project.

At this point, we should probably admit that it's almost impossible to verify totally a model for a complex system. We have seen verified models, which have already been used for extensive analysis, suddenly produce flawed results. This is typically caused by a unique combination of circumstances that was never originally considered being imposed on the model. Unfortunately, you need at least to consider that all of your previous analysis may be inaccurate. We should also point out that there's no magic to verification, nor is there a single method accepted by all. The key is that you, and your client, become totally convinced that the model is working as intended. Having done that, you are now ready to consider trying to validate the model. Although we are treating verification and validation as two separate topics or tasks, the difference between the two is often blurred.

In order to validate a simulation model, you should compare the results from your model to the results from the real system. If the system does not yet exist, you're in trouble right from the start. Even if the system does exist, such a comparison may be a difficult task. It's not uncommon for organizations to keep extensive metrics on past system performance, but often they do not keep the information that tells you what the system was doing, making, or subject to during that time period. Even if the data exist, they may be inaccurate or misleading.

Years ago, there was a large and complex simulation model developed for a facility that produced heavy-duty transmissions. The facility had approximately 1,400 machines and resources grouped into about 30 departments. The model had been developed to determine the sensitivity of product mix on the total throughput and the potential effect of introducing a new product line into the system. The primary performance measures were all related to the system's capacity. All of the process plans and process times were downloaded from the facility's databases. In addition, all the order releases were available for the last year of production. It looked like it would be an easy task to validate the model. Having already verified the model, the validation was undertaken.

The first set of runs produced results that totally mystified the clients. There were bottlenecks in the model that did not occur in the real system and bottlenecks in the real system that did not show up in the model results. In fact, a detailed comparison of resource-utilization statistics from the model to the historical records showed that there were major differences. This problem was resolved rather quickly when it became apparent that the processing times maintained in the facility's databases were based on standard times used for costing. They in no way represented the actual processing times observed on the shop floor. Be aware that this is a fairly common problem. Data are often kept for the convenience of the accountant, not the systems engineer. Luckily, someone had developed a set of conversion tables that converted the standard times to accurate processing times. These tables were included in the model and the validation process continued.

Everything looked good until a simulation run was made for the entire year, and the results were compared to the historical records. Eleven of the 12 months produced almost identical results. There was, however, one month near the middle of the simulated year where the records indicated that the output was approximately 40% larger than that predicted by the simulation model. A lot of time was spent trying to figure out what had caused this discrepancy. It was assumed that there was a problem with the simulation model—wrong! Finally, the analysis group started to suspect that maybe the historical data were flawed. By looking at the detailed statistics for the month in question, it became apparent that the resource statistics from the simulation closely matched the recorded statistics. In fact, it was quickly determined that the only difference was in the total number of transmissions produced. In addition, everyone was convinced that the system did not have the capacity to produce that many units. It took several weeks of part-time sleuthing before the answer emerged.

Someone finally realized that the month in question was the last month of the company's fiscal year. As it turned out, there was a small amount of product that was rejected for various reasons through the first 11 months of the year. These products were simply set aside until the final month when they suddenly became acceptable and were reported as part of that month's output. The reasoning was that it improved the performance of the system for the year. Weeks later these products were formally rejected, but the previous year's records were never adjusted. Removing these rejected items from the monthly throughput finally allowed the model to be validated.

If accurate records on the actual system do not exist, then it may be impossible to validate the model. In that case, concentrate on the verification and use the best judgment of individuals who are the most familiar with the system's capability. Sometimes individuals from the production floor have an uncanny ability to predict the performance of a new system accurately. Remember, the key is for both the analyst and the client to have confidence in the results from the model.

If you've gotten this far with your simulation project, you've cleared most of the major hurdles. You're now ready to use the model to answer some questions.

13.7 Experimentation and Analysis

We've already covered most, if not all, of the statistical issues associated with simulation analysis, and we'll not repeat that material here. However, there are several practical implications that we would like you to be aware of. Ideally, before you start any analysis, you would design a complete set of experiments that you intend to conduct. You would also decide on the types of analysis tools that you would be using. Although this may be ideal, it's far from reality. In some cases, you just don't have the luxury of sufficient time; in other cases, you don't really know where you're going with the analysis until you get there.

For example, if your objective is to improve the system throughput, you may not know what changes are required to achieve that goal. Sometimes the best analysis method is a group of knowledgeable individuals sitting around a computer suggesting and trying alternatives. This normally means that almost all the alternatives that are investigated will, at least, be feasible. One problem with this type of approach is that you may be tempted to reduce run times drastically in order to reduce the time you have to wait for the next set of results. Or worse yet, you may want to evaluate the system based only on a view of the animation. Don't yield to this pressure as it can lead to a disaster (and disasters can alter the direction of your career).

You might also consider structuring your experimentation based on the type of analysis that needs to be performed. For the sake of this discussion, let's identify three different types of analysis: candidate, comparative, and predictive.

Candidate analysis is normally done during the early design phases of a system. You are generally trying to identify the best candidate systems from a much larger group of potential designs that merit additional study. Models for these types of analysis are normally lacking in detail. You might think of these as rough-cut capacity models. You're trying to weed out the obvious losers and identify the potential winners. When you're performing this type of analysis, you can't put much faith in the accuracy of the true system's performance. You still need to make sure that you have a sufficient number of replications to provide good estimates of the system's performance (or your run times are of sufficient length), but there is very little value in increasing the number of replications (or the run time) to obtain tighter confidence limits on your results.

Comparative analysis would normally be the next logical step in selecting the final system design. You have a finite set of designs, and you want to compare them to identify the best design. This type of analysis typically requires a more detailed model, but we're only concerned with comparing one system to another. For example, there may be system activities that will affect the performance of the system, but the amount of the effect is common across all systems under consideration. For example, it may not be necessary to incorporate preventive maintenance or operator schedules into these types of models. The activities that are common to all systems under consideration and have the same effect do not have to be included in the models. For these types of systems, there may be value in increasing run times (or number of replications) to boost your confidence in selecting the best system.

Predictive analysis typically deals with only a few systems—often only one. By this time, you've selected what appears to be the best system, and you're now interested in estimating the actual performance of that system. This type of analysis requires that you include all activities that will affect the system's ability to achieve the predicted performance. At this point, you've selected the best system, and you're going forth with a recommendation to build it, provided that it meets the required objectives. These types of models need to be very detailed, and you also need to be confident in the data that are being used.

Regardless of the type of analysis you're conducting, be careful not to make strong judgments based on limited information. Be sure that your results are based on sound statistical practice, and when you conclude that one system is better than another, be sure that there is a significant statistical difference between them. More details on these issues are in Sections 6.4-6.6, 7.2, and 12.4-12.6.

Before proceeding, we find it necessary to point out again that the activities that constitute a simulation project appear to follow a logical time pattern. In reality, it's not uncommon to jump back and forth between activities, often re-visiting activities that you thought were complete. In practice, you often find that a good project will force you through several iterations of these activities.

13.8 Presenting and Preserving the Results

When you get to this stage, you've completed at least the initial study, and you're ready to go forth with your results. In many instances, only a written report is required or expected. In other cases, you may have to go before a group of decision makers and give an informal or formal presentation of your results. This may actually be the most important phase of the project because it may determine whether your results are accepted. We're not going to go into great detail about how to write or give a report. If you feel that your skills are lacking, we suggest you find a book devoted to these subjects or take a formal course to sharpen your skills. You'll find that you'll need these types of skills in many other areas as well! However, there are a few obvious points to be made.

First, be sure you're addressing the correct questions, and be sure to provide concise answers. Always include the equivalent of an executive summary that states your recommendations clearly—and the major reasons for them—in one page (or slide) or less. If possible, try to avoid presenting numbers in absolute terms; use ranges or confidence intervals. Understand your audience before you proceed. If you have volumes of material to support your recommendations, put them in an appendix or hold them in reserve (to be used only if requested). Don't be tempted to answer questions about features of the system that were not included in your model. Finally, be prepared to receive instructions to go back and look at additional alternatives.

Preserving the results should be an ongoing task throughout the duration of the project. Normally, we would call this the documentation process. It should include not only the final recommendations, but also documentation of the model and the details of the analysis. It is true that most simulation projects are not revisited. But there are exceptions, and it is not uncommon to have to dust off an old model years later to perform more analysis. We're aware of one simulation model developed in 1983, as a SIMAN model, that's still being used periodically, even though the model is basically the same.

Most practitioners agree that there's a law of nature, or maybe even a theorem, about documentation. The more documentation that you have on a project, the lower the probability of ever having to use the model again. It always seems to be those crash projects that leave no time for documentation that you're asked to resurrect years later. However, if you've followed our advice in this chapter, most of your documentation is already available and in electronic form. If you store all of this information (simulation model, specification, report, and presentation) in the same folder, you've covered most of the bases. An alternative is to use the methods described in Chapter 10 to embed these items directly into the simulation file.

You might also admit that in most cases the documentation will be used by someone else, if it's used at all. Individuals are promoted, change positions or change organizations, so your best approach to this task is to ask yourself what you'd want or need to know if someone else had performed the original study. This may not help you, but the next person will certainly be grateful.

13.9 Disseminating the Model

During the course of a simulation study, you will certainly share your simulation model and animation with your client. If the clients have a copy of Arena, you only need to send them the latest model (*.doe*) file, and they can view and alter your model at any time. If the clients (or anyone else, for that matter) do not have the software, this obviously won't work.

Arena does have a run-time mode, which will allow the user to open, run, and view the animation and the results of any model. It allows the user to edit any existing data and re-save the model. However, it does not allow the user to add or alter any logic or data modules. If the run-time version is required, you should contact Rockwell Automation for more information.

APPENDIX A

A Functional Specification for The Washington Post

A.1 Introduction

This appendix contains functional specification material provided to *The Washington Post* as part of a Systems Modeling simulation modeling consulting project (in the mid-1990s, which was prior to acquisition by Rockwell Automation). This project was delivered to Gary Lucke, manager of manufacturing systems engineering, and Olivier Girod, manager of industrial engineering, of *The Washington Post*. This file has been modified slightly to retain confidentiality of certain proprietary information. Some of the technology will appear dated, but we've left its original form from the time of this project because the process is unaffected by it.

A.1.1 Document Organization

This document is provided to describe the mailroom operations at *The Washington Post's* Springfield, Va., and proposed Maryland facilities. The description will include the detail necessary to develop accurately an Arena simulation model of both operations.

This document is divided into six sections. The first section defines the objectives of the simulation project, the purpose of this document, the use of the Arena model, and the software and hardware required to run the Arena model. The second section describes the physical components of the mailroom operation, as well as the modeling approaches for each component. The third section describes the animation of the simulation model. The fourth section summarizes the user input requirements for the Arena model and the desired output generated from the Arena model. The fifth section describes the project deliverables. Finally, the sixth section contains the agreement and acceptance signatures required to proceed with the project.

A.1.2 Simulation Objectives

The objective of the simulation study is to provide *The Washington Post* with a decision support tool that will assist in evaluating the truck-loading operation at the Springfield plant as well as the proposed Maryland facility. The simulation will aid in assessing the impact of press output, tray utilization, tray trip rate, and truck arrival patterns on the loading operations.

In order to obtain the simulation objectives, two simulation models will be developed under this contract. These models will incorporate actual operational data and information collected from the Springfield and Maryland facilities. The simulation will utilize actual production sequences and decision logic required to represent facility operations accurately. In addition, the simulation will incorporate the information and outputs generated by the AGV Roll Delivery simulation currently under development by Systems Modeling (SM), a material-handling vendor, and the newspaper.

The development of this model requires a number of tasks. Initially, a thorough understanding of the newspaper's Springfield and Maryland facilities is necessary to conceptualize and develop an accurate representation of this system. This process has been, and will continue to be, a joint effort between *The Washington Post* and Systems Modeling. This procedure defines the user's conceptualization of the system that SM will use to develop the Arena simulation model. Included in this process is defining the inputs and outputs of the model, verifying that the Arena model has been implemented accurately, and validating that the model accurately represents the facility.

Upon completion of the project, the newspaper will have a tool that will allow an analyst to specify various operating scenarios for the mailroom facility, execute simulation experiments of the scenarios, and perform statistical analysis of the scenarios.

A.1.3   Purpose of the Functional Specification

This functional specification serves four purposes. First and foremost, this document describes the mailroom operation at the newspaper's Springfield and Maryland facilities, at the level of detail required for modeling purposes. This description includes process flow, equipment functionality, operating procedures and rules, system interactions, and logistical issues. These systems must be thoroughly understood before they can be represented in a computer simulation.

Second, the user input required to perform the simulation analysis is defined. The input required includes such things as the press behavior, dock allocations, truck load time, and other operational characteristics.

Third, the output generated by the computer simulation is defined. Output is generally in the form of system performance measures from which the simulation analysis is performed. These output statistics include such things as dock utilization, truck load times, tray utilization, etc.

Finally, the project deliverables are described. The deliverables will be contained in a three-ring binder and will include computer diskettes containing the Arena model, input data files, hard copies of the data files, a user's manual to describe how to use the software, and the final report.

This document explicitly defines issues that are part of the quotation and other unofficial documents exchanged by SM and The Washington Post. Therefore, for those issues that are discussed in other documents, this document supersedes all correspondence in defining the project.

A.1.4   Use of the Model

The use of the simulation model will be detailed completely in the user's manual provided at the completion of the project. Use of the model includes initializing the Arena model and input files with the desired input parameters, running the model on the computer, and generating summary statistics and reports. The summary reports generated from various runs can be compared to evaluate the impact of specific parameters on system performance.

A.1.5   Hardware and Software Requirements

SM will develop the Arena simulation model under the Microsoft® Windows® operating system environment. The software and hardware required to run the model include:

- Arena Standard Edition 1.25 or higher
- IBM-compatible 486 PC or higher
- Windows 3.1 or 3.11
- 8 MB RAM (16 MB recommended)
- 30 MB hard disk space

The above-mentioned software is not included with this project, but can be purchased under a separate contract.

A.2   System Description and Modeling Approach

The following sections define the flow of newspapers from the presses through the mailroom to the loading docks. The various system components will be described for both the Springfield and Maryland facilities since the two plants have different layouts and modes of operation. Any operational differences will be addressed in the detail needed for this modeling effort. Overall, the logic defined will be similar for both facilities.

The model will include the production of headsheets, movement of headsheet product via the tray system to the docks, and the palletizing and loading of the bundles onto the trucks at the docks. It will also include the loading of previously produced advance product onto the trucks at the docks.

A.2.1   Model Timeline

The model will be able to simulate mailroom activities ranging from one day to a complete week.

A.2.2   Presses

The presses will be the starting point within the simulation model. Product generated from a press is sent to a stacker where it is bundled and then sent to the tray system. Neither the presses nor the stackers will be modeled in detail for this project. Press product is produced on each of four identical presses at a constant press rate during normal operations. Assuming a constant, user-defined bundle size at the stacker, bundles will be modeled entering the system at a constant rate for each press. The production rate for each press will be defined on an hourly basis over a one-week time horizon. Most of the time, press runs should start between 12:15 AM and 12:30 AM and finish between 4:30 AM and 4:40 AM.

At the conclusion of the AGV Roll Delivery simulation project, logic will be added so that the AGV simulation generates a press schedule that includes all uptimes and downtimes for each press. These press schedules can then be used as input files for the mailroom simulation. The mailroom simulation will be designed so that the analyst can choose to use either (1) "late run" press output schedules generated from the AGV simulation, or (2) user-defined press parameters and press downtime distributions. Both options are described below.

Table A-1. Late-Run Press Output Generated by AGV Simulation (Press 1)

Press 1 Sequence	Up/Down	Time (minutes)	Press Speed (copies/hr)	Ramp Up (copies)	Ramp-Up Speed (copies/hr)
1	Up	30	70,000	5,000	56,000
2	Down	5	0	0	0
3	Up	45	70,000	5,000	56,000
4	Down	10	0	0	0

A.2.2.1   AGV Simulation Press Output Schedule

Late-run press output schedules generated by the AGV simulation model will be imported into the mailroom model as ASCII files. For each press, these files will consist of a sequence of uptimes, downtimes, and speeds. An example of a late-run press output schedule generated by the AGV simulation is presented in Table A-1.

A.2.2.2   User-Defined Press Schedule

User-defined press parameters will be read into the simulation model from one ASCII file and will include, for each press: (1) total work order size, (2) press speeds, (3) ramp-up counts, and (4) ramp-up speeds. An example of user-defined press parameters is provided in Table A-2.

Besides replating downtime, each press also experiences random downtime. This downtime is caused by either (1) bad paper roll, or (2) newsprint web breaks. For each failure type, the frequency of occurrence will be based on headsheet count, while the time to repair distribution will be expressed in minutes.

A.2.2.3   Replating

For each press, replating will occur according to a user-defined schedule, and its associated downtimes will be modeled using a time-based distribution. An example of the replating schedule is given in Table A-3. In this example, it is assumed that the presses are turned on between 12:15 AM and 12:30 AM and are off between 4:30 AM and 4:40 AM. For the purpose of the simulation, it will be assumed that there may be as many as ten replatings in a given run.

Table A-2. User-Defined Press Output Schedule

Press Number	Work Order (copies)	Press Speed (copies/hr)	Ramp Up (copies)	Ramp-Up Speed (copies/hr)
1	120,000	70,000	5,000	56,000
2	120,000	65,000	3,000	49,000
3	120,000	55,000	5,000	49,000
4	120,000	70,000	5,000	56,000

Table A-3. User-Defined Replating-Induced Downtime

Replating	Replating Occurrence (minutes)	Replating Downtime (minutes)
1st	1:15 to 1:30	7 to 10
2nd	2:15 to 2:30	10 to 15
3rd	2:45 to 3:00	7 to 10
4th	:	:

Most of the time, replating should induce the following sequence: Presses 1 and 2 go down, Press 1 goes back up, Press 3 goes down, Press 2 goes back up, Press 4 goes down, Press 3 goes back up, and Press 4 goes back up.

A.2.3   Product Types

Presses produce two types of product—advance and headsheet. Advance product is the part of the newspaper that is not time sensitive and can therefore be produced earlier in the day. The headsheet is typically the first few sections of the newspaper that contain time-sensitive news. For purposes of this simulation, it will be assumed that all of the advance product needed for a given day is available when needed to load the delivery trucks. Only the headsheet production will be modeled explicitly.

A.2.4   Press Packaging Lines

Each press feeds three press packaging lines and each press packaging line is connected to the tray system. Press output can be regulated through each of the three press packaging lines. These lines input headsheets and produce bundles. A press packaging line has three operational modes: (1) backup, (2) regular, and (3) manual insertion. In the backup mode, the press packaging line is idle and does not produce any bundles. In the regular mode, a user-defined percentage of press output is fed to the line and mechanically transformed into bundles. Bundle size will be user defined. In the manual insertion mode, the line functions in the regular mode except that bundles are manually reworked to satisfy additional product requirements. These requirements involve merging advance products with headsheets. The manual insertion process will be described in A.2.9.

A.2.5   Tray System

Bundles entering the tray system from the press packaging lines have no pre-defined destination. At the determination point in the tray system, the bundles will be given a final destination (dock or palletizer) based on the tray trip rate and the trucks currently waiting for product.

A.2.5.1   Springfield, Va., Tray System

The Springfield tray system consists of two identical tray conveyors that transport bundles from the presses and stackers to the loading docks and palletizers. The two tray conveyors are differentiated and identified by their color, green or yellow. The output from each press is dedicated to one of the two conveyors, with the output from presses 1 and 3 sent to the green tray conveyor, and the output from presses 2 and 4 sent to the

Table A-4. Truck Arrival Schedule

Number	Bucket	TruckID	Draw Qty	Truck Type
1	1	0906	12294	2
2	1	0907	5050	3
⋮	⋮	⋮	⋮	⋮
n	8	9451	650	1

yellow tray conveyor. The green tray conveyor has exactly 263 trays, while the yellow tray conveyor has exactly 266 trays. Both conveyors move at a velocity of 150 trays/min and can deliver bundles to any of the bundle docks or any of the palletizers. The tray trip rate determines how often product may be diverted to a given location. At the determination point in the tray system, the bundle will look at the truck with the highest priority, currently at a dock. If the trip rate is not exceeded, the bundle will be diverted to that location. If the trip rate would be exceeded, the bundle will look at trucks with lower and lower priorities until a valid destination is found. If a bundle cannot find a valid destination, it is recirculated on the tray conveyor. The conveyor velocity and trip rate are user defined.

A.2.5.2 Maryland Tray System

The tray system in the Maryland facility will function in a similar manner. However, because the Maryland facility is one story, there is only one conveyor in the tray system. In addition, each of the packaging lines is connected directly to one to four docks. This enables bundles to be sent directly to a dock, bypassing the tray system. A packaging line will only divert a bundle to the tray system if there are no trucks ready for loading at one of its docks. The physical layout of the conveyor and the exact number of trays in the conveyor must be defined by the newspaper before the model may be built. The conveyor velocity and trip rate are user defined.

A.2.6 Truck Arrivals

Truck arrivals are controlled from a user-defined schedule file (Table A-4). Each truck has a unique truck ID, a time bucket, a draw quantity, and a truck type. The time bucket determines the time segment in which the truck arrives. The draw quantity is the number of copies to be loaded onto that truck. The truck type is the type of service for that truck. Truck types can be either highway, home delivery, or newsstand.

Trucks arrive in the model in time bucket (n-1) to begin loading the advance product at an advance dock. When the advance product is completely loaded, the truck may move to a dispatch dock where it will receive the headsheet needed for completion. Trucks are serviced based on their arrival time at a dock.

A.2.6.1 Home Delivery

Home delivery (HD) trucks receive their advance product at one of the dedicated advance docks. Pallets of advance product are disassembled and conveyed in bundles on a dedicated conveyor to the truck. The loading rate in bundles per minute and the dock setup time in minutes are user defined. After receiving all of the advance product, the HD truck can proceed to the regular dispatch docks where the same amount of headsheet product is loaded. Headsheet bundles arrive via the tray system. The loading rate in bundles per minute and the dock setup time in minutes are user defined. A user-defined percentage of the HD trucks are deemed to be "slow" and as such have a longer load time and setup time.

A.2.6.2   Highway

Highway (HWY) trucks are serviced at dedicated highway docks. Entire pallets of advance product are loaded by fork trucks onto the HWY trucks. The loading rate in pallets per minute and the wrap-up time (for last pallet) in minutes are user defined. Headsheet product destined for a HWY truck is sent to the palletizer dedicated to that HWY truck. Pallets are formed and are then loaded onto the truck by fork trucks. The loading rate and wrap-up time are the same as for the advance pallets. In addition, there is a dock change time between trucks. A HWY truck remains at a dock until all of its advance and headsheet product have been loaded.

A.2.6.3   Newsstand/Street Sales

Newsstand/street sales (NS/SS) trucks receive no advance product. Instead, upon arrival at a dispatch dock, bundles are generated from the manual insertion process. These bundles travel via the tray system to the NS/SS truck. Load times and wrap-up times are similar to those for the HD trucks.

A.2.7   Docks

In the model, there are three types of docks: (1) pallet, (2) advance bundle, and (3) headsheet bundle. Pallet docks are designed for forktruck access and service HWY trucks. Advance bundle docks are used to load the advance product onto the trucks during time bucket (n-1). Headsheet bundle docks are used to load the headsheet product delivered to the trucks via the tray system. A truck receives press product based on its arrival time (FIFO) at a given dock area and not its dock location. For each dock type, the number of active docks will be user defined.

A.2.7.1   Springfield, Va., Docks

The Springfield facility has exactly 12 dedicated pallet docks, exactly 10 advance bundle docks, and exactly 14 dedicated headsheet bundle docks.

A.2.7.2   Maryland Docks

The proposed Maryland facility has exactly 10 combination pallet and advance bundle docks and 18 dedicated headsheet bundle docks.

A.2.8   Palletizers

For purposes of the simulation, all of the palletizers are assumed to be identical. A palletizer is used to package headsheet bundles destined for a HWY truck at a pallet dock. Each HWY truck at a dock has a dedicated palletizer. For purposes of the model, a palletizer acts as a possible destination for bundles on the tray system. The palletizing of advance product will not be modeled in this project. Palletizers operate at a constant rate and form pallets of a constant size. The last pallet loaded will be a partial pallet with as much product as necessary to finish the load. The palletizer rate and pallet size are user defined. Palletizers experience various downtimes at random intervals. These downtimes can either be modeled as one downtime by combining all of the failure modes or by modeling each individual failure mode separately.

If a bundle arriving at a palletizer must wait because the equipment is busy or down, it will accumulate in a queue. The capacity of the palletizer queues will be user defined. If the queue at a palletizer reaches the user-defined limit, any additional bundles arriving at that palletizer will be turned away. Bundles turned away will be recirculated on the tray conveyor and will try to exit at the same palletizer on the next circuit.

A.2.8.1   Springfield, Va., Palletizers
The Springfield facility contains five palletizers, located on the first floor.

A.2.8.2   Maryland Palletizers
The proposed Maryland facility contains four palletizers.

A.2.9   Manual Insertion Process
Manual insertion begins when the first NS/SS truck arrives at a dock and continues until all NS/SS trucks are serviced. As the press packaging lines are changed over to manual insertion mode, hand inserters from a pool of workers are assigned to assemble bundles of advance product and headsheets. These bundles, called NS/SS bundles, are dispatched to the NS/SS trucks. Each packaging line is staffed by a maximum of 25 workers. Each worker can insert 500 copies per hour. Bundle size, worker pool size, and worker capacity are user defined.

Manual insertion is a critical component of the mailroom simulation because it conditions press packaging line operations. For each production day, the press packaging line operation experiences three sequential phases: (1) the home delivery (HD) phase, (2) the combined home delivery and newsstand/street sales (HD+NS/SS) phase, and (3) the newsstand/street sales (NS/SS) phase. During the first phase, all bundles produced by the press packaging lines are assembled mechanically and dispatched to either HD or HWY trucks. During the second phase, bundles may be dispatched to either HD, HWY, or NS/SS trucks. During the third phase, all bundles are manually assembled and dispatched to NS/SS trucks. The following section contains descriptions of these phases for a one-press configuration. The example is provided to illustrate the concept of a press packaging line operation.

Example

- *HD Phase.* Two packaging lines operate in the regular mode and share press output (50%/50%). One packaging line is in backup mode.
- *HD+NS/SS Phase.* Two packaging lines operate in the regular mode and share 80% of press output (40%/40%). One packaging line operates in manual insertion mode and is fed 20% of press production. On the hand insertion line, headsheet bundles that are not set aside by a hand inserter are down-stacked on a pallet (user-defined pallet size) before the bundles reach the tray system. These

headsheet bundles are stored until the NS/SS phase. The headsheet bundles that are set aside are reworked and transformed into NS/SS bundles.

- *NS/SS Phase*. While the press is still running, the three packaging lines operate in the manual insertion mode and share press output (33%/33%/33%). On these hand insertion lines, headsheet bundles that are not set aside and reworked by a hand inserter are down-stacked on a pallet before reaching the tray system. These bundles are stored until the press completes its work order. Once the press is turned off, all down-stacked headsheet bundles are re-fed into hand insertion lines for NS/SS production.

In the simulation, the user will define the length of a phase by defining the amount of product to be produced by the press during that phase. When that amount of product has been produced, the next phase will begin. The NS/SS phase will start when all of the HD trucks are serviced and end when all of the NS/SS trucks have been serviced. The user will also define the mode of operation for each packaging line during each phase. If a packaging line is working in manual insertion mode, the user will define the number of hand inserters on the line and the hand insertion rate.

A.3   Animation

The Arena simulation model will include an animation that will capture the general flow of bundles through the mailroom to the docks and palletizers. The animation will be a top-down, two-dimensional view of the system. The animation will show the movement of bundles through the tray system to the loading docks and palletizers. In addition, various system statistics will be displayed on the animation to show dynamic performance criteria.

The newspaper will supply a scaled layout of both the Springfield facility and the proposed Maryland facility in CAD format that will provide the static background for the animation.

A.4   Summary of Input and Output

A.4.1   Model Input

The data items read into the simulation model from ASCII files include but are not restricted to:

Presses
- Production Rate
- Production Schedule (1 day or 1 week)
- MTBF, MTTR

Note: The production schedule output from the AGV Roll Delivery Simulation may be used instead of these inputs.

Press Packaging Line
- Line Activation Patterns (Modes, Phases)
- Tray System
- Velocity
- Trip Rate (Docks)
- Trip Rate (Palletizers)

Palletizers
- Production Rate
- Bundles per Pallet
- MTBF, MTTR

Bundles
- Copies per Bundle (Advance)
- Copies per Bundle (Headsheet)
- Copies per Bundle (Hand Insertion)

Process Time
- Truck Load Rates
  - HD-Headsheet (Reg)
  - HD-Headsheet (Slow)
  - HD-Advance
  - HWY (Last Pallet)
  - HWY (Other Pallets)
- Dock Change Times
  - HD-Headsheet (Reg)
  - HD-Headsheet (Slow)
  - HD-Advance
  - HWY

Truck Schedule (for Each Truck)
- Bucket
- Truck ID
- Draw Qty
- Truck Type

Hand Insertion
- Hand Inserters per Line
- Hand Inserter Service Rate

A.4.2   Model Output

The following performance measures will be written to one or more ASCII data files at the conclusion of the simulation run. These statistics will be collected and output on a daily and weekly basis.

Input Parameters
- Process Times
- Equipment and Resource Levels
- Failure Rates

Press Activity (for Each Press)
- Available Time (Up)
- Total Production
- Average Production Rate
- Breakdown History

Hand Insert Activity (for Each Line)
- Total Production
- Average Production Rate
- Down-Stacked Headsheet Bundles

Tray Activity
- Utilization (Green & Yellow)

Bundles
- Destination Statistics

Dispatch Statistics
- Trucks Early
- Trucks Late
- Time to Complete Pct of Trucks

Bucket Loading (per Bucket)
- Total Units Loaded (Actual)
- Total Units Loaded (Scheduled)

Dock Statistics (per Dock)
- Dock Utilization
- Total Units Loaded
- Utilization per Bucket

Truck Statistics (per Truck)
- Dock Used
- Advance Load Time
- Wait Time
- Idle Time
- Load Time
- Change Time
- Summary for Each Truck Type

A.5   Project Deliverables

The following sections discuss the project deliverables. Upon completion of the project, *The Washington Post* will be supplied with all the computer files that were developed under the contract. The simulation model and all supporting data files are the sole and exclusive property of the buyer. This does not include any Arena software unless specifically stated and a quote has been provided. SM will keep backup copies for maintenance and recordkeeping in the event that *The Washington Post* desires changes to the model in the future.

A.5.1   Simulation Model Documentation

Model documentation will be a continuous process throughout the development of the model. This documentation will be contained within the Arena model and will include detailed comments describing all major sections of the model logic. Also included will be a complete variable listing containing descriptions of all variables, entity attributes, stations, queues, etc., used in the model.

A.5.2   User's Manual

The user's manual will include all of those items for which SM is responsible. This manual will only include information that is specific to this project. The user is referred to the *Arena User's Guide* for items that are specific to the Arena software. The contents of the user's manual will be:

1. The functional specification.
2. A copy of the model file on diskette.
3. A copy of all data input files on diskette.
4. Instructions on how to use the model.

A.5.3   Model Validation

Validation is the process of establishing that the model accurately represents the real system. The actual performance data required as inputs to the simulation model will be essential for this validation. The amount of data available, in the proper format, determines the level of detail of the validation process. SM and *The Washington Post* will run initial validation tests of the models to confirm system performance and the logic and algorithms implemented.

A.5.4   Animation

The Arena model will include a two-dimensional animation designed to allow the analyst insight into the dynamic features of the model. The animation will closely resemble the facility layout supplied by the newspaper in the form of a CAD file.

A.6   Acceptance

The estimate for the above-described modeling project is six person-weeks of effort to commence after this functional specification document has been signed.

The cost of this effort is $ xx. Billing will occur based on the following schedule:

ACCEPTANCE OF THE FUNCTIONAL SPECIFICATION	$ xx
COMPLETION OF MODEL DEVELOPMENT	$ xx
FINAL ACCEPTANCE	$ xx

This cost does not include software required to run the simulation model.

The Washington Post Mailroom Simulation Model
Functional Specification Furnished By:

_____

Scott A. Miller, Project Manager
Simulation and Consulting Services
Systems Modeling Corporation

Agreed and Accepted By:

_____

The Washington Post representatives:
Gary Lucke, Manager of Manufacturing Systems Engineering
Olivier Girod, Manager of Industrial Engineering

APPENDIX B

IIE/RA Contest Problems

This appendix contains information on the first 12 IIE/Rockwell Automation (Systems Modeling) student simulation competitions. Teams of three undergraduate students from universities worldwide compete in this competition annually. The winners for the first 12 contests were the University of Calgary, Kansas State University, co-winners University of Pittsburgh and Virginia Tech, University of South Queensland, Southern Polytechnic Institute & State University, University of Central Florida, University of Central Florida, University of Arizona, ITESM Sonora Norte, Universidad de los Andes, Kyong-gii University, and North Carolina State University.

Detailed statements for each problem, together with necessary tables, figures, and in some cases, separate data files, are available on the book's Web site, with no login or password needed (see the book's Preface for the Web site URL). For each year's contest, there is a single downloadable *.zip* file containing the description and any additional files (e.g., input data); see http://www.winzip.com if you need the WinZip® utility for decompressing these *.zip* files.

Here are the titles of the individual contests, in chronological order:

1. SM Superstore
2. SM Market
3. Sally Model's SM Pizza Shop
4. SM Office Repair
5. SM Rental
6. SM Theme Parks
7. SM Testing
8. SM Travel
9. SM Electronics
10. SM Paints
11. Rockland Steel Company
12. Rocksoft City Hospital

APPENDIX C

A Refresher on Probability and Statistics

The purpose of this appendix is to provide a brief refresher on selected topics in probability and statistics necessary to understand some of the probabilistic foundations of simulation, as well as to design and analyze simulation experiments appropriately. While this material underlies many parts of the book (including every time we use a probability distribution in a model to represent some random input quantity like a time duration), it's particularly relevant to Section 2.6, Section 4.6, Chapter 6, Section 7.2, and Chapter 12.

We intend this appendix to serve as a very brief tutorial, and not as anything like a comprehensive treatment of these topics. There are many excellent texts on probability and statistics in general, such as Anderson, Sweeney, and Williams (2006), Devore (2004), and Hogg and Craig (2005); for a free online "book" on the subject, which links to additional online resources, see Lane (2005) at http://davidmlane.com/hyperstat/index.html.

Though we'll start pretty much from scratch on probability and statistics, we do assume that you're comfortable with algebraic manipulations, including summation notation. For a complete understanding of continuous random variables (Section C.2.3), you'll need to know some calculus, particularly integrals.

In Section C.1, we go over the basic ideas and terminology of probability. Section C.2 contains a discussion about random variables in general, describing discrete and continuous random variables as well as joint distributions. The notion of sampling, and the associated probabilistic structure, is discussed in Section C.3. Statistical inference, including point estimation, confidence intervals, and hypothesis testing, is covered in Sections C.4–C.6. Throughout, we'll make the discussion relevant to simulation by way of examples.

C.1 Probability Basics

An *experiment* is any activity you might undertake whose exact outcome is uncertain (until you do it and see what happens). While the term might conjure up images of high school chemistry lab, its interpretation in probability is much broader. For instance:

- Flip a "fair" coin. Will it come up tails?
- Throw a "fair" die (that's singular for "dice"). Will it be 4? Will it be an odd number? Will it be more than 2 but no more than 5?
- Drive to work tomorrow. How long will it take? How long will you be delayed because of construction? Will you be hit by an asteroid?

- Run a call center for a week. How many calls will be handled? What will be the average duration of your customers' wait on hold? How many customers will be turned away because the hold queues are full?
- Run a *simulation* of a call center (rather than operate the real center) for a *simulated* week. Ask the same questions as we just did, except now for what happens in the simulation rather than in reality; if your simulation model is valid, you hope to get the same answers, or at least close.

The *sample space* of an experiment is the complete list of all the individual outcomes that might occur when you do your experiment. For something like flipping a coin or throwing a die, it's easy to write down the sample space. In other cases, though, the sample space could have an infinite number of possibilities, like how long it will take you to drive to work tomorrow. Fortunately, it's often possible to understand an experiment and its probabilistic structure without writing down explicitly what the sample space is.

An *event* is a subset[1] of the sample space. An event can sometimes be described by just listing out the individual outcomes defining the event, if it's simple enough. Usually, though, an event is defined by some condition on what happens in the experiment; for example, in the call center mentioned above, an event of interest might be that at least 500 calls are handled in a day. Events are often denoted as capital letters like E, F, E_1, E_2, etc. The usual set operations apply to events, such as *union* ($E \cup F$), *intersection* ($E \cap F$), and *complementation* (E^c = the set of possible outcomes *not* in E), as represented in Figure C-1.

The *probability* of an event is the relative likelihood that it will occur when you do the experiment. By convention (and for convenience of arithmetic), probabilities are always between 0 and 1. $P(E)$ denotes the probability that the event E will occur. While it may be impossible to know or derive the probability of an event, it can be interpreted as the proportion of time the event occurs in many independent repetitions of the experiment (even if you can't actually repeat the experiment). Here are some properties (among many others) of probabilities:

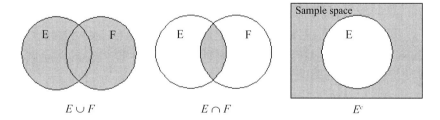

$$E \cup F \qquad\qquad E \cap F \qquad\qquad E^c$$

Figure C-1. Union, Intersection, and Complementation

---

[1] In advanced treatments of probability, an event is not allowed to be *any* subset of the sample space, but only a particular kind of subset, called a *measurable* subset. For our purposes, though, all the subsets we'll consider will be measurable, so will be events.

- If S is the entire sample space, then $P(S) = 1$.
- If $\varnothing$ is the *empty* (*null*) event, then $P(\varnothing) = 0$.
- $P(E^C) = 1 - P(E)$.
- $P(E \cup F) = P(E) + P(F) - P(E \cap F)$.
- If E and F are *mutually exclusive* ($E \cap F = \varnothing$), then $P(E \cup F) = P(E) + P(F)$.
- If E is a subset of F (i.e., the event E implies the event F), then $P(E) \le P(F)$.
- If $o_1, o_2, o_3, \ldots$ are the individual outcomes in the sample space (finite or infinite), then $\sum_{\text{all } i} P(o_i) = 1$.

Sometimes, the knowledge that one event occurred can alter the probability that another event also occurred. The *conditional probability* of E given F is defined as

$$P(E \mid F) = P(E \cap F) / P(F)$$

(assuming that $P(F) > 0$). The intuition for this is that knowing that F occurred reduces the "world" from the whole sample space down to F, and so the only relevant part of E is that part that intersects with F. This probability is then measured relative to the "size" of the reduced world, $P(F)$.

Events E and F are called *independent* if $P(E \cap F) = P(E)\,P(F)$. If this is so, then $P(E \mid F) = P(E)$, and $P(F \mid E) = P(F)$, by definition of conditional probability. In other words, knowing that one of two independent events has occurred tells you nothing about whether the other one occurred.

There are a lot of different kinds of events and probabilities that come up in simulation. For example, you might want to ask about the probability that:

- a part passes inspection.
- an arriving part is of priority 3.
- a service time will be between 2 and 6.
- no customers will arrive during a five-minute time interval.
- the maximum queue length during a simulation will exceed 10.
- the average time in system of parts is less than four hours.

C.2 Random Variables

Events can be defined in many different ways, and can be very complex. One way of quantifying and simplifying events is by defining *random variables* relating to them. In this section, we'll discuss the basic ideas of random variables, their two basic forms (discrete and continuous), and then consider multiple random variables defined together and their possible relationships.

C.2.1 Basics

A *random variable* is a number whose value is determined by the outcome of an experiment, so can be thought of as a quantification of an experiment. Technically, a random variable is a function defined from the sample space to the real numbers. As such, it's a rule or mapping that assigns a number to each possible outcome of the experiment. While you can sometimes define a random variable in this way, by going back to the sample space and assigning the mapping, you often just start by defining the random variable without bothering with the sample space. A reasonable way to think about a random variable is that it's a number whose value you don't know for sure before doing the experiment, but you will generally know *something* about it, like its range of possible values or its probability of being equal to something or falling in some range. Random variables are typically denoted by capital letters like X, Y, W_1, W_2, etc.

Random variables come in two basic "flavors": *discrete* and *continuous*. A discrete random variable can take on only certain separated values. For instance, the number of defective items in a shipment of 50 items would have to be an integer between 0 and 50. Another example of a discrete random variable is the number of times a part has to undergo inspection in order to pass; without further information, this random variable would be a positive integer with no upper bound. Thus, a discrete random variable could have a finite or infinite range of possible values.

A continuous random variable, on the other hand, can take on any real value, possibly bounded on the left or the right. Continuous random variables typically represent physical measurements like time or distance. There are always infinitely many possible values for a continuous random variable, even if there are limits on its value on both ends. For instance, if X is the time to process a part on a machine, the range would be $[0, \infty)$ unless we assumed that no part was allowed to stay on the machine for longer than a certain time a, in which case the range would be $[0, a]$.

In simulation, random variables are used for several different purposes. They often serve as "models" for input quantities like uncertain time durations (service or interarrival times), the number of customers in an arriving group, or which of several different part types a given arriving part is. Random variables are also used to represent output quantities like the average time in system, the number of customers served, or the maximum length of a buffer.

The probabilistic behavior of a random variable is described by its *probability distribution*. Since the nature of this distribution is somewhat different for discrete and continuous random variables, we'll consider them separately; we'll also define some basic properties of random variables, like expected value and variance.

C.2.2   Discrete

For a discrete random variable X, there will be a list x_1, x_2, . . . (finite or infinite) of possible values it can take on. Note that the x_i's are fixed, non-random values, but the random variable X is, well, random. The *probability mass function* (PMF) is simply a function that gives the probability that X will take on each of the possible values:

$$p(x_i) = P(X = x_i)$$

for all i. Note that the statement "$X = x_i$" is an event that may or may not happen, and the PMF gives the probability that it does. The PMF may be expressed in a variety of different ways—a numerical list or table, a graph, or some kind of mathematical formula. Since the complete list of the x_i's is supposed to represent all the different possible values of X, $\sum_{\text{all } i} p(x_i) = 1$. Usually, the PMF is estimated from data, or simply assumed.

The *cumulative distribution function* (CDF) of a discrete random variable X is a function that gives the probability that X will be *less than or equal to* its argument:

$$F(x) = \sum_{\substack{\text{all } i \text{ such that} \\ x_i \le x}} p(x_i).$$

This summation is taken over all possible values x_i that are $\le$ the argument x of F. Note that $0 \le F(x) \le 1$ for all x, that $F(x) \to 0$ as $x \to -\infty$, and that $F(x) \to 1$ as $x \to +\infty$. Thus, $F(x)$ is a non-decreasing function going from 0 up to 1 as x goes from left to right. For a discrete random variable, $F(x)$ is a "step" function that's flat between adjacent possible values x_i, and takes a "jump" of height $p(x_i)$ above x_i.

The probability of an event involving a discrete random variable X generally can be found by adding up the appropriate values of the PMF. For instance,

$$P(a \le X < b) = \sum_{\substack{\text{all } i \text{ such that} \\ a \le x_i < b}} p(x_i).$$

This just says to add up the probabilities of those x_i's that are at least a but (strictly) less than b. Note that with discrete random variables, you need to be careful about weak vs. strong inequalities.

Just as data sets have a "center" measured by the average of the data, random variables have a "center" in a certain sense. The *expected value* of the discrete random variable X is defined as

$$E(X) = \sum_{\text{all } i} x_i p(x_i)$$

(this is also called the *mean* or *expectation* of X and is often denoted by μ or, if there's need to identify the random variable, μ_x). This is a weighted average of the possible values x_i for X, with the weights' being the respective probabilities of occurrence of each x_i. In this way, those x_i's with high probability of occurrence are counted more heavily than are those that are less likely to occur. If there are finitely many x_i's and each is equally likely to occur, then $E(X)$ is just the simple average of the x_i's since they all "count" the same. Despite the name, it's important to understand that $E(X)$ is *not* to be interpreted as the value of X you "expect" to get when you do the experiment defining X. Indeed, $E(X)$ might not even be a possible value of a discrete random variable X (the x_i's might be integers but, depending on the situation, $E(X)$ need not be an integer). Instead, interpret the expected value like this: do the experiment many times (technically, infinitely many times), observe a value of X each time, and compute the average of all these values of X you observe—this average will be the expectation of X.

And just as data sets have a measure of variability, so too do random variables. The *variance* of the discrete random variable X is defined as

$$Var(X) = \sum_{\text{all } i} (x_i - \mu)^2 p(x_i)$$

(often denoted σ^2 or σ_X^2), where μ is the expected value of X. This is a weighted average of the squared deviation of the possible values x_i around the expectation, with the weights' being the probability of occurrence of each x_i. The variance is a measure of the "spread" of the random variable about its mean. The units of the variance are the squares of the units of X, so people often use the positive square root of the variance (denoted σ or σ_X) as a measure of spread; this is called the *standard deviation* of X.

As mentioned earlier, there are different ways to define the PMF of a discrete random variable. Arena supports several common discrete random variables for modeling input quantities, and these are defined and described in Appendix D.

C.2.3 Continuous

A continuous random variable can take on any real value in some range. The range can be limited or unlimited on either or both ends. No matter how narrow the range may be, a continuous random variable can always take on an infinite[2] number of real values (i.e., any value in a continuum). Thus, it doesn't make sense to talk about the probability that a continuous random variable *equals* (exactly) some fixed number x; technically, this probability will always be 0 even if x is within the range of X.

Instead, the probabilistic behavior of a continuous random variable is described in terms of its falling *between* two fixed values, which can be far apart or close together. The *probability density function* (PDF) of a continuous random variable X is defined to be a function $f(x)$ with the following properties and interpretation:

- $f(x) \geq 0$ for all real values x.
- The total area under $f(x)$ is 1. In calculus terminology, the total integral of $f(x)$ is 1:

$$\int_{-\infty}^{+\infty} f(x)dx = 1.$$

- For any fixed real values a and b, with $a \leq b$, the probability that X will fall between a and b is the area under $f(x)$ between a and b. In calculus terminology,

$$P(a \leq X \leq b) = \int_a^b f(x)dx.$$

This last property says that if we slide a slim interval left and right along the x axis (keeping the width of the interval constant), we "pick up" more area (probability) in those regions where the density is high, so we're more likely to observe lots of values of X where the density is high than where the density is low. If you think of repeating the experiment many times and making a dot on the x axis where the value of the random variable X lands each time, your dots will be highly dense where the density function is high and of low density where the density function is low (get it?). Note that the height (value) of $f(x)$ is itself *not* the probability of *anything*. Indeed, we don't require that $f(x)$ be ≤ 1, and some PDFs can rise above 1 at some points; what's required is that the total

---

[2] More precisely, an *uncountably* infinite number of values, for those of you who are concerned with the different sizes of infinity.

area under the PDF be equal to 1. For example, consider a uniform distribution between 3.0 and 3.1. In order for the total area under the PDF to be equal to 1, the height $f(x)$ for all possible values of x (between 3.0 and 3.1) needs to be equal to 10. Also, we could specify that $f(x) = 0$ for values of x in or outside some range, which would mean that these ranges are impossible for the random variable X to fall in. Unlike discrete random variables, we can be sloppy with whether the endpoints of the ranges over which we want probabilities are defined by weak or strong inequalities (i.e., $<$ is the same as $\leq$, and $>$ is the same as $\geq$).

The cumulative distribution function of a continuous random variable X has the same basic definition and interpretation as for discrete random variables: $F(x) = P(X \leq x)$ for all real values x. Thus, $F(x)$ is the probability that X will land on or to the left of x. However, computing it requires getting the area under the density function to the left of x:

$$F(x) = \int_{-\infty}^{x} f(t)dt.$$

Depending on the form of the PDF $f(x)$, the CDF $F(x)$ may or may not be expressible as a closed-form formula involving x. For instance, the exponential and Weibull distributions (see Appendix D for definitions) do have simple formulas for the CDF, but the general gamma and normal distributions do not. If the CDF is not expressible as a formula, its evaluation must be left to some numerical method or a table (which is why every statistics book has a table of normal-distribution areas). As in the discrete case, the CDF $F(x)$ for a continuous random variable rises from 0 at the extreme left to 1 at the extreme right, but in this case is a continuous function rather than a step function. Since $f(x)$ is the slope (derivative) of $F(x)$, those regions on the x axis where $F(x)$ rises steeply are those regions where we'd get a lot of observations on X; conversely, where $F(x)$ is relatively flat, we won't see many observations on X.

The expected value of a continuous random variable is, as in the discrete case, one measure of the "center" of the distribution and is the average of infinitely many observations on X. It's defined as

$$E(X) = \int_{-\infty}^{+\infty} xf(x)dx$$

and often denoted as μ or μ_X. Roughly, this is a weighted "average" of the x values, using the density as the weighting function to count more heavily those values of x around which the density is high. The variance of X, measuring its "spread," is

$$Var(X) = \int_{-\infty}^{+\infty} (x - \mu)^2 f(x)dx$$

and often denoted σ^2 or σ_X^2; the positive square root of the variance (denoted σ or σ_X) is the standard deviation.

Arena supports several different continuous random variables for use in modeling random input quantities; these are defined and discussed in Appendix D.

C.2.4 Joint Distributions, Covariance, Correlation, and Independence

So far we've considered random variables only one at a time. But sometimes they naturally come in pairs or triples or even longer ordered sequences (having the wonderful name *tuples*), which are called *jointly distributed* random variables or *random vectors*. For instance, in the input modeling for a job-shop simulation, an arriving order might have random variables representing the part type (dictating its route through the shop), priority, and processing times at the steps along its route. On the output side, the simulation might generate a sequence $W_1, W_2, W_3, \ldots$ representing the times in system of the finished parts in order of their exit. One issue that naturally arises, and which can affect how we model and analyze such random vectors, is whether the random variables composing them are related to each other, and if so, how.

To address this, we'll start with the complete probabilistic representation of random vectors. In order to keep things at least partially digestible, we'll consider just a pair (two-tuple) of random variables (X_1, X_2), but things extend in the obvious way to higher dimensions. The *joint CDF* of (X_1, X_2) is a function of two variables defined as

$$F(x_1, x_2) = P(X_1 \leq x_1 \text{ and } X_2 \leq x_2)$$

for all pairs (x_1, x_2) of real numbers. Often the word "and" is replaced by just a comma:

$$F(x_1, x_2) = P(X_1 \leq x_1, \ X_2 \leq x_2).$$

If the random variables are both discrete, the *joint PMF* is

$$p(x_1, x_2) = P(X_1 = x_1, X_2 = x_2).$$

If the random variables are both continuous, the *joint PDF* is denoted $f(x_1, x_2)$, which you can visualize as some kind of surface floating above the (x_1, x_2) plane, and which has total volume under it equal to one. The interpretation of the joint PDF in the continuous case is this: the probability that X_1 will fall between a_1 and b_1, and, simultaneously, that X_2 will fall between a_2 and b_2, is the volume under the joint PDF above the rectangle $[a_1, b_1] \times [a_2, b_2]$ in the (x_1, x_2) plane. In calculus notation, this interpretation is expressed as

$$P(a_1 \leq X_1 \leq b_1, a_2 \leq X_2 \leq b_2) = \int_{a_1}^{b_1} \int_{a_2}^{b_2} f(x_1, x_2) dx_2 \, dx_1.$$

The joint distribution (expressed as either the CDF, PMF, or PDF) contains a *lot* of information about the random vector. In practice, it's usually not possible to know, or even estimate, the complete joint distribution. Fortunately, we can usually address the issues we need to without having to know or estimate the full-blown joint distribution.

Given a joint distribution of two random variables, we can derive the individual, or *marginal*,[3] distributions of each of the random variables on their own. In the jointly discrete case, the marginal PMF of X_1 is, for all possible values x_{1i} of X_1,

---

[3] The term "marginal" is not to suggest that these distributions are of questionable moral integrity. Rather, it simply refers to that fact that, in two dimensions for the jointly discrete case, if the joint probabilities are arranged in a table, then summing the rows and columns results in values on the margins of the table, which will be the individual "marginal" distributions.

$$p_{X_1}(x_{1i}) = P(X_1 = x_{1i}) = \sum_{\text{all } x_{2j}} p(x_{1i}, x_{2j})$$

and the marginal CDF of X_1 is, for all real values of x,

$$F_{X_1}(x) = \sum_{\substack{\text{all } i \text{ such that} \\ x_{1i} \leq x}} p_{X_1}(x_{1i})$$

(symmetric definitions apply for X_2). In the jointly continuous case, the marginal PDF of X_1 is

$$f_{x_1}(x_1) = \int_{-\infty}^{+\infty} f(x_1, x_2) dx_2$$

and the marginal CDF of X_1 is

$$F_{X_1}(x_1) = \int_{-\infty}^{x_1} f_{X_1}(t) dt$$

(symmetrically for X_2). Note that we can get the marginal distributions in this way from the joint distribution, but knowledge of the marginal distributions is generally not sufficient to determine what the joint distribution is (unless X_1 and X_2 are independent random variables, as discussed below).

The *covariance* between the components X_1 and X_2 of a random vector is defined as

$$Cov(X_1, X_2) = E\big[(X_1 - E(X_1))(X_2 - E(X_2))\big].$$

Note that the quantity inside the [ ] is a random variable with its own distribution, etc., and the covariance is the expectation of this random variable. The covariance is a measure of the (linear) relationship between X_1 and X_2, and can be positive, zero, or negative. If the joint distribution is shaped so that, when X_1 is above its mean, then X_2 tends to be above its mean, then the covariance is positive; this implies that a small X_1 tends to be associated with a small X_2 as well. On the other hand, if large X_1 is associated with small X_2 (and vice versa), then the covariance will be negative. If there is no tendency for X_1 and X_2 to occur jointly in agreement or disagreement over being big or small, then the covariance will be zero. Thus, the covariance tells us whether the two random variables in the vector are (linearly) related or not, and if they are, whether the relationship is positive or negative.

However, the covariance's magnitude is difficult to interpret since it depends on the units of measurement. To rectify this, the *correlation* between X_1 and X_2 is defined as

$$Cor(X_1, X_2) = \frac{Cov(X_1, X_2)}{\sigma_{X_1} \sigma_{X_2}}.$$

Clearly, the correlation has the same sign as the covariance (or is zero along with the covariance), so the direction of any relationship is also indicated by the sign of the correlation. Also, the correlation is a dimensionless quantity (that is, it has no units of measurement and will be the same regardless of what units of measurement you choose for X_1 and X_2). What's perhaps not obvious (but it's true) is that the correlation will always fall between -1 and $+1$, giving its magnitude universal emotional impact. Without knowing anything about the situation, you can say that a correlation of 0.96 or -0.98 is quite strong, whereas a correlation of 0.1 or -0.08 is pretty weak. Thus, the correlation provides a very meaningful way to express both direction and strength of linear[4] relationship between random variables.

The random variables X_1 and X_2 are called *independent* if their joint CDF always factors into the product of their marginal CDFs:

$$F(x_1, x_2) = F_{X_1}(x_1)\, F_{X_2}(x_2) \text{ for all } (x_1, x_2).$$

Equivalently, independence can be defined in terms of similar factorization of the joint PMF into the product of the marginal PMFs, or factorization of the joint PDF into the product of the marginal PDFs. Here are some properties of independent random variables X_1 and X_2:

- In words, independence means that knowing the value that X_1 hit tells you nothing about where X_2 may have landed.
- If two random variables are independent, then they will also be uncorrelated. The converse, however, is generally not true (unless the random variables have a joint normal distribution). Admittedly, the counterexamples are pathological, but they're there.
- For independent random variables, $E(X_1 X_2) = E(X_1)\, E(X_2)$.
- Independence of random variables is a pretty big deal in probability and statistics since, believe it or not, the factorization property in the above definition renders a whole lot of derivations possible where they would be totally impossible otherwise.
- Maybe because independence is so important analytically, it's awfully tempting just to assume it when you're not too sure it's justified.[5] All we can do is alert you to the fact that this assumption is there, and violating it usually has unknown ramifications.

In the case of more than two random variables in the random vector, independence means that the joint CDF (or PMF or PDF) factors into the product of the single marginal counterparts; this implies pairwise independence, but pairwise independence does not necessarily imply independence of all the random variables.

The issues of correlation and independence of random variables comes up in at least a couple of places in simulation. On the input side, we usually model the various random quantities driving the simulation as being independent, and simply generate them accordingly. But there sometimes might be some kind of dependence present that we

---

[4] The reason we keep hedging the language with this "linear" qualifier for the covariance and correlation is that these measures may fail to pick up nonlinear relationships that might be present. However, in most modeling applications the relationships will be fairly close to linear, at least over a restricted range.

[5] In this case, some people have been known to refer to this as the Declaration of Independence since that's pretty much all it is.

need to capture for the sake of model validity; Section 4.6.7 discusses this briefly and gives references. On the output side, a run of a simulation over time typically produces a sequence of output random variables that may be correlated, perhaps heavily, among themselves. This complicates proper statistical analysis of such data, and care must be taken to design the runs appropriately and use the output properly; Section 7.2 gets into some of the particular problems and remedies.

C.3   Sampling and Sampling Distributions

The main purpose of statistical analysis is to estimate or infer something concerning a large *population*, assumed to be too large to look at completely, by doing calculations with a *sample* from that population. Sometimes it's more convenient to think of sampling from some ongoing process rather than from a static population. The mathematical basis for this is that there is a random variable (or maybe random vector) with some distribution, which governs the behavior of the population; in this sense, the population can be thought of as the random variable and its distribution, and a sample is just a sequence of independent and identically distributed (IID) observations, or *realizations*, on this random variable. Whichever the case, you don't know the parameters of the population or its governing distribution, and you need to take a sample to make estimates of quantities or test hypotheses.

There's been a lot of statistical theory worked out to do this—we'll describe just a little of it, pertinent to the simulation examples in the book, and refer you to the references mentioned at the beginning of this appendix for more. This statistical theory assumes that the sample has been taken *randomly*—that is, so that every possible sample of whatever size you're taking had the same chance of being the sample actually chosen. The link between sampling for statistical inference and the probability theory discussed so far in this appendix is that "the experiment" referred to earlier is in this case the act of taking a random sample. The outcome is a particular data set (one of many possible) from which various quantities are computed, which clearly depend on the sample that happened to have been obtained.

In simulation, sampling boils down to making some runs of your model, since random input makes for random output. Assuming that the random-number generator is operating properly and being used correctly, the randomness of the sample is guaranteed; this is to be contrasted with physical or laboratory experiments where great pains are often taken to ensure that the sample obtained is really random.

Let $X_1, X_2, \ldots, X_n$ be a random sample observed from some population. Equivalently, these data are IID observations on some underlying random variable X whose distribution governs the population. In simulation, this arises on the input side where the data are observations from the real system to which some distribution is to be fitted to serve as an input to the simulation (as in Section 4.6). It also arises on the output side where the data points are summary statistics across n IID replications of the simulation run (as in Sections 6.2-6.6 and 7.2.2), or are perhaps batch means from within a single long run of a steady-state simulation (as in Section 7.2.3). Let $\mu = E(X)$, $\sigma^2 = Var(X)$, and $p = P(X \in B)$ where B is a set defining some "distinguishing characteristic" of X (like being more than 25). As we'll formalize in Section C.4, it's reasonable to "estimate" these three quantities, respectively, by:

- The sample mean:
$$\overline{X} = \frac{\sum\limits_{i=1}^{n} X_i}{n} .$$

- The sample variance:
$$s^2 = \frac{\sum\limits_{i=1}^{n} (X_i - \overline{X})^2}{n-1} .$$

- The sample proportion:
$$\hat{p} = \frac{\text{number of } X_i\text{'s that are in } B}{n} .$$

(Other population/distribution parameters, and estimates of them, are certainly possible, but these three will serve our purposes.) Note that each of these quantities can be computed from knowledge of the sample data only (i.e., there are no population/distribution parameters involved); such quantities are called (*sample*) *statistics*. The important thing to remember about statistics is that they are based on a random sample and, as such, are themselves random—you got your sample and we got ours (of the same size n), which were probably different samples so probably produced different numerical values of the statistics. Thus, statistics are actually random variables themselves, relative to the "experiment" of taking a sample.

Accordingly, statistics have their own distributions (sometimes called *sampling distributions*), expectations, variances, etc. Here are some results about the distributions of the above three statistics:

- $E(\overline{X}) = \mu$ and $Var(\overline{X}) = \sigma^2/n$. If the underlying distribution of X is normal (see Appendix D for distribution definitions), then the distribution of $\overline{X}$ is also normal, written $\overline{X} \sim N(\mu, \sigma/\sqrt{n})$; we'll use the convention that the second "argument" of the normal-distribution notation is the standard deviation rather than the variance. Even if the underlying distribution of X is not normal, the central limit theorem says that, under fairly mild conditions, the distribution of $\overline{X}$ will be approximately normal for large n.
- $E(s^2) = \sigma^2$. If the underlying distribution is normal, then the quantity $(n-1)s^2/\sigma^2$ has a chi-square distribution with $n-1$ degrees of freedom (DF), denoted χ^2_{n-1}. Any standard statistics book will have a definition and discussion of this distribution.
- $E(\hat{p}) = p$ and $Var(\hat{p}) = p(1-p)/n$. For large n, the distribution of $\hat{p}$ is approximately normal.

The importance of sampling distributions is that they provide the basis for estimation and inference about the population/distribution parameters.

C.4 Point Estimation

Quantities that are characteristic of the population/distribution, such as μ, σ^2, and p, are called *parameters*. Unless you somehow know (or assume) everything about the population/distribution, you won't know the values of parameters. Instead, the best you can usually do is to *estimate* these parameters with sample statistics, as described in Section C.3. Since we're estimating a parameter by just a single number (rather than an interval), this is called *point estimation*. While point estimates on their own frankly aren't worth much (since you don't know how close or stable or generally good they are), they're a start and can have some properties worth mentioning.

A statistic serving as a point estimator is called *unbiased* for some population/distribution parameter if E(point estimator) = parameter. In words, this says that if we took a lot of samples and computed the point estimator from each sample, the average of these estimators would be equal to the parameter being estimated. Clearly, this is a comforting property, and, from the sampling-distribution results cited in Section C.3, is one enjoyed by $\overline{X}$ for m, s^2 for σ^2, and $\hat{p}$ for p.

While unbiasedness is nice, it doesn't speak to the stability of the estimator across samples. Other things (like unbiasedness) being equal, we'd prefer an estimator that has low variance since it's more likely to be close to the parameter being estimated. The lower-variance estimator is called more *efficient*; this term is actually analogous to economic efficiency in sampling since a more efficient estimator will require a smaller sample for its variance to come down to a tolerable level.

Related to efficiency is the notion of *consistency* of an estimator. While there are several different kinds of consistency, the basic idea is that, as the sample size n grows, the estimator gets "better" in some sense (lack of this property would certainly be upsetting). For instance, we'd like the variance of an estimator to decline, hopefully to zero and hopefully quickly, as the sample size is increased. Taking a glance at the expressions for the variances of $\overline{X}$, s^2, and $\hat{p}$ in Section C.3 shows that they all satisfy this property.

C.5 Confidence Intervals

Most of the commonly used point estimators have good properties. But they all have variability associated with them, so will generally "miss" the parameter they're estimating. A *confidence interval* provides one way of quantifying this imprecision. The goal of a confidence-interval procedure is to form an interval, with endpoints determined by the sample, that will contain, or "cover" the target parameter with a prespecified (high) probability called the *confidence level*. The usual notation is that the confidence level is $1 - \alpha$, resulting in a $100 (1 - \alpha)$ percent confidence interval.

Using the sampling-distribution results, as well as similar results found in statistics books like those referenced at the beginning of this appendix, the following confidence intervals for several common parameter-estimation problems have been derived:

- The population/distribution expectation μ: The confidence interval is:

$$\overline{X} \pm t_{n-1,\,1-\alpha/2}\,\frac{s}{\sqrt{n}}$$

where $t_{n-1,1-\alpha/2}$ is the point that has below it probability $1 - \alpha/2$ for Student's t distribution with $n - 1$ DF (this point is called the upper $1 - \alpha/2$ *critical point* for this distribution and is tabled as part of Exercise C-4 at the end of this appendix and in any statistics book). Proper coverage probability for this interval assumes that the underlying distribution is normal, but the central limit theorem ensures at least approximately correct coverage for large n.

- The population/distribution variance σ^2: The confidence interval is

$$\left(\frac{(n-1)s^2}{\chi^2_{n-1,1-\alpha/2}}, \frac{(n-1)s^2}{\chi^2_{n-1,\alpha/2}} \right)$$

where $\chi^2_{n-1,1-\alpha/2}$ is the upper $1 - \alpha/2$ critical point for the chi-square distribution with $n - 1$ DF (tabled in statistics books). This interval assumes a normal population/distribution.

- The population/distribution standard deviation σ: Due to the definition and interpretation of confidence intervals, we can simply take the square roots of the endpoints of the preceding interval:

$$\left(\sqrt{\frac{(n-1)s^2}{\chi^2_{n-1,1-\alpha/2}}}, \sqrt{\frac{(n-1)s^2}{\chi^2_{n-1,\alpha/2}}} \right)$$

- The difference between the expectations of two populations/distributions, $\mu_A - \mu_B$: There are different approaches and resulting formulas for this, discussed in Section 12.4.1. One important issue in deciding which approach to use is whether the sampling from the two populations/distributions was done independently or not. The important interpretation is that if this interval contains 0, the conclusion is that we cannot discern a statistically significant (at level α) difference between the two expectations; if the interval misses 0 then there appears to be a significant difference between the expectations.

- The ratio of the variances of two populations/distributions, σ_A^2/σ_B^2: The confidence interval is

$$\left(\frac{s_A^2 / s_B^2}{F_{n_A-1,n_B-1,1-\alpha/2}}, \frac{s_A^2 / s_B^2}{F_{n_A-1,n_B-1,\alpha/2}} \right)$$

where the subscripts A and B on the sample variances and sample sizes indicate the corresponding population/distribution, and $F_{k_1,k_2,1-\alpha/2}$ denotes the upper $1 - \alpha/2$ critical point of the F distribution with (k_1, k_2) DF (tabled in statistics books). A normal population/distribution is assumed for this interval. We conclude that there is a statistically significant difference between the variance parameters if and only if this interval does not contain 1.

- The ratio of the standard deviations of two populations/distributions, σ_A/σ_B: Just take the square root of the endpoints in the preceding confidence interval; conclude that the standard-deviation parameters differ significantly if and only if this interval misses 1.

- The population/distribution proportion p: The confidence interval is

$$\hat{p} \pm z_{1-a/2} \sqrt{\frac{\hat{p}(1-\hat{p})}{n}}$$

where $z_{1-\alpha/2}$ is the upper $1 - \alpha/2$ critical point of the standard (mean = 0, standard deviation = 1) normal distribution (tabled in statistics books). This is an approximate-coverage interval, valid for large n (one definition, among many, of "large n" is that both $n\hat{p}$ and $n(1-\hat{p})$ be at least 5 or so).

- The difference between the proportions of two populations/processes $p_A - p_B$: The confidence interval is

$$\hat{p}_A - \hat{p}_B \pm z_{1-a/2} \sqrt{\frac{\hat{p}_A(1-\hat{p}_A)}{n_A} + \frac{\hat{p}_B(1-\hat{p}_B)}{n_B}}$$

with the obvious interpretation of the subscripts A and B. Both sample sizes need to be "large," as described in the preceding point.

C.6   Hypothesis Tests

In addition to estimating, either via a point or an interval, population/distribution parameters, you might want to use the data to "test" some assertion made about the population/distribution. These questions and procedures are called *hypothesis tests*, and there are many, many such tests that people have devised for a wide variety of applications. We won't make any attempt to give a complete treatment of hypothesis testing, but only describe the general idea and give a couple of simulation-specific examples. For derivation and specific formulas for doing hypothesis tests, please see a statistics book like those we mentioned at the beginning of this appendix.

The assertion to be tested is called the *null hypothesis*, usually denoted H_0. Often, it represents the status quo or the historical situation, or what is being claimed by somebody else. The denial (opposite) of the null hypothesis is the *alternate hypothesis*, which we'll denote H_1. The intent of hypothesis testing is to develop a decision rule for using the data to choose either H_0 or H_1 and be as sure as we can that whichever we declare to be true really *is* the truth.

Barring complete information about the population/distribution, though, we can never be 100% sure that we're making the right choice between H_0 and H_1. If H_0 is really the truth yet we reject it in favor of H_1, we've committed a *Type I Error*. But if H_1 is really the truth yet we don't reject H_0, we've made a *Type II Error*. Hypothesis tests are set up to allow you to specify the probability α of a Type I Error, while doing the best to minimize the probability β of a Type II Error. If you demand a really small α, you'll get it but at the cost of a higher β (though the relationship between the two is not simple) unless you can go collect some more data. In hypothesis testing, H_0 and H_1 are not given equal treatment—the benefit of the doubt is given to H_0, so if we reject H_0, we're making a fairly strong and confident decision that H_1 is true. But if we can't mount enough evidence against H_0 to reject it, we haven't necessarily "proved" that H_0 is the truth— we've just failed to find evidence against it. The reason for failing to reject H_0 could, of course, be that it's really true. But another reason for failing to reject H_0 is that we just don't have enough data to "see" that H_0 is false.

Another way to set up and carry out a test is not to make a firm yes/no decision, but rather quantify how "certain" you are about which hypothesis is correct. The *p-value* of a data set in a test is the probability of getting a data set, if H_0 is true, that's more in favor of H_1 than the one you got. So if the *p*-value is tiny, you're saying that it's very hard to get information more in favor of H_1 than the information you already have in your data set, so that the evidence for H_1 is strong. If the *p*-value is large, say 0.3 or 0.6, then it's entirely possible that just by chance you'd get data more in favor of H_1, so there's no particular reason to suspect H_0. If the *p*-value is "on the edge," say something like 0.1, you're left with an inconclusive result, which is sometimes all you can say with your data.

One place in simulation that hypothesis tests come up is in fitting input probability distributions to observed data on the input quantity being modeled (Section 4.6). Here, H_0 is the assertion that a particular candidate fitted distribution adequately explains the data. If H_0 is not rejected, then you have insufficient evidence that this distribution is wrong, so you might go with it. The Arena Input Analyzer has two different tests for this built in, the chi-square test and the Kolmogorov-Smirnov test. These tests basically ask how close the fitted distribution is to the empirical distribution defined directly by the data; for details on how these and other *goodness-of-fit tests* work, see any standard statistics book.

Another place in simulation that a hypothesis test comes up is on the output side. If there are several (more than two) models you're comparing on the basis of some selected output performance measure, a natural question is whether there's any difference at all among the means of this measure across the different models. A collection of specific problems and techniques, called *analysis of variance* (ANOVA), is a standard part of any statistics book, so we won't go into its inner workings here. The null hypothesis is that all the means across the different models are the same; if you do not reject this, you have insufficient evidence of any difference on this measure that the different models make. But if you reject H_0, you're saying that there *is* some difference somewhere among the means, though not that they're all unique and different. A natural question in this case is then precisely which means differ from which other ones, sometimes called *multiple comparisons* in ANOVA. There have been several different methods developed for attacking this problem, three of which are due to Bonferroni, Scheffé, and Tukey. The Arena Output Analyzer (Section 6.4) has built-in facility for carrying out an ANOVA test, including these multiple-comparisons methods.

Related to this goal, but more useful in simulation output analysis, is the question of which among several variants (sometimes called *scenarios*) is best according to some overall performance measure. The Arena Process Analyzer, or PAN (Section 6.5) has a built-in facility for identifying best scenarios and does so in a way related to hypothesis tests.

C.7 Exercises

C-1   Suppose that X is a discrete random variable with probability mass function given by $p(1) = 0.1$, $p(2) = 0.3$, $p(3) = 0.3$, $p(4) = 0.2$, and $p(5) = 0.1$.

(*a*)   Plot $p(x)$ for all x in the range.

(*b*)   Compute numerically and plot the cumulative distribution function $F(x)$.

(*c*)   Compute $P(1.4 \le X \le 4.2)$.

(*d*)   Compute $P(1.4 < X < 4.2)$.

(*e*)   Compute the expected value $E(X)$.

(*f*)   Compute the variance $Var(X)$.

(*g*)   Compute the standard deviation of X.

C-2   Suppose that X has a (continuous) uniform distribution between 10 and 20 (see Appendix D).

(*a*)   Compute $E(X)$.

(*b*)   Compute $Var(X)$.

(*c*)   Compute $P(X < 12)$.

(*d*)   Compute $P(X > 12)$.

(*e*)   Compute $P(X < 8 \text{ or } X > 22)$.

C-3   Suppose that X is an exponential random variable with mean 5. (The cumulative distribution function is $F(x) = 1 - e^{-x/5}$ for $x \ge 0$, and $F(x) = 0$ for $x < 0$; see also Appendix D.)

(*a*)   Compute $P(X > 5)$.

(*b*)   Compute $P(1.4 \le X \le 4.2)$.

(*c*)   Compute $P(1.4 < X < 4.2)$.

C-4   Suppose that 7.3, 6.1, 3.8, 8.4, 6.9, 7.1, 5.3, 8.2, 4.9, and 5.8 are ten observations from a normal distribution with unknown mean and unknown variance. Compute the sample mean, sample variance, and a 95 percent confidence interval for the population mean. Here's part of a t table that contains what you'll need (and then some):

degrees of freedom	1 − α/2				
	0.900	0.950	0.975	0.990	0.995
1	3.078	6.314	12.706	31.821	63.656
2	1.886	2.920	4.303	6.965	9.925
3	1.638	2.353	3.182	4.541	5.841
4	1.533	2.132	2.776	3.747	4.604
5	1.476	2.015	2.571	3.365	4.032
6	1.440	1.943	2.447	3.143	3.707
7	1.415	1.895	2.365	2.998	3.499
8	1.397	1.860	2.306	2.896	3.355
9	1.383	1.833	2.262	2.821	3.250
10	1.372	1.812	2.228	2.764	3.169
11	1.363	1.796	2.201	2.718	3.106
12	1.356	1.782	2.179	2.681	3.055
13	1.350	1.771	2.160	2.650	3.012
14	1.345	1.761	2.145	2.624	2.977
15	1.341	1.753	2.131	2.602	2.947
16	1.337	1.746	2.120	2.583	2.921
17	1.333	1.740	2.110	2.567	2.898
18	1.330	1.734	2.101	2.552	2.878
19	1.328	1.729	2.093	2.539	2.861
20	1.325	1.725	2.086	2.528	2.845
25	1.316	1.708	2.060	2.485	2.787
30	1.310	1.697	2.042	2.457	2.750

APPENDIX D

Arena's Probability Distributions

Arena contains a set of built-in functions for generating random variates from the commonly used probability distributions. These distributions appear on drop-down menus in many Arena modules where they're likely to be used. They also match the distributions in the Arena Input Analyzer (except for the Johnson distribution). This appendix describes all of the Arena distributions.

Each of the distributions in Arena has one or more parameter values associated with it. You must specify these parameter values to define the distribution fully. The number, meaning, and order of the parameter values depend on the distribution. A summary of the distributions (in alphabetical order) and parameter values is given in Table D-1.

Table D-1. Summary of Arena's Probability Distributions

Distribution			Parameter Values
Beta	BETA	BE	Beta, Alpha
Continuous	CONT	CP	$CumP_1, Val_1, \ldots CumP_n, Val_n$
Discrete	DISC	DP	$CumP_1, Val_1, \ldots CumP_n, Val_n$
Erlang	ERLA	ER	ExpoMean, k
Exponential	EXPO	EX	Mean
Gamma	GAMM	GA	Beta, Alpha
Johnson	JOHN	JO	Gamma, Delta, Lambda, Xi
Lognormal	LOGN	RL	LogMean, LogStd
Normal	NORM	RN	Mean, StdDev
Poisson	POIS	PO	Mean
Triangular	TRIA	TR	Min, Mode, Max
Uniform	UNIF	UN	Min, Max
Weibull	WEIB	WE	Beta, Alpha

The distributions can be specified by using one of two formats: you can select a single format, or you can mix formats within the same model. The format is determined by the name used to specify the distribution. The primary format is selected by using either the variable's full name or a four-letter abbreviation of the name consisting of the first four letters. For example, UNIFORM or UNIF specifies the uniform distribution in the primary format. The secondary format is selected by specifying the distribution with a two-letter abbreviation. For example, UN specifies the uniform distribution in the secondary format. The names are not case-sensitive.

In the primary format, you explicitly enter the parameters of the distribution as arguments of the distribution. For example, UNIFORM(10, 25) specifies a uniform distribution with a minimum value of 10 and a maximum value of 25. In the alternative format, you indirectly define the parameters of the distribution by referencing a parameter set within the Parameters module from the Elements panel. For example, UN(DelayTime) specifies a uniform distribution with the minimum and maximum values defined in the parameter set named DelayTime. The main advantage of the indirect method of defining the parameters provided by the alternative format is that the parameters of the distribution can be modified from within the Parameters module.

The random-number stream, which is used by Arena in generating the sample, can also be specified in both formats. In the primary format, you enter the stream number as the last argument following the parameter value list. For example, UNIFORM(10,25,2) specifies a sample from a uniform distribution using random-number stream 2. In the secondary format, you enter the random-number stream as a second argument following the identifier for the parameter set. For example, UN(DelayTime,PTimeStream) specifies a sample from a uniform distribution using random-number stream PTimeStream (which must be an expression defined to be the stream you want, and is not an entry in the Parameters module).

In the following pages, we provide a summary of each of the distributions supported by Arena, listed in alphabetical order for easy reference. The summary includes the primary and secondary formats for specifying the distribution and a brief description of the distribution. This description includes the density or mass function, parameters, range, mean, variance, and typical applications for the distribution. If you feel in need of a brief refresher on probability and statistics, see Appendix C.

Beta(β, α)   **BETA(Beta, Alpha) or**
BE(ParamSet)

Probability Density Function

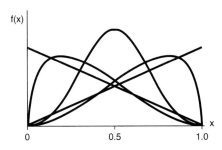

$$f(x) = \begin{cases} \dfrac{x^{\beta-1}(1-x)^{\alpha-1}}{B(\beta,\alpha)} & \text{for } 0 < x < 1 \\[2mm] 0 & \text{otherwise} \end{cases}$$

where B is the complete beta function given by

$$B(\beta,\alpha) = \int_0^1 t^{\beta-1}(1-t)^{\alpha-1}\,dt$$

Parameters   Shape parameters Beta (β) and Alpha (α) specified as positive real numbers.

Range   [0, 1] (Can also be transformed to a general range $[a,b]$ as described below.)

Mean $\dfrac{\beta}{\beta+\alpha}$

Variance $\dfrac{\beta\alpha}{(\beta+\alpha)^2(\beta+\alpha+1)}$

Applications   Because of its ability to take on a wide variety of shapes, this distribution is often used as a rough model in the absence of data. Because the range of the beta distribution is from 0 to 1, the sample X can be transformed to the scaled beta sample Y with the range from a to b by using the equation $Y = a + (b - a)X$. The beta is often used to represent random proportions, such as the proportion of defective items in a lot. It can also be used as a general and very flexible distribution to represent many input quantities that can be assumed to have a range bounded on both ends.

Continuous	CONTINUOUS(CumP$_1$, Val$_1$, . . ., CumP$_n$, Val$_n$) or
$(c_1, x_1, . . ., c_n, x_n)$	CONT(CumP$_1$, Val$_1$, . . ., CumP$_n$, Val$_n$) or CP(ParamSet)

Probability Density Function

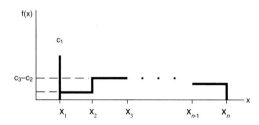

Cumulative Distribution Function

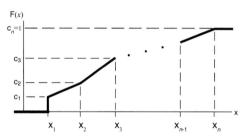

$$f(x) = \begin{cases} c_1 & \text{if } x = x_1 \text{ (a mass of probability } c_1 \text{ at } x_1) \\ c_j - c_{j-1} & \text{if } x_{j-1} \leq x < x_j, \text{ for } j = 2, 3, \ldots, n \\ 0 & \text{if } x < x_1 \text{ or } x \geq x_n \end{cases}$$

Parameters   The CONTINUOUS function in Arena returns a sample from a user-defined empirical distribution. Pairs of cumulative probabilities c_j (= CumP$_j$) and associated values x_j (= Val$_j$) are specified. The sample returned will be a real number between x_1 and x_n, and will be less than or equal to each x_j with corresponding cumulative probability c_j. The x_j's must increase with j. The c_j's must all be between 0 and 1, must increase with j, and c_n must be 1.

The cumulative distribution function $F(x)$ is piecewise linear with "corners" defined by $F(x_j) = c_j$ for $j = 1, 2, \ldots, n$. Thus, for $j \geq 2$, the returned value will be in the interval $(x_{j-1}, x_j]$ with probability $c_j - c_{j-1}$; given that it is in this interval, it will be distributed uniformly over it.

You must take care to specify c_1 and x_1 to get the effect you want at the left edge of the distribution. The CONTINUOUS function will return (exactly) the value x_1 with probability c_1. Thus, if you specify $c_1 > 0$ this actually results in a mixed discrete-continuous distribution returning (exactly) x_1 with probability c_1, and with probability $1 - c_1$ a continuous random variate on $(x_1, x_n]$ as described above. The graph of $F(x)$ above depicts a situation where $c_1 > 0$. On the other hand, if you specify $c_1 = 0$, you will get a (truly) continuous distribution on $[x_1, x_n]$ as described above, with no "mass" of probability at x_1; in this case, the graph of $F(x)$ would be continuous, with no jump at x_1.

   As an example use of the CONTINUOUS function, suppose you have collected a set of data $x_1, x_2, \ldots, x_n$ (assumed to be sorted into increasing order) on service times, for example. Rather than using a fitted theoretical distribution from the Input Analyzer (Section 4.5), you want to generate service times in the simulation "directly" from the data, consistent with how they're spread out and bunched up, and between the minimum x_1 and the maximum x_n you observed. Assuming that you don't want a "mass" of probability sitting directly on x_1, you'd specify $c_1 = 0$ and then $c_j = (j - 1)/(n - 1)$ for $j = 2, 3, \ldots, n$.

Range $[x_1, x_n]$

Applications   The continuous empirical distribution is used to incorporate empirical data for continuous random variables directly into the model. This distribution can be used as an alternative to a theoretical distribution that has been fitted to the data, such as in data that have a multimodal profile or where there are significant outliers.

Discrete	**DISCRETE(CumP$_1$, Val$_1$, . . ., CumP$_n$, Val$_n$) or**
$(c_1, x_1, \ldots, c_n, x_n)$	**DISC(CumP$_1$, Val$_1$, . . ., CumP$_n$, Val$_n$) or DP(ParamSet)**

Probability Mass Function

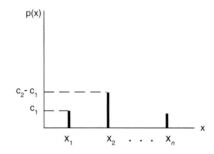

$$p(x_j) = c_j - c_{j-1}$$

where $c_0 = 0$

Cumulative Distribution Function

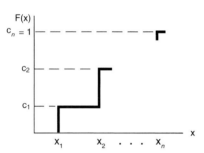

Parameters   The DISCRETE function in Arena returns a sample from a user-defined discrete probability distribution. The distribution is defined by the set of n possible discrete values (denoted by $x_1, x_2, \ldots, x_n$) that can be returned by the function and the cumulative probabilities (denoted by $c_1, c_2, \ldots, c_n$) associated with these discrete values. The cumulative probability (c_j) for x_j is defined as the probability of obtaining a value that is less than or equal to x_j. Hence, c_j is equal to the sum of $p(x_k)$ for k going from 1 to j. By definition, $c_n = 1$.

Range $\{x_1, x_2, \ldots, x_n\}$

Applications   The discrete empirical distribution is used to incorporate discrete empirical data directly into the model. This distribution is frequently used for discrete assignments such as the job type, the visitation sequence, or the batch size for an arriving entity.

Erlang(β, k)   **ERLANG(ExpMean, k) or ERLA(ExpMean, k) or ER(ParamSet)**

Probability Density Function

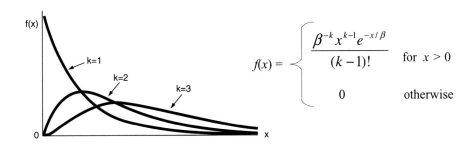

$$f(x) = \begin{cases} \dfrac{\beta^{-k} x^{k-1} e^{-x/\beta}}{(k-1)!} & \text{for } x > 0 \\ \\ 0 & \text{otherwise} \end{cases}$$

Parameters   If $X_1, X_2, \ldots, X_k$ are IID exponential random variables, then the sum of these k samples has an Erlang-k distribution. The mean (β) of each of the component exponential distributions and the number of exponential random variables (k) are the parameters of the distribution. The exponential mean is specified as a positive real number, and k is specified as a positive integer.

Range $[0, +\infty)$

Mean $k\beta$

Variance $k\beta^2$

Applications   The Erlang distribution is used in situations in which an activity occurs in successive phases and each phase has an exponential distribution. For large k, the Erlang approaches the normal distribution. The Erlang distribution is often used to represent the time required to complete a task. The Erlang distribution is a special case of the gamma distribution in which the shape parameter, α, is an integer (k).

Exponential(β)   **EXPONENTIAL(Mean) or EXPO(Mean) or EX(ParamSet)**

Probability Density Function

$$f(x) = \begin{cases} \dfrac{1}{\beta} e^{-x/\beta} & \text{for } x > 0 \\[2mm] 0 & \text{otherwise} \end{cases}$$

Parameters   The mean (β) specified as a positive real number.

Range $[0, +\infty)$

Mean β

Variance β^2

Applications   This distribution is often used to model interevent times in random arrival and break-down processes, but it is generally inappropriate for modeling process delay times.

Gamma(β, α)   **GAMMA(Beta, Alpha) or GAMM(Beta, Alpha) or GA(ParamSet)**

Probability Density Function

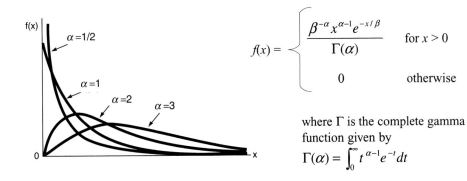

$$f(x) = \begin{cases} \dfrac{\beta^{-\alpha}\, x^{\alpha-1}\, e^{-x/\beta}}{\Gamma(\alpha)} & \text{for } x > 0 \\[2ex] 0 & \text{otherwise} \end{cases}$$

where Γ is the complete gamma function given by

$$\Gamma(\alpha) = \int_0^\infty t^{\alpha-1} e^{-t}\, dt$$

Parameters   Scale parameter (β) and shape parameter (α) specified as positive real values.

Range $[0, +\infty)$

Mean $\alpha\beta$

Variance $\alpha\beta^2$

Applications   For integer shape parameters, the gamma is the same as the Erlang distribution. The gamma is often used to represent the time required to complete some task (for example, a machining time or machine repair time).

Johnson   **JOHNSON(Gamma, Delta, Lambda, Xi) or JOHN(Gamma, Delta, Lambda, Xi)
or JO(ParamSet)**

***Probability
Density
Function***

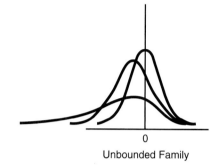
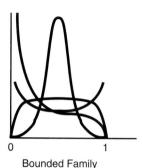

Unbounded Family   Bounded Family

Parameters   Gamma shape parameter (γ), Delta shape parameter ($\delta > 0$), Lambda scale parameter
($\lambda > 0$), and Xi location parameter (ξ).

Range $(-\infty, +\infty)$ Unbounded Family

$[\xi, \xi + \lambda]$ Bounded Family

Applications   The flexibility of the Johnson distribution allows it to fit many data sets. Arena can
sample from both the unbounded and bounded form of the distribution. If Delta (δ) is
passed as a positive number, the bounded form is used. If Delta is passed as a negative
value, the unbounded form is used with $|\delta|$ as the parameter. (At present, the Input
Analyzer does not support fitting Johnson distributions to data.)

Lognormal(μ, σ)   **LOGNORMAL(LogMean, LogStd) or LOGN(LogMean, LogStd) or RL(ParamSet)**

Probability Density Function

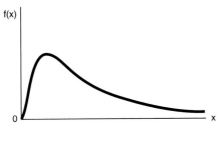

Denote the user-specified input parameters as LogMean = μ_l and LogStd = σ_l. Then let $\mu = \ln(\mu_l^2 / \sqrt{\sigma_l^2 + \mu_l^2})$ and $\sigma = \sqrt{\ln\left[(\sigma_l^2 + \mu_l^2)/\mu_l^2\right]}$. The probability density function can then be written as

$$f(x) = \begin{cases} \dfrac{1}{\sigma x \sqrt{2\pi}} e^{-(\ln(x)-\mu)^2/(2\sigma^2)} & \text{for } x > 0 \\ 0 & \text{otherwise} \end{cases}$$

Parameters   Mean LogMean ($\mu_l > 0$) and standard deviation LogStd ($\sigma_l > 0$) of the lognormal random variable. Both LogMean and LogStd must be specified as strictly positive real numbers.

Range $[0, +\infty)$

Mean   LogMean = $\mu_l = e^{\mu + \sigma^2/2}$

Variance $(\text{LogStd})^2 = \sigma_l^2 = e^{2\mu + \sigma^2}(e^{\sigma^2} - 1)$

Applications   The lognormal distribution is used in situations in which the quantity is the product of a large number of random quantities. It is also frequently used to represent task times that have a distribution skewed to the right. This distribution is related to the normal distribution as follows. If X has a Lognormal (μ_l, σ_l) distribution, then $\ln(X)$ has a Normal(μ, σ) distribution. Note that μ and σ are *not* the mean and standard deviation of the lognormal random variable X, but rather the mean and standard deviation of the normal random variable $\ln X$; the mean LogMean = μ_l and variance $(\text{LogStd})^2 = \sigma_l^2$ of X are given by the formulas earlier on this page.

Normal(μ, σ)   **NORMAL(Mean, StdDev) or NORM(Mean, StdDev) or RN(ParamSet)**

Probability Density Function

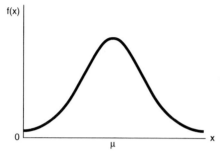

$$f(x) = \frac{1}{\sigma\sqrt{2\pi}}e^{-(x-\mu)^2/(2\sigma^2)} \quad \text{for all real } x$$

Parameters   The mean (μ) specified as a real number and standard deviation (σ) specified as a positive real number.

Range $(-\infty, +\infty)$

Mean μ

Variance σ^2

Applications   The normal distribution is used in situations in which the central limit theorem applies—that is, quantities that are sums of other quantities. It is also used empirically for many processes that appear to have a symmetric distribution. Because the theoretical range is from $-\infty$ to $+\infty$, the distribution should not be used for positive quantities like processing times.

Poisson(λ)   **POISSON(Mean) or POIS(Mean) or
                        PO(ParamSet)**

***Probability
Mass
Function***

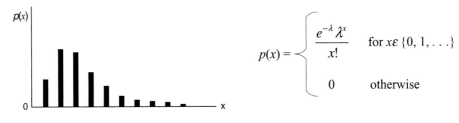

$$p(x) = \begin{cases} \dfrac{e^{-\lambda}\,\lambda^x}{x!} & \text{for } x\varepsilon\,\{0,\,1,\,\ldots\} \\[2mm] 0 & \text{otherwise} \end{cases}$$

Parameters   The mean (λ) specified as a positive real number.

Range   {0, 1, ...}

Mean   λ

Variance   λ

Applications The Poisson distribution is a discrete distribution that is often used to model the number
                    of random events occurring in a fixed interval of time. If the time between successive
                    events is exponentially distributed, then the number of events that occur in a fixed time
                    interval has a Poisson distribution. The Poisson distribution is also used to model random
                    batch sizes.

Triangular(*a*, *m*, *b*)   TRIANGULAR(Min, Mode, Max) or TRIA(Min, Mode, Max) or TR(ParamSet)

Probability Density Function

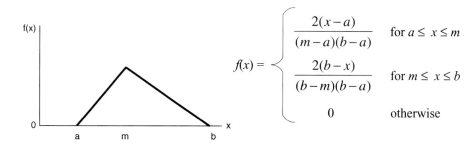

$$f(x) = \begin{cases} \dfrac{2(x-a)}{(m-a)(b-a)} & \text{for } a \leq x \leq m \\[2ex] \dfrac{2(b-x)}{(b-m)(b-a)} & \text{for } m \leq x \leq b \\[2ex] 0 & \text{otherwise} \end{cases}$$

Parameters   The minimum (*a*), mode (*m*), and maximum (*b*) values for the distribution specified as real numbers with $a < m < b$.

Range $[a, b]$

Mean $(a + m + b)/3$

Variance $(a^2 + m^2 + b^2 - ma - ab - mb)/18$

Applications   The triangular distribution is commonly used in situations in which the exact form of the distribution is not known, but estimates (or guesses) for the minimum, maximum, and most likely values are available. The triangular distribution is easier to use and explain than other distributions that may be used in this situation (e.g., the beta distribution).

Uniform(*a*, *b*)   **UNIFORM(Min, Max) or UNIF(Min, Max) or**
                              UN(ParamSet)

Probability
Density
Function

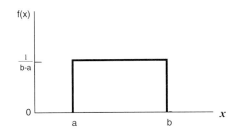

$$f(x) = \begin{cases} \dfrac{1}{b-a} & \text{for } a \le x \le b \\[2ex] 0 & \text{otherwise} \end{cases}$$

Parameters   The minimum (*a*) and maximum (*b*) values for the distribution specified as real numbers with *a* < *b*.

Range $[a,\ b]$

Mean $(a + b)/2$

Variance $(b - a)^2/12$

Applications The uniform distribution is used when all values over a finite range are considered to be equally likely. It is sometimes used when no information other than the range is available. The uniform distribution has a larger variance than other distributions that are used when information is lacking (e.g., the triangular distribution).

Weibull(β, α)   **WEIBULL(Beta, Alpha) or WEIB(Beta, Alpha) or**
                                         WE(ParamSet)

Probability
Density
Function

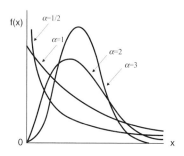

$$f(x) = \begin{cases} \alpha\beta^{-\alpha}x^{\alpha-1}e^{-(x/\beta)^{\alpha}} & \text{for } x > 0 \\ \\ 0 & \text{otherwise} \end{cases}$$

Parameters   Scale parameter (β) and shape parameter (α) specified as positive real numbers.

Range $[0, +\infty)$

Mean $\dfrac{\beta}{\alpha}\Gamma\left(\dfrac{1}{\alpha}\right)$, where Γ is the complete gamma function (see gamma distribution).

Variance $\dfrac{\beta^2}{\alpha}\left\{2\Gamma\left(\dfrac{2}{\alpha}\right) - \dfrac{1}{\alpha}\left[\Gamma\left(\dfrac{1}{\alpha}\right)\right]^2\right\}$

Applications   The Weibull distribution is widely used in reliability models to represent the lifetime of a device. If a system consists of a large number of parts that fail independently, and if the system fails when any single part fails, then the time between successive failures can be approximated by the Weibull distribution. This distribution is also used to represent non-negative task times.

APPENDIX E

Academic Software Installation Instructions

Arena requires Microsoft® Windows® 98, Windows 98 SE, Windows ME, Windows 2000 (Service Pack 3 or later), Windows Server 2003, or Windows XP (Service Pack 1 or later). Under Windows 2000 and Windows XP, you must have Administrator privileges to install the software. [1]

E.1 Authorization to Copy Software

This academic software can be installed on any university computer, as well as on students' computers. It is intended for use in conjunction with this book for the purpose of learning simulation and Arena. You have the right to use and make copies the software for academic use for teaching and research purposes only. Commercial use of the software is prohibited. Software support is supplied only to the registered instructor.

This textbook describes Arena 10.0 and uses it for compiling and running the examples, but it is expected that future versions of Arena can be used equally well with this book. Since new releases of Arena typically outpace updates to this textbook, academic institutions are encouraged to keep their lab and research software up to date and to make copies of their latest install CDs available to students to replace what is included in this book. However, it is important that students use the same software version as used in the labs, because models are forward compatible, but not backward compatible (e.g., an Arena 10.0 model can be loaded into Arena 11.0, but an Arena 11.0 model cannot be loaded into Arena 10.0).

E.2 Installing the Arena Software

Follow this sequence to install your Arena software. Please note that you cannot merely accept all the defaults; there are some specific steps you must follow during the installation process:

1. Insert the Arena CD to initiate the autorun program, which displays the Arena installation screen. If it does not run automatically, browse the CD directory to locate `autorun.exe` and double-click it to start the installation.
2. From the installation dialog, select *Install Arena 10.00.00 (CPR 7)*. When prompted for a serial number, enter `STUDENT` to activate the academic version. Doing so customizes the install, provides access to the examples referenced in the textbook, and allows you to build larger models.

---

[1] It is not necessary to have Administrator privileges to run Arena after installation. For more information or help, please consult your Systems Administrator.

3. When choosing a location to install Arena on the PC's hard drive, please note that Arena will be placed in the Arena 10.0 subfolder of the folder you specify.
4. After Arena installs, reboot your computer if requested.
5. If you have further questions, please refer to the User Zone section of our Web site at www.ArenaSimulation.com.

License activation is not supplied with, or required for, the STUDENT version of Arena. If you see an option to install the activation for Arena, this option should be cleared (unchecked). If you are installing the Arena PE Educational Lab Package or for more information on license activation or any other aspect of installation, select *Installation Notes* from the Arena installation screen.

E.3   System Requirements

The minimum requirements/recommendations for running the Arena software are:

- Microsoft® Windows® 98, Windows 98 SE, Windows ME, Windows 2000 (Service Pack 3 or later), Windows Server 2003, or Windows XP (Service Pack 1 or later)
- Microsoft Internet Explorer 5.01 or later. (This does not need to be your active browser, but some IE components are necessary for Arena to work.)
- Adobe Acrobat Reader 7.0 or later to view documentation
- Hard drive with 75-250 MB free disk space (depending on options installed)
- 64MB RAM (recommended 128MB RAM or higher, depending on operating system)
- Minimum Pentium processor, 300Mhz or higher
- The running and animation of simulation models can be calculation-intensive, so a faster processor with additional memory may result in significantly improved performance. In addition, a larger monitor and a screen resolution of at least 1024 x 768 is recommended for improved animation viewing.
- **Under Windows 2000 and Windows XP, you must have Administrator privileges to install the software.**
- **Some Windows 98 SE systems may not execute the Autorun.exe or Setup.exe correctly. If this problem occurs, create a directory on the local hard drive and copy the contents of the Arena 10.0 installation CD into that directory. Then load either the Autorun.exe or Setup.exe from the newly created directory.**

References

Anderson, D. R., D. J. Sweeney, and T. A. Williams, (2006), *Essentials of Statistics for Business and Economics*, 4th ed., Thomson South-Western, Cincinnati, OH.

Ashour, S. and S. G. Bindingnavle, (1973), "An Optimal Design of a Soaking Pit Rolling Mill System," *Simulation*, vol. 20, pp. 207-14.

Balci, O., (1990), "Guidelines for Successful Simulation Studies," *Proceedings of the 1990 Winter Simulation Conference*, O. Balci et al. (eds.), pp. 25-32.

Balci, O., (1995), "Principles and Techniques of Simulation Validation, Verification, and Testing," *Proceedings of the 1995 Winter Simulation Conference*, C. Alexopoulos et al. (eds.), pp. 147-154.

Banks, J. and R. R. Gibson, (1996), "Getting Started in Simulation Modeling," *IIE Solutions*, vol. 28, pp. 34-39.

Bauer, K. W. and J. R. Wilson, (1992), "Control-Variate Selection Criteria," *Naval Research Logistics 39*, pp. 307-321.

Bratley, P., B. L. Fox, and L. E. Schrage, (1987), *A Guide to Simulation*, 2d ed., Springer-Verlag, New York, NY.

Cinlar, E., (1975), *Introduction to Stochastic Processes*, Prentice-Hall, Englewood Cliffs, NJ.

Decisioneering, Inc., (2006), Crystal Ball, http://www.decisioneering.com/, Denver, CO.

Devore, J. L., (2003), *Probability and Statistics for Engineering and the Sciences*, 6th ed., Wadsworth Inc, Belmont, CA.

Devroye, L, (1986), *Non-Uniform Random Variate Generation*, Springer-Verlag, New York, NY.

Farrington, P. A. and J. J. Swain, (1993), "Design of Simulation Experiments with Manufacturing Applications," *Proceedings of the 1993 Winter Simulation Conference*, G. W. Evans et al. (eds.), pp. 69-75.

Fishman, G. S. (1978), "Grouping Observations in Digital Simulation," *Management Science 24*, pp. 510-521.

Forgionne, G. A., (1983) "Corporate Management Science Activities: An Update," *Interfaces*, vol. 13, pp. 20-23.

Gass, Saul I. and Donald R. Gross, (2000), "In Memoriam: Carl M. Harris, 1940-2000," *INFORMS Journal on Computing 12*, pp. 257-260.

Glover, F., J. Kelly, and M. Laguna, (1999), "New Advances for Wedding Optimization and Simulation," Proceedings of the 1999 Winter Simulation Conference, 255-260.

Goldsman, D., (1992), "Simulation Output Analysis," *Proceedings of the 1992 Winter Simulation Conference*, J. J. Swain et al. (eds.), pp. 97-103.

Gross, D. and C.M. Harris, (1998), *Fundamentals of Queueing Theory*, 3d ed., Wiley, New York, NY.

Harpell, J. L., M. S. Lane, and A. H. Mansour, (1989), "Operations Research in Practice: A Longitudinal Study," *Interfaces*, vol. 19, pp. 65-74.

Harrison, J. M. and C. H. Loch, (1995), "Operations Management and Reengineering," working paper, Graduate School of Business, Stanford University, Stanford, CA.

Hogg, R. V. and A. T. Craig, (2005), *Introduction to Mathematical Statistics*, 6th ed., Macmillan, New York, NY.

Johnson, M. A., S. Lee, and J. R. Wilson, (1994), "Experimental Evaluation of a Procedure for Estimating Nonhomogeneous Poisson Processes Having Cyclic Behavior," *ORSA Journal on Computing 6*, pp. 356-368.

Kelton, W. D., (1996), "Statistical Issues in Simulation," *Proceedings of the 1996 Winter Simulation Conference*, J. M. Charnes et al. (eds.), pp. 47-54.

Kleindorfer, G. B. and R. Ganeshan, (1993), "The Philosophy of Science and Validation in Simulation," *Proceedings of the 1993 Winter Simulation Conference*, G. W. Evans et al. (eds.), pp. 50-57.

Kuhl, M. E. and J. R. Wilson (2000), "Least squares estimation of nonhomogeneous Poisson processes," *Journal of Statistical Computation and Simulation 67*, pp. 75-108.

Kuhl, M. E. and J. R. Wilson (2001), "Modeling and simulating Poisson processes having trends on nontrigonometric cyclic effects," *European Journal of Operational Research 133*, pp. 566-582.

Kulwiec, R. A., (1985), *Materials Handling Handbook*, 2d ed., John Wiley & Sons, Inc., New York, NY.

L'Ecuyer, P., (1996), "Combined Multiple Recursive Generator," *Operations Research 44*, pp. 816–822.

L'Ecuyer, P., (1999), "Good Parameter Sets for Combined Multiple Recursive Random Number Generators," *Operations Research 47*, pp. 159–164.

L'Ecuyer, P., R. Simard, E. J. Chen, and W. D. Kelton, (2002), "An Object-Oriented Random-Number Package with Many Long Streams and Substreams," *Operations Research 50*, pp. 1073-1075 (with accompanying Online Companion at http://or.pubs.informs.org/Media/L%27Ecuyer.pdf).

Lane, David M., (2005), *HyperStat Online Textbook*, http://davidmlane.com/hyperstat/index.html.

Lane, M. S., A. H. Mansour, and J. L. Harpell, (1993), "Operations Research Techniques: A Longitudinal Update 1973-1988," *Interfaces*, vol. 23, pp. 63-68.

Law, A. M. and W. D. Kelton, (2000), *Simulation Modeling and Analysis,* 3d ed., McGraw-Hill, New York, NY.

Leemis, L. M., (1991), "Nonparametric Estimation of the Intensity Function for a Nonhomogeneous Poisson Process," *Management Science 37*, pp. 886-900.

Lindley, D. V., (1952), "The Theory of Queues with a Single Server," *Proceedings of the Cambridge Philosophical Society 48*, pp. 277-289.

Marsaglia, G., (1968), "Random Numbers Fall Mainly in the Planes," *National Academy of Science Proceedings*. vol. 61, pp. 25-28.

Morgan, B. J. T., (1984), *Elements of Simulation*, Chapman & Hall, London.

Morgan, C. L., (1989), "A Survey of MS/OR Surveys," *Interfaces*, vol. 19, pp. 95-103.

Musselman, K. J., (1993), "Guidelines for Simulation Project Success," *Proceedings of the 1993 Winter Simulation Conference*, G. W. Evans et al. (eds.), pp. 58-64.

Nance, R. E., (1996), *A History of Discrete Event Simulation Programming Languages, History of Programming Languages*, T. J. Bergin and R. J. Gibson (eds.), ACM Press and Addison-Wesley Publishing Company, pp. 369-427.

Nance, R. E. and R. G. Sargent, (2003), "Perspectives on the Evolution of Simulation," *Operations Research 50*, pp. 161-172.

Nelson, B. L., (1990), "Control-Variate Remedies," *Operations Research 38*, pp. 974-992.

Nelson, B. L., J. Swann, D. Goldsman, and W. Song, (2001), Simple Procedures for Selecting the Best Simulation System when the Number of Alternatives is Large, *Operations Research 49,* pp. 950-963.

Palisade Corporation, (2006), @RISK, http://www.palisade.com/, Ithaca, NY.

Pegden, C. D., R. E. Shannon, and R. P. Sadowski, (1995), *Introduction to Simulation Using SIMAN*, 2d ed., McGraw-Hill, New York, NY.

Rasmussen, J. J. and T. George, (1978), "After 25 Years: A Survey of Operations Research Alumni, Case Western Reserve University," *Interfaces*, vol. 8, pp. 48-52.

Sadowski, R. P., (1989), "The Simulation Process: Avoiding the Problems and Pitfalls," *Proceedings of the 1989 Winter Simulation Conference*, E. A. MacNair et al. (eds.), pp. 72-79.

Sadowski, R. P., (1993), "Selling Simulation and Simulation Results," *Proceedings of the 1993 Winter Simulation Conference*, G. W. Evans et al. (eds.), pp. 65-68.

Sargent, R. G., (1996), "Verifying and Validating Simulation Models," *Proceedings of the 1996 Winter Simulation Conference*, J. M. Charnes et al. (eds.), pp. 55-64.

Sargent, R. G., K. Kang, and D. Goldsman, (1992), "An Investigation of Finite-Sample Behavior of Confidence Interval Estimators," *Operations Research 40*, pp. 898-913.

Schmeiser, B. W., (1982), "Batch Size Effects in the Analysis of Simulation Output," *Operations Research 30*, pp. 556-568.

Schriber, T. J., (1969), *Fundamentals of Flowcharting*, John Wiley & Sons, Inc., New York, NY.

Seila, A. F., (1990), "Output Analysis for Simulation," *Proceedings of the 1990 Winter Simulation Conference*, O. Balci et al. (eds.), pp. 49-54.

Shannon, R. E., S. S. Long, and B. P. Buckles, (1980), "Operations Research Methodologies in Industrial Engineering," *AIIE Trans.*, vol. 12, pp. 364-367.

Swart, W. and L. Donno, (1981), "Simulation Modeling Improves Operations, Planning, and Productivity for Fast Food Restaurants," *Interfaces*, vol. 11, pp. 35-47.

Thomas, G. and J. DaCosta, (1979), "A Sample Survey of Corporate Operations Research," *Interfaces*, vol. 9, pp. 102-111.

Wysk, R. A., J. S. Smith, D. T. Sturrock, S. E. Ramaswamy, G. D. Smith, and S. B. Joshi, (1994), "Discrete Event Simulation for Shop Floor Control," in *Proceedings of the 1994 Winter Simulation Conference,* J. D. Tew et al. (eds.), pp. 962-969.

Symbols

A

T